C0-ALU-729

Translational Regulation of Gene Expression 2

Translational Regulation of Gene Expression 2

Edited by
Joseph Ilan
Case Western Reserve University
Cleveland, Ohio

Plenum Press • New York and London

Library of Congress Cataloging-in-Publication Data

Translational regulation of gene expression 2 / edited by Joseph Ilan.
p. cm.
Includes bibliographical references and index.
ISBN 0-306-44374-0
1. Genetic translation. 2. Genetic regulation. 3. Gene expression. I. Ilan, Joseph.
QH450.5.T733 1993
574.87'3223--dc20

93-17794
CIP

ISBN 0-306-44374-0

A Division of Plenum Publishing Corporation
233 Spring Street, New York, N.Y. 10013

Printed in the United States of America

Contributors

DONALD D. ANTHONY • Department of Biochemistry, School of Medicine, Case Western Reserve University, Cleveland, Ohio 44106-4935

G. ARROYO • Department of Biology, University of Puerto Rico, Rio Piedras, Puerto Rico 00931

ROSTOM BABLANIAN • Department of Microbiology and Immunology, SUNY, Health Science Center at Brooklyn, Brooklyn, New York 11203

G. C. CANDELAS • Department of Biology, University of Puerto Rico, Rio Piedras, Puerto Rico 00931

C. CARRASCO • Department of Biology, University of Puerto Rico, Rio Piedras, Puerto Rico 00931

E. CARRASQUILLO • Department of Biology, University of Puerto Rico, Rio Piedras, Puerto Rico 00931

JANE-JANE CHEN • Harvard–Massachusetts Institute of Technology Division of Health Sciences and Technology, Massachusetts Institute of Technology, Cambridge, Massachusetts 02139

A. MARK CIGAN • Section on Molecular Genetics of Lower Eukaryotes, Laboratory of Molecular Genetics, National Institute of Child Health and Human Development, National Institutes of Health, Bethesda, Maryland 20892

BANSIDHAR DATTA • Department of Chemistry, University of Nebraska, Lincoln, Nebraska 68588-0304

THOMAS E. DEVER • Section on Molecular Genetics of Lower Eukaryotes, Laboratory of Molecular Genetics, National Institute of Child Health and Human Development, National Institutes of Health, Bethesda, Maryland 20892

THOMAS F. DONAHUE • Department of Biology, Indiana University, Bloomington, Indiana 47405

MARK E. DUMONT • Department of Biochemistry, School of Medicine and Dentistry, Rochester, New York 14642

LAN FENG • Department of Biology, Indiana University, Bloomington, Indiana 47405

ROBERT M. FREDERICKSON • Department of Biochemistry and McGill Cancer Centre, McGill University, Montréal, Québec, Canada H3G 1Y6

GHANASHYAM D. GHADGE • Department of Microbiology and Immunology, and Robert H. Lurie Cancer Center, Northwestern University Medical School, Chicago, Illinois 60611

MARIANNE GRUNBERG-MANAGO • Institut de Biologie Physico-Chemique, 75005 Paris, France

NABA K. GUPTA • Department of Chemistry, University of Nebraska, Lincoln, Nebraska 68588-0304

BOYD HARDESTY • Department of Chemistry and Biochemistry, and Clayton Foundation Biochemical Institute, University of Texas at Austin, Austin, Texas 78712

J. WOODLAND HASTINGS • Department of Cellular and Developmental Biology, Harvard University, Cambridge, Massachusetts 02138

G. WESLEY HATFIELD • Department of Microbiology and Molecular Genetics, College of Medicine, University of California, Irvine, California 92717

ALAN G. HINNEBUSCH • Section on Molecular Genetics of Lower Eukaryotes, Laboratory of Molecular Genetics, National Institute of Child Health and Human Development, National Institutes of Health, Bethesda, Maryland 20892

ARA G. HOVANESSIAN • Unité de Virologie et Immunologie Cellulaire (UA CNRS 1157), Institut Pasteur, 75015 Paris, France

JOSEPH ILAN • Institute of Pathology, Case Western Reserve University School of Medicine, Cleveland, Ohio 44106

M. IRIZARRY • Department of Biology, University of Puerto Rico, Rio Piedras, Puerto Rico 00931

ROGER L. KASPAR • Department of Biochemistry, University of Washington, Seattle, Washington 98133

GISELA KRAMER • Department of Chemistry and Biochemistry, and Clayton Foundation Biochemical Institute, University of Texas at Austin, Austin, Texas 78712

WIESLAW KUDLICKI • Department of Chemistry and Biochemistry, and Clayton Foundation Biochemical Institute, University of Texas at Austin, Austin, Texas 78712

DONG-HEE LEE • Department of Cellular and Developmental Biology, Harvard University, Cambridge, Massachusetts 02138

SUSAN LINDQUIST • Howard Hughes Medical Institute and Department of Molecular Genetics and Cell Biology, University of Chicago, Chicago, Illinois 60637

WILLIAM C. MERRICK • Department of Biochemistry, School of Medicine, Case Western Reserve University, Cleveland, Ohio 44106-4935

MARIA MITTAG • Department of Cellular and Developmental Biology, Harvard University, Cambridge, Massachusetts 02138

RICHARD P. MOERSCHELL • Department of Biochemistry, School of Medicine and Dentistry, Rochester, New York 14642; *present address*: Institute for Protein Research, Osaka University, Suita, Osaka 565, Japan.

DAVID R. MORRIS • Department of Biochemistry, University of Washington, Seattle, Washington 98133

KNUD H. NIERHAUS • Max-Planck-Institut für Molekulare Genetik, D-1000 Berlin 33, Germany

A. PLAZAOLA • Department of Biology, University of Puerto Rico, Rio Piedras, Puerto Rico 00931

CLAUDE PORTIER • Institut de Biologie Physico-Chemique, 75005 Paris, France

PRITHI RAJAN • Department of Microbiology and Immunology, and Robert H. Lurie Cancer Center, Northwestern University Medical School, Chicago, Illinois 60611

MANAS K. RAY • Department of Chemistry, University of Nebraska, Lincoln, Nebraska 68588-0304

ANANDA L. ROY • Department of Chemistry, University of Nebraska, Lincoln, Nebraska 68588-0304

ALEXEY G. RYAZANOV • Institute of Protein Research, Academy of Sciences of Russia, 142292, Pushchino, Moscow Region, Russia; and Department of Pharmacology, UMDNJ–Robert Wood Johnson Medical School, Piscataway, New Jersey 08854

KATHERINE T. SCHMEIDLER-SAPIRO • Department of Biological Sciences, California State University—Long Beach, Long Beach, California 90840

ROBERT J. SCHNEIDER • Department of Biochemistry and Kaplan Cancer Center, New York University Medical Center, New York, New York 10016

STEVEN SCZEKAN • Department of Cellular and Developmental Biology, Harvard University, Cambridge, Massachusetts 02138

FRED SHERMAN • Departments of Biochemistry and Biophysics, School of Medicine and Dentistry, Rochester, New York 14642

NAHUM SONENBERG • Department of Biochemistry and McGill Cancer Centre, McGill University, Montréal, Québec, Canada H3G 1Y6

ALEXANDER S. SPIRIN • Institute of Protein Research, Academy of Sciences of Russia, 142292, Pushchino, Moscow Region, Russia

ROLF STERNGLANZ • Department of Biochemistry, State University of New York at Stony Brook, Stony Brook, New York 11794

SATHYAMANGALAM SWAMINATHAN • Department of Microbiology and Immunology, and Robert H. Lurie Cancer Center, Northwestern University Medical School, Chicago, Illinois 60611

DIETER TECHEL • Department of Cellular and Developmental Biology, Harvard University, Cambridge, Massachusetts 02138

BAYAR THIMMAPAYA • Department of Microbiology and Immunology, and Robert H. Lurie Cancer Center, Northwestern University Medical School, Chicago, Illinois 60611

FRANCISCO TRIANA • Max-Planck-Institut für Molekulare Genetik, D-1000 Berlin 33, Germany

SUSUMU TSUNASAWA • Institute for Protein Research, Osaka University, Suita, Osaka 565, Japan

WILLIAM E. WALDEN • Department of Microbiology and Immunology, University of Illinois, Chicago, Illinois 60612

RONALD C. WEK • Section on Molecular Genetics of Lower Eukaryotes, Laboratory of Molecular Genetics, National Institute of Child Health and Human Development, National Institutes of Health, Bethesda, Maryland 20892

MICHAEL W. WHITE • Department of Veterinary Molecular Biology, Montana State University, Bozeman, Montana 59717

YAN ZHANG • Department of Biochemistry and Kaplan Cancer Center, New York University Medical Center, New York, New York 10016

Preface

This book, which results from the dramatic increase in interest in the control mechanism employed in gene expression and the importance of the regulated proteins, presents new information not covered in *Translational Regulation of Gene Expression*, which was published in 1987. It is not a revision of the earlier book but, rather, an extension of that volume with special emphasis on mechanism.

As the reader will discover, there is enormous diversity in the systems employing genes for translational regulation in order to regulate the appearance of the final product—the protein. Thus, we find that important proteins such as protooncogenes, growth factors, stress proteins, cytokines, lymphokines, iron-storage and iron-uptake proteins, and a panorama of prokaryotic proteins, as well as eukaryotic viral proteins, are translationally regulated. Since for some gene products the degree of control is greater by a few orders of magnitude than their transcription, we can state that for these genes, at least, the expression is translationally controlled.

Translational regulation of gene expression in eukaryotes has emerged in the last few years as a major research field. The present book describes mechanisms of translational regulation in bacteria, yeast, and eukaryotic viruses, as well as in eukaryotic genes. In this book we try to provide in-depth coverage by including important examples from each group rather than systematically including all additional systems not described in the previous volume.

The first paper on translational regulation of gene expression in eukaryotes appeared in 1970 (Ilan, J., Ilan, J., and Patel, N., 1970, Mechanism of gene expression in *Tenebrio molitor*—Juvenile hormone determination of translational control through transfer ribonucleic acid and enzyme, *J. Biol. Chem.* **245**:1275–1281). In this publication, translational control, mRNA stability during development, codon bias, and hormonal involvement in translational regulation were described for the first time. This paper was greeted with skepticism since the thinking at that time was based on the model for transcriptional control of gene expression suggested by F. Jacob and J. Monod emphasizing that mRNA must be a short-lived intermediate. The reviewers of the 1970 paper considered the described mechanism with disbelief, disregarding the experimental evidence. Our description of the role of isoacceptor tRNA in translational regulation in insects has been rediscovered 20 years later by the groups of Lazrini and Candelas (see Chapter 12 in this volume).

I would like to thank the contributors for their considerable patience and labor. I think the final product will satisfy contributors and readers alike.

Joseph Ilan

Cleveland, Ohio

Contents

Chapter 5
Regulation of *GCN4* Expression in Yeast: Gene-Specific Translational Control by Phosphorylation of eIF-2α

Alan G. Hinnebusch, Ronald C. Wek, Thomas E. Dever, A. Mark Cigan, Lan Feng, and Thomas F. Donahue

Chapter 6
Co- and Posttranslational Processes and Mitochondrial Import of Yeast Cytochrome *c*

FRED SHERMAN, RICHARD P. MOERSCHELL, SUSUMU TSUNASAWA, ROLF STERNGLANZ, AND MARK E. DUMONT

Chapter 7
eIF-4E Phosphorylation and the Regulation of Protein Synthesis

ROBERT M. FREDERICKSON AND NAHUM SONENBERG

Chapter 10

Translational Control by Adenovirus-Associated RNA I

Bayar Thimmapaya, Ghanashyam D. Ghadge, Prithi Rajan, and Sathyamangalam Swaminathan

Chapter 11
Translational Regulation in Adenovirus-Infected Cells

ROBERT J. SCHNEIDER AND YAN ZHANG

Chapter 12
Concerted Gene Expressions in Elicited Fibroin Synthesis

G. C. CANDELAS, G. ARROYO, C. CARRASCO, E. CARRASQUILLO, A. PLAZAOLA, AND M. IRIZARRY

Chapter 13
Translational Control in the Circadian Regulation of Luminescence in the Unicellular Dinoflagellate *Gonyaulax polyedra*

STEVEN SCZEKAN, DONG-HEE LEE, DIETER TECHEL, MARIA MITTAG, AND J. WOODLAND HASTINGS

Chapter 16
Control of Ribosomal Protein Synthesis in Eukaryotic Cells

Roger L. Kaspar, David R. Morris, and Michael W. White

Chapter 17
Translational Regulation in Reticulocytes: The Role of Heme-Regulated eIF-2α Kinase

Jane-Jane Chen

Chapter 18
Regulation of Reticulocyte eIF-2α Kinases by Phosphorylation

GISELA KRAMER, WIESLAW KUDLICKI, AND BOYD HARDESTY

Chapter 19
Initiation Mechanisms Used in the Translation of Bicistronic mRNAs

WILLIAM C. MERRICK AND DONALD D. ANTHONY

Chapter 20

Protein Synthesis Initiation in Animal Cells: Mechanism of Ternary and Met-tRNA$_f$·40S·mRNA Complex Formation and the Regulatory Role of an eIF-2-Associated 67-kDa Polypeptide

NABA K. GUPTA, BANSIDHAR DATTA, MANAS K. RAY, AND ANANDA L. ROY

Chapter 21

Phosphorylation of Elongation Factor 2: A Mechanism to Shut Off Protein Synthesis for Reprogramming Gene Expression

ALEXEY G. RYAZANOV AND ALEXANDER S. SPIRIN

Chapter 1

A Two-Ribosome Model for Attenuation

G. Wesley Hatfield

1. INTRODUCTION

The first insights into the mechanisms of gene regulation were provided by the work of François Jacob and Jacques Monod in early 1960s. The results of their elegant genetic studies concerning the regulation of the lactose utilization operon of *Escherichia coli* suggested the presence of a repressor molecule that acted in *trans* to repress the expression of the genes of the operon.[1,2] Later, Ellis Englesberg and his colleagues showed that *trans*-acting factors could affect gene expression in a positive manner as well.[3] Subsequently, positive and negative *trans*-acting effectors of gene expression were demonstrated for many metabolic pathways in bacteria, bacteriophage, and eukaryotic organisms. In each of these systems, however, regulation was exerted at the level of transcription initiation. It was not until the late 1970s that Charles Yanofsky and his colleagues demonstrated a fundamentally different type of gene regulation. They showed that, in addition to regulation by repression, the genes of the tryptophan operon of *E. coli* are subject to regulation by transcription termination. The regulation of transcription termination at a site preceding the structural genes of an operon is called attenuation. Transcription attenuation is now recognized as a common mechanism for the

G. Wesley Hatfield • Department of Microbiology and Molecular Genetics, College of Medicine, University of California, Irvine, California 92717.

Translational Regulation of Gene Expression 2, edited by Joseph Ilan. Plenum Press, New York, 1993.

regulation of gene expression in bacteria and more general forms of attenuation have been described in eukaryotic cells and their viruses (reviewed in refs. 4–6).

The common features of attenuation that have been elucidated for amino acid biosynthetic operons in bacteria are described in Section 2 and the temporal mechanics of attenuation for the tryptophan (*trp*) operon are described in Section 3. In Section 4 the structural and functional differences between the attenuators of the well-studied *trp* and isoleucine–valine (*ilvGMEDA*) operons are discussed, and in Section 5 the temporal mechanics of attenuation for the *ilvGMEDA* operon are described. In Section 6 the evidence in support of the idea that the rate of ribosome release from the leader RNA of the *trp* and *ilvGMEDA* operons is important for setting the basal level of attenuation for these operons is presented. In the final sections (Sections 7 and 8) I argue that the functional and structural differences between the *trp* and *ilvGMEDA* attenuators cannot be accommodated by a single model for attenuator function and propose that the mechanisms for regulating the levels of transcriptional readthrough at the these attenuators are fundamentally different.

2. THE COMMON FEATURES OF ATTENUATION

Attenuation was proposed as a regulatory mechanism for the regulation of the tryptophan operon of *E. coli* by Yanofsky and his co-workers in 1976.[7–9] Since that time, attenuation has been demonstrated to regulate the expression of operons required for the biosynthesis of isoleucine and valine (the *ilvGMEDA* and *ilvBN* operons[10–13]), threonine (the *thrABC* operon[14]), leucine (the *leuABCD* operon[15]), histidine (the *hisGDCBHAFIE* operon[16,17]), and phenylalanine (the *pheA* operon[18]).

The leader–attenuator regions of each of these operons (Fig. 1) have several features in common. The leader region is the DNA sequence between the site of transcription initiation and the first structural gene of each operon. The attenuator is a factor-independent transcription termination site in the distal portion of the leader (reviewed in ref. 19). Termination of transcription at the attenuator results in the production of a leader RNA. The leader region of each operon encodes a short polypeptide, the leader polypeptide. The leader polypeptide coding sequence contains multiple codons for the amino acid(s) synthesized by the gene products of the operon. These are the regulatory codons. If the aminoacylated transfer RNA(s) [tRNA(s)] for the regulatory codons are in ample supply within the cell, then the leader polypeptide can be synthesized without interruption and transcription will terminate at the attenuator before the structural genes of the operon are expressed. If, on the other hand, the cells are starving for the amino acid(s) produced by the gene products of the operon, then the aminoacylated tRNA(s) for the regulatory codons will be in short supply and transcription will proceed through the attenuator into the structural genes. Consequently, the catalysts for the biosynthesis of the limiting amino acids(s) will be produced and its supply will be replenished.

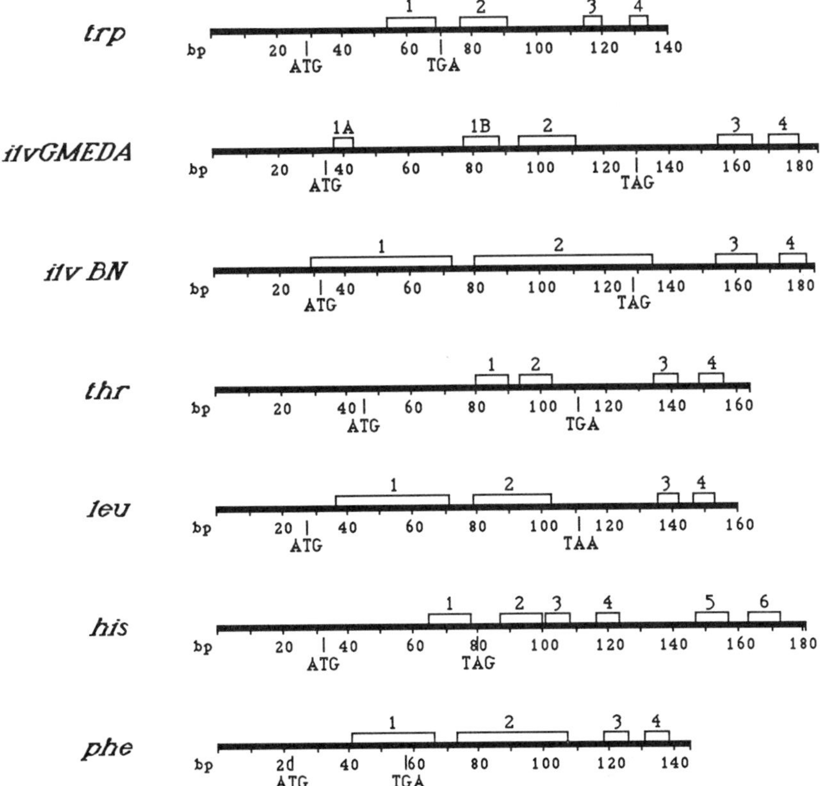

Figure 1. The leader–attenuator regions of the amino acid biosynthetic operons regulated by attenuation. The bold horizontal lines represents the DNA encoding the leader RNA of each operon. The numbered open boxes above each bold horizontal line identify the DNA sequences in the leader region that encode the alternative base-pairing regions of the leader RNA. Regions 1 and 2 base pair to form the pause structure, stem–loop 1:2. Regions 2 and 3 base pair to form the antiterminator structure, stem–loop 2:3. Regions 3 and 4 base pair to form the attenuator structure, stem–loop 3:4. For the *his* operon, regions 1 and 2 base pair to form the pause structure and regions 5 and 6 base pair to form the attenuator structure. Regions 2 and 3, and 4 and 5, base pair to form stem–loop 2:3, and stem–loop 4:5, for antitermination. Reviewed in ref. 4.

How does the translational efficiency of the regulatory codons in the leader polypeptide coding region effect transcription termination or readthrough by a transcribing RNA polymerase at the downstream attenuator site? This coupling of translation and transcription functions is brought about by alternative secondary structures that the leader RNA can assume. The alternative secondary structures of the *trp* and *ilvGMEDA* leader RNAs are shown in Fig. 2. One structure results in the formation of a rho-independent transcription termination signal, a stem–loop structure followed by a tract of uridine residues (the 3:4 stem–loop),

A

Excess Amino Acids

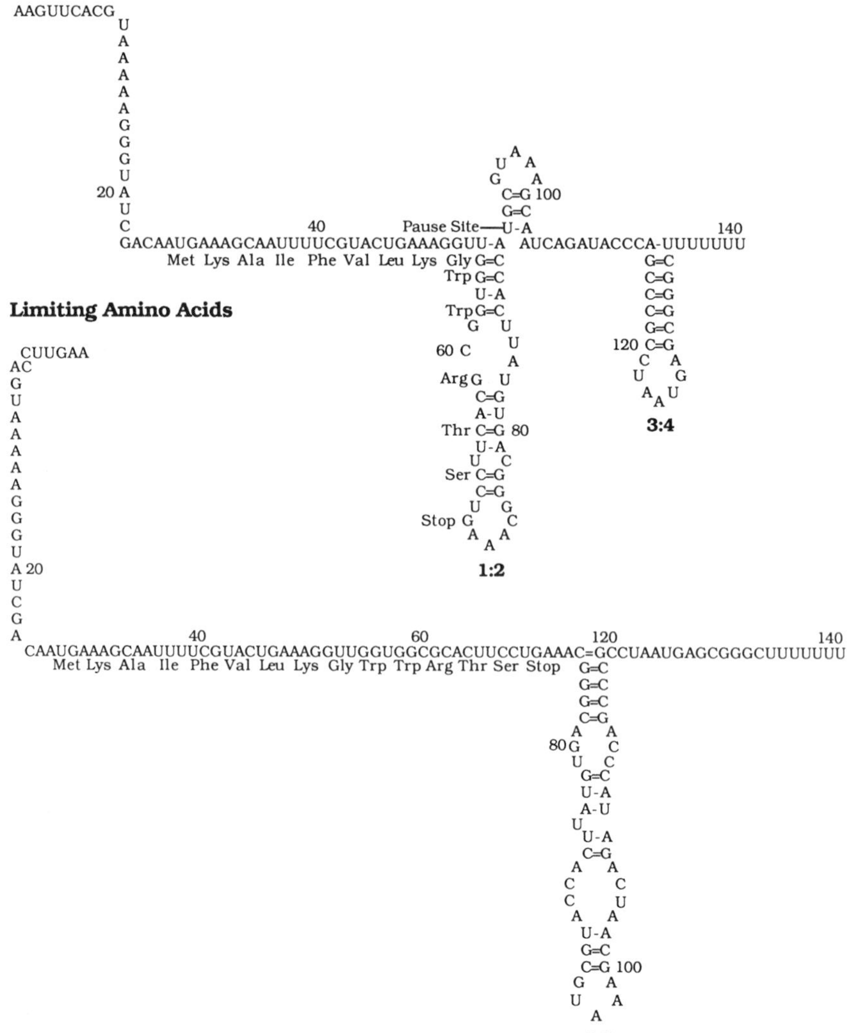

Figure 2. Alternative secondary structures of the leader RNA transcripts of the **(A)** *trp* and **(B)** *ilvGMEDA* operons. The structures labeled excess amino acids are in the terminated (attenuated), stem–loops 1:2 and 3:4, configuration. The structures labeled limiting amino acids are in the deattenuated, stem–loop 2:3, configuration. The amino acids encoded by the leader polypeptide coding region and the transcriptional pause sites are identified.

B

Excess Amino Acids

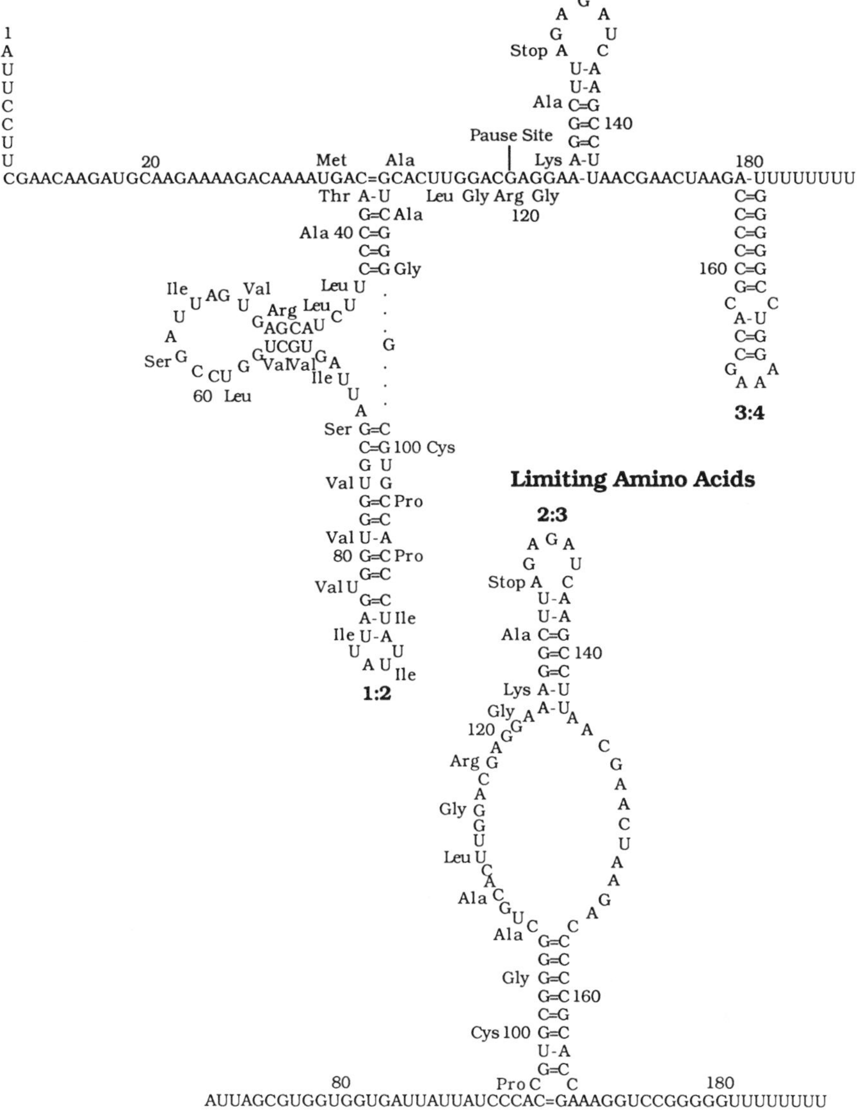

Figure 2. (*Continued*)

the attenuator, at the 3′ end of the leader RNA.[19] Another structure (the 1:2 stem–loop) results in the formation of a transcriptional pause signal. The formation of this structure in the nascent leader RNA causes the transcribing RNA polymerase to pause after the synthesis of the stem 2 region of the stem–loop structure. Stem 2 of stem–loop 1:2 and stem 3 of stem–loop 3:4 can form yet a third, alternative structure, the antiterminator stem–loop 2:3. The formation of the antiterminator preempts the formation of the attenuator structure and effects deattenuation, transcription through the attenuator. The formation of these alternative structures (stem–loops 1:2 and 3:4 or stem–loop 2:3) is directed by the temporal positioning of a translating ribosome in the leader polypeptide coding region in relation to the transcribing RNA polymerase. The coupling of the movement of ribosome and RNA polymerase molecules through the leader–attenuator region is ensured by the presence of the transcriptional pause site. The transcribing RNA polymerase pauses at the base of the stem–loop 1:2 coding sequence until a ribosome attaches to the leader RNA and begins to translate through the leader polypeptide coding region. The paused RNA polymerase is released to resume transcription when the translating ribosome enters the stem 1 coding region and disrupts the stem–loop 1:2 pause structure. At this point transcription and translation are synchronized. If the cellular levels of aminoacylated tRNAs are high, then translation will proceed unimpeded and translation and transcription will remain synchronized until the ribosome reaches the stop codon of the leader polypeptide coding region. The stop codon of the *trp* leader polypeptide coding region is in the loop region of the stem–loop 1:2 structure. The presence of a ribosome at this site disrupts the stem–loop 1:2 structure. The basal attenuated level of transcriptional readthrough at the attenuator is determined by the time it takes the ribosome to release the leader RNA template relative to the transcriptional speed of the RNA polymerase. If the ribosome were to release before the synthesis of stem 3, then stem–loop 1:2 would be free to form, followed by the synthesis, and the unchallenged formation, of the stem–loop 3:4 attenuator structure. This situation would result in a maximal level of transcription termination at the attenuator. Since the average ribosome release time is thought to be about 0.6 sec, however, and the average transcription rate is about 50 nucleotides per second, it is likely that the released RNA polymerase would have synthesized stem 3 and be in stem–loop 4 at the time the ribosome releases. In this case, the formation of the antiterminator structure, stem–loop 2:3, could compete with the formation of the alternative stem–loop 1:2 and 3:4 structures to set an average basal level of transcriptional readthrough at the attenuator. Thus, the basal level of attenuation is thought to be determined by the location of the transcribing RNA polymerase when the ribosome dissociates from the leader RNA template and by the relative probabilities for the formation of the competing RNA structures (reviewed in ref. 4).

Regulatory codons for each attenuator are located in the stem 1 region of the leader RNA. If the cell is starving for the regulatory amino acid and the cellular level of the cognate aminoacylated tRNA is low, then the translating ribosome will pause at the hungry codons in the stem 1 region. At the same time, the RNA polymerase continues its synthesis of the leader RNA. If it completes the synthesis of stem 3 before the ribosome moves out of the stem 1 region, then the antiterminator structure (stem–loop 2:3; Fig. 2) will form. This will, of course, result in deattenuation, that is, transcription through the attenuator site into the structural genes of the operon. When the intracellular level of the amino acid(s) is replenished, transcription through the attenuator is restored to its basal level (reviewed in ref. 4).

3. THE TEMPORAL MECHANICS OF *trp* ATTENUATION

It should be clear from the above explanation of attenuation that "timing is everything." That is, the decision to attenuate or to deattenuate is made on the basis of the relative rates of translation and transcription through the leader–attenuator region. In order to illustrate this point, I shall describe the temporal events of transcription and translation through the leader–attenuator region of the *trp* operon. These events are displayed in Fig. 3 in the order that they are described in the following discussion.

The attenuated leader RNA of the *trp* operon is 140 nucleotides long. The leader polypeptide coding region contains 14 codons located between nucleotide positions 27 and 69, with the stop codon centered on nucleotide 71. For the purpose of this analysis I shall assume that the transcription rate through the *trp* leader is the same as the average transcription rate for *E. coli*, 50 base pairs per second,[20] and that the rate of synthesis of the leader polypeptide is equal to the average translation rate for *E. coli*, 50 nucleotides per second.[21–23] At these rates, an RNA polymerase initiates transcription of the *trp* leader and arrives at the transcriptional pause site, at base 92, 1.84 sec later (1.3 sec after the completion of the synthesis of sequences necessary for the initiation of translation of the leader polypeptide coding region; Fig. 3B). If we assume that a ribosome initiates translation of leader polypeptide synthesis about once every second,[4] then a ribosome will attach to the leader RNA when the transcribing RNA polymerase is at base pair 77, 0.30 sec before the RNA polymerase reaches the transcriptional pause site (Fig. 3A). The RNA polymerase will, however, reach the pause site (Fig. 3B) and wait for about 0.1 sec before the translating ribosome is centered on nucleotide 48 (six nucleotides before the beginning of the stem–loop 1:2 structure) and releases the paused RNA polymerase (Fig. 3C) by disrupting stem–loop 1:2. (A ribosome masks about nine nucleotides on either side of the codon on which it is centered.[24] For the purpose of this discussion, it is assumed that the ribosome must disrupt three base pairs of stem–loop 1:2 to release the paused RNA polymerase.) At this point, the translating ribosome and the transcribing RNA polymerase are synchronized. If an ample supply of aminoacylated tRNAs is present, translation and transcription will continue in synchrony until the translat-

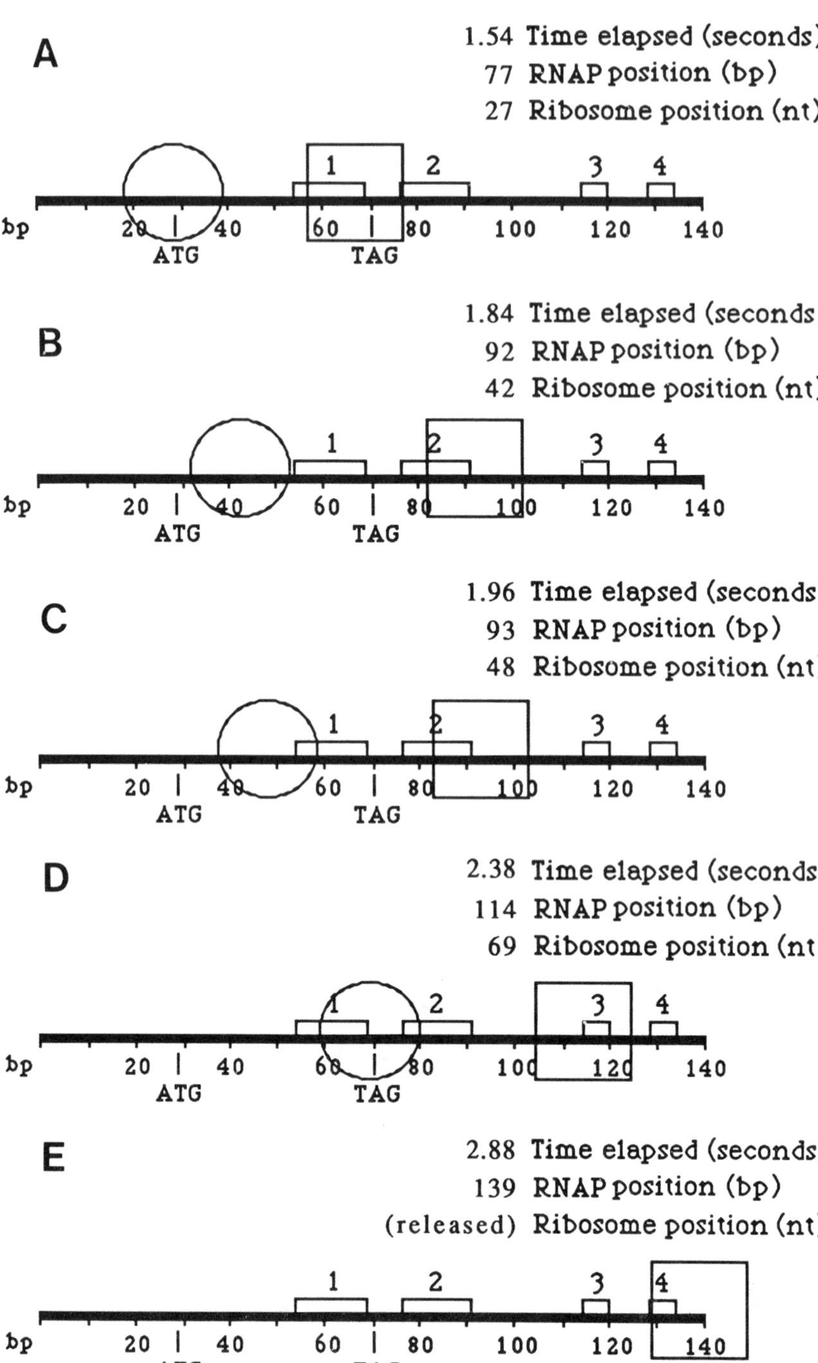
A
1.54 Time elapsed (seconds)
77 RNAP position (bp)
27 Ribosome position (nt)
1
2
3
4
bp
20
40
60
80
100
120
140
ATG
TAG
B
1.84 Time elapsed (seconds)
92 RNAP position (bp)
42 Ribosome position (nt)
C
1.96 Time elapsed (seconds)
93 RNAP position (bp)
48 Ribosome position (nt)
D
2.38 Time elapsed (seconds)
114 RNAP position (bp)
69 Ribosome position (nt)
E
2.88 Time elapsed (seconds)
139 RNAP position (bp)
(released) Ribosome position (nt)

ing ribosome arrives at the translation stop codon centered at nucleotide position 69 (Fig. 3D). At this time, the RNA polymerase will be at base pair position 114 in the region between the DNA sequences encoding stem 2 and stem 3. If, as described above, the ribosome were to release immediately upon its arrival at the stop codon in the middle of the stem–loop 1:2 structure, this structure would be free to form prior to the synthesis of stem 3. This would leave stem 3 free to base pair with stem 4 as these regions are synthesized and cause maximal termination at the attenuator. However, since the average ribosome release time in *E. coli* is about 0.6 sec,[25] the transcribing RNA polymerase would usually be near the end of the leader RNA when the ribosome releases (Fig. 3E). This situation would set up a competition between the alternative secondary structures for their formation. The more frequent formation of the antiterminator structure under this condition would produce a higher basal level of transcription through the attenuator and a correspondingly higher basal level of expression of the structural genes of the operon.

A second transcriptional pause site has been revealed in the leader–attenuator region of the *trp* operon of *Serratia marcesens*.[26] This second pause site, at the distal region of stem 3, was detected in synchronized *in vitro* transcription assays. Transcriptional pausing during the synthesis of stem 3 of the *Serratia* operon would delay transcription of stem 4. Such a delay would permit the translating ribosome time to reach the stop codon of the leader polypeptide coding region, to release, and when it does, to permit the alternative formation of either stem–loop 1:2 (attenuation) or antiterminator stem–loop 2:3 (deattenuation). The schematic in Fig. 3E shows that, in *E. coli*, the ribosome is expected to be at base pair 139 in the *trp* leader region if the ribosome releases 0.5 sec after it arrives at the stop codon of the leader polypeptide. In order for the RNA polymerase to be around

←

Figure 3. The temporal mechanics of attenuation in the tryptophan operon. The bold horizontal line represents the leader DNA and RNA region of the tryptophan operon. The numbered open boxes identify the DNA sequences in the leader region that encode the alternative base-pairing regions of the leader RNA. Leader RNA regions 1 and 2 base pair to form the pause structure, stem–loop 1:2, regions 2 and 3 base pair to form the antiterminator structure, stem–loop 2:3; and regions 3 and 4 base pair to form the attenuator structure, stem–loop 3:4 (see Fig. 2A for details). The transcriptional pause site is located at base pair position 92. An open square represents a transcribing RNA polymerase and an open circle represents a translating ribosome. Each panel, **A–E**, is a "stopped-frame" image generated by a HyperCard animated attenuation simulator program written for the Macintosh computer. The total time elapsed since the initiation of transcription and the positions of the transcribing RNA polymerase and translating ribosome are given and indicated schematically in each panel. The variable parameters were set as follows: transcription rate, 50 base pairs/sec; translation rate, 50 nucleotides/sec; translation initiation frequency, 1.0 sec; ribosome release time, 0.5 sec. **(A)** The RNA polymerase and ribosome positions when the ribosome initiates translation of the leader RNA. **(B)** The RNA polymerase and ribosome positions when the RNA polymerase reaches the transcriptional pause site. **(C)** The RNA polymerase and ribosome positions when the translating ribosome releases the paused RNA polymerase. **(D)** The RNA polymerase and ribosome positions when the ribosome reaches the stop codon of the leader RNA. **(E)** The RNA polymerase position when the ribosome releases the leader RNA.

base pair 124 at the base of stem 3 when the ribosome releases, either the polymerase must pause at a second pause site in stem 3 when the ribosome releases, either the polymerase must pause at a second pause site in stem 3 for about 0.33 sec or the ribosome must release after 0.27 sec instead of the average 0.6 sec. Thus, if a second pause site were present in the stem 3 region of the *E. coli trp* leader–attenuator region, the coordination of stem 4 synthesis and ribosome release could also contribute to the mechanism for setting the basal level of expression of the tryptophan operon.

For deattenuation under conditions of limiting intracellular concentrations of aminoacylated tryptophanyl-tRNA, a ribosome must stall at the regulatory *trp* codons in stem 1 for at least 0.66 sec to allow the transcribing RNA polymerase time to complete the synthesis of stem 3 in order to ensure the formation of the antiterminator, stem–loop 2:3, structure.

4. STRUCTURAL AND FUNCTIONAL DIFFERENCES BETWEEN THE *trp* AND *ilvGMEDA* ATTENUATORS

While the common structural features of the leader–attenuator regions are conserved in all amino acid-regulated attenuators (Fig. 1), there are nevertheless significant structural and functional differences between the *ilvGMEDA* and *trp* operons. Functionally, the *ilvGMEDA* operon is regulated by a multivalent attenuation mechanism. That is, the attenuator of the *ilvGMEDA* operon responds to the limiting intracellular concentration of any one of the three branched-chain amino acids, isoleucine, valine, or leucine, while the *trp* attenuator responds to the intracellular level of only one amino acid, tryptophan. Structurally, the leader RNA transcript of the *ilvGMEDA* operon is much longer, 186 nucleotides, than the leader RNA of the *trp* operon, 140 nucleotides (Figs. 1 and 2). The leader polypeptide coding region of the *ilvGMEDA* operon is also longer, 32 codons, than the leader polypeptide coding region of the *trp* operon, 14 codons (Fig. 2). The stem 1 region of the leader RNA of the *trp* operon contains two tandem Trp regulatory codons. The leader polypeptide coding region of the *ilvGMEDA* operon contains 14 regulatory amino acid codons and some of these codons are located in the distal end of stem 1 and in the proximal half of stem 2 at positions that would not be predicted to affect the attenuation mechanism (Fig. 2). Another couple of significant structural differences between the leader RNAs of these operons is that stem–loop 1:2 of the *ilvGMEDA* operon is bifurcated with an extra stem–loop in the stem 1 region that separates the RNA sequences that participate in the stem 1:2 base pairings. These separated sequences are designated stem 1A and stem 1B (Fig. 2). Whereas the leader polypeptide coding region of the *trp* leader RNA ends at the base of stem 1, the leader polypeptide coding region of the *ilvGMEDA* leader RNA extends to a site beyond the end of stem 2 in yet another short stem–loop structure located between the stem–loop 1:2 and 3:4 structures in the loop of the antiterminator stem–loop 2:3 (Fig. 2). Finally, the leader polypeptide coding region of the *trp* leader RNA begins nine codons before the beginning of the stem–loop 1:2 structure. Consequently, a ribosome translating the leader polypeptide coding region of the *trp* leader RNA must travel six codons (0.36 sec at 50 nucleotides per second) before it disrupts the stem 1 region to release the pause RNA polymerase at the base of the stem–loop 1:2 structure. In contrast, the leader polypeptide coding region of the *ilvGMEDA* leader RNA begins at the proximal end of the stem–loop 1:2 structure (Fig. 2). This means that as soon as a ribosome attaches to the leader RNA a paused RNA polymerase at the distal base of the stem–loop 1:2 structure will be released. Thus, the structural properties and the positions of the polypeptide-coding regions of the *trp* and *ilvGMEDA* operons are quite different. How do these differences serve to differentiate the basic mechanisms of these two operons?

5. THE TEMPORAL MECHANICS OF *ilvGMEDA* ATTENUATION

In Section 3, I demonstrated that the temporal mechanics of attenuation for the *trp* operon are adequately explained with a few basic assumptions about the relative rates of transcription and translation and the average rates of translation initiation and ribosome release. If the basic mechanism of attenuation is the same for all amino acid-regulated attenuators, then these assumptions should suffice to explain the temporal mechanics of other attenuators. In this section, I suggest that, in fact, it is *not* possible to use these same assumptions to explain the regulation of the *ilvGMEDA* attenuator without the addition of another major assumption. Namely, that the structure of the *ilvGMEDA* attenuator requires the participation of *two*, rather than one, ribosomes in the leader polypeptide coding region. I suggest that it is the relative positions and translation rates of both of these ribosomes in relation to the position and transcription rate of the transcribing RNA polymerase that set the basal and deattenuated levels of transcriptional readthrough at the attenuator of this operon.

In order to analyze how the structural differences between the *ilvGMEDA* and *trp* leader–attenuator regions are thought to affect the attenuation mechanism, it is instructive to examine the temporal events of transcription and translation through the leader–attenuator region of the *ilvGMEDA* operon in the same way that these events were analyzed for the *trp* operon in Section 3. These events are displayed in Fig. 4 in the order that they are described in the following discussion. The attenuated leader RNA of the *ilvGMEDA* operon is 186 nucleotides long. The leader polypeptide coding region contains 32 codons located between nucleotides 32 and 129. The stop codon of the leader polypeptide coding region is centered on nucleotide 130 in the loop region of a seven-nucleotide-long stem–loop structure located between the stem 2 and stem 3 regions of the leader RNA. This stem–loop is not affected by the alternative base-pairing properties of the leader RNA;

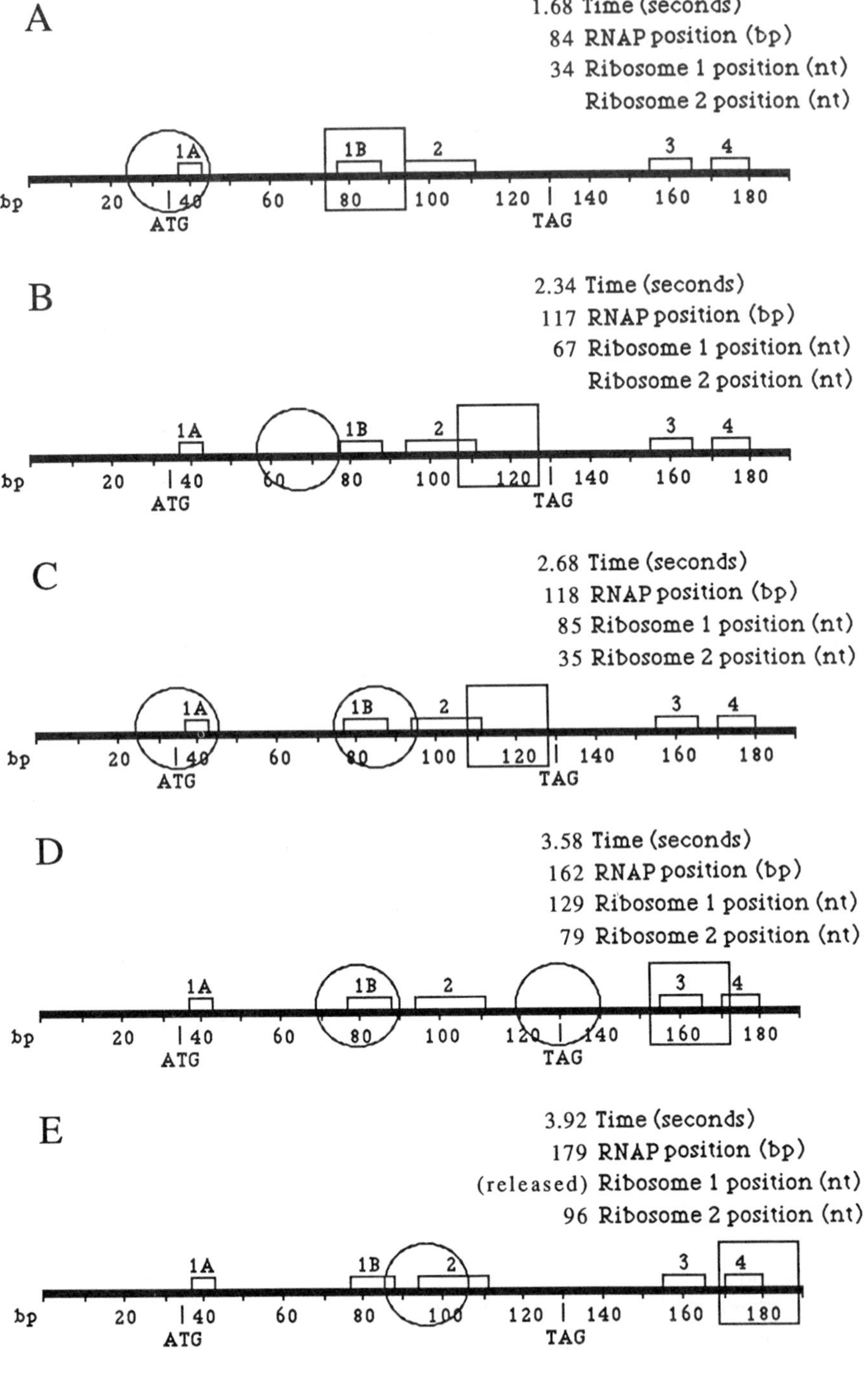
A
1.68 Time (seconds)
84 RNAP position (bp)
34 Ribosome 1 position (nt)
Ribosome 2 position (nt)
1A
1B
2
3
4
bp
20
40
60
80
100
120
140
160
180
ATG
TAG
B
2.34 Time (seconds)
117 RNAP position (bp)
67 Ribosome 1 position (nt)
Ribosome 2 position (nt)
C
2.68 Time (seconds)
118 RNAP position (bp)
85 Ribosome 1 position (nt)
35 Ribosome 2 position (nt)
D
3.58 Time (seconds)
162 RNAP position (bp)
129 Ribosome 1 position (nt)
79 Ribosome 2 position (nt)
E
3.92 Time (seconds)
179 RNAP position (bp)
(released) Ribosome 1 position (nt)
96 Ribosome 2 position (nt)

however, this stem–loop structure does stabilize the stem–loop 2:3 antiterminator structure. It will be referred to, therefore, as the antiterminator stabilizer (Fig. 2).

Assuming the same rate parameters that were used for the temporal analysis of the *trp* attenuator (50 bases per second for translation and transcription rates and a translation initiation frequency of 1 sec), one would expect a ribosome to initiate translation of the nascent leader RNA 1.68 sec after the initiation of transcription (Fig. 4A). At this point, the transcribing RNA polymerase is at base pair 84 in the leader region, 33 base pairs from the transcriptional pause site at base pair position 117. By the time the RNA polymerase arrives at the pause site at the base of the stem 1A–stem 2 base-paired region, the ribosome has passed through the stem 1A region and is entering the stem 1B region (Figs. 2 and 4B). Therefore, the base pairings between the stem 1A and stem 2 regions of the leader RNA can reform and cause the RNA polymerase to pause at the distal end of this RNA duplex. The pausing of the RNA polymerase at this site is also enhanced by the binding of integration host factor to a site in the leader–attenuator region immediately downstream of the transcriptional pause site[27,28] and by the action of the *nusA* protein.[29] The RNA polymerase is presumed to remain at the transcriptional pause site until it is released by the disruption of the base pairing between stems 1A and 2 by the initiation of translation by a second ribosome. This is assumed to occur 1 sec after the first ribosome-initiated translation of the leader polypeptide coding region.[4] Thus, the RNA polymerase is not released from its transcriptional pause site in the leader region by ribosome 1 because the RNA polymerase will not have arrived at the pause site while ribosome 1 is in a position of the leader RNA that will mask the base pairings between stems 1A and 2 that are required for the polymerase to pause at nucleotide 117. At 0.34 sec after the RNA polymerase arrives at the pause site, however, while ribosome 1 is in the stem 1B region, it

Figure 4. The temporal mechanics of attenuation in the *ilvGMEDA* operon. The bold horizontal line represents the leader DNA and RNA region of the *ilvGMEDA* operon. The numbered open boxes identify the DNA sequences in the leader region that encode the alternative base-pairing regions of the leader RNA. Leader RNA regions 1A and 1B base pair with region 2 to form the pause structure, stem–loop 1:2; regions 2 and 3 base pair to form the antiterminator structure, stem–loop 2:3; and regions 3 and 4 base pair to form the attenuator structure, stem–loop 3:4 (see Fig. 2B for details). The transcriptional pause site is located at base pair position 117. An open square represents a transcribing RNA polymerase and open circle(s) represents a translating ribosome(s). Each panel, **A–E**, is a "stopped-frame" image generated by a HyperCard animated attenuation simulator program written for the Macintosh computer. The total time elapsed since the initiation of transcription and the positions of the transcribing RNA polymerase and translating ribosomes 1 and 2 are given and indicated schematically in each panel. The variable parameters were set as follows: transcription rate, 50 base pairs/sec; translation rate, 50 nucleotides/sec; translation initiation frequency, 1.0 sec; ribosome release time, 0.34 sec. **(A)** The RNA polymerase and ribosome positions when ribosome 1 initiates translation of the leader RNA. **(B)** The RNA polymerase and ribosome 1 positions when the RNA polymerase reaches the transcriptional pause site. **(C)** The RNA polymerase and ribosome 1 positions when ribosome 2 initiates translation and releases the paused RNA polymerase. **(D)** The RNA polymerase and ribosome 2 positions when ribosome 1 reaches the stop codon of the leader RNA. **(E)** The RNA polymerase and ribosome 2 positions when ribosome 1 releases the leader RNA.

would be released to resume transcription of the leader region by the attachment of a *second* ribosome, ribosome 2, to the leader RNA (Fig. 4C). At this point, all three molecules, RNA polymerase, ribosome 1, and ribosome 2, are synchronized and, if the intracellular concentrations of aminoacyled tRNAs for the branched-chain amino acids remain high, this synchrony will be maintained until ribosome 1 reaches the stop codon of the leader polypeptide coding region (Fig. 4D). The stalled ribosome at the stop codon disrupts the antiterminator stabilizer (Fig. 2 and ref. 10). The longer ribosome 1 remains at the stop codon, the greater is the probability for the formation of stem–loops 1:2 and 3:4. When ribosome 1 first arrives at the stop codon, however, ribosome 2 is still in the stem 1 region antagonizing the formation of stem–loop 1:2 and the transcribing RNA polymerase is at base pair position 162 near the middle of the DNA sequence encoding stem 3 of the leader RNA. Therefore, stem–loop 3:4 cannot form, since the synthesis of stem 3 of the leader RNA is incomplete. Thus, the entire system is poised. If ribosome 1 were to release from the leader immediately upon its arrival at the stop codon, then ribosome 2 would inhibit the formation of stem–loop 1:2 and the subsequent formation of the antiterminator stem–loop 2:3 would be favored. This would result in a high basal level of transcriptional readthrough at the attenuator. A slower release time for ribosome 1 will allow more time for ribosome 2 to exit stem 1. At the same time, ribosome 2 will move from the stem 1 region into the stem 2 region and progressively inhibit the formation of the antiterminator stem–loop 2:3. Also, the longer it takes ribosome 1 to release, the more time there is for the synthesis of stem 4 and, therefore, the more effectively attenuator, stem–loop 3:4, formation can compete with antiterminator, stem–loop 2:3, formation. Thus, it is proposed that the basal level expression of the structural genes of the *ilvGMEDA* operon is set by the position of ribosome 2 and the position of the transcribing RNA polymerase at the time ribosome 1 releases from the stop codon of the leader RNA (Fig. 4E).

This analysis discloses several features of the mechanisms for setting the basal level of transcription through the attenuators of the *ilvGMEDA* and *trp* operons that are fundamentally different. The most obvious difference is, of course, that the proper function of the *ilvGMEDA* operon might require the participation of *two* ribosomes. Another difference is that the paused ribosome at the stop codon in the *trp* leader RNA masks the base-pairing regions of stems 1 and 2, thus favoring the formation of the attenuator stem–loop 3:4. Early release of the ribosome from the *trp* leader RNA increases the probability of antiterminator formation and, therefore, deattenuation. Conversely, delayed release of the ribosome from the *trp* leader RNA will facilitate attenuation. In contrast, the ribosome waiting to release from the stop codon in the antiterminator stabilizer region of the *ilvGMEDA* leader RNA inhibits the formation of the antiterminator structure, but it does not mask the RNA sequences involved in the base pairings between stems 1 and 2. Instead, it is suggested that ribosome 2 inhibits the formation of the stem–loop 1:2 structure. Nevertheless, the functional consequences of these differences are the same. The early release of the ribosome at the translation stop codon of these leader RNAs will facilitate decreased termination (deattenuation) and the delayed release of the ribosome will facilitate increased termination (attenuation) of transcription through the attenuator sites of both operons. The delayed release of ribosome 1 from the *ilvGMEDA* leader RNA would be predicted to increase transcription termination at the attenuator because it is in a region of the leader RNA where it inhibits antiterminator formation but does not interfere with the formation of stem–loop 1:2 or 3:4. This basic difference in attenuator function is the consequence of the fact that the leader polypeptide coding region for the *trp* operon ends in the loop of the stem–loop 1:2 region, whereas the leader polypeptide coding region of the *ilvGMEDA* operon ends in the antiterminator stabilizer structure between stems 2 and 3 of the antiterminator structure. According to the two-ribosome model, this is an important feature for setting the basal level of transcriptional readthrough at the *ilvGMEDA* attenuator. The slower ribosome 1 is to release from the stop codon of the leader RNA, the more attenuation is favored, both because there is more time for stem 4 synthesis and because ribosome 2 will enter stem 2 to interrupt the antiterminator and to promote attenuator formation.

6. RIBOSOME RELEASE TIME DETERMINES THE BASAL LEVEL OF TRANSCRIPTIONAL READTHROUGH AT THE ATTENUATORS OF THE *trp* AND *ilvGMEDA* OPERONS

The addition of an extra ribosome allows for an interpretation of the attenuation mechanism of the *ilvGMEDA* operon that follows the same assumptions used to interpret the *trp* attenuator mechanism. That is, the rates of transcription and translation through the leader–attenuator regions of both operons are the same, the rates of translation initiation and ribosome release are the same, and the timing of ribosome release sets the basal level of transcription through the attenuator. While it is difficult to ascertain the relative kinetic parameters of transcription and translation through these leader–attenuator regions, the importance of the timing of ribosome release for setting the basal levels of attenuation for each of these operons can be compared. Roesser and Yanofsky[30,31] demonstrated the importance of the rate of ribosome release from the *trp* leader RNA by showing that mutations in *prfB*, encoding release factor 2 (UGA- and UAA-specific[32]), increase transcription termination at the attenuator of the *trp* operon approximately twofold. On the other hand, these *prfB* mutations exert no effect on basal level expression in strains in which the naturally occuring *trp* leader polypeptide stop codon, UGA, is replaced with UAG or UAA. Transcription through the attenuator is increased in these strains, however, when they contain mutations in *prfA*, which encodes release factor 1 (UAG- and UAA-specific[32]). They also showed that preventing the synthesis of the *trp* leader polypeptide by changing the start codon from AUG to AUA increases transcription at the attenuator fourfold (superattenuation) and, as predicted, neither the *prfA* nor the *prfB* mutations affect transcriptional readthrough at the attenuator in superattenuated strains.

These experiments, together with experiments that affect the relative stabilities of the alternative secondary structures of the *trp* leader RNA, provide the basis of the model for basal level expression of the *trp* operon.[4,31] Recently, similar experiments have been performed to determine if leader polypeptide synthesis and ribosome release influence the basal level of expression of the *ilvGMEDA* operon.[33] The promoter–leader region was transcriptionally fused to the β-galactosidase gene, with (*ilv*P_G1P_G2attenuator::*lacZ*) or without (*ilv*P_G1P_G2::*lacZ*) the attenuator, and single-copy integrants of these constructs in the chromosome of strains containing wild-type or mutant alleles of *prfA* or *prfB* (release factor 1, UAG- and UAA-specific; and release factor 2, UGA- and UAA-specific; respectively) were isolated. The naturally occurring stop codon for the *ilvGMEDA* leader polypeptide is UAG. The results of these experiments (Table I) show that mutations in *prfA*, encoding release factor 1 (UAG- and UAA-specific), increase transcription termination at the attenuator greater than twofold, whereas *prfB* mutations, encoding release factor 2 (UGA- and UAA-specific), do not affect transcription termination at the attenuator. The results in Table I also show that these mutations are attenuator-specific, since they do not affect transcription initiation from the tandem promoter region of the *ilvGMEDA* operon. Furthermore, a mutation in the leader–attenuator region that changes the start codon of the leader polypeptide coding region from AUG to AUA increases transcription termination at the attenuator fivefold (data not shown). Thus, like the *trp* operon,

Table I. The Effects of Ribosome Release Factor Mutants on Transcription Through the Attenuator of the *ilvGMEDA* Operon

E. coli strain	Relevant genotype	β-Galactosidase specific activity[a]	
		30°C	37°C
IH550	pRS550::*lacZ*	ND	ND
IH450	*ilv*P_G1P_G2::*lacZ*	9640	10830
IH454	*ilv*P_G1P_G2::*lacZ*, *RF-1*	9990	9740
IH455	*ilv*P_G1P_G2::*lacZ*, *RF-2*	9820	10100
IH211	*ilv*P_G1P_G2attenuator::*lacZ*	1100	1090
IH842	*ilv*P_G1P_G2attenuator::*lacZ*, *RF-1*	1330	515
IH853	*ilv*P_G1P_G2attenuator::*lacZ*, *RF-2*	1280	1180

[a]β-Galactosidase-specific activity (nmole ONP/min per mg protein) was assayed as described by Miller.[34] Each β-galactosidase-specific activity measurement is the average of assays performed at three extract concentrations with four assay times each. Each strain was grown at 30 or 37°C in glucose minimal salts medium supplemented with L-isoleucine, L-valine, and L-leucine. Wild-type strains grew with an average doubling time of 51 min and release factor mutant strains grew with an average doubling time of 59 min. ND, none detected.

the basal level of expression of the *ilvGMEDA* operon is influenced by the timing of the release of a ribosome from the leader RNA.

Any model for the *ilvGMEDA* attenuator mechanism that includes the observation that a delayed ribosome release time from the leader RNA results in superattenuation of transcription into the structural genes of the *ilvGMEDA* operon must acknowledge that a ribosome must reach the stop codon of the leader polypeptide coding region well in advance of the attenuation decision. If a single-ribosome model is assumed, the ribosome must release the RNA polymerase at base pair 117 in the leader attenuator region and get to the stop codon, 96 nucleotides downstream, and wait about 0.6 sec to dissociate before the released RNA polymerase finishes the synthesis of the stem 3 region of the leader RNA. At a translation rate of 50 nucleotides per second, it takes the ribosome 1.92 sec to translate the leader polypeptide coding region. It takes the RNA polymerase only 1 sec, however, to transcribe to the middle of the stem–loop 3:4 coding region, the polymerase position around which the attenuation decision is made. If the 0.6 sec average ribosome release time is added to the leader polypeptide translation time, then the ribosome must translate the leader polypeptide coding region at a minimum rate of 240 nucleotides per second. This translation rate, which corresponds to a polypeptide chain synthesis rate of 80 amino acids per second, is unprecedented. Alternatively, a slow translation initiation rate (greater than 2.34 sec; Fig. 4B) must be assumed, since the attachment of a ribosome to the leader RNA is the event that is presumed to release the paused RNA polymerase. Neither of these assumptions appears as attractive as the two-ribosome model.

7. DEATTENUATION AND THE SECOND RIBOSOME

I favor a two-ribosome model for *ilvGMEDA* attenuation not only because a one-ribosome model demands either an unacceptably rapid translation elongation rate or a very slow translation initiation rate, but also because deattenuation by starvation for isoleucine and valine is more easily explained with a two-ribosome model. In fact, we have previously proposed such a model to satisfy this problem.[10] In order to illustrate the importance of the second ribosome for deattenuation, it is constructive to consider the leader RNA secondary structures that can form when a ribosome(s) is stalled at different regulatory codons.

If the nascent leader RNA is not translated, then complementary sequences will base pair as they are synthesized (Figs. 2B and 5A). The uninterrupted formation of the attenuator in the absence of translation probably explains the efficient termination observed in *in vitro* transcription assays.[10] In the presence of translation, however, a ribosome stalled at the tandem leucine codons at the fourth and fifth positions of the leader polypeptide coding region will mask nine nucleotides on either side of the codon in the A-site.[24] A ribosome stalled at this position will inhibit the attachment of a second ribosome to the leader RNA and

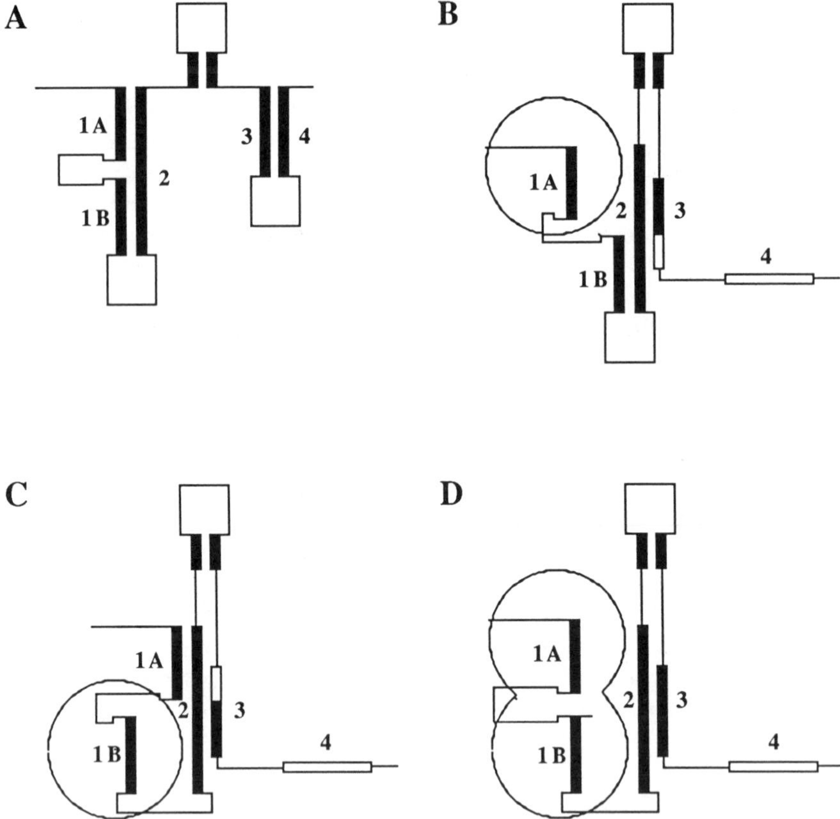

Figure 5. The postulated effects of ribosome stalling in the leader RNA of the *ilvGMEDA* operon. Black bars represent base-paired regions of the leader RNA. Open bars represent nonpaired regions. The large open circles represent ribosomes. **(A)** No translation. **(B)** Ribosome stalled at the tandem leucine codons. **(C)** Ribosome stalled at the tandem valine or isoleucine codons. **(D)** Ribosome stalled at the tandem valine or isoleucine codons with the attachment of a second ribosome to the leader RNA.

disrupt base pairings between stem 1A and the 3′ half of stem 2. This will facilitate the formation of the antiterminator by freeing the 3′ half of stem 2 to base pair with complementary sequences in stem 3. These base pairings will, in turn, preempt the subsequent base pairing of complementary sequences in the 5′ half of stem 3 and the 3′ half of stem 4 (Fig. 5B). This partial disruption of the attenuator leads to a 4- to 5-fold decrease in transcription termination at the attenuator (Fig. 5B).[19] While there is another leucine codon at position 10 of the leader polypeptide coding region, this codon is not important for attenuation, since replacing the tandem leucine codons at positions 4 and 5 with codons that do not encode leucine, and do not affect the basal level of transcription through the attenuator, completely eliminates the effect of leucine starvation.[35]

Twelve of the 15 codons between codon positions 5 and 20 of the leader polypeptide coding region encode isoleucine and valine. It is not known which of these codons are regulatory codons. Since ribosomes mask about three codons on either side of the codon being read,[24] however, the isoleucine, valine, and leucine codons at positions 7, 8, and 10 would not be expected to mask RNA sequences that directly affect attenuator formation, although it is quite possible that these codons are important for controlling the overall transit time of ribosomes through the stem 1 region during conditions of mild branched-chain amino acid starvation. For example, the presence of ribosome 2 in this side loop of the bifurcated stem–loop 1:2 structure would be expected to destabilize stem–loop 1:2 and counteract the effect of the late arrival, and late release, of ribosome 1 at the stop codon.

The tandem valine codons at positions 11 and 12 and the three valine codons at positions 15–17 are situated in places where ribosome stalling will disrupt base pairings between stem 1B and the 5′ half of stem 2. These base pairings will, in turn, preempt subsequent base pairings between complementary sequences between the 3′ half of stem 3 and the 5′ half of stem 4 of the attenuator (Fig. 5C). Thus, the tandem leucine and valine codons in the stem 1 region of the leader RNA are in positions where the stalling of a single ribosome (ribosome 1) can effect deattenuation. On the other hand, the tandem isoleucine codons are situated in the leader RNA where the effect of ribosome stalling is more problematic. While the stalling of a ribosome on these codons in the loop region of stem–loop 1:2 would weaken the stability of base pairings between stem 1A and stem 2, it would not directly facilitate the formation of the antiterminator stem–loop 2:3 structure in the way effected by the stalling of a ribosome on the tandem leucine and valine codons near the bifurcation in stem 1A. However, if a second ribosome were to initiate translation of the leader 1 sec after the first (the exact time that ribosome 1 reaches the first isoleucine codon), then the second ribosome would queue up approximately 30 nucleotides behind the first[36] over the isoleucine codon at position 8 in the leader polypeptide coding region. The position of this second ribosome would disrupt the base pairings between stem 1A and stem 2 and effect deattenuation as described above for starvation by leucine (Fig. 5D).

8. SUMMARY

Attenuation is a complicated regulatory mechanism. Despite its complexities, however, Charles Yanofsky and his colleagues have provided us with a thorough understanding of the attenuation mechanism responsible for the regulation of the genes of the tryptophan operon of *E. coli*. This understanding is based on the results of rigorous and comprehensive experimental documentation gathered over the past 15 years. Following the initial description of attenuation for the tryptophan operon, it was discovered that other amino acid biosynthetic operons are also regulated by attenuation. Due to the remarkable functional and structural similarities among the attenuators of these operons it has been assumed that the fundamental mechanisms for attenuator regulation are also similar. Indeed, this has proven, in general, to be the case. The basic model that describes attenuation in terms of alternative leader RNA structures that form in response to the relative positions of a ribosome and an RNA polymerase in the leader region has been proven for *trp* attenuation, and the existence of similar alternative base-pairing structures important for attenuator function in other amino acid operons has been described (reviewed in refs. 4 and 19). Nevertheless, many functional and structural variations exist among the leader–attenuator regions of the amino acid operons regulated by attenuation. The analysis of these differences between the *ilvGMEDA* and *trp* operons suggests a fundamental difference in the attenuation mechanism employed by these two operons. That is, whereas the *trp* leader RNA is translated by only one ribosome per attenuation event, the *ilvGMEDA* operon might require the participation of two ribosomes to effect an attenuation decision. There are at least two reasons why this difference might exist. First, the *ilvGMEDA* attenuator must respond to the intracellular depletion of any one of three amino acids, while the *trp* attenuator monitors the level of only one amino acid. The physical presence of multiple codons for three amino acids obviously requires more coding sequence. Second, the placement of multiple tandem codons for three different amino acids at a place in the leader RNA that would effect deattenuation by ribosome stalling presents a logistical problem. Both of these problems are satisfied, however, by the two-ribosome model described in Sections 5–7.

The leader RNA of the *trp* operon is "hard wired" in two ways so that it will not be translated by more than one ribosome per attenuation event. First of all, it takes the released RNA polymerase only about 1 sec to transcribe from base pair 92, where it is released by the translating ribosome, to the end of the leader RNA at base pair 140. This is probably an insufficient time for a second ribosome to initiate translation before the attenuation decision is made. Second, there are sequences in the 3′ end of the *trp* leader RNAs of *E. coli*, *S. marcesens*, and *Salmonella typhimurium* that base pair with RNA sequences in the ribosome attachment site.[37] The base pairing of these sequences inhibits any subsequent translation initiation of these *trp* leader RNAs.[38] On the other hand, there are no sequences in the leader RNA of the *ilvGMEDA* operon that are complementary to the ribosome attachment site and, as pointed out in Section 4, there should be ample time for a second ribosome to initiate translation and effect the attenuation decision in the leader region of this operon.

A basic assumption used to compare the *ilvGMEDA* and *trp* operons in this chapter is that the temporal events associated with the translation and transcription of the leader–attenuator regions of these two operons are comparable. It should be emphasized that this assumption has not been verified. Indeed, it is quite possible that codon context effects could cause the translation rate through the leader polypeptide coding region of these operons to vary or that additional transcriptional pause sites are created during coupled transcription and translation that are not observed with *in vitro* transcription assays. Nevertheless, the average rate parameters used in the analyses presented here are sufficient to explain the attenuation mechanism of the well-defined tryptophan operon. Large changes in these parameters of an unprecedented magnitude would be required to accommodate a single-ribosome model for the *ilvGMEDA* operon and, even if these temporal differences were assumed, the organization of the mutually exclusive base-pairing regions of the leader RNA of this operon demands two ribosomes for efficient deattenuation during isoleucine starvation (Section 7).

Another assumption that is critical for the two-ribosome model concerns the spacing between queued ribosomes on a messenger RNA. While I am unaware of any measurements of this spacing in a prokaryote, Wolin and Walter[36] have performed elegant experiments with a eukaryotic translation system which demonstrate that ribosomes are spaced 27–29 nucleotides apart behind a translational pause site. It does not seem unreasonable, therefore, to assume, as I have here, that smaller prokaryotic ribosomes can stack as tightly as 30 nucleotides apart, although, the ability of the *ilvGMEDA* leader RNA to accommodate translation by two ribosomes at one time must, of course, be verified experimentally.

References

1. Jacob, F., and Monod, J., 1961, *J. Mol. Biol.* **3:**318.
2. Jacob, F., and Monod, J., 1961, *Cold Spring Harbor Symp. Quant. Biol.* **26:**193.
3. Englesberg, E., Irr, J., Power, J., and Lee, N., 1965, *J. Bacteriol.* **90:**946.
4. Landick, R., and Yanofsky, C., 1987, in: *Escherichia coli and Salmonella typhimurium, Cellular and Molecular Biology* (F. C. Neidhardt, J. L. Ingraham, K. B. Low, B. Magasanik, M. Schaechter, and H. E. Umbarger, eds.), p. 1276, American Society for Microbiology, Washington, D. C.
5. Yanofsky, C., 1988, *J. Biol. Chem.* **263:**609.
6. Aloni, Y., and Hay, N., 1985, *CRC Crit. Rev. Biochem.* **18:**327.
7. Bertrand, K., Squires, C., and Yanofsky, C., 1976, *J. Mol. Biol.* **103:**319.
8. Lee, F., Squires, C. L., Squires, C., and Yanofsky, C., 1976, *J. Mol. Biol.* **103:**383.
9. Bertrand, K., Korn, L. J., Lee, F., and Yanofsky, C., 1977, *J. Mol. Biol.* **117:**227.
10. Lawther, R., and Hatfield, G. W., 1980, *Proc. Natl. Acad. Sci. USA* **77:**1862.
11. Nargang, F. E., Subrahmanyam, C. S., and Umbarger, H. E., 1980, *Proc. Natl. Acad. Sci. USA* **77:**1823.
12. Friden, P., Newman, T., and Freundlich, M., 1982, *Proc. Natl. Acad. Sci. USA* **79:**6156.
13. Hauser, C. A., and Hatfield, G. W., 1984, *Proc. Natl. Acad. Sci. USA* **81:**76.
14. Gardner, J. F., 1979, *Proc. Natl. Acad. Sci. USA* **76:**1706.
15. Keller, E. B., and Calvo, J. M., 1979, *Proc. Natl. Acad. Sci. USA* **76:**6186.
16. Barnes, W. M., 1978, *Proc. Natl. Acad. Sci. USA* **75:**4281.
17. DiNocera, P. P., Blasi, F., DiLauro, R., Frunzio, R., and Bruni, C. B., 1978, *Proc. Natl. Acad. Sci. USA* **75:**4276.
18. Zurawski, G., Elseviers, D., Stauffer, G., and Yanofsky, C., 1978, *Proc. Natl. Acad. Sci. USA* **75:**4271.
19. Sharp, J. A., and Hatfield, G. W., 1987, in: *Translational Regulation of Gene Expression* (J. Ilan, ed.), p. 447, Plenum Press, New York.
20. Gausing, K., 1972, *J. Mol. Biol.* **71:**529.

21. Kassavetis, G. A., and Chamberlin, M. J., 1981, *J. Biol. Chem.* **256:**2777.
22. Sørensen, M., Kurland, C. G., and Pedersen, S., 1989, *J. Mol. Biol.* **297:**365.
23. Jensen, K. F., 1988, *Eur. J. Biochem.* **175:**587.
24. Oxender, D. L., Zurawski, G., and Yanofsky, C., 1979, *Proc. Natl. Acad. Sci. USA* **76:**5524.
25. Curran, J. F., and Yarus, M., 1988, *J. Mol. Biol.* **203:**75.
26. Roesser, J. R., and Yanofsky, C., 1990, *J. Biol. Chem.* **265:**6055.
27. Pagel, J. M., and Hatfield, G. W., 1991, *J. Biol. Chem.* **266:**1985.
28. Tsui, P., and Freundlich, M., 1989, *J. Mol. Biol.* **203:**817.
29. Hauser, C. A., Sharp, J. A., Hatfield, L. K., and Hatfield, G. W., 1985, *J. Biol. Chem.* **260:**1765.
30. Roesser, J. R., and Yanofsky, C., 1988, *J. Biol. Chem.* **263:**14251.
31. Roesser, J. R., and Yanofsky, C., 1989, *J. Biol. Chem.* **264:**12284.
32. Caskey, C. T., 1980, *Trends Biochem. Sci.* **54:**234.
33. Parekh, B., Pagel, J. M., and Hatfield, G. W., (in preparation).
34. Miller, J. H., 1972, In: *Experiments in Molecular Genetics*, Cold Spring Harbor Laboratory, Cold Spring Harbor, New York.
35. Chen, J.-W., Harris, E., and Umbarger, H. E., 1991, *J. Bacteriol.* **173:**2328.
36. Wolin, S. L., and Walter, P., 1988, *EMBO J.* **7:**3569.
37. Yanofsky, C., 1984, *Mol. Biol Evol.* **1:**143.
38. Das, A., Crawford, I. P., and Yanofsky, C., 1982, *J. Biol. Chem.* **257:**8795.

Chapter 2

Regulation of Ribosomal Protein mRNA Translation in Bacteria

The Case of S15

Claude Portier and Marianne Grunberg-Manago

1. INTRODUCTION

Ribosomal proteins (r-proteins) bind to specific sites on ribosomal RNA (rRNA) to assemble a ribosome. Except for L12/L7, there is always a stoichiometric amount of all r-proteins and rRNA and the amount of free ribosomal proteins is low in the cell. Thus a coordinated regulation must exist. For the vast majority of ribosomal proteins, the mechanism of coordinate expression is at the translational autoregulation level. Some exceptions to this rule are observed, e.g., in the S10 operon (regulated both by transcription attenuation and translation inhibition[1]), the *spc* operon,[2] and the *trmD* operon (nonautoregulated).[3,4] In most cases, each operon encoding a ribosomal protein encodes a ribosomal protein repressor (Fig. 1). This repressor binds generally specifically to the 5′ region of the messenger and stops the translation of all downstream cistrons. This phenomenon, called translational coupling, has been demonstrated for the L11–L1 operon. When the ribosome binding site for L11, the upstream gene, is mutated, the L1 cistron is no longer

Claude Portier and Marianne Grunberg-Manago • Institut de Biologie Physico-Chemique, 75005 Paris, France.

Translational Regulation of Gene Expression 2, edited by Joseph Ilan. Plenum Press, New York, 1993.

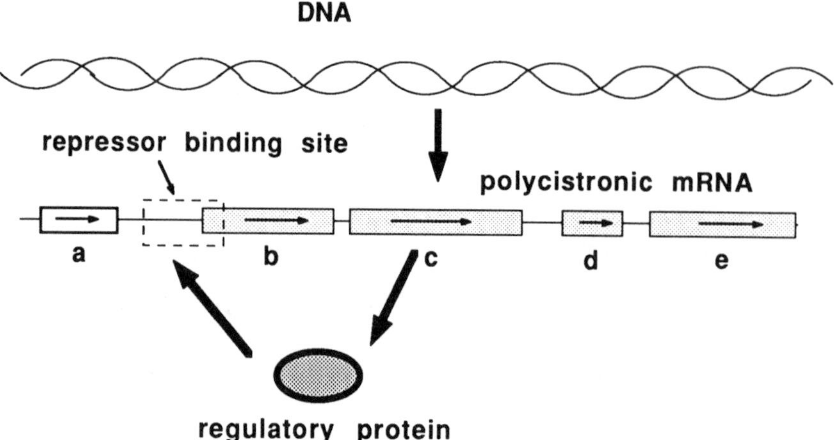

Figure 1. General regulatory scheme for translational regulation. The messenger encoding a ribosomal gene carries a repressor binding site generally located at close proximity to a ribosome loading site. This overlap induces competition between the ribosome and the ribosomal repressor protein for binding at this site. After binding of the repressor, translational coupling blocks the expression of the downstream reading frames.

translated.[5] This suggests that the ribosome, after initiating on a polycistronic message, is able to translate each coding region without leaving the messenger or by allowing a simultaneous entry of a new initiating ribosome after translation termination of a cistron. This has been established for several messenger RNAs (mRNAs) with overlapping messages or containing short intercistronic regions.[6,7] Long-range translational coupling has also been described.[8,9] In Eubacteria, a bidirectional scanning movement of the ribosome allows, after translation termination of a first message, for the selection of another initiation codon if it is located in a radius of about 40 nucleotides around the termination signal.[10] Often, a more or less complex "slippery" region is present to allow for efficient frameshifting.[11,12] When the intercistronic region is large, the presence of a ribosome binding site upstream of each coding frame is occluded by the formation of various secondary structures forbidding the access to the external ribosomes. It is the translating ribosomes which will open these sequences and either translate the downstream message or allow reinitiation.[13,14]

In translational control, the binding of the repressor to the mRNA occurs only when the repressor is in excess in the cell, that is, when its binding site on the ribosomal RNA is saturated. Thus, the ribosomal protein synthesis rate is tightly coupled to the rRNA synthesis rate. This type of regulation raises several questions: Where is the repressor binding site located on the mRNA? How does ribosomal protein bind to it? Do structural similarities exist between messenger and ribosomal protein binding sites? What are the mechanisms used to stop translation? Some answers to these questions have been obtained through the detailed study of only a limited number of examples, particularly the α, S10, *rpsO–pnp*, *spc*, and L11-*rif* operons.

When measured, the binding affinity of a ribosomal protein for its own messenger is practically identical to the value observed for rRNA[15,16]: 10^7 M^{-1}. This is considerably lower than the affinity for protein-DNA complexes, which reaches 10^{12} M^{-1}. Consequently, protein–RNA complexes have short lifetimes which fit with rapid equilibrium and high turnover. In the ribosomes however, interactions between neighboring proteins probably increase the RNA binding affinity, which accounts for ribosome stability. It is quite different for messenger binding, where presumably one protein binds to the repressor binding site, excluding any cooperativity in this association. This lack of cooperativity could explain why the ribosomal repressor proteins bind to rRNA with an apparent higher affinity when compared to the value experimentally measured and why these repressors bind to mRNA and inhibit its translation only when in excess in the cell, that is, when no more ribosomal RNA sites are available. Another problem is to understand how a repressor protein binds to its RNA site. In other words, what is the structure recognized or, more precisely, what are the determinants involved? And finally, how does this binding modify the translation rate of the messenger? Two mechanisms have been proposed, the displacement model and the entrapment model (Fig. 2). In the displacement model, the repressor binding site generally overlaps the ribosome loading site on the mRNA. Thus, there is a competition between the repressor and the ribosome for this site and any increase in the ribosome binding efficiency decreases the repressor binding capacity. Fixation of the ribosomal repressor protein on this site hinders ribosome loading and blocks initiation. In the entrapment model, there are two independent sites, one for ribosome loading and another for repressor binding. At first approximation, ribosome binding is unaffected by repressor binding.[17] In this case, the repressor stops the ribosome movement, either by stabilizing the ribosome binding to the Shine–Dalgarno sequence, by slowing its dissociation, or in any other way. Almost all examples of translational autoregulation mechanisms belong to the displacement model. Only two systems seem different: autoregulation of S4 in the α operon and of S15 in the *rpsO–pnp* operon. Since these systems are exceptional and new information is available, especially for S15, this chapter is entirely devoted to S15 autocontrol.

2. THE *rpsO–pnp* OPERON

Besides *rpsO*, the gene encoding the small ribosomal protein S15, this operon carries *pnp*, the gene for polynucleotide phosphorylase. These two genes are located at 69 min on the *Escherichia coli* chromosome and are surrounded by several genes involved in translation. Among the identified genes, *metY*, *nusA*, and *infB* are located upstream and are part of another operon. They code for a minor form of $tRNA_{Metf'}$, NusA, a transcription termination factor, and IF2, a translation

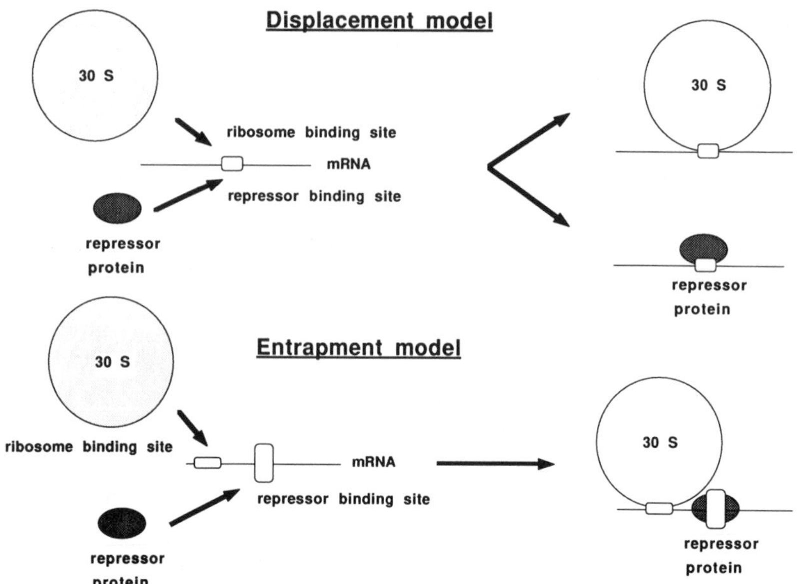

Figure 2. The two models for translational control according to Draper. The displacement model is described in Fig. 1. It appears to be a general mechanism for control. In the entrapment model, the ribosome and the ribosomal protein bind at two different sites on the mRNA. The repressor binding site is located downstream of the Shine-Dalgarno sequence in order to inhibit the ribosome progression. Interaction between the ribosome and the repressor protein is predicted and might play a role in the stabilization of the complex.

initiation factor, respectively. Downstream of the *rpsO–pnp* operon, the *deaD* gene has recently been identified. It shows similarity with the ATP-dependent RNA helicase family and might play a role in translation.[18]

The *rpsO* gene is expressed from a promoter located about 100 nucleotides upstream (Fig. 3). This transcript does not terminate entirely at the *rpsO* terminator. About 10% of the messengers cross the barrier and extent into the *pnp* gene. This bicistronic message is cleaved by RNase III in the intercistronic space.[19] Moreover, there are two other endonucleolytic cleavages by RNase E on either side of the terminator.[20] These cuts remove the terminator and probably initiate the chemical degradation of S15 mRNA. The necessity for a second cut, just downstream of the terminator, is not clear.

3. THE EXPRESSION OF S15 IS CONTROLLED AT THE TRANSLATIONAL LEVEL

Previous studies have shown that S15 expression is very tightly regulated. Its concentration *in vivo* increases only twofold, even when *rpsO* is carried by a

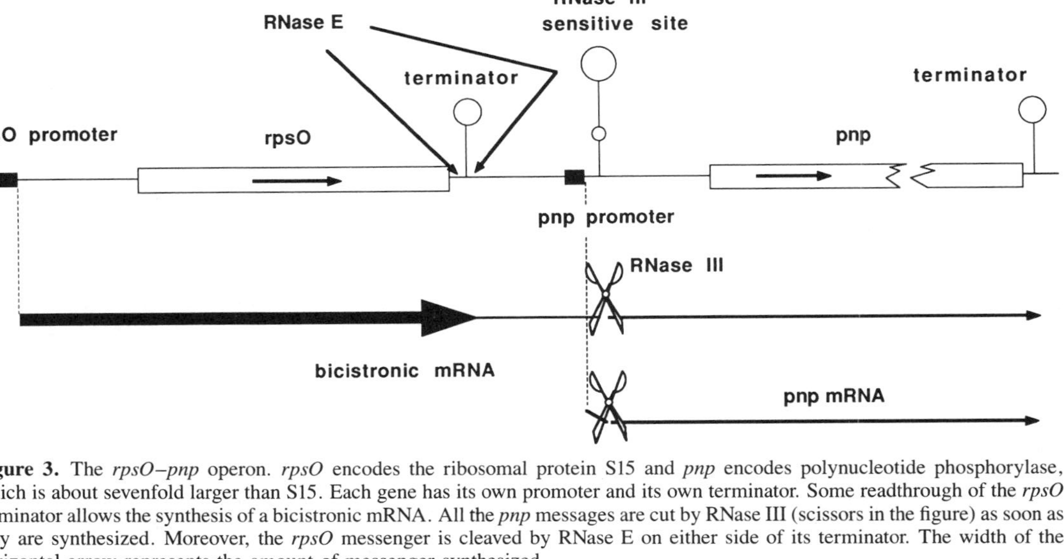

Figure 3. The *rpsO–pnp* operon. *rpsO* encodes the ribosomal protein S15 and *pnp* encodes polynucleotide phosphorylase, which is about sevenfold larger than S15. Each gene has its own promoter and its own terminator. Some readthrough of the *rpsO* terminator allows the synthesis of a bicistronic mRNA. All the *pnp* messages are cut by RNase III (scissors in the figure) as soon as they are synthesized. Moreover, the *rpsO* messenger is cleaved by RNase E on either side of its terminator. The width of the horizontal arrow represents the amount of messenger synthesized.

multicopy plasmid.[21] Since the vast majority of ribosomal proteins are translationally autoregulated, a model can be proposed. Considering the mass action law, the amount of S15 synthesized must increase, even if tightly regulated, because the equilibrium between the different constituents is modified:

$$\frac{[\mathrm{R}] + [\mathrm{mRNA}]}{[\mathrm{R{-}mRNA}]} = K$$

Here R is the repressor (here S15), mRNA = free S15 mRNA, and R–mRNA = S15 mRNA fraction bound to S15. If the concentration of S15 mRNA increases, the concentration of S15 and that of the complex must also increase to maintain the equilibrium. When S15 is expressed from a multicopy plasmid, however, the total repressor amount comes from the contribution of several *rpsO* genes. Then, although the total concentration of S15 increases, the S15 synthesized from only one copy should decrease to account for the moderate increase of S15 in the cell. To analyze the variations in the synthesis rate of S15 under different conditions, a translational fusion between the 5′ upstream part of *rpsO* and the distal part of *lacZ* was constructed and used as a reporter gene.[22] In this system, the fusion is inserted into the chromosome and plasmids expressing S15 are introduced in the cell (Fig. 4). In the presence of a translational control, the level of the S15–β-galactosidase fusion protein, which indicates the expression of one copy of the *rpsO* gene, should decrease. Table I shows that, in the presence of *rpsO* in *trans*, the level of the chimeric protein is decreased about ninefold. Moreover, this repression effect is not observed in the presence of only *pnp*, the second gene of the operon. Thus, S15 synthesis is regulated by its own product. The transcription rate of the fusion was determined in the presence and absence of *rpsO* in *trans*. The values are quite similar (Table II), indicating that the control is at the translational step. Thus, it can be concluded that S15 expression is autoregulated at the translational level.

4. THE CONTROL OF *rpsO* AND THAT OF *pnp* EXPRESSION ARE INDEPENDENT

Since the *pnp* gene is coexpressed with *rpsO*, at least partially, is expression of the *pnp* gene also regulated by S15? In other words, is its expression translationally coupled to that of S15? By constructing an inframe translational fusion between *pnp* and *lacZ*, the same kind of experiment was performed. A strain carrying this fusion in the chromosome was transformed with plasmids expressing *rpsO*. No effects were observed when compared to a control with plasmids without *rpsO*. This indicates that translational coupling does not occur and that the two genes are controlled entirely independently at the translational level. This result is not quite unexpected, given the existence of the two types of endonucleolytic cleavage in the intercistronic domain of the cotranscripts (see above). Moreover,

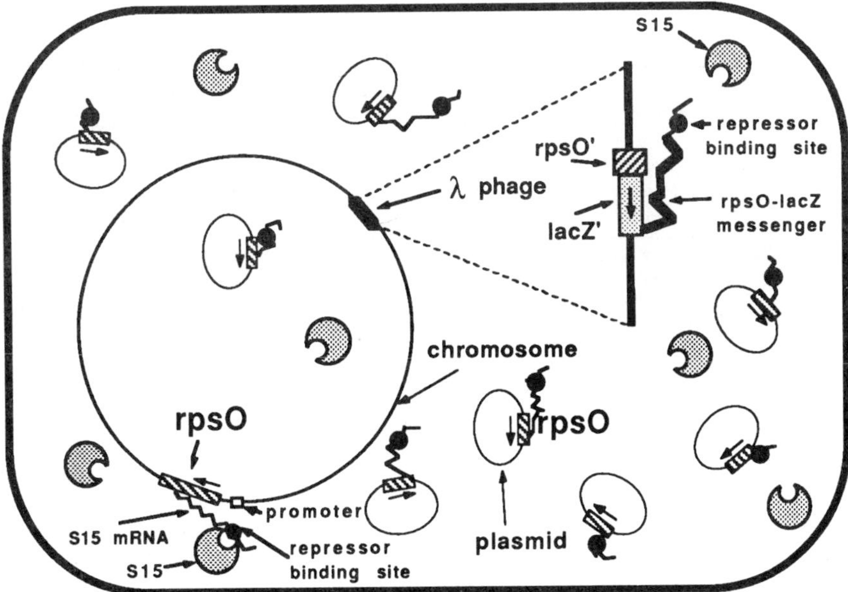

Figure 4. System used to study translational repression of S15. The chromosome is shown as a large circle and plasmids by small ovals. Both carry *rpsO* genes which are actively transcribed into S15 mRNA. The repressor binding site of the mRNA is indicated by a black circle. The protein S15 is represented by a shaded, partial circle. The λ phage integrated in the chromosome is carrying a translational fusion between the 5′ region of *rpsO*(*rpsO′*) and the distal part of *lacZ* (*lacZ′*) (hatched rectangle). The correspondinɡ ' ybrid messenger exhibits the S15 binding site and its translation is controlled by the free conc‹ ıtration of S15. Thus, the β-galactosidase concentration produced from this fusion mirrors the synthesis rate of S15.

Table 1. Repression of the Activity of a *rpsO–lacZ* Fusion by a Plasmid Carrying *rpsO* in *Trans*[a]

Plasmid	Activity of the fusion (β-gal units)	Repression ratio
None (control)	116	—
pBP Δ10 (+*pnp*)	118	1.0
pBP Δ7 (+*rpsO*)	24	4.8

[a]When this fusion is introduced into the chromosome, the β-galactosidase level is not changed when a plasmid expressing *pnp* is present, but only if the plasmid expresses *rpsO*. This is illustrated by the repression ratio (the value without repressor divided by the value in the presence in *trans* of a gene carried by the operon studied). The higher the repression ratio, the higher the effect.

Table II. Synthesis Rate of the Fusion *rpsO–lacZ* in the Presence of S15 Expression in *Trans*[a]

Plasmid	ThrSase mRNA (control) (% input)	s15–βgal mRNA (% input)	Percent of the control
None (control)	0.020	0.080	100
pBP Δ10 (+*pnp*)	0.019	0.080	105
pBP Δ7 (+*rpsO*)	0.022	0.058	67

[a]The *rpsO–lacZ* mRNA synthesis rate was determined by hybridizing one-min-pulse radiolabeled total mRNA to specific single stranded probes. The radioactivity hybridized on filters was counted and standardized to an internal control (threony-l-tRNA-synthetase mRNA) for any loss during extraction or alteration in hybridization yields, and expressed as a percentage of the input. In the presence of *rpsO* in *trans*, there is only about a 30% decrease, probably induced by an increased mRNA degradation rate. If the control were transcriptional, the value would have to decrease about fivefold (see Table I).

recent work on *pnp* expression has shown that polynucleotide phosphorylase is also autoregulated.[23] This situation is similar to that for the *rif* operon (Fig. 5): the first two genes encode the ribosomal proteins L10 and L12, and the two distal genes encode the β and β′ subunits of RNA polymerase. In this operon, expression of the first two genes is autocontrolled independently of the downstream genes.[24–27] Very curiously, the intercistronic space between L12 and β′ carries a terminator for L12 followed by an RNase III-sensitive site.[28] Moreover, there is cotranscription between the ribosomal genes and β, β′.[29] The significance, if any, of the structural and functional similarities between these operons is not known.

5. POSITION OF THE REPRESSOR BINDING SITE

If S15 regulates its expression by binding to its messenger, where is this binding site located? Since the fusion is regulated, it appears that the S15 binding site is in the 5′ end of the message. A more precise location was obtained by isolation of mutants deregulated for the expression of *rpsO*.[22] Under repressed conditions, the hybrid β-galactosidase protein concentration is not enough to support growth on lactose media. Deregulated mutants, thus, are colonies that grow under these conditions. The spontaneous mutations obtained are all located in the beginning of the coding phase (Fig. 6). The effect of the mutations on regulation might be accounted for by the formation of a hairpin. In this model, the Shine–Dalgarno sequence would be paired in the stem. Each mutation, by destabilizing this structure, would liberate the ribosome binding site and increase expression. Secondary structure probes of the S15 mRNA leader using chemical and enzymatic treatment detected two stable helices upstream of the presumed hairpin. The expected hairpin showed considerable heterogeneity,[30] probably

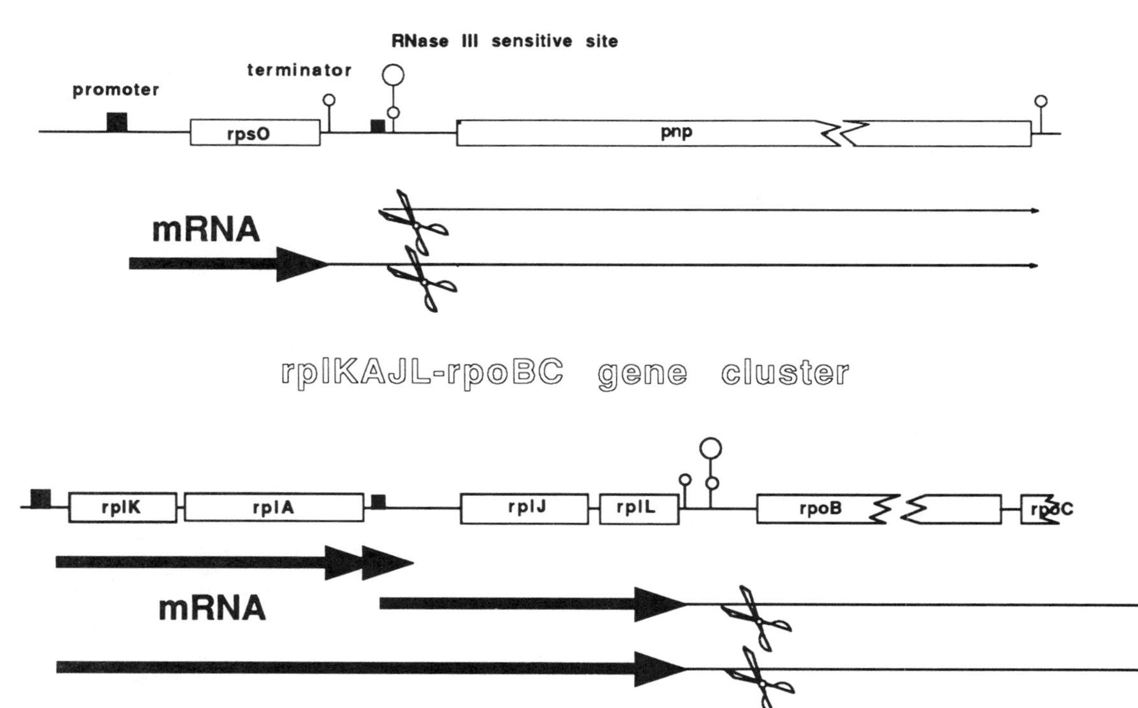

Figure 5. Comparison between the *rpsO–pnp* operon and the *rplKAJL–rpoBC* cluster. Same explanation as in legend to Fig. 3. In both cases, there is readthrough after the terminator of the ribosomal gene and the polycistronic message is cleaved by RNase III at a site located in the intercistronic space. Moreover, both have translational autocontrol for the expression of genes located upstream and downstream of the RNase III site.

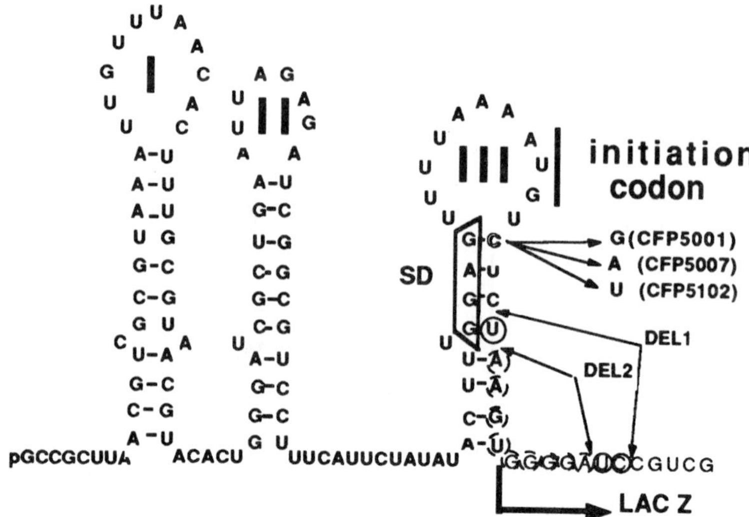

Figure 6. Localization of the spontaneous mutations in the S15 mRNA leader. The stem–loops in the leader mRNA are numbered from 5′ and 3′. The Shine–Dalgarno sequence (SD) is boxed and the initiation codon underlined. Three point mutations were located at the same nucleotide site (outlined) in the S15 coding phase. Two deletions were isolated downstream (DEL1 and DEL2) of this nucleotide and the deleted nucleotides are circled. The junction of the fusion is indicated by an arrow.

because of its low stability. Further studies, however, do not entirely support these conclusions. Deletion of the second stem–loop, for example, totally abolishes the autoregulation, but deletion of the first stem–loop has no effect. This result indicates that the repressor binding site overlaps the ribosome loading site and extends through two stem–loops.

6. PSEUDOKNOT STRUCTURE OF THE REPRESSOR BINDING SITE

The presence of mutations in the two different stem–loops suggests that the two domains may interact. The existence of perfect complementarity between two sequences of seven nucleotides located in different hairpins indicates that a pseudoknot could form. Several compensatory mutations were created in these sequences to ascertain the existence of a pseudoknot: when a single mutation is created, the control is lost, and after introducing a compensatory base change, control is restored.[31] These results suggest that the nucleotides tested base pair and that a pseudoknot is formed. In this case, helix 1 is presumed to stack with helix 2, corresponding to the base pairs formed between the nucleotides of the loop and a complementary sequence outside (Fig. 7). The handedness of the double helix

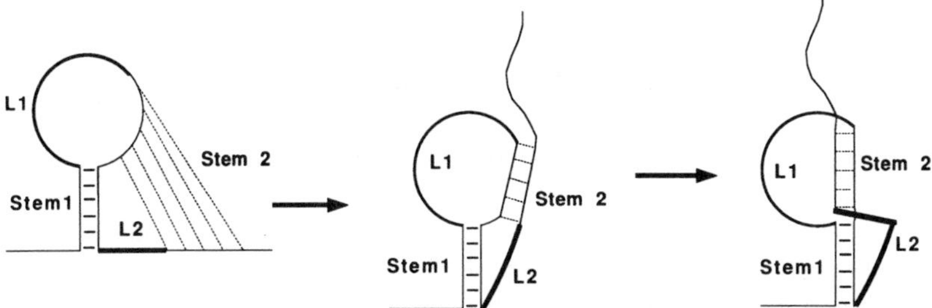

Figure 7. Formation of a pseudoknot. In the presence of a stem–loop, interaction between a part of the loop and its complementary sequence outside creates a helix (stem 2) which will stack on top of another (stem 1), forming a quasicontinuous regular double helix. Due to the handedness of the double helix and the polarity of the chain, the connecting loops L1 and L2 are different: L1 cross the deep major groove and L2 crosses the minor shallow groove.

and the polarity of the polymer create two different connecting loops. Loop 1 (L1) crosses the deep major groove and loop 2 (L2) the shallow minor groove.[32] Recent studies have shown that stacking between the stems induces the formation of a helix exhibiting a fairly undistorted A-form helix on one side and carrying both loop regions and the junction of the loops with the stems on the other side.[33] Moreover, a pseudoknot structure is stable, even in the absence of further interaction with RNA outside. The stability, however, is strongly affected by the length of the loops 1 and 2.[32,34] Theoretical considerations suggest that at least one nucleotide is enough for L1 stabilization and two for L2.[32] In fact, the number of nucleotides forming the loops is dependent upon the number of bases they must cross because the distances between one phosphate atom on one strand and another on the opposite strand vary in a regular RNA double helix and because this variation is quite different depending upon whether the major or minor groove is crossed. In our example, theoretically, one nucleotide could be the minimum number for loop 1 if stem 2 is constituted by seven nucleotides, with two nucleotides in loop 2, and if stem 1 is one to three base pairs. Experimental data are necessary to confirm these predictions. This hypothesis is demonstrated by the observations of Wyatt *et al.*,[34] showing that if two nucleotides for loop 1 is the minimum number necessary to cross five nucleotides in stem 2, three (and not two) is necessary for loop 2 to cross three nucleotides in stem 1. They also show that magnesium and high ionic strength stabilize pseudoknot formation.[34]

Concerning the pseudoknot of S15 mRNA, stem 2 has seven base pairs and one nucleotide crosses the deep major groove. This was demonstrated by creating compensatory mutations in either end of stem 2 and analyzing their effects on the autocontrol.[31] In either case, single mutations disrupt the control and double mutations restore it. Thus, the crossing nucleotide must be located in the deep major groove with the ribose phosphate backbone in an extended configuration, presumably by changes in the "ribose puckering" (C_3' endo to C_2' endo). The structure created is a rather regular helix of 17 base pairs with a bulged U in stem 1 and loop 2 protruding outside (Fig. 8). Additional information was gathered by performing chemical and enzymatic probe sensitivity experiments. The results confirms the genetic data except for one nucleotide, A(−47). This nucleotide is highly reactive toward dimethylsulfoxide (DMS) and 1-cycloexyl-3(2-(1-methyl-morpholino)-ethyl)-carbodiimide) CMCT, indicating that it should not be in a Watson–Crick position in the pseudoknot.[30] This discrepancy could reflect a slight distorsion at the junction of the two helices as suggested by Puglisi *et al.*[33]

7. SPECIFICITY OF THE RECOGNITION BY S15

The pseudoknot structure of the repressor binding site has been clearly established. Mutants show that this structure is necessary for regulation of S15 expression. S15 binds to the pseudoknot and stops translation. Direct evidence for this binding was shown by *in vitro* protection experiments. Sensitivity of nucleo-

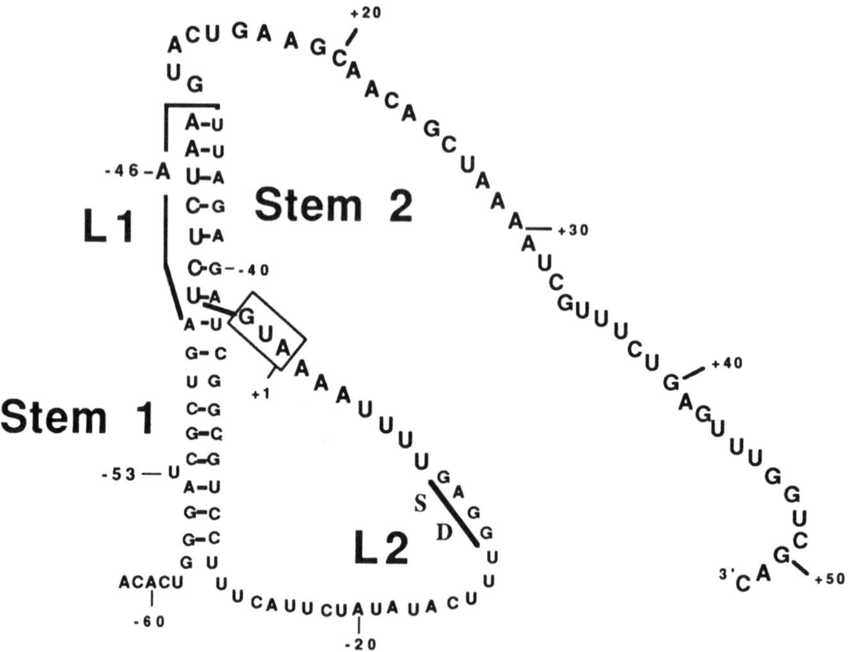

Figure 8. Structure of the pseudoknot of the S15 repressor binding site. The identification of stems and loops (L) is as for Fig. 7. The initiation codon is boxed and the Shine–Dalgarno sequence (SD) is underlined. Note that the loop L1 is composed of only one nucleotide [A at the (−46) position].

tides toward chemical reagents like DMS or CMCT indicates that any nucleotide reactive toward these probes is not engaged in Watson–Crick base pairing. If nucleotides no longer react in the presence of S15, it suggests that they are protected by the protein from the chemical probes. Two major pitfalls must be avoided: first, the observed shielding effect could result from steric hindrance. This problem is generally limited with chemical probes, given their small radii. Second, the binding of the protein may induce a change in the RNA conformation, modifying the secondary structure. For S15, the equilibrium between the hairpins and the pseudoknot is displaced in favor of the pseudoknot by S15 binding as shown by *in vitro* RNA structure probing in the presence and in the absence of S15. When S15 is bound, the paired nucleotides located in the loop are not modified by the chemical probes. This result does not imply a protection by the protein, but local structural modifications. With these limitations in mind, the protection experiments suggest that three points are protected from the chemical probes.[30] One point is the nucleotide crossing the deep major groove at the (−46) position. The adjacent A at (−47) was shown to be engaged with U(−38) and does not react (Fig. 8). The bulged U at (−53), strongly protected, is another point, and the initiation codon is the third. Some of the protection observed for U(−12) can be accounted for by base pairing between the Shine–Dalgarno region and a sequence located from (−20) to (−26) as supported by *in vitro* reactivities in these regions.

To determine to what extent these nucleotides are involved in the specific recognition by S15 to the mRNA, these nucleotides were modified by site-directed mutagenesis and the effect on the control of deletions, insertions, or substitutions in the pseudoknot on the control measured. (Insertions of triplets downstream of the initiation codon were necessary in order to maintain the reading frame.) To avoid mutations affecting only the translational efficiency, only mutants with altered repression ratios were considered. Insertion of a triplet just downstream of the initiation condon has variable effects depending on the triplet inserted. It should be kept in mind that the mutations created may induce the formation of alternate pseudoknot structures like those drawn in Fig. 9 and that the modifications of geometry induced may be sufficient, in some cases, to decrease or abolish control. This would imply that the determinants carried by the pseudoknot can no longer be recognized, as they would not occupy the correct position, although the pseudoknot is present. An illustration of this possibility might be given when A(−46) was changed to U. This mutation abolishes control. On the other hand, when A(−46) was substituted with G, the effect was minor. Thus, if mutations have some effect, the magnitude of the effect on control is strongly base dependent. Only one change entirely abolishes the control, suggesting that the nucleotide involved plays a particular role in S15–mRNA recognition. However, it must be ascertained that the substitution with another nucleotide has not disrupted the pseudoknot structure before concluding that there is a direct interaction between the mRNA and the protein. If it is assumed that these mutations have no influence on pseudoknot formation, then the different effects observed between

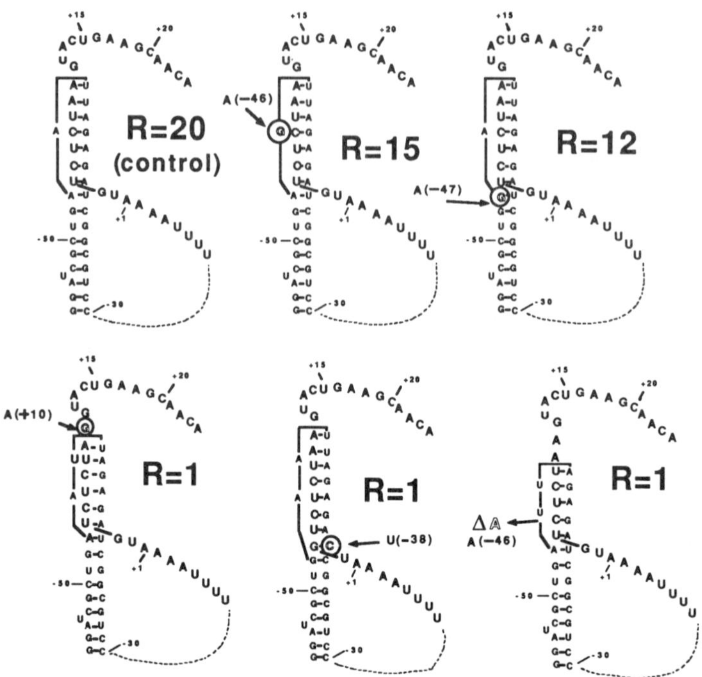

Figure 9. Putative alternate structures of the pseudoknot induced by some mutations. The nucleotide which is mutated (or deleted) is indicated in outlined characters. *R* corresponds to the repression ratio. Note that in this model any change in the length of the stem or of the loop L1 abolishes the control ($R = 1$).

the G and U substitutions at (−46) remain to be explained. Is the loss of autocontrol induced solely by a pyrimidine? In the two other examples of this phenomenon, two different effects have been observed. For glutaminyl-tRNA-synthetase, a pyrimidine instead of a purine in position 73 prevents the formation of a hairpin turn.[35] In the translational operator of the R17 replicase gene, the bulged purine residue is needed for intercalation into the helix.[36] For the S15 translational operator, it is possible that the recognition is not base-specific, but purine-specific. The existence of quasi-identical binding of A (at N_1 and N_6 or G (at N_1 and O_6) to $O_{\epsilon 1}$ and $N_{\epsilon 2}$ of glutamine or to $O_{\delta 1}$ and $N_{\delta 2}$ of asparagine might give some support to these observations.[37] Another possibility could be that any purine at this position allows stacking between the bases in the pseudoknot and an aromatic residue of the protein. The stack would be favored by the extended configuration of the (−46) nucleotide: ribose puckering from C_3' endo to C_2' endo would be induced by the large (7 Å instead of 5.9 Å) phosphate–phosphate distance across the deep groove and drastic changes in P–O torsion angles α and ζ, a consequence of the abrupt change in the polynucleotide chain direction. In fact, the phosphodiester backbone could be recognized directly when distorted, due to some peculiar structural features as suggested by the interaction between TAR and Tat in human immunodeficiencies virus (HIV)[38] or between the translational operator of the replicase of bacteriophage R17 and the coat protein of this bacteriophage.[39] It is striking that, in the pseudoknot described here, the presumed distance between the phosphate residues for A(−46) in C_2' endo is 7.0 Å, very close to the 7.1 Å of the bulging U(23) of TAR. An "arginine fork" could result which would account for the RNA–S15 contact. It is also possible that the effect of pyrimidine substitution results from the impossibility of formation of a triple base interaction in the deep major groove when A(−46) is changed to U.

Whatever the precise interaction between the mRNA and the protein, the existence of this extended, "open" configuration is a general structural feature of nucleotides involved in binding sites of proteins. Thus, it is tempting to postulate that A(−46) is involved in S15 recognition. More difficult to account for is the minor effect observed for either the bulge or the initiation codon. Perhaps the effect is smaller because the strength of interactions is less and is dispersed over several points. More data must be collected before the pattern of nucleotides recognized by S15 is defined.

8. COMPARISON BETWEEN THE REPRESSOR BINDING SITE AND THE RIBOSOME BINDING SITE

The binding site of S15 on the ribosomal RNA was identified several years ago.[40–43] It is located in the central domain of the 16S RNA. It covers helix 655–672/734–751 and a part of the dissymetrical interior loop 672–676/714–733 (Fig. 10). The protected nucleotides are at positions 741 and 742 in the long helix and in a small adjacent sequence (principally residues 724, 727, 729, and 730).[43,44] The two adjacent Gs (741, 742) in the helix are paired with As in the other strand at 663 and 665, thus creating a bulged G at 664. All but one of the other protected nucleotides are located in the loop of a two-base-pair helix formed in the interior loop. These nucleotides are protected by more than one protein (e.g., in addition to S15, S4, and S6 + S18). Thus, S15 protection could arise from local rearrangements of the RNA after binding of these other proteins at their sites. If so, the only specific protection points are located at 741 and 742 in the helix. What are the similarities between the site on the rRNA and the repressor binding site in the S15 messenger? Both sites exhibit the same long stem of 17 nucleotides. Both helices present a nucleotide bulge located just after the fourth base pair from the bottom. The question of whether the noncanonical A–G base pair and the bulge at this position mimic the discontinuities induced by the insertion of loop 2 into stem 2 in the pseudoknot cannot be adressed now, but must be kept in mind, especially considering the Gs protected in rRNA and the initiation codon involved in the recognition process (see above). The most important point, when comparing

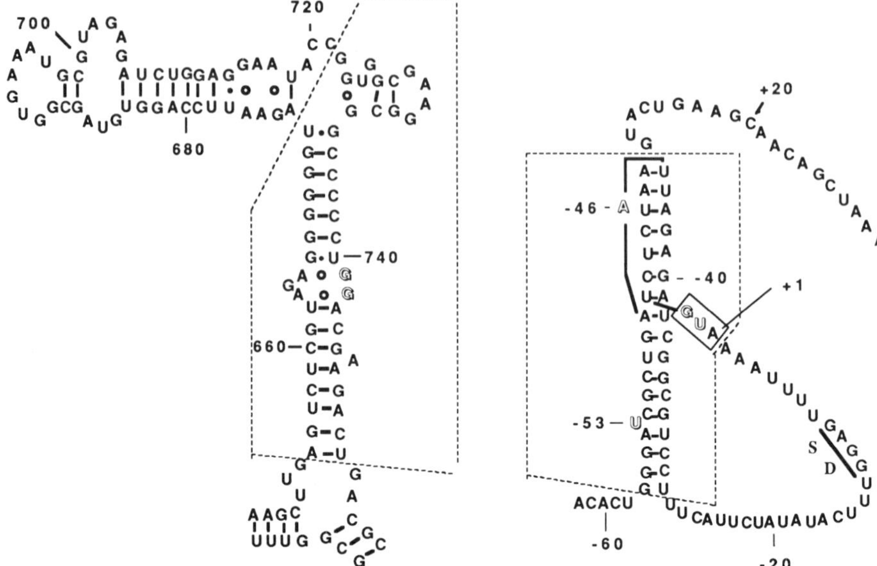

Figure 10. Comparison between the S15 ribosomal binding site and the S15 repressor binding site. *Left*: The secondary structure of the ribosomal RNA known to be bound by S15 is surrounded by dashed lines. *Right*: The dashed lines surround the helix created by the pseudoknot recognized by S15. Outlined characters represent the nucleotides specifically protected by S15 against chemical probes. Same explanation as in legend to Fig. 8.

the two binding sites is that the existence in rRNA of a nucleotide analogous to A(−46) is not clear. Presumably, it should not have a bulged conformation, since the crossing stretches it through the major groove and, consequently, the position of this A is different when compared to the ribosomal bulged nucleotide. The precise contact point in rRNA remains to be determined and no clear analogy between the repressor binding site and the rRNA can be identified except, possibly, the long helix.

9. MECHANISM OF TRANSLATIONAL AUTOREGULATION

Two mechanisms, displacement and entrapment, have been proposed to account for translational repression.[17] In the displacement model, any increase in translational expression must derive from a better translational efficiency or a decrease in repressor affinity. A modification in translational efficiency could result from alteration of the ribosome binding affinity, or from a change in elongation rate, or from a variation in the inactivation rate of the messenger. Any effect on the translation rate which does not affect the repressor binding capacity is easily detected in our system because, in all these cases, the repression ratio is not modified. But if the protein affinity for repressor binding site is modified, the repression ratio should be changed. Any decrease in the repressor binding affinity should induce an increase in the efficiency of translation.

Therefore, any mutation increasing the β-galactosidase level of the *rpsO–lacZ* fusion under repressed conditions, i.e., those partially or fully disrupting the autocontrol, should cause a decrease in the repression ratio. This is not the case if one compares the expression of the two different *rpsO–lacZ* fusions.[31] Although the expression of the two messages is autoregulated, in one of these fusions both the expression and the repression capacity are strongly increased. This means that an increase in the ribosome binding affinity does not necessarily correspond to a decrease in repression capacity as predicted by a competition mechanism. Rather, in this case, the reverse effect is observed.

Interestingly, *in vitro* studies cast some light on this unexpected phenomenon.[45] The 3′ position of the initiating ribosome can be located by its capacity to inhibit the extension of a primer annealed at the 3′ end of the messenger by reverse transcriptase (toe-printing experiments). Using this method, the *in vitro* studies showed that the binding of S15 to mRNA, by stabilizing the pseudoknot structure, hinders the formation of the initiation complex, but not the binary complex (30S subunit-mRNA). Moreover, the binding of S15 to mRNA is increased by the formation of the binary (30S-mRNA) initiation complex, as shown by a strong stop of the reverse transcriptase movement at the 3′ edge of the pseudoknot. Conversely, the binary complex stabilizes the binding of S15 to its repressor site.[45] In addition, the presence on the same messenger of one 30S subunit and of the S15 repressor is confirmed by RNase T1 footprinting experiments. These results show that 30S subunits bind preferentially to the pseudoknotted structure. Also, the ribosomal particles do not compete with the S15 repressor for binding but rather promote the S15 binding, suggesting a 30S-dependent S15 binding. Thus the regulation of S15 expression is not the result of a competition between the initiating ribosome and the repressor for an overlapping site, but the consequence of an entrapment mechanism,—a model predicted by Draper.[17] In this model (Fig. 11), the 30S subunit, after binding to the message in the Shine-Delgarno region, is likely stopped by the pseudoknot stabilized by S15. In the absence of S15, the structure can be opened to allow the formation of the ternary initiation complex and then translation. That the 30S subunit binds more efficiently to the S15 mRNA when the pseudoknot structure is formed might come as a suprise. Is the Shine-Dalgarno sequence, when carried by loop L2 of the pseudoknot, better recognized by the 30S subunit? On the other hand, there is an equilibrium between this structure and the formation of two independent stem-loops. This apparent contradiction raises a question: what does this equilibrium mean? *In vivo* and *in vitro* studies of different mutants suggest that kinetic parameters are likely to be very important in regulation. Perhaps the two RNA conformers, one active and the other inactive, are exchanged at a rate that could be modulated by a factor *in trans*. The

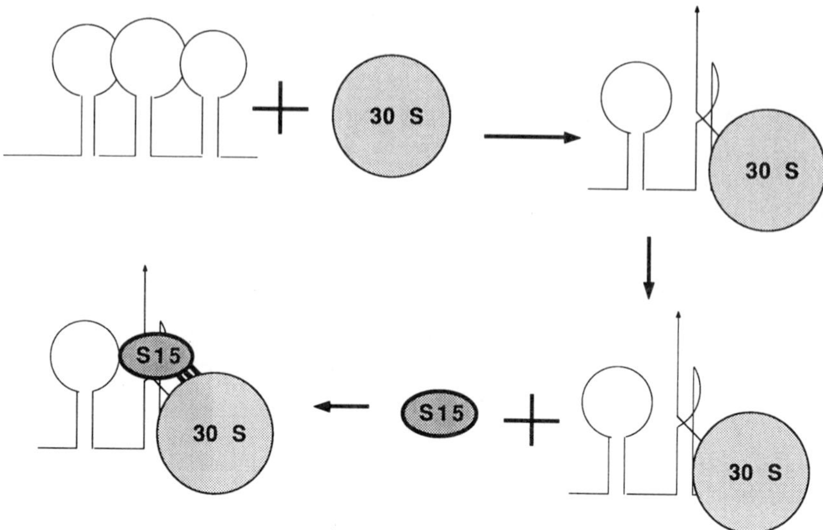

Figure 11. Presumed mechanism for S15 autocontrol. The 30 S ribosomal subunit binds to the Shine–Dalgarno sequence of the S15 messenger (represented by three stem–loops), presumably preferentially when the structure is pseudoknotted and translation occurs. In the presence of excess S15 in the medium, S15 binds to the pseudoknot and interacts with the initiating ribosome to stop its translocation to the initiating site. Translation is then blocked.

elongating ribosome, by disrupting the pseudoknot, could induce the formation of two transitory structures with stem loops. Then, the message would take the pseudoknot conformation and would be readily recognized by another 30S subunit, which would bind to the Shine-Dalgarno sequence. In the absence of S15, the translation initiation complex would form because the pseudoknot structure is not very stable. In the presence of excess S15, the pseudoknot, bound by S15, would be stabilized and would hinder the bound ribosome (preternary complex) from opening the structure to form the ternary complex and to initiate translation. Whether the bound ribosome is then blocked at the preinitiation stage or if it is released in the medium is not known. If the affinity of the 30S for binding mRNA relies mainly on an unfolded stretch of RNA,[46] and if the stability of the binary complex essentially is given by the Shine-Dalgarno sequence, there could be a better complementarity between the rRNA and mRNA sequences, which, while decreasing the dissociation rate, would also increase the translation rate and decrease the repression ratio. Conversely, a decrease in 30S affinity would induce the reverse effect: a lower translation rate and an increased repression ratio. This hypothesis would not necessarily contradict the 30S-dependent S15 binding model if the modification of the 30S affinity does not negatively affect the pseudoknot formation rate and the S15 repressor binding site, as this is the structure recognized by S15. Paradoxically, *in vivo* these effects would mimic a competition model.

10. THE S4 MODEL OF TRANSLATIONAL REPRESSION

The expression of the α operon is autoregulated at the translational level by the ribosomal protein S4. This regulatory protein is encoded by the third gene of the operon. It binds to a repressor site overlapping the initiation site of the S13 messenger and stops the translation of the ribosomal messages downstream. The structural features of the S4 repressor site are known[47]: it is a complex pseudoknot (Fig. 12). In a recent analysis, the binding constants of a set of mutants were compared with the translational repression levels.[48] For several mutants, no repression was observed, although the S4 binding affinity for the mRNA was

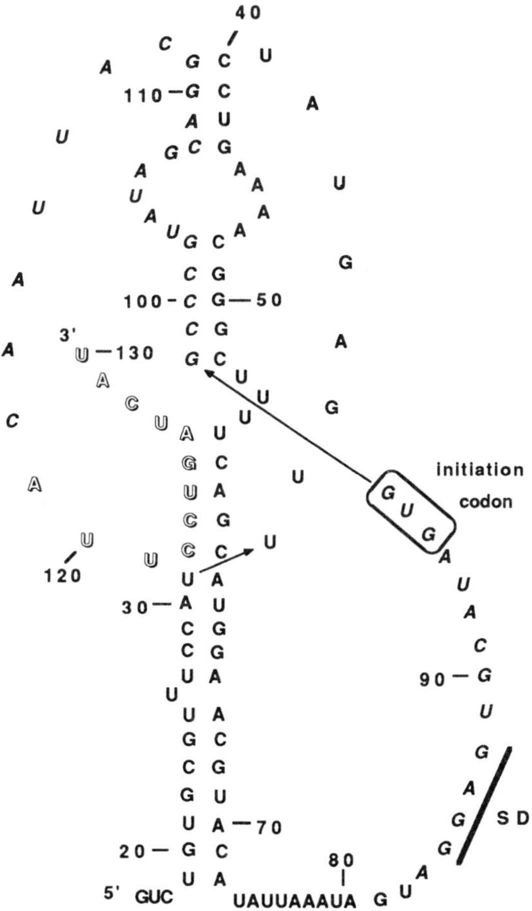

Figure 12. Pseudoknotted structure of the S4 repressor binding site. Italicized and outlined characters are used to facilitate the schematic of the structure. Arrows indicate the polarity of the chain and the bond between two adjacent nucleotides. Same explanation as in the legend to Fig. 8.

unmodified. Moreover, for other mutants, no correlation between the repression ratio and the translation level exists, suggesting that the displacement model does not apply. Tang and Draper propose that some interaction exists between the ribosome and the repressor binding sites and that an allosteric mechanism could account for the effects of these mutations. The messenger would mediate the interaction between repressor and ribosome, that is, the binding of S4 would induce, through the messenger, an alteration in the ribosome binding capacity. This RNA–RNA interaction could be short range, i.e., S4 repressor binding, or long-range, i.e., L10 binding. Very recent data support an entrapment mechanism, and an allosteric model for translational repression has been proposed.[49,50] These authors showed that there is no competition between 30S subunits and S4 for mRNA binding. Moreover, the S4 repressor binding site exhibits two conformations, which, although trapped by 30S subunits, differ in their abilities to bind S4 and $tRNA^{met}_{f}$. Formation of the binary complex (mRNA-30S) is promoted by the pseudoknot structure, (as observed for the S15 mRNA pseudoknot), which is thus stabilized. The inactive-to-active transition of the mRNA structure is expected to affect a kinetic step in initiation complex formation as well as the S4 binding affinity. In addition, temperature seems to play an important role in this kinetic step: S4 binding is increased at low temperature (30°C), while the $tRNA^{met}_{f}$ binding is increased at high temperature (40°C). The switch between active and inactive complexes is likely to be much faster *in vivo* than *in vitro*, because the initiation factors alter the kinetics of initiation. Compared with S15 autoregulation, the two systems share several structural and functional similarities: a pseudoknot in the repressor binding site, an unpaired purine (A) as the potential determinant for protein recognition,[31,48] a rare codon at the fourth (S4) or third (S15) position after the initiation codon, the same relative position of the initiation codon (juxtaposed at the foot of a stem), the presence of the Shine-Dalgarno sequence in the center of the same loop (L2), and probably the same mechanism of ribosome entrapment, that is, 30S binding promoted by the pseudoknot structure and an equilibrium between two mRNA conformations, active or inactive. A point of dissimilarity can be added: The lack of evidence for structural analogies between the ribosomal and repressor binding sites. This situation is quite different for S8,[51,52] L1, or L10 ribosomal proteins,[53,54] where similarities between the two binding sites have been shown. In each case, short sequence identities are present.

11. CONCLUDING REMARKS

The studies of translational control in bacteria have illuminated important aspects of the physiology of the cell, the molecular interactions between proteins and RNA, and the mechanisms of translation. Some of the results can be generalized.

Translational control is a general mechanism used to coordinate the synthesis of several ribosomal proteins encoded in many operons dispersed on the chromosome. Ribosomal RNA is the conductor by synchronizing the protein production rates. Constitutive proteins could be subject to this type of control because the autocontrol mechanism allows variation in the amount synthesized in response to the concentration of the targets, themselves modulated by other factors. In this sense, this type of mechanism is a way of metabolism integration. Thus, a full understanding of these mechanisms is quite necessary for unraveling the connections between the different biochemical pathways in the cell.

The complex phenomenon of translation initiation is very important, but very difficult to study. A detailed analysis of autoregulation is a way to identify the different steps involved in this process. In fact, formation of the ternary complex is the crucial event because, after it is formed, nothing seems able to stop translation.[55] Therefore, translation initiation can be stopped either by competition for 30S ribosome subunit binding or by blocking the access of the ribosome to the initiation codon. Overlapping the ribosome loading site with the repressor site creates a competition between ribosome and repressor for occupying the site. Another way to stop the ribosome is physically to hinder its progression by inserting a protein on its path before the formation of the ternary complex. The first mechanism has been shown in some cases, but the second, the entrapment model, remains to be fully demonstrated. Recent studies suggest that it is not the protein, but rather the stabilization of the RNA structure through the repressor binding which is able to stop the formation of the initiation complex. It should be possible, through isolation of a series of mutants, to identify the different steps involved in this mechanism and to define how the ribosome movement is stopped by the repressor protein. Also, what, if any, is the effect of initiation factors on translational control? Extragenic mutations could be tools to analyze these contacts. These mutations should allow the precise identification of all the components involved in the initiation process.

A third point, the study of molecular interactions between protein and RNA, is difficult because RNA molecules are often in equilibrium between different structures and because these molecules can have more varied overall three-dimensional shapes than DNA. In the system described here, the mimicry between rRNA and mRNA binding sites observed for regulatory ribosomal proteins like S8, L1, or L10 is not found, except for the presence of a regular long helix. The specificity of recognition seems to be restricted to unpaired or single-stranded residues, thus confirming previous studies.[56] It is known that alterations of a regular double-stranded helix by bulges, loops, or noncanonical base pairs can provide a basis for protein recognition.[36,38,57] Moreover, although only the minor groove of the helix is accessible to α-helices and antiparallel β-strands, the major groove could also be involved at its extremities or when distorted, for example, by a bulge.[58] The atoms recognized in these contacts can be quite limited in number and can be located on the ribose, phosphate, or base, or mediated by a water molecule or metal ion. Additional weak contacts could also account for an increased stability and, of course, for increased specificity.

A special comment should be made regarding pseudoknots, which are novel important structures involved in molecular interactions. Unexpectedly, they are involved in translational control, either directly as repressor binding sites (for S4 and S15 ribosomal proteins in *E. coli* or for protein involved in plasmid replication), or indirectly by allowing frameshifting (in several retroviruses and avian bronchitis virus). Their general properties seem linked to the formation of the structure and/or to a pseudoknot-dependent presentation of nucleotide determinants involved in protein recognition. The first point is best illustrated by a recent study of the coronavirus ribosomal frameshifting signal.[59] The pseudoknot involved in this phenomenon is essential, as is the heptanucleotide "slippery" sequence located upstream of it. In this sequence, generally X XXY YYZ, X may be the same as Y and the triplet represents the initial reading frame. However, the pseudoknot per se does not carry the specificity. Shortening of the helices or connecting loops or extensive changes in the sequence does not modify the frameshifting properties. Moreover, a hairpin with the same stem length cannot replace it. Since the presence of the pseudoknot is necessary, it must create either some specific but very localized interaction with the translating ribosome, necessary for frameshifting, or, more likely, an unusual geometry which would be more resistant to the secondary structure melting activity of the ribosome, inducing a more efficient pausing of the ribosome.[60] Kinetic parameters would be prominent in this mechanism.

In other cases, the recognition of specific determinants carried by the pseudoknot seems to be the rule. This is the case for the 3′ terminal tRNA-like structures of many single-stranded plant RNA viruses involved in viral RNA multiplication and packaging and which also serve as substrates for tRNA-specific enzymes. There are some exceptions to this rule, however, for example, in the tobacco mosaic virus (TMV) or the pea early browning virus (PEBV), which cannot be aminoacylated despite their 3′ terminal tRNA-like structures. These observations hinder any generalization on the identity of the determinants involved. For example, the different abilities to be aminoacylated can be extended to susceptibility to RNase P cleavage of turnip yellow mosaic virus (TYMV) and TMV. TYMV, which is aminoacylated, is cleaved by RNase P, as opposed to TMV, which is neither aminoacylated nor cleaved. Brome mosaic virus (BMV), however, can be aminoacylated,[61] but is resistant to RNase P.[62] The presence of specific determinants different from those of TYMV could account for this discrepancy. A more complex tertiary structure could well be involved in the multiplication of TMV, as suggested by the role of other adjacent pseudoknots in the TMV sequence.[63] In this case, the signal seems localized to a double-helical segment immediately upstream of the tRNA-like structure. Thus, if a particular configuration carrying specific determinants seems to be recognized, ongoing experiments are necessary to assess their identities.

Concerning the pseudoknots involved in translational control, if the presence of specific determinants is no longer questioned, their precise identification and the relationship to their cognate ribosomal binding sites remains an open problem.

In addition to S4 or S15 regulation, could some information concerning the protein recognition motif be derived from the T4 phage protein 32? This protein binds primarily to single-stranded DNA and, secondarily, cooperatively to its mRNA to stop translation. A first molecule of protein 32 binds to a pseudoknot located about 40 nucleotides from the initiation codon, nucleating the binding of other molecules of this protein.[64] Double-stranded pseudoknot RNA recognition could be mediated by a zinc domain (three cysteines and one histidine) present in protein 32, as in plant virus nucleocapsid proteins.[65] If this is true, the S15 translational operator could constitute an exception to this model: S15 has no cysteine in its sequence and recognition of its pseudoknot must involve another motif.

The difficulties encountered for the identification of the RNA determinants underline the value of the results obtained for S4 and S15. In both cases a single base appears to be involved, a purine. In both cases substitution by a pyrimidine abolishes the autocontrol. The number of these determinants is thus very limited and presumably their spacing is essential. If the specificity is not directly linked to the bases, the limited identity between the ribosomal and messenger sites could be revealed only by three-dimensional analysis of these sites or by genetic analysis. In this view, context effects could be important and should be tested whenever possible to discriminate between the presence of positive or negative determinants. Then, mutational analyses of the small RNA fragments that constitute the repressor binding sites must be taken with caution until the determinants identified can be substituted into the ribosomal site and shown to positively affect the binding activity of S15. Nevertheless, this type of analysis remains a powerful tool to gain useful information for solving these problems.

REFERENCES

1. Lindhal, L., Archer, R. H., and Zengel, J. M., 1983, *Cell* **33**:241.
2. Mattheakis, L., Vu, L., Sor, F., and Nomura,M., 1989, *Proc. Natl. Acad. Sci. USA* **86**:448.
3. Wikström, P. M., Byström, A. S., and Björk, G. R., 1988, *J. Mol. Biol.* **203**:141.
4. Byström, A. S., von Gabain, A., and Björk, G. R., 1989, *J. Mol. Biol.* **208**:575.
5. Baughman, G., and Nomura, M., 1983, *Cell* **34**:979.
6. Oppenheim, D. S., and Yanovsky, C., 1980, *Genetics* **95**:785.
7. Takiff, H. E., Chen, S.-M., and Court, D. L., 1989, *J. Bacteriol.* **171**:2581.
8. Petersen, C., 1989, *J. Mol. Biol.* **206**:323.
9. Berkhout, B., and van Duin, J., 1985, *Nucleic Acids Res.* **13**:6955.
10. Adhin, M. R. and van Duin, J., 1990, *J. Mol. Biol.* **213**:811.
11. Jacks, T., Madhani, H. D., Masiarz, F. R., and Varmus, H. E., 1988, *Cell* **55**:447.
12. Brierley, I., Digard, P., and Inglis, S. C., 1989, *Cell* **57**:537.
13. Saito, H., and Richardson, C. C., 1981, *Cell* **27**:533.
14. Hellmuth, K., Rex, G., Surin, B., Zinck, R., and McCarthy, J. E. G., 1991, *Mol. Microbiol.* **5**:813.
15. Deckman, I. C., and Draper, D. E., 1985, *Biochemistry* **24**:7860.
16. Schwarzbauer, J., and Craven, G., 1981, *Nucleic Acids Res.* **9**:2223.
17. Draper, D. E., 1988, In: *Translational Regulation of Gene Expression* (J. Ilan, ed.), p. 1, Plenum Press, New York.

18. Toone, W. M., Rudd, K. E., and Friesen, J. D., 1991, *J. Bacteriol.* **173:**3291.
19. Régnier, P., and Portier, C., 1986, *J. Mol. Biol.* **187:**23.
20. Régnier, P., and Hajnsdorf, E., 1991, *J. Mol. Biol.* **217:**283.
21. Takata, R., Aoyagi, M., and Mukai, T., 1982, *Mol. Gen. Genetic.* **188:**334.
22. Portier, C., Dondon, L., and Grunberg-Manago, M., 1990, *J. Mol. Biol.* **211:**417.
23. Robert-Le Meur, M., and Portier, C., 1992, *EMBO J.* **11:**2633.
24. Climie, S. C., and Friesen, J. D., 1987, *J. Mol. Biol.* **198:**371.
25. Dennis, P. P., Nene, V., and Glass, R. E., 1985, *J. Bacteriol.* **161:**803.
26. Yates, J. L., Arfsten, A. E., and Nomura, M., 1980, *Proc. Natl. Acad. Sci. USA* **77:**1837.
27. Meek, D. W., and Hayward, R. S., 1986, *Mol. Gen. Genet.* **202:**500.
28. Barry, G., Squires, C., and Squires, C. L., 1980, *Proc. Natl. Acad. Sci. USA* **77:**3831.
29. Dennis, P. P., 1984, *J. Biol. Chem.* **259:**3202.
30. Philippe, C., Portier, C., Mougel, M., Grunberg-Manago, M., Ebel, J. P., Ehresmann, B., and Ehresmann, C., 1990, *J. Mol. Biol.* **211:**415.
31. Benard, L., Dondon, L., Philippe, C., Grunberg-Manago, M., Ehresmann, B., Ehresmann, C., and Portier, C. (in preparation).
32. Pleij, C. W. A., Rietfveld, K., and Bosch, L., 1985, *Nucleic Acids Res.* **13:**1717.
33. Puglisi, J. D., Wyatt, J. R., and Tinoco, I., 1990, *J. Mol. Biol.* **214:**437.
34. Wyatt, J. R., Puglisi, J. D., and Tinoco, I., 1990, *J. Mol. Biol.* **214:**455.
35. Jahn, M., Rogers, M. J., and Söll, D., 1991, *Nature* **352:**258.
36. Wu, H.-N., and Uhlenbeck, O. C., 1987, *Biochemistry* **26:**8221.
37. Sänger, W., 1984, In: *Principles of Nucleic Acid Structure* (C.R. Cantor, ed.), p. 385, Springer-Verlag, New York.
38. Calnan, B. J., Tidor, B., Biancalana, S., Hudson, D., and Frankel, A. D., 1991, *Science* **252:**1167.
39. Milligan, J. F., and Uhlenbeck, O. C., 1989, *Biochemistry* **28:**2849.
40. Zimmermann, R. A., Mackie, G. A., Muto, A., Garrett, R. A., Ungewickell, E., Ehresmann, C., Stiegler, P., Ebel, J. P., and Fellner, P., 1975, *Nucleic Acids Res.* **2:**279.
41. Maly, P., and Brimacombe, R., 1983, *Nucleic Acids Res.* **11:**7263.
42. Stark, M. J. R., Gregory, R. J., Gourse, R. L., Thurlow, D. L., Zwieb, C., Zimmermann, R. A., and Dahlberg, A. E., 1984, *J. Mol. Biol.* **178:**303.
43. Mougel, M., Philippe, C., Ebel, J. P., Ehresmann, B., and Ehresmann, C., 1988, *Nucleic Acids Res.* **16:**2825.
44. Svensson, P., Changchien, L. M., Craven, G. R., and Noller, H. F., 1988, *J. Mol. Biol.* **200:**301.
45. Philippe, C., Eyermann, F., Bénard, L., Portier, C., Ehresmann, B., and Ehresmann, C. 1993, *Proc. Natl. Acad. Sci. USA*, in press.
46. de Smit, M. H., and van Duin, J. (1990), In: *Progress in Nucleic Acid Research and Molecular Biology* (W. E. Cohn, and K. Moldave, eds.) pp 1–35, Academic Press, New York.
47. Tang, C. K., and Draper, D. E., 1989, *Cell* **57:**531.
48. Tang, C. K., and Draper, D. E., 1990, *Biochem.* **29:**4434.
49. Spedding, G., Gluick, T. C., and Draper, D. E., 1993, *J. Mol. Biol.* in press.
50. Spedding, G., and Draper, D. E., 1993, *Proc. Natl. Acad. Sci. USA.* in press.
51. Ceretti, D. P., Mattheakis, L. C., Kearney, K. R., Vu, L., and Nomura, M., 1988, *J. Mol. Biol.* **204:**309.
52. Gregory, R. J., Cahill, P. B. F., Thurlow, D.L., and Zimmerman, R. A., 1988, *J. Mol. Biol.* **204:**295.
53. Thomas, M. S., and Nomura, M., 1987, *Nucleic Acids Res.* **15:**3085.
54. Saïd, B., Coles, J. R., and Nomura, M., 1988, *Nucleic Acids Res.* **16:**10529.
55. Gold, L., 1988, *Ann. Rev. Biochem.* **57:**199.
56. Romaniuk, P. J., Lowary, P., Wu, H. N., Stormo, G., and Uhlenbeck, O. C., 1987, *Biochemistry* **26:**1563.
57. Schimmel, P., 1989, *Biochemistry* **28:**2747.

58. Weeks, K. M., and Crothers, D. M., 1991, *Cell* **66:**577.
59. Brierley, I., Rolley, N .J. R., Jenner, A. J., and Inglis, S. C., 1991, *J. Mol. Biol.* **220:**889.
60. Draper, D. E., 1990, *Curr. Op. Cell Biol.* **2:**1099.
61. Dreher, T. W., and Hall, T.C. (1988) *J. Mol. Biol.* **201:**31.
62. Mans, R. M. W., Guerrier-Takada, C., Altman, S., and Pleij, C., 1990, *Nucleic Acids Res.* **18:**3479.
63. Takamatsu, N., Watanabe, Y., Meshi, T., and Okada, Y., 1990, *J. Virol.* **64:**3686.
64. McPheeters, D. S., Stormo, G. D., and Gold, L., 1988, *J. Mol. Biol.* **201:**517.
65. Shamoo, Y., Webster, K. R., Williams, K. R., and Konigsberg, W. H., 1991, *J. Biol. Chem.* **266:**796.

Chapter 3

Role of Elongation Factors in Steering the Ribosomal Elongation Cycle

Knud H. Nierhaus and Francisco Triana

1. INTRODUCTION

The detection and characterization of the third transfer RNA (tRNA)-binding site on the ribosome, the E site, in addition to the classical A and P sites has led to the allosteric three-site model (ref. 1 and references therein), which has provided fresh impetus for discussion of and insights into the ribosomal elongation mechanism. It allows, for example, for the first time the identification of a common inhibition mechanism for aminoglycoside antibiotics,[2] which do not exert their antibiotic activity by inducing misreading[3] as previously assumed. The implications of the allosteric three-site model for the selection of the correct aminoacyl-tRNA and for the role of the elongation factors will be surveyed here. We start with a brief description of the main features of the allosteric three-site model, then address the problem of recognition involved in the selection of cognate aminoacyl-tRNAs and describe a surprising solution to this problem, which might be related to such fundamental structural features as the two-subunit nature of all ribosomes. We close the chapter with a first attempt to describe the mechanism of both elongation factors.

Knud H. Nierhaus and Francisco Triana • Max-Planck-Institut für Molekulare Genetik, D-1000 Berlin 33, Germany.

Translational Regulation of Gene Expression 2, edited by Joseph Ilan. Plenum Press, New York, 1993.

2. THE TWO MAIN FEATURES OF THE ALLOSTERIC THREE-SITE MODEL FOR THE ELONGATING RIBOSOME

It is now well established that ribosomes in all living cells contain three binding sites for tRNA: the A site (A for aminoacyl-tRNA), where the decoding takes place; the P site (P for peptidyl-tRNA); and the E site (E for exit), which is specific for deacylated tRNA (for review see ref. 4).

The allosteric three-site model is characterized by two main features:

1. The first and third sites, A and E, respectively, are coupled by an allosteric interaction in the sense of a negative cooperativity. A deacylated tRNA at the E site induces a low affinity at the A site, and conversely an aminoacyl-tRNA at the A site induces a low affinity at the E site.[1,5] It follows that two out of the three sites always have a high affinity for tRNA and that consequently two tRNAs are always found on the elongating ribosome. In the pretranslocational state the tRNAs are present at the A and P sites, in the posttranslocational state at the P and E sites. A further consequence is that deacylated tRNA is released from the E site upon A-site occupation, but not during the translocation reaction as formerly believed (see Fig. 1).

2. The second important feature is the fact that both tRNAs present on the elongating ribosome simultaneously undergo codon–anticodon interaction.[1,6,7] Figure 2 illustrates this feature with a simple experiment.

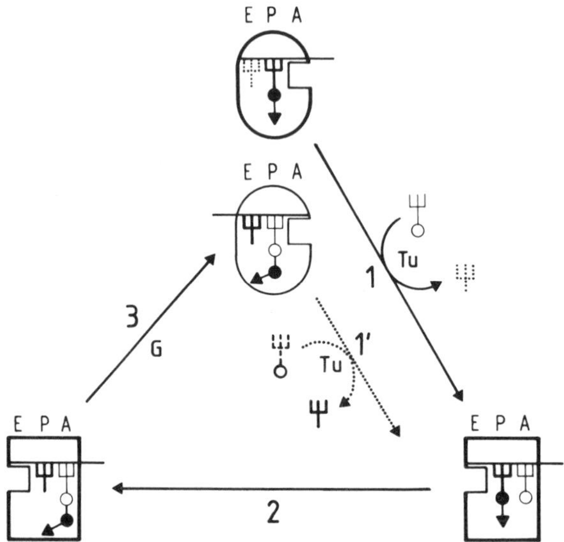

Figure 1. The three essential reactions of the ribosomal elongation cycle in the framework of the allosteric three-site model. (1) Binding of aminoacyl-tRNA to the A site; (2) peptidyl transfer; (3) translocation. The pretranslocational state of the ribosome is indicated by a rectangular ribosome, the posttranslocational conformer by a round ribosome. Tu and G mean EF-Tu and EF-G. Taken from ref. 1.

Deacylated tRNAs have considerably different intrinsic affinities for nonprogrammed ribosomes. If, for example, [^{14}C]tRNA^{Glu} is bound to nonprogrammed, tightly-coupled ribosomes, then nonlabeled tRNA^{Phe} can effectively chase the [^{14}C]tRNA^{Glu}, but nonlabeled tRNA^{Lys} cannot (Fig. 2, left side). In other words, tRNA^{Phe} has a high intrinsic affinity for ribosomes in the absence of the cognate codon, and tRNA^{Lys} a low affinity.[1] We now take advantage of this fact and test for the presence or absence of codon–anticodon interaction at the various tRNA-binding sites on the ribosome in the presence of messenger RNA (mRNA).

We use poly(A) as mRNA, which codes for tRNA^{Lys}. If a tRNA present at a distinct site undergoes codon–anticodon interaction, the cognate chasing substrate should be more efficient than the noncognate one, i.e., nonlabeled deacylated tRNA^{Lys} should be a better chasing substrate than tRNA^{Phe}. If there is no codon–anticodon interaction, however, then tRNA^{Phe} would be more effective in chasing than tRNA^{Lys} due to its higher intrinsic affinity (Fig. 2, left side). As a control, we bind [^{14}C]tRNA^{Lys} to the P site, where codon–anticodon interaction is known to occur.[8–10] As can be seen in Fig. 2 (right side), here the chasing effect of nonlabeled, cognate tRNA^{Lys} exceeds by far that of noncognate tRNA^{Phe}, thus indeed indicating codon–anticodon interaction. If we now establish the pretranslocational state by binding Ac[^{3}H]Lys-tRNA^{Lys} to the A site, or if we establish the posttranslocational state by translocating both tRNAs to E and P sites, respectively, the results of the chasing experiments clearly indicate codon–anticodon interactions in all cases (Fig. 2). This means that both tRNAs simultaneously undergo codon–anticodon interaction, regardless of whether they are at the A and P sites (pretranslocational state) or at the P and E sites (posttranslocational state).

Codon–anticodon interaction at the ribosomal E site is one of the most unexpected aspects of the allosteric three-site model, since the decoding process occurs exclusively at the ribosomal A site in the course of the elongation cycle. One attractive explanation is that the ribosome requires the formation of six base pairs (i.e., two adjacent codon–anticodon interactions) in order to maintain and secure the reading frame. A second possibility, which might be valid in addition, is that a tight connection between *both* tRNAs and the mRNA is essential for the translocation reaction, i.e., the movement of the tRNA–mRNA complex (involving both tRNAs) relative to the ribosome. A third reason has already been evaluated experimentally, namely codon–anticodon interaction at the E site is important for both the accuracy and rate of protein synthesis[11]; this surprising connection is discussed in Section 4.

3. THE TWO MAIN STATES OF THE ELONGATING RIBOSOME

The activation energies required for the formation of the various states of the ribosome have been assessed in a systematic analysis.[12] The fact that we derived the activation energies from first-order reactions cogently indicates that the rate-

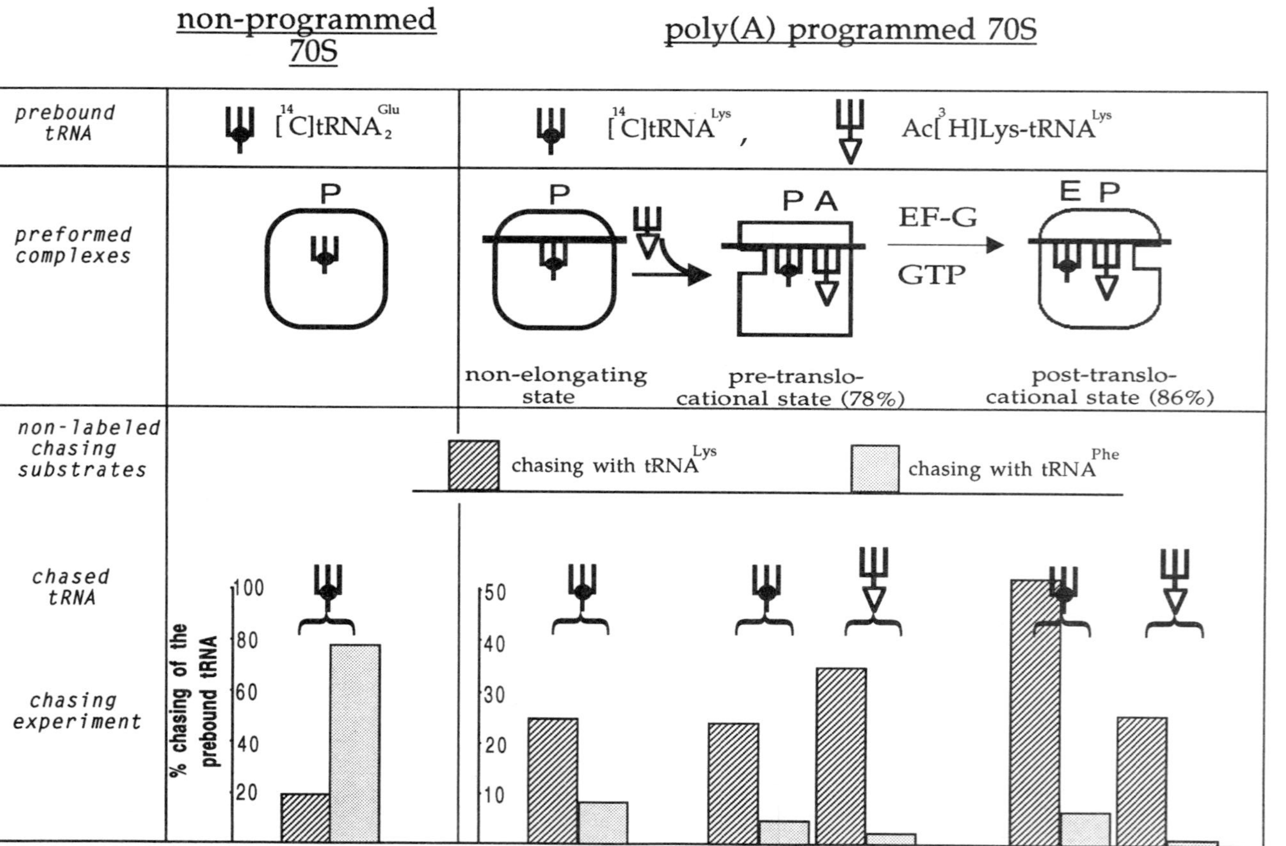

non-programmed 70S
poly(A) programmed 70S
prebound tRNA
[14C]tRNA2 Glu
[14C]tRNA Lys ,
Ac[3H]Lys-tRNA Lys
preformed complexes
P
P
P A
EF-G
GTP
E P
non-elongating state
pre-translocational state (78%)
post-translocational state (86%)
non-labeled chasing substrates
chasing with tRNA Lys
chasing with tRNA Phe
chased tRNA
chasing experiment
% chasing of the prebound tRNA
100
80
60
40
20
50
40
30
20
10

limiting step of the corresponding reaction is a conformational change, whereas the reverse argument is not true; the absence of a measurable activation energy does not exclude a conformational change.

Using this criterion, we have to distinguish at least four different conformers, two of which carry two tRNAs and thus represent states of the elongation cycle (Fig. 3; the ribosomes drawn above each other are considered as being equivalent conformers). The empty, but programmed ribosome can efficiently bind an aminoacyl-tRNA or the corresponding ternary complex aminoacyl-tRNA·EF-Tu·GTP (EF, elongation factor) to the A site (note that the P site is not occupied). Since a high-affinity A site is indicative of the pretranslocational state as discussed in Section 2, we conclude that empty ribosomes (or at least a significant population among them) are in a related but not identical conformation to the pretranslocational state. When the P sites of empty, but programmed ribosomes become occupied, a surprisingly high activation energy of 72 kJ/mole was observed, indicating that the ribosome switches to a second state. The ribosomes can bind a deacylated tRNA to the E site without activation energy, and since a high-affinity E site is indicative of the posttranslocational state, we conclude that the second state is related but not identical to the posttranslocational state. The difference between the second state and the posttranslocational state becomes evident when the activation energies for the transitions to the pretranslocational state (A-site binding) are compared; from the second state (E site free) 47 kJ/mole is required, from the posttranslocational state 87 kJ/mole. The first case (E site free) occurs only once during protein synthesis, namely just after the formation of the initiation complex, and therefore this kind of A-site occupation is termed the "i type" ("i" for initiation; Fig. 4).[2] The next and all subsequent A-site occupations start from the posttranslocational complex (E site occupied) and are termed "e type" ("e" for elongation). An important consequence of this observation is that the second state (only one tRNA at the P site) does not represent a conformer of the elongation cycle and is certainly different from the posttranslocational state.

The essential point of Fig. 3 is that the activation energies for the allosteric transitions from the post- to the pretranslocational state (A-site occupation of the

←

Figure 2. Codon–anticodon interaction at the three ribosomal binding sites analyzed by chasing experiments (for explanations see text). *Control of nonprogrammed 70S:* 7.7 pmole [^{14}C]tRNA$_2^{Glu}$ (specific activity 110 cpm/pmole) was bound to 18 pmole 70S ribosomes. For chasing, 108 pmole nonlabeled tRNA^{Phe} or tRNA^{Lys} was added and the residual binding assessed. For example, 80% chasing means that 80% of the prebound, labeled tRNA was lost upon addition of the chasing substrate. Data were taken from ref. 1, where more experimental details can be found. *Poly(A)-programmed 70S:* 15–17.8 pmole of [^{14}C]tRNA^{Lys} (60 cpm/pmole) was bound to 24 pmole 70 S in the three subsequently established states shown, before the chasing substrates were added. The corresponding numbers for the binding of Ac[^{3}H]Lys-tRNA^{Lys} (50 cpm/pmole) were 10 and 11 pmole in the pre- and posttranslocational states, respectively. According to the puromycin reaction, 78% of the AcLys-tRNA was found at the A site in the pretranslocational state, and after translocation induced by EF-G, 86% was found at the P site in the posttranslocational state. For chasing, 120 pmole of nonlabeled tRNA^{Phe} (noncognate) or nonlabeled tRNA^{Lys} (cognate) was added. Data were taken from ref. 7.

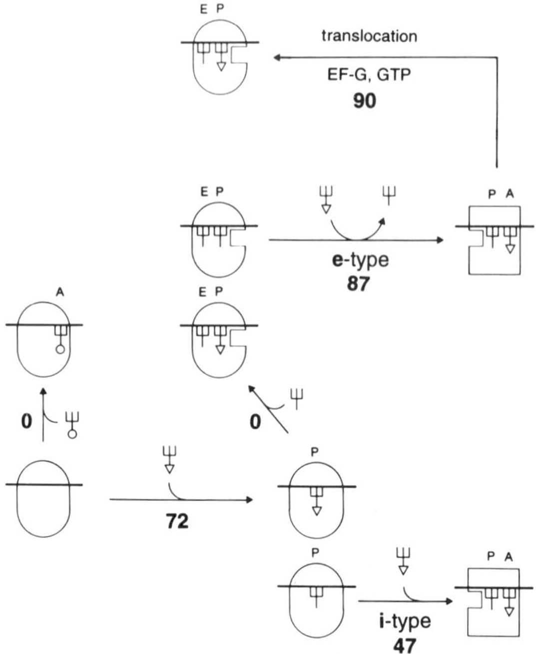

Figure 3. Four different conformations of 70S ribosomes. Ribosomes drawn about each other are considered to be in equivalent states. Only ribosomes carrying two tRNAs represent states of the elongation cycle. The numbers give the respective activation energies in kJ/mole measured at 15 mM Mg^{2+}. For details see text. Taken from ref. 12.

e type) and *vice versa* (translocation) are both rather high (about 90 kJ/mole) and that the corresponding rates are very slow.[12] In contrast, the reactions which follow A-site occupation (EF-Tu-dependent GTP cleavage, peptide-bond formation) are fast and occur efficiently even at O°C. It follows that the latter two reactions represent substrates of the pretranslocational state, which is separated by a high activation energy barrier from the posttranslocational state. Thus, the elongating ribosome oscillates between two main states, the pre- and the posttranslocational state (Fig. 4). Due to the high activation energy barrier either state can be prepared *in vitro* with high homogeneity.

Recently, a model for the elongation cycle has been proposed according to which the peptidyl residue moves to the P site of the peptidyl-transferase center before translocation, whereas the anticodon loop is still located at the A site of the 30S subunit. The model was derived from patterns of protection against some base-modifying reagents, the protection of bases in the 23S ribosomal RNA (rRNA) being solely dependent on an intact –CCA end of the tRNA ("hybrid-site model").[13] However, the evidence presented did not conclusively support the

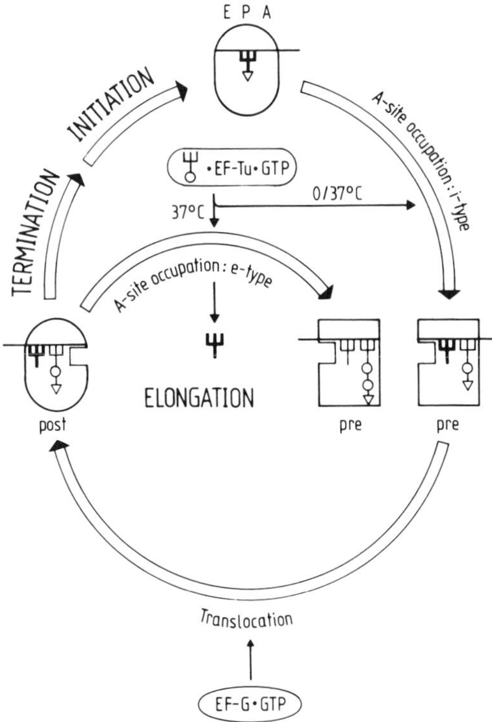

Figure 4. Functional scheme of the ribosome showing the three main states of the protein-synthesizing ribosome. The 70S ribosome with one tRNA indicates the initiation complex; that with two tRNAs indicates the elongation complex. Taken from ref. 2.

conclusions derived, and three major objections can be made: (1) A nonelongating state (one tRNA at the P site, E site free) was compared with the pretranslocational state, which does not allow — as mentioned above — a "pre–post" comparison. (2) The Mg^{2+} concentration was uncritically changed between 5 and 25 mM, which severely effects the binding properties of the ribosomal sites[14] (for discussion see ref. 4). (3) The flexible –CCA end is not a well-suited reporter structure for the site location of the rather rigid tRNA molecule. Furthermore, a hybrid-site character A/P and P/E cannot easily be reconciled with the allosteric linkage between A and E sites, which has been observed under a variety of experimental conditions.[1,6,7] Due to these inconsistencies we should be careful before accepting one implication of the hybrid-site model which is of great consequence: namely, the hybrid-site model claims that in the A/P state the peptidyl residue lies at the P site of the peptidyl-transferase center although it does not react with puromycin. This view of the P site confuses the operational definitions of A and P sites which have been used during the last two to three decades and according to which a peptidyl residue at the P-site region of the peptidyl-transferase center can react with puromycin, whereas a peptidyl residue at the A-site region cannot.

4. RECOGNITION OF THE COGNATE AMINOACYL-tRNA: THE IMPORTANCE OF THE ALLOSTERIC COUPLING BETWEEN A AND E SITES

Precise recognition is a basic property of living processes. Enzymes must recognize their substrate with high precision (one error in 10^3–10^6 reactions) in order to prevent the metabolism from lapsing into chaos.

Roughly three types of molecular interaction can be distinguished during specific recognition. Type I is characterized by a relatively small contact area between the substrate and the recognizing molecule, and here the contact area is largely identical with the recognition area, i.e., the binding energy is nearly identical with the discrimination or recognition energy. The contact area is small because, for example, the substrate is a small molecule, and this is valid for most of the enzymatic reactions in metabolism. But interactions between large molecules may also belong to this class, e.g., the specific antigene–antibody interaction, where the contact region (≈recognition area) is restricted to the tips of the antibody molecules made by the highly variable domains.

Type II is only quantitatively different from type I, in that here the contact area is large (e.g., the whole surface area of some ribosomal proteins) and also practically identical with the recognition area. The resulting enormous discrimination energy allows the precise assembly of highly complicated structures, the most prominent examples of which are the ribosomes, However, the assembly of virus heads, nucleosomes, some subunit-structured enzymes, etc., also has to be mentioned here.

Type III is qualitatively different from the two other types. In this case a large contact area contains only a small recognition site. Consequently the binding energy ($\Delta G°$) by far exceeds the recognition energy $\Delta\Delta G°$ ($= \Delta G°_{[\text{correct substrate}]} - \Delta G°_{[\text{wrong substrate}]}$). This generates the problem of efficient discrimination between similar, but wrong substrates, which have a significant affinity for the contact area of the recognizing molecule. The higher these affinities are, the longer are the "sticking times" (reciprocal of the dissociation rate constant) of wrong substrates. This means that the binding equilibrium is reached only after a prolonged time. Since the recognition energy ($\Delta\Delta G°$), which determines the highest level of accuracy, is fully exploited only in the equilibrium state, type III reactions will be either very slow or prone to error.

The decoding at the ribosomal A site is an example of a type III recognition process, because the ribosome has to select the one cognate aminoacyl-tRNA out of a large number of aminoacyl-tRNAs, which all have a remarkable codon-independent affinity for the A site. In addition to the specific codon–anticodon interaction which provides the selection of the cognate aminoacyl-tRNA, this common codon-independent affinity is also necessary for the ribosomal function, because the aminoacyl residue of the acylated tRNA, which is about 76 Å away from the anticodon,[15] must be precisely positioned in the peptidyl-transferase center.

Due to the large contribution of the nonselective component relative to the total binding energy, reaching the equilibrium state of A-site occupation should be extremely time-consuming. As discussed for type III recognition in general, occupation of the ribosomal A site and hence protein biosynthesis should be either slow or work with low accuracy. In contrast, it can be shown *in vivo* and *in vitro* that incorporation of amino acids into proteins is a rather fast process, but nevertheless translation fidelity is very high (error frequency less than 10^{-3}).

The enormous problem the ribosome has to solve is that on the one hand it needs a nonselective affinity of all tRNAs to maintain general ribosomal functions such as peptidyl transfer, but on the other hand precisely this nonselective affinity causes trouble with respect to the recognition problem. This problem becomes even more severe if one takes into account that it is not the already large aminoacyl-tRNA which is the substrate for the A site, but rather the ternary complex aminoacyl-tRNA·EF-Tu·GTP with a molecular mass of about 73 kDa (in this chapter we use the data from *Escherichia coli* as an example). On the other hand, it is only the anticodon, with a molecular mass of 1 kDa, i.e., 1–2% of the mass of the complex, which forms the recognition area. Large portions of the complex apart from the anticodon participate in A-site binding, e.g., EF-Tu contributes significantly and directly to the very high A-site affinity of the ternary complex (the apparent K_a is two orders of magnitude larger than the corresponding K_a of *N*-acetyl-aminoacyl-tRNA[16]).

The allosteric interplay between A and E sites is the key event for the solution to this problem. It has been shown that the occupied E site induces a low-affinity state of the A site in the posttranslocational state.[1] We assume that, at this low-affinity A site, contacts between ribosome and ternary complex outside the recognition area are abolished, but the interaction with the anticodon region is unaffected (or might even be optimized). Thus, the ribosome has reduced the recognition problem to that of type I (a small contact area which is identical to the recognition area). Of course the recognition energy ($\Delta\Delta G°$) is not affected, but it is no longer superimposed on a large basis of nondiscriminatory binding energy. This has even deeper consequences for the rate and accuracy of protein biosynthesis, which have been experimentally explored and will be briefly outlined.

The 41 different species[17] of ternary complexes can be grouped into three classes with respect to the codon present at the ribosomal A site (Fig. 5). The first class is the cognate aminoacyl-tRNA, and comprises only the one species with an anticodon complementary to the codon. The second class of near-cognate anticodons contains about four aminoacyl-tRNA species, with an anticodon similar to that of the cognate aminoacyl-tRNA. Most of the aminoacyl-tRNAs (about 90%)

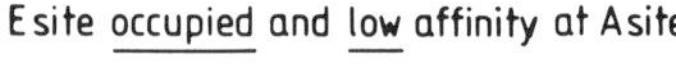

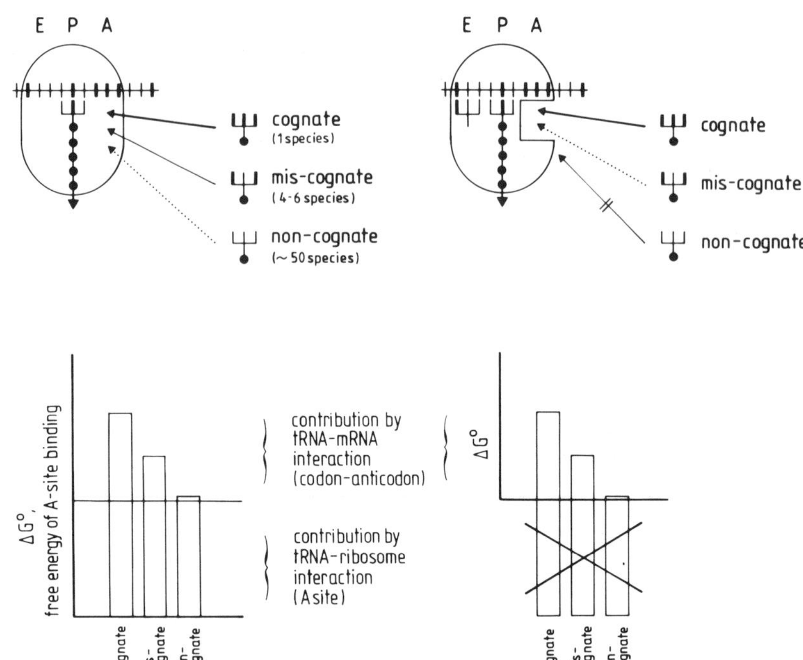

Figure 5. The importance of the E-site-induced low affinity of the A site for the accuracy of protein synthesis. For details see text. Taken from ref. 11.

belong to the third, noncognate class. If the A site were to be in a high-affinity state throughout the elongation cycle, then even the noncognate aminoacyl-tRNA would be able to bind outside the anticodon region (e.g., with the aminoacyl stem), and this would occasionally trigger the formation of a peptide bond and thus lead to the incorporation of the noncognate amino acid. Since this could happen with any amino acid and not only with the (often chemically similar) near-cognate amino acids, the effect would be frequently disastrous for the function of the corresponding protein. Not only would the accuracy of protein synthesis be severely affected, but also its rate, since due to the large binding energy the sticking time of any aminoacyl-tRNA (including the noncognate species) would be relatively long.

In contrast, if the nondiscriminatory contacts are abolished in the low-affinity state of the A site and the transition to the high-affinity state needs correct Watson–Crick base pairing as a prerequisite, then the transition can only be triggered by the cognate and, to a much lesser extent, the near-cognate aminoacyl-tRNAs. However, the noncognate aminoacyl-tRNAs could no longer interfere with the accuracy or rate of protein synthesis, since no contacts would be possible at the recognition area (dissimilar anticodon) or at the nondiscriminatory area (low-affinity state). It follows that the A site practically does not exist for noncognate aminoacyl-tRNA.

These predictions were tested with poly(U)-programmed ribosomes carrying an AcPhe-tRNA at the P site.[11] A mixture of ternary complexes was added, which contained either [^{14}C]Phe-tRNA or [^{3}H]Asp-tRNA (codon: GAC/U), respectively, and the dipeptides formed were analyzed by high-performance liquid chromatography (HPLC) techniques. In one case the A site was in the low-affinity state (E site occupied, posttranslocational state), whereas in the second case the A site was in a state of higher affinity (E site free). The latter state is not the pretranslocational state (see Section 3 and Fig. 3), but our idea was that if we observed an effect under this condition, the effect must be even more extensive in the pretranslocational state, where the A site achieves its full high-affinity state.

Figure 6 shows the results. When the E site was free (A site with increased affinity), the noncognate Asp was incorporated significantly into dipeptides. In contrast, when the E site was occupied (A site with low affinity), the formation of incorrect dipeptides was depressed to background levels, whereas the incorporation of the correct amino acid Phe was hardly affected. Most interestingly, only a

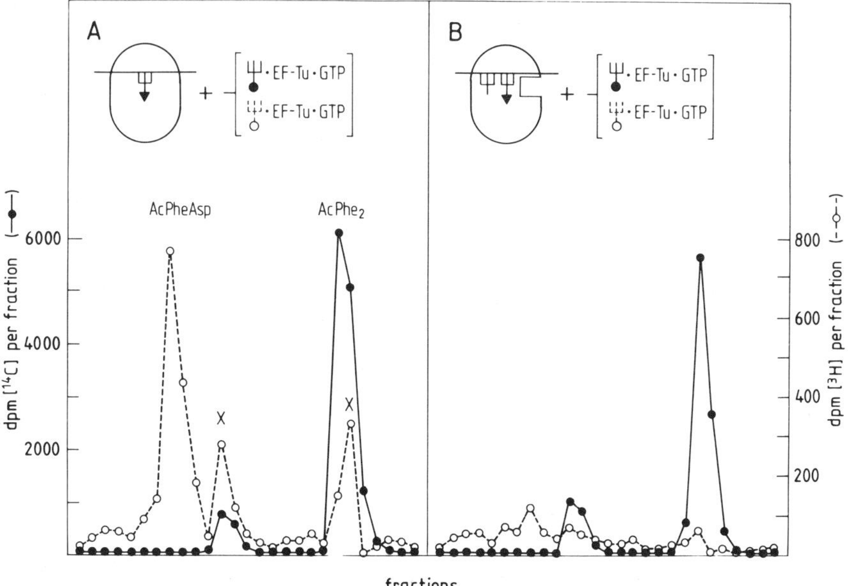

Figure 6. HPLC analysis of dipeptide formation by poly(U)-programmed ribosomes with E site free (left half) or occupied (right half). Taken from ref. 11.

tRNA cognate to the codon present at the E site prevented the incorporation of the wrong amino acid, whereas a near-cognate deacylated tRNA did not.[11] Codon–anticodon interaction at the E site apparently is the signal for keeping the ribosome in the posttranslocational state, underlining the functional importance of this interaction at the E site.

It follows that the low-affinity state of the A site prevents any interference of the noncognate aminoacyl-tRNAs, which reduces the ribosomal problem of aminoacyl-tRNA selection by an order of magnitude: the ribosome only has to select 1 out of about 6 aminoacyl-tRNAs (cognate plus near-cognate) instead of 1 out of 41 aminoacyl-tRNAs.

In addition, a second feature might contribute to the solution of the ribosomal recognition problem, namely that the value of the recognition energy ($\Delta\Delta G°$) is possibly higher than is generally estimated. The ribosome might recognize a canonically formed partial Watson–Crick structure between codon and anticodon,[18] which represents a prerequisite for the allosteric transition to the pretranslocational state. The view that the ribosome recognizes the correctness of the partial Watson–Crick structure and does not "measure" number and strength of hydrogen bonds between codon and anticodon has at least two important advantages over alternative views. (1) It explains why codon rich in A and U are recognized as precisely as those rich in G and C.[19] (2) Recognition of the correctness of the partial Watson–Crick structure means that the correct positions of the sugar pucker and the phosphorus atoms participate in the discrimination, which enlarges the potential of the discrimination energy enormously. A large discrimination energy allows a high precision of the binding step and makes a proofreading mechanism for the selection of aminoacyl-tRNAs dispensable. Note that the assumption of proofreading mechanisms in ribosomal translation was invoked because of a lack of discrimination energy during the binding step was assumed.[20–22]

The allosteric three-site model requires a sharply different allosteric response of the ribosomal subunits upon tRNA binding to the A site or translocation. In the posttranslocational state the recognition center of the A site, where codon–anticodon interaction takes place, should be fully active, while the additional contact area is in a nonbinding state. The recognition area certainly resides on the 30S subunit, whereas the nondiscriminatory area lies at least partially in the neighborhood of the peptidyl-transferase center on the 50S subunit. During the transition to the pretranslocational state the affinity of the nondiscriminative area is restored. It follows that the 30S subunit is relatively inert during this transition, whereas the 50S subunit seems to undergo a gross conformational change. Likewise, in both transitions (post → pre and pre → post) the E site oscillates between high- and low-affinity states. Since it is known that the E site is largely located on the 50S subunits,[23] these facts indicate that in the course of elongation the large ribosomal subunit severely and actively changes its conformation, whereas the small ribosomal subunit seems to behave more passively. A strikingly different behavior of the two subunits during elongation provides a clue to the explanation of the universal subunit structure of ribosomes.

5. A UNIVERSALLY CONSERVED SEQUENCE OF 23S-TYPE rRNA IS RELATED TO THE ROLE OF ELONGATION FACTORS

The α-sarcin stem–loop structure of 23S-type rRNA (helix 95, *E. coli* numbering) contains a 12-mer, which is among the longest universally conserved sequences of all rRNAs. The name is derived from the RNase α-sarcin, a 17-kDa protein, which cleaves with remarkable specificity a single phosphodiester bond after position G2661 in ribosomes and large subunits from all kingdoms, resulting in a complete block of protein biosynthesis.[24–26] The cleavage specificity of α-sarcin is lost with naked RNA,[27] indicating that the α-sarcin stem–loop structure must be prominently exposed in the ribosome.

Both elongation factors EF-Tu and EF-G (EF-1 and EF-2 in eukaryotes) bind to overlapping sites, since their binding is mutually exclusive.[28,29] The overlapping region of the binding sites is probably related to the common function of both factors, namely the reduction of the activation energy barrier between pre- and posttranslocational states.

Evidence is accumulating that the α-sarcin stem–loop structure is part of the overlapping binding region of both factors. The specific cleavage after G2661 severely impedes the binding of both factors, but has no effect on intrinsic ribosomal functions such as nonenzymatic tRNA binding, peptide-bond formation, or spontaneous translocation.[25] Furthermore, both factors are in contact with the α-sarcin loop, as shown by protection patterns against modifying reagents.[30] Finally, hybridization of oligodeoxynucleotides complementary to the α-sarcin stem–loop structure within tightly coupled ribosomes indeed triggered a conformational change of the large subunits, resulting in a loss of functional competence, dissociation of 70S ribosomes (seen in sucrose-gradient runs) and a substantial inactivation of the resulting 50S subunits in contrast to the 30S subunits. The inactivation largely persisted after removal of the hybridized antisense DNA.[31] Interestingly, the antisense DNA has to cover not only part of the loop including the α-sarcin cleavage site, but also the complete 3′ branch of the stem structure. Clearly, a melting of the stem structure is related to the disastrous effects.

The conformational change triggered by the antisense DNA might correspond to the common function of both elongation factors, namely to reduce the activation energy barrier for a conformational change, thus facilitating a unidirectional allosteric transition from one main state to the other. The fact that the antisense DNA leads to a functional "dead end" of the 50S subunit is probably related to the second function, which characterizes the elongation factors and not the antisense DNA, in that the elongation factors together with the appropriate tRNAs control the transition to either state. The lack of this control in the case of the antisense DNA results in a disastrous conformational change of the 50S subunit.

At this level we can combine the individual pieces of information by proposing a model mechanism for the action of the elongation factors. This model mechanism is composed of at least two steps (Fig. 7). In the first step, the elongation factors bind probably loosely to the α-sarcin domain, melt the α-sarcin stem, and thus reduce the activation energy barrier. A special requirement for EF-Tu over that of EF-G with respect to this first step is the preceding double-helix formation between the codon and anticodon of the newly incoming aminoacyl-tRNA. The reduction of the activation energy barrier now allows the allosteric transition to the new main state, which is trapped in the second step by tight binding of the respective elongation factor, i.e., EF-Tu binds tightly to the pretranslocational state, and EF-G to the posttranslocational state. The main step trapped by the tight binding of an elongation factor triggers the factor-dependent GTPase activity, the factor switches to the GDP-conformer, which does not bind to the ribosome, and dissociates from the ribosome. The α-sarcin stem reforms and again erects the activation energy barrier. The main state achieved is thus stabilized until the complementary factor catalyzes and promotes the reverse allosteric transition. It is clear that the GTP energy is not used for the allosteric transition itself, but is rather required for an effective removal of the factors after they have fulfilled their functions on the ribosome.

The action of the elongation factors shows similarities and dissimilarities with that of a standard enzyme. In the first step the function of the elongation factors corresponds to the action of standard enzymes, i.e., to accelerate a reaction by reducing the activation energy. The difference is that factors do not form or cleave covalent bonds like enzymes, but catalyze allosteric transitions, i.e., conformational changes. The second step is strikingly different from the action of an enzyme, in that the factors determine the direction of reaction; EF-Tu promotes the transition to the pretranslocational state, and EF-G that to the posttranslocational state. This view is supported by the observation that a dramatic increase of the rate of protein biosynthesis is only observed when both elongation factors are present, but not in the presence of only one.[32]

In this scenario the effects of elongation factors are based on the initial melting of the α-sarcin stem (helix 95). Melting and rejoining has to occur easily, and one would therefore predict that not only is the loop sequence universally conserved, but probably also a structural weakness of the helix. That is precisely what we found.

In Table I we present the predicted free energy change of helix formation (determined according to an updated version of the energy parameters for secondary structure prediction[33,34]) for the α-sarcin stem-and-loop motif in a wide variety of organisms, and compare it with the corresponding value from another universally conserved loop–helix motif with 11 bases in the loop and 6 base pairs in the stem (helix 81). Since the α-sarcin stem length varies between 5 and 8 base

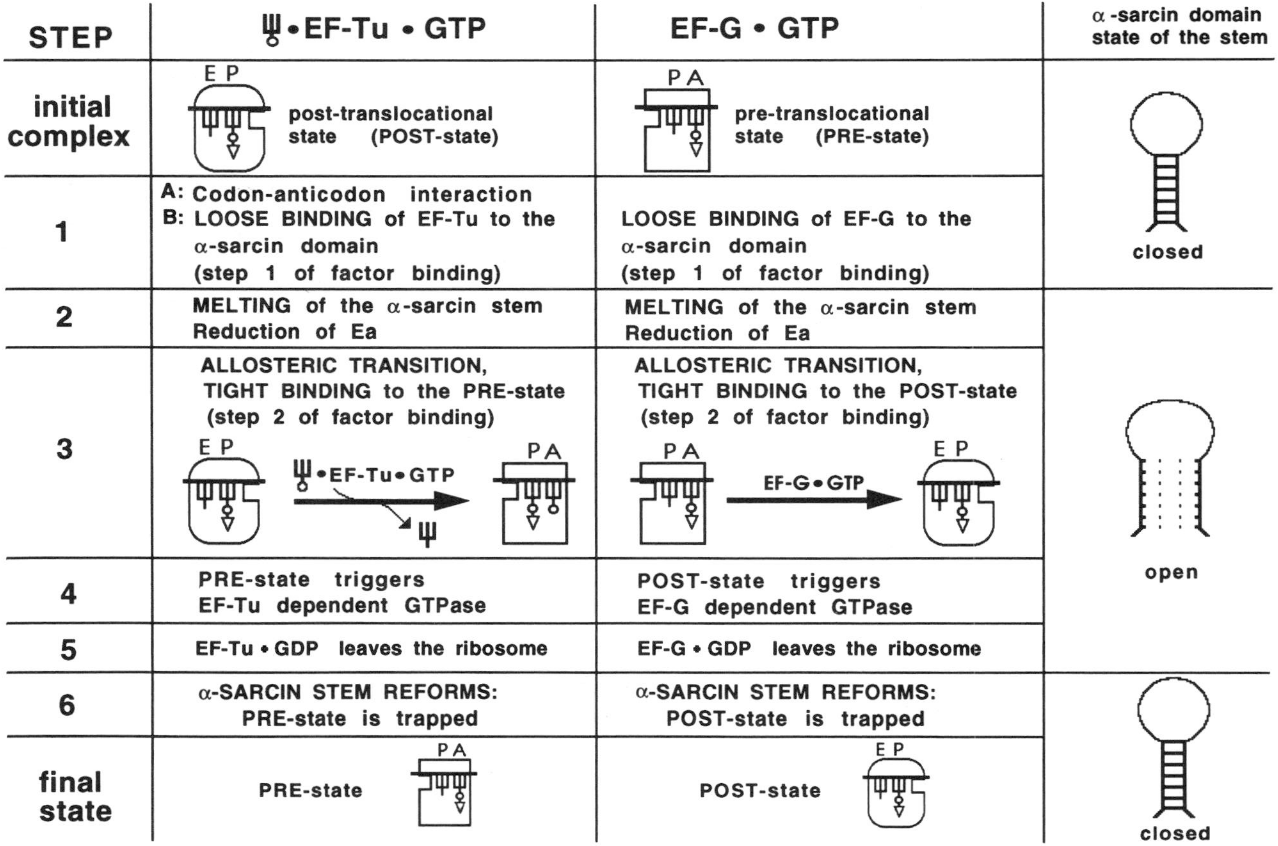

STEP	•EF-Tu • GTP	EF-G • GTP	α-sarcin domain state of the stem
initial complex	E P; post-translocational state (POST-state)	P A; pre-translocational state (PRE-state)	closed
1	A: Codon-anticodon interaction B: LOOSE BINDING of EF-Tu to the α-sarcin domain (step 1 of factor binding)	LOOSE BINDING of EF-G to the α-sarcin domain (step 1 of factor binding)	
2	MELTING of the α-sarcin stem Reduction of Ea	MELTING of the α-sarcin stem Reduction of Ea	open
3	ALLOSTERIC TRANSITION, TIGHT BINDING to the PRE-state (step 2 of factor binding) E P → •EF-Tu•GTP → P A	ALLOSTERIC TRANSITION, TIGHT BINDING to the POST-state (step 2 of factor binding) P A → EF-G•GTP → E P	
4	PRE-state triggers EF-Tu dependent GTPase	POST-state triggers EF-G dependent GTPase	
5	EF-Tu • GDP leaves the ribosome	EF-G • GDP leaves the ribosome	
6	α-SARCIN STEM REFORMS: PRE-state is trapped	α-SARCIN STEM REFORMS: POST-state is trapped	closed
final state	PRE-state (P A)	POST-state (E P)	

Figure 7. Outline of the proposed mechanism of ribosomal elongation factors.

Table I. Secondary Structure Stability of α-Sarcin Domain and Helix 81[a]

	α-Sarcin domain helix 95		Helix 81	
	ΔG°_{37}	$\Delta G^\circ_{37,bp}$	ΔG°_{37}	$\Delta G^\circ_{37,bp}$
Eubacteria				
Anacystis nidulans	−5.7	−0.81	−8.6	−1.43
Bacillus stearothermophilus	−3.5	−0.50	−8.6	−1.43
Bacillus subtilis	−3.5	−0.50	−8.6	−1.43
Chlorobium limicola	−4.9	−0.70	−8.6	−1.43
Escherichia coli	***−4.9***	***−0.70***	***−7.5***	***−1.25***
Flexibacter flexilis	−3.1	−0.44	−8.6	−1.43
Leptospira interrogans	−4.9	−0.70	−8.6	−1.43
Micrococcus luteus	−7.3	−1.04	−8.6	−1.43
Pirellula marina	−1.3	−0.22	−7.5	−1.25
Pseudomonas aeruginosa	−4.9	−0.70	−7.5	−1.25
Pseudomonas cepacia	−4.9	−0.70	−7.5	−1.25
Rhodobacter capsulatus	−5.7	−0.81	−8.6	−1.43
Rhodobacter sphaeroides	−5.7	−0.81	−8.6	−1.43
Ruminobacter amyhlophilus	−3.9	−0.56	−7.5	−1.25
Streptomyces ambofaciens	−7.3	−1.04	−8.6	−1.43
Thermus thermophilus	−4.7	−0.67	−8.6	−1.43
Average		−0.68 ± 0.21		−1.37 ± 0.09
Archaebacteria				
Halobacterium halobium	−6.6	−0.83	−6.4	−1.07
Haloarcula marismortui[b]	−6.3	−0.79	−8.2	−1.37
Halococcus morrhuae	−4.5	−0.56	−6.8	−1.13
Methanobacterium thermoautothrophicum	−4.2	−0.53	−6.8	−1.13
Methanococcus vannielli	−6.5	−0.81	−6.9	−1.15
Thermoplasma acidophilum	−1.6	−0.23	−9.2	−1.53
Subtotal average		−0.63 ± 0.23		−1.23 ± 0.18
Thermoproteus tenax	−11.1	−1.39	−7.7	−1.28
Thermophilum pendens	−10.3	−1.29	−7.9	−1.32
Desulfurococcus mobilis	−11.2	−1.40	−5.9	−1.18
Subtotal average		−1.36 ± 0.06		−1.26 ± 0.07
Average		−0.87 ± 0.41		−1.24 ± 0.15
Eukaryotes				
Caenorhabditis elegans	−5.1	−0.85	−5.7	−0.95
Homo sapiens	−3.4	−0.68	−5.8	−0.97
Mus musculus (mouse)	−3.4	−0.68	−5.8	−0.97
Oryza sativa (rice)	−0.7	−0.14	—	—
Rattus norvegicus (rat)	−3.4	−0.68	−5.8	−0.97
Saccharomyces cerevisiae	−2.9	−0.48	−5.8	−0.97
Tetrahymena thermophila	−1.7	−0.28	−7.1	−1.18
Xenopus laevis	−3.4	−0.68	−5.8	−0.97
Average		−0.60 ± 0.22		−1.00 ± 0.08

(Continued)

Table I. *(Continued)*[a]

	α-Sarcin domain helix 95		Helix 81	
	ΔG°_{37}	$\Delta G^\circ_{37,bp}$	ΔG°_{37}	$\Delta G^\circ_{37,bp}$
Plastids				
Astasia longa	−1.7	−0.24	−8.6	−1.43
Chlamydomonas reinhardtii	−4.6	−0.66	−8.6	−1.43
Chlorella ellipsoidea	−6.0	−0.86	−8.6	−1.43
Euglena gracilis	−3.7	−0.53	−8.6	−1.43
Marchantia polymorpha	−5.1	−0.73	−7.1	−1.18
Nicotiana tabacum	−5.1	−0.73	−8.6	−1.43
Oryza sativa	−5.1	−0.73	−8.6	−1.43
Pisum sativum (pea)	−3.6	−0.60	−8.6	−1.43
Zea mays	−0.8	−0.13	−8.6	−1.43
Average		−0.58 ± 0.24		−1.40 ± 0.08
Mitochondria				
Antilocapra americana	−1.7	−0.28	−6.6	−1.10
Bos taurus	−1.7	−0.28	−6.6	−1.10
Cervus unicolor	−1.7	−0.28	−6.6	−1.10
Capra hircus	−1.7	−0.28	−6.6	−1.10
Gallus gallus	−1.9	−0.32	−5.0	−0.83
Homo sapiens	+2.2	—	−6.6	−1.10
Hydroptes inermis	−1.7	−0.28	−6.6	−1.10
Muntiacus reevesi	−1.7	−0.28	−6.6	−1.10
Mus musculus	−1.7	−0.28	−6.6	−1.10
Odocoileus virginianus	−3.9	−0.56	−6.6	−1.10
Rattus norvegicus	−1.7	−0.28	−6.6	−1.10
Paracentrotus lividus	−1.1	−0.18	−7.7	−1.28
Schizosaccharomyces pombe	−0.5	−0.08	−6.0	−1.00
Tragulus napu	−1.7	−0.28	−6.6	−1.10
Average		−0.28 ± 0.10		−1.09 ± 0.09

[a]The free energy change calculations were done using and updated version of the energy parameters of Freier *et al.*[33,34] and the sequence data were taken from a recent compilation of 23S and 23S-like rRNA sequences.[35] ΔG°_{37} is the free energy change of the helix formation at 37°C (kcal/mole), and ΔG°_{37},bp = ΔG°_{37} per base pair in the stem (kcal/mole).

[b]Formerly *Halobacterium marismortui*.

pairs, we defined the term $\Delta G^\circ_{37,bp}$, which is the standard free energy change per base pair formation (see Table I), as a normalized energy parameter that we can use for comparison of analogous structures. The data presented in Table I clearly show that for cytoplasmic ribosomes of all kingdoms the $\Delta G^\circ_{37,bp}$ value from the α-sarcin domain is significantly higher than that of helix 81 (−0.64 ± 0.28 versus −1.40 ± 0.22 kcal/mole, respectively). Exceptions to the general tendency are three thermophilic archaebacteria ($\Delta G^\circ_{37,bp}$ = −1.36 ± 0.06 kcal/mole), where the thermophily might require an increased stability for appropriate function. Chloroplast ribosomes behave like mesothermic eubacteria in this structural respect, whereas the mitochondrial ribosomes show a particularly low stability of helix 95.

Comparison with other helices yields the same picture; the stability of the α-sarcin stem is unusually low (not shown). This striking weakness is also evident when we compare the average α-sarcin stem (7 base pairs; $\Delta G^{\circ}_{37,\mathrm{bp}} = -0.64$ kcal/mole) with the weakest canonical helix of the same size, namely A7:U7, which is predicted to be even more stable ($\Delta G^{\circ}_{37,\mathrm{bp}} = -0.77$ kcal/mole). This weak character of the secondary structure in the α-sarcin domain is caused by both the base-pair composition of the stem and the destabilizing effect of the loop.

We conclude that the α-sarcin stem–loop structure is characterized by two universally conserved features, the dodecamer sequence in the loop and the structural weakness of the stem, which is well suited to a repeated melting and rejoining during various functional phases.

Although the mechanism for the elongation factors outlined in this chapter still contains speculative elements, numerous and sometimes puzzling observations can be combined to give a coherent picture which allows us to begin to understand the action of elongation factors.

ACKNOWLEDGMENTS. We are grateful to J. Belart, R. Brimacombe, N. Burkhardt, R. Jünemann, M. Rühl, and E. Philippi for help and support, and to D. Beyer, C. Lapke, and A. P. Potapov for discussion.

REFERENCES

1. Gnirke, A., Geigenmüller, U., Rheinberger, H.-J., and Nierhaus, K. H., 1989, The allosteric three-site model for the ribosomal elongation cycle: Analysis with a heteropolymeric mRNA, *J. Biol. Chem.* **264:**7291–7301.
2. Hausner, T.-P., Geigenmüller, U., and Nierhaus, K. H., 1988, The allosteric three-site model for the ribosomal elongation cycle: New insights into the inhibition mechanisms of aminoglycosides, thiostrepton, and viomycin, *J. Biol. Chem.* **263:**13103–13111.
3. Fast, R., Eberhard, T. H., Ruusala, R., and Kurland, C. G., 1987, Does streptomycin cause an error catastrophe? *Biochimie* **69:**131–136.
4. Nierhaus, K. H., 1990, The allosteric three-site model for the ribosomal elongation cycle: Features and future, *Biochemistry* **29:**4997–5008.
5. Rheinberger, H.-J., and Nierhaus, K. H., 1986, Allosteric interactions between the ribosomal transfer RNA-binding sites A and E, *J. Biol. Chem.* **261:**9133–9139.
6. Rheinberger, H.-J., Sternbach, H., and Nierhaus, K. H., 1986, Codon–anticodon interaction at the ribosomal E site, *J. Biol. Chem.* **261:**9140–9143.
7. Rheinberger, H.-J., and Nierhaus, K. H., 1986, Adjacent codon–anticodon interactions of both tRNAs present at the ribosomal A and P or P and E sites, *FEBS Lett.* **204:**97–99.
8. Wurmbach, P., and Nierhaus, K. H., 1979, Codon–anticodon interaction at the ribosomal P (peptidyl-tRNA) site, *Proc. Natl. Acad. Sci. USA* **76:**2143–2147.
9. Lührmann, R., Eckhard, H., and Stöffler, G., 1979, Codon–anticodon interaction at the ribosomal peptidyl site, *Nature* **280:**423–425.

10. Peters, M., and Yarus, M., 1979, Transfer RNA selection at the ribosomal A and P sites, *J. Mol. Biol.* **134:**471–491.
11. Geigenmüller, U., and Nierhaus, K. H., 1990, Significance of the third tRNA binding site, the E site, on *E. coli* ribosomes for the accuracy of translation: An occupied E site prevents the binding of non-cognate aminoacyl-tRNA to the A site, *EMBO J.* **9:**4527–4533.
12. Schilling-Bartetzko, S., Bartetzko, A., and Nierhaus, K. H., 1992, Kinetic and thermodynamic parameters for tRNA binding to the ribosome and for the translocation reaction, *J. Biol. Chem.* **267:**4703–4712.
13. Moazed, D., and Noller, H. F., 1989, Intermediate states in the movement of transfer RNA in the ribosome, *Nature* **342:**142–148.
14. Rheinberger, H.-J., and Nierhaus, K. H., 1987, The ribosomal E site at low Mg^{2+}: Coordinate inactivation of ribosomal functions at Mg^{2+} concentrations below 10 mM and its prevention by polyamines, *J. Biomol. Struct. Dynam.* **5:**435–446.
15. Quigley, G. J., Wang, A. H. J., Seeman, N. C., Suddath, F. L., Rich, A., Sussman, J. L., and Kim, S. H., 1975, Hydrogen bonding in yeast phenylalanine transfer RNA, *Proc. Natl. Acad. Sci. USA* **72:**4866–4870.
16. Schilling-Bartetzko, S., Franceschi, F. J., and Nierhaus, K. H., 1992, Apparent association constants of tRNAs for the ribosomal A, P and E sites, *J. Biol. Chem.* **267:**4633–4702.
17. Komine, Y., Adachi, T., Inokuchi, H., and Ozeki, H., 1990, Genomic organization and physical mapping of the transfer RNA genes in *Escherichia coli* K12, *J. Mol. Biol.* **212:**579–598.
18. Potapov, A. P., 1982, A stereospecific mechanism for the aminoacyl-tRNA selection at the ribosome, *FEBS Lett.* **146:**5–8.
19. Andersson, S. G. E., Buckingham, R. H., and Kurland, C. G., 1984, Does codon composition influence ribosome function? *EMBO J.* **3:**91–94.
20. Kurland, C. G., 1980, On the accuracy of elongation, in: *Ribosomes* (G. Chambliss, G. R. Craven, J. Davies, K. Davis, L. Kahan, and M. Nomura, eds.), pp. 597–614, University Park Press, Baltimore.
21. Hopfield, J. J., and Yamane, T., 1980, The fidelity of protein synthesis, in: *Ribosomes* (G. Chambliss, G. R. Craven, J. Davies, K. Davis, L. Kahan, and M. Nomura, eds.), pp. 585–596, University Park Press, Baltimore.
22. Nierhaus, K. H., 1982, Structure, assembly, and function of ribosomes, in: *Current Topics in Microbiology and Immunology*, Volume 97 (W. Henle, P. H. Hofschneider, H. Koprowski, F. Melchers, R. Rott, H. G. Schweiger, and P. K. Vogt, eds.), pp. 81–155, Spinger-Verlag, Berlin.
23. Gnirke, A., and Nierhaus, K. H., 1986, tRNA binding sites on the subunits of *Escherichia coli* ribosomes, *J. Biol. Chem.* **261:**14506–14514.
24. Endo, Y., and Wool, I. G., 1982, The site of action of α-sarcin on eukaryotic ribosomes: The sequence at the α-sarcin cleavage site in 28S ribosomal ribonucleic acid, *J. Biol. Chem.* **257:** 9054–9060.
25. Hausner, T.-P., Atmadja, J., and Nierhaus, K. H., 1987, Evidence that the G^{2661} region of 23S rRNA is located at the ribosomal binding sites of both elongation factors, *Biochimie* **69:**911–923.
26. Wool, I. G., 1984, The mechanism of action of the cytotoxic nuclease α-sarcin and its use to analyse ribosome structure, *Trends Biochem. Sci.* **9:**14–17.
27. Endo, Y., Huber, P. W., and Wool, I. G., 1983, The nuclease activity of the cytotoxin α-sarcin: The characteristics of the enzymatic activity of α-sarcin with ribosomes and ribonucleic acids as substrates, *J. Biol. Chem.* **258:**2662–2667.
28. Miller, D. L., 1972, Elongation factors EF-Tu and EF-G interact at related sites on ribosomes, *Proc. Natl. Acad. Sci. USA* **69:**753–755.
29. Richter, D., 1973, Competition between the elongation factors 1 and 2, and phenylalanyl transfer ribonucleic acid for the ribosomal binding sites in a polypeptide-synthesizing system from the brain, *J. Biol. Chem.* **248:**2853–2857.

30. Moazed, D., Robertson, J. M., and Noller, H. F., 1988, Interaction of elongation factors EF-G and EF-Tu with a conserved loop in 23S RNA, *Nature* **334:**362–364.
31. Twardowski, T., and Nierhaus, K. H., 1993, The α-sarcin stem–loop structure of 23S rRNA: Hybridization of antisense probes provokes drastic effects on the large ribosomal subunit, submitted.
32. Gavrilova, L. P., Perminova, I. N., and Spirin, A. S., 1981, Elongation factor Tu can reduce translation errors in poly(U)-directed cell-free systems, *J. Mol. Biol.* **149:**69–78.
33. Freier, S. M., Kierzek, R., Jaeger, J. A., Sugimoto, N., Caruthers, M. H., Neilson, T., and Turner, D. H., 1986, Improved free-energy parameters for predictions of RNA duplex stability, *Proc. Natl. Acad. Sci. USA* **83:**9373–9377.
34. Jaeger, J. A., Turner, D. H., and Zuker, M., 1989, Improved predictions of secondary structures for RNA, *Proc. Natl. Acad. Sci. USA* **86:**7706–7710.
35. Gutell, R. R., Schnare, M. N., and Gray, M. W., 1990, A compilation of large subunit (23S-like) ribosomal RNA sequences presented in a secondary structure format, *Nucleic Acids Res.* **18:**2319–2330.

Chapter 4

Genetics of Translation Initiation Factors in *Saccharomyces cerevisiae*

Lan Feng and Thomas F. Donahue

1. INTRODUCTION

The first genetic mutant affecting the translation initiation process of the yeast *Saccharomyces cerevisiae* was isolated almost 25 years ago as a temperature-sensitive mutant. But it was not until recent years that genetic analysis in yeast proved worthy as an effective approach to study the mechanism of eukaryotic translation. As a result of these recent genetic studies, yeast is now accepted as a eukaryotic organism with a similar mechanism for initiating translation as observed in mammalian cells. The similarities between the mammalian and yeast initiation processes, now well established, enables us to assess the mechanism of initiation by combining both biochemical and genetic data, a very powerful means to dissect the eukaryotic translation machinery. The ability to do genetics provides additional dimensions to the analysis of translation initiation. As discussed below, it affords direct selection for suppressor mutations in genes which encode factors. These suppressor genes help to define those factors that function at particular steps in the initiation pathway. This has always been one of the rate-limiting

LAN FENG AND THOMAS F. DONAHUE • Department of Biology, Indiana University, Bloomington, Indiana 47405.

Translational Regulation of Gene Expression 2, edited by Joseph Ilan. Plenum Press, New York, 1993.

problems associated with the biochemical analysis of translation initiation, as the complexity of the process was not compatible, in many cases, with developing specific assays for factors at defined steps in the pathway. Genetic suppressor analysis can also be a very sensitive method for detecting factor function. It has the potential to define new factors that may not be realizable by biochemical approaches, especially if these factors are not abundant proteins. Aside from suppressor analysis, genetics has also allowed the ability to establish the biological significance of factors for translation initiation and to define translational control mechanisms which operate in response to different physiological conditions, and may be providing insights into cell cycle control mechanisms. The purpose of this chapter is to provide an overview of these genetic studies in yeast with special emphasis on studies that relate to factors involved in the eukaryotic translation initiation process.

2. *SACCHAROMYCES CEREVISIAE* UTILIZES A MECHANISM OF TRANSLATION INITIATION THAT IS CONSISTENT WITH THE SCANNING MODEL

The first genetic attempt to determine the signals for directing translation initiation at a yeast messenger RNA (mRNA) was performed by Sherman and colleagues over 20 years ago.[1] Using reversion analysis of *CYC1* initiator codon mutants, they found that the most 5′ proximal AUG serves as the start codon for translation initiation when present at any site within a 37-nucleotide region.[1,2] These genetic data provided the initial evidence that AUG was the primary signal for translation initiation in yeast mRNA and also contributed to the establishment of the scanning model.[3] These findings were corroborated by other genetic studies in yeast.[4] Our lab designed a genetic selection scheme to identify mutations that have a negative effect on translation initiation at the AUG start codon in the *HIS4* message. The scheme was designed in such a way so that yeast cells that failed to initiate translation at the first AUG initiator codon could be detected by their ability to initiate at a downstream AUG codon. All mutations identified that altered initiation at the first AUG were mutations within the AUG itself. No mutations in other regions of the leader were identified.

The sequence context neighboring the AUG initiator codon of a mammalian gene has been shown by mutational analysis to contribute significantly to the efficiency of initiation.[5] The consensus derived from this and comparative studies, 5′-CACCAUGG-3′, is believed to be optimal for initiation. Comparative analysis of yeast mRNAs indicates that yeast leader regions are A rich and, similar to mammalian genes, do not have a Shine and Dalgarno ribosomal binding site.[6] The sequence context of yeast initiator regions is biased toward having an A nucleotide at the -3 position and a U nucleotide at the $+4$ position (the A of the AUG as $+1$). Mutational analysis of the sequence context at both the *CYC1* and *HIS4* genes, however, showed that changing the sequence surrounding the AUG start codon had no more than a twofold effect on the efficiency of initiation at the AUG start codon.[7,8] Thus, in contrast to the finding in mammalian studies, AUG appears to be the primary signal for translation initiation.

Comparison of yeast leader regions also indicated that the majority of yeast mRNAs have leader lengths that range between 25 and 95 nucleotides, with 52 nucleotides being the average length.[6] Mutational analysis at the *HIS4* locus showed that an elongated leader does not affect the efficiency of translation initiation.[7] A similar mutational analysis at the *PGK* locus indicated that a leader length of 21 nucleotides or as short as 7 nucleotides only decreased translation initiation at an AUG by about 50%.[9] Observations on these short leader mutants further point to the fact that the ribosome can bind near the 5′ end of mRNA and that no sequence information (signals) appears to reside in the leader region that is required for translation initiation. This is in agreement with similar studies in mammalian systems, which show that a minimum length of 15 nucleotides is required for efficient initiation and no specific sequence element is needed to promote initiation (reviewed in ref. 10).

Finally, as observed in studies with mammalian genes,[11,12] insertion of secondary structures (stem–loop) in the leader region of yeast mRNAs abolishes gene expression.[7,13] These studies have been interpreted to be consistent with the scanning model, in light of the assumption that stem–loop structure in a leader region abolishes translation initiation either by affecting the accessibility of the 5′ end of mRNA to ribosomes or by blocking the ability of the ribosome to scan toward an AUG.

3. GENETICS OF INITIATION FACTORS IN YEAST

The data discussed above suggest a mechanism of initiation in yeast that has 5′-to-3′ directionality, whereby ribosomes (1) bind near the 5′ end of mRNA, (2) scan the leader region, and (3) select an AUG codon nearest the 5′ end of the message. For simplicity we will present the genetic studies that relate to translation initiation factors in yeast as if this mechanism occurs as a three-step pathway. In order to relate the genetic studies to the biochemistry of initiation, we have prefaced each section with a brief introduction which defines the factors that are believed to be involved at each step and their possible mechanistic function. It should be noted that most of the factorology mentioned is limited to that which has been obtained from studies of mammalian systems. For more extensive reviews on the biochemistry of mammalian translation initiation we refer to some recent articles.[14,15]

3.1. Ribosomal Binding of mRNA

The mechanism by which ribosomes bind mRNA is not clear. Based on biochemical studies, two models have been proposed as to how this might occur (reviewed in ref. 16). The first model, cap-dependent initiation, is generally considered to have greater application to basic mechanisms for initiation of translation at the majority of mRNAs and involves eIF-4F. eIF-4F is composed of three subunits: eIF-4E, the cap-binding protein; eIF-4A, the prototype DEAD-box-containing RNA helicase protein; and p220, a subunit for which no specific function has been proposed. eIF-4F is believed to bind mRNA through the cap-binding protein and brings eIF-4A to the 5′ end of mRNA. The eIF-4A protein is then thought to remove secondary structure in the mRNA to facilitate ribosomal binding near the 5′ end of the message. eIF-3, a large multisubunit factor, is reported to bind both the ribosome and eIF-4F; eIF-3 could be considered as a factor that might play a role in linking ribosome binding to mRNA. The second model, cap-independent initiation, is much more speculative in nature, and, based on intuition, might only apply to a limited subset of mRNAs. In this model eIF-4A is proposed to bind mRNA internally and in conjunction with eIF-4B unwinds mRNA to provide a single-stranded region for the ribosome to bind near an AUG start codon. It has been suggested that two other proteins, p52[17] and p57,[18] might be involved in this process, although neither protein is a known initiation factor nor have they been directly demonstrated to be important for translation initiation.

Some of the factor genes that are believed to be important for ribosomal binding to mRNA have been isolated and the manipulation of these genes in yeast has proved quite useful in demonstrating that their encoded gene products play an important *in vivo* role in the translation initiation process. Unfortunately, the genetics of yeast has not yet provided any insight or clarity with regard to the mechanism by which ribosomes bind mRNA. Having these genes in hand should lend to some experimental testing of these models.

3.1.1. Identification of the eIF-4E Gene

The Trachsel lab isolated a 24-kDa protein from a yeast postribosomal supernatant by m^{7}GTP agarose affinity chromatography.[19] This protein binds m^{7}GTP, m^{7}GDP, but not GDP, and can be cross-linked to the 5′ cap structure of oxidized reovirus mRNA. The gene encoding this 24-kDa protein was subsequently cloned and sequenced and the derived amino acid sequence is 33% identical to the mammalian eIF-4E sequence.[20,21] These data suggested that this protein could be the yeast homologue of the mammalian cap-binding protein, eIF-4E. The genetics of yeast seems to prove this point. Gene disruption experiments showed that this yeast gene is essential for cell viability, as expected if this gene encoded an important initiation factor.[20] Alleles were constructed which conferred a temperature-sensitive (ts) phenotype to yeast and extracts made from these strains were shown to have a defect in *in vitro* translation.[22] In addition, it was shown that the mammalian eIF-4E cDNA can substitute for the yeast gene homologue *in vivo*,[21] demonstrating that the yeast gene encodes eIF-4E. These data clearly indicate that eIF-4E is biologically relevant to the translation initiation process.

The eIF-4E gene in yeast was also identified as a cell cycle mutant, *cdc33*.[23]

The *cdc33* mutant was originally isolated as a temperature-sensitive mutant that arrested at G_1 upon shifting cells from 25 to 35°C.[24] eIF-4E isolated from this strain was shown to have a defect in cap-binding activity.[25] It has been speculated[23] that the G_1 arrest of the *cdc33* mutant could represent a specific involvement of eIF-4E in controlling the cell cycle, as cells with a ts defect in protein synthesis might be expected to arrest at random points in the cell cycle. These results are quite interesting, as they suggest some link between the translation initiation machinery and the cell cycle control process. Other observations suggest that eIF-4E may play an important role in controlling rates of protein synthesis. Sequence comparison shows that the phosphorylated serine-53 position and its flanking sequence in mammalian eIF-4E are conserved in yeast eIF-4E.[21,26] The phosphorylated form of the mammalian protein has been speculated to be important for stimulating translation initiation. This similarity could imply that phosphorylation of eIF-4E may have functional significance in controlling translation initiation in yeast.

3.1.2. eIF-4A Genes Are Identified as Genetic Suppressors of a Mitochondrial Missense Mutation

A yeast eIF-4A gene was isolated by a genetic suppression scheme, but not in such a way that might have been predictable based on the proposed function of this protein in translation initiation. A yeast strain containing a missense mutation in the *mitochondrially encoded OXI2* gene was transformed with a yeast genomic bank inserted in a high-copy vector.[27] The *oxi2* mutation causes respiratory insufficiency due the inability to make the mitochondrial cytochrome oxidase III subunit and this mutant phenotype can be detected by the inability of the strain to grow on a nonfermentable carbon source such as glycerol.[28] One clone, which could grow on glycerol medium despite a mutation in the *oxi2* gene, was isolated by transformation of this strain with the high-copy bank. Thus, the eIF-4A gene was isolated as a high-copy suppressor of a mitochondrial missense mutation.

The basic question is why did this occur or how does this relate to the proposed function of eIF-4A in translation initiation? The answer lies in the fact that *OXI2* expression is dependent on the expression of nuclear encoded genes which are translated in the cytoplasm and transported into the mitochondria.[29] One possible explanation is that suppression of *oxi2* by overexpressed eIF-4A is a direct effect. Perhaps eIF-4A is important to both the cytoplasmic and mitochondrial translation initiation processes. In this case increased levels of eIF-4A are transported into the mitochondria, which results in increased levels of mutant Oxi2 protein that partially corrects the mutant phenotype. However, the eIF-4A gene lacks features, such as a mitochondrial targeting sequence, that are usually found associated with a protein that is shared between the cytoplasm and the mitochondria.[27] Alternatively, eIF-4A is only involved in cytoplasmic translation initiation and suppression of *oxi2* is an indirect effect. In this case, increased levels of eIF-4A would lead to an increase in translation of a protein which functions in *oxi2* expression in the mitochondria. When this protein is abundant in the mitochondria it can correct the defect conferred by the *oxi2* gene by making more mutant Oxi2 protein. This latter explanation is more compatible with biochemical observations that eIF-4A is important for initiation of cellular mRNAs. Finding the gene(s) whose expression is rate limiting by the amount of eIF-4A in the cell is of the utmost importance. The ability to define the inherent features of this mRNA which confer rate-limiting eIF-4A expression would provide an important clue into the function of eIF-4A in the translation initiation process.

3.1.3. eIF-4A Is an Essential Gene Required for *in Vivo* Translation Initiation

There are two copies of the eIF-4A gene in the yeast haploid genome: *TIF1* (translation initiation factor), which was identified as a high-copy suppressor of the *oxi2* mutant; and *TIF2*, which was identified by probing with the *TIF1* gene.[27] The DNA sequence of *TIF1* and *TIF2* showed that they encode an identical 395-amino acid protein which is 65% identical (81% similar when conserved residues are considered) to the mammalian eIF-4A protein.[27,30] *TIF2* is more highly expressed than *TIF1* at the level of transcription.[31] Gene disruption experiments demonstrated, however, that yeast cells are viable when either *TIF1* or *TIF2* is the sole copy of eIF-4A. Cells are inviable only when the expression of both *TIF1* and *TIF2* is disrupted.[27]

The yeast eIF-4A genes in hand have provided the ability to assess the function of this protein *in vivo*. Promoter fusion constructs have enabled the preparation of cell-free extracts which are depleted of yeast eIF-4A. These extracts are defective in promoting translation, but the activity can be restored by the addition of purified eIF-4A.[32] Again, this demonstrates the biological relevance of this protein for *in vivo* translation. In light of the high degree of similarity between the yeast and mammalian proteins, it is surprising that the mammalian eIF-4A gene cannot substitute for the yeast eIF-4A genes *in vivo*.[31] The yeast protein can function, however, in conjunction with the mammalian eIF-4B protein in *in vitro* assays that are designed to measure the displacement of RNA:RNA duplexes as an indication of helicase activity.[33] The yeast gene has also been analyzed extensively by mutation and most of the mutations that interfere with its ability to function map to the motifs that are considered to be the conserved signatures of RNA helicase.[34]

3.1.4. eIF-3

The first mutants that affected global protein synthesis were identified almost 25 years ago among temperature-sensitive (ts) yeast mutants.[35] One of these mutants, *prt1*, has been characterized extensively by biochemical analysis and is believed to encode one of the subunits of yeast eIF-3 (for review see ref. 36). The major criteria for this assignment are that: (1) the Prtl protein copurifies with a high-molecular-weight complex analogous to that of mammalian eIF-3; (2) cell extracts derived from a *prtl*, ts mutant are defective in translation initiation at the nonpermissive temperature; (3) these cell extracts are also defective in converting 40S to 45S ribosomal particles, which represents a 40S–eIF-3 complex, as well as defective in ternary complex (eIF-2–$tRNA_i^{met}$–GTP) association with the 40S, two activities associated with eIF-3.

The *PRT1* gene has been cloned and sequenced and encodes a 763-amino acid protein.[37] Interestingly, the *PRT1* gene appears to have also been identified independently as a temperature-sensitive cell cycle mutant, *cdc63*, that arrests at the start position when cells are shifted to the nonpermissive temperature[38] and similar observations have been made with the ts^-, *prtl* mutant strain.[39] Perhaps eIF-4E and eIF-3 both play a role in cell proliferation in addition to their roles in translation initiation.

3.1.5. Observations on Cap-Independent Initiation in Yeast

There have been some observations made with yeast cell-free extracts that are depleted of eIF-4E that the leader of the tobacco mosaic virus (TMV) mRNA still promotes translation initiation at a downstream CAT reporter coding region.[40,41] The TMV leader is believed to be unstructured and therefore probably promotes translation initiation in the absence of eIF-4E, by bypassing a need for eIF-4E function in removing secondary structure from mRNA as explained in Section 3.1.

We have taken a different approach to determine if a mechanism of cap-independent translation initiation occurs in yeast. We have put the *HIS4* gene under PolI transcriptional control in an attempt to make "capless" *HIS4* transcripts in yeast (H. J. Lo and T. F. Donahue, unpublished observations). Transcription of *HIS4* occurs in this strain and initiates at the *Pol*I start site. In addition, the level of *HIS4* transcript is comparable to that normally observed in wild-type *HIS4* strains. Although this strain does not have a wild-type $His4^+$ phenotype, it does grow in the absence of histidine. This could mean that yeast has some capacity to translate a capless mRNA, although other interpretations for these effects can be imagined. We have also identified revertants of this strain that can grow better on synthetic dextrose-histidine plates. These revertants contain extragenic suppressor mutations. These suppressor mutants fall into two classes: Class I revertants lead to an increase in transcript levels and may encode mutated forms of transcriptional factors; and class II revertants do not change the level of the transcript nor affect the initiation site. This latter class of suppressor mutants could define initiation factors that promote capless translation initiation events in yeast.

3.2. Ribosomal Scanning

The ribosomal scanning model proposes that once bound to the 5′ end of the message the ribosome scans the leader in search of an AUG start codon.[3] We do not think that even the slightest clue exists as to how the ribosome scans or what factors are directly involved in this process. Mutational studies do appear to be quite strong in implying that ribosomes have 5′-to-3′ directionality in their ability to select a start site for translation (see Section 2). One can only speculate as to how scanning might occur. The ribosome could be propelled as originally proposed in the model. One variation on the model is that eIF-4F in conjunction with eIF-4A and 4B, which are proposed to have mRNA unwinding activity,[42] remove secondary structure in a 5′-to-3′ direction and pave a path of least resistance for the ribosome to follow passively.

Recently we have initiated a genetic reversion analysis of stem–loop structure mutations that have been inserted at different positions in the leader region of *HIS4* and presumably block either the ability of the ribosome to bind mRNA or scan toward an AUG (see Section 2; T. F. Donahue, unpublished observations). The strategy of the selection scheme is to identify extragenic suppressors that can now restore *HIS4* expression despite this stem–loop structure mutation in the leader. These suppressors presumably would represent mutations in genes that encode factors that function during the ribosomal binding or scanning steps of translation initiation. As a result of a mutation in these genes the factor can now perform this function better and melts out the stem–loop inserted at *HIS4* to allow translation initiation to occur. Preliminary studies indicate that at least six unlinked suppressor genes have been identified. One gene, *SSL1*, encodes an approximately 50-kDa protein. Cell extracts prepared from conditional *SSL1* mutants exhibit a defect in translation initiation.[43] A second gene characterized, *SSL2*, has protein sequence motifs that are characteristic of RNA helicases.[44] Neither of these two genes is an allelic form of eIF-4E or eIF-4A nor do the derived amino acid sequences of *SSL1* and *SSL2* resemble the mammalian eIF-4B sequence;[45] we anticipated to identify the eIF-4A, 4B, and 4E genes as suppressors based on the selection strategy and the proposed function of these proteins in the translation initiation process. It is too early to speculate about these findings and the other suppressor genes are currently being characterized.

3.3. Ribosomal Recognition of an AUG Start Codon

The biochemistry of translation initiation defined eIF-2 as an essential component of the preinitiation complex whose function was to bind $tRNA_i^{met}$ in a GTP-dependent fashion.[15] This ternary complex (eIF-2–$tRNA_i^{met}$–GTP) then associated with the 40S ribosome, which in turn binds mRNA and scans the leader looking for an AUG start codon. At the time of 80S complex formation eIF-2 is released as a GDP–eIF-2 binary complex. In order for eIF-2 to recycle back into the translation initiation process the GDP-bound form must be converted to a GTP-bound form that is necessary for $tRNA_i^{met}$ binding or ternary complex formation.

As discussed below, the genetics of yeast defined eIF-2 and $tRNA_i^{met}$ as components of the preinitiation complex that function in directing the ribosome to the start site of translation. Because of a unique blend of information obtained by biochemical and genetic analyses, we now understand more about ribosomal recognition of an AUG start codon than any other step in translation initiation.

3.3.1. $tRNA_i^{met}$

In light of the fact that in yeast the AUG initiator codon solely serves as the start signal for translocation, we speculated that a base-pair interaction between the anticodon of $tRNA_i^{met}$ and the AUG was an important component for ribosomal recognition of a start site. To test this idea, we initiated genetic studies in which the anticodon of one of the $tRNA_i^{met}$ genes from yeast was mutated from 3′-UAC-5′ to 3′-UCC-5′ and also mutated the initiator codon at the *HIS4* gene to AGG to be complementary with UCC–$tRNA_i^{met}$.[46] Both genes when present in single copy in yeast cells would not allow translation initiation to occur at *HIS4*, as noted by a His$^-$ phenotype, a result anticipated, as the anticodon mutation was expected to result in the inability of this mutant tRNA to be charged by the methionyl-tRNA synthetase.[47] The UCC–$tRNA_i^{met}$, however, could restore translation at *his4* that was specific for the AGG codon in either of two ways. First, when we selected a mutation in the methionyl-tRNA synthetase which now had the ability to charge this UCC–$tRNA_i^{met}$, and second, when we overexpressed either the wild-type synthetase gene or the UCC-initiator-tRNA. Presumably overexpression of either gene led to increased levels of charged mutant tRNA that could now support an initiation event at the AGG codon at *his4*.

The ability to put this genetic system together enabled us to test the importance of the anticodon–codon interaction in directing the ribosome to the site of initiation.[46] A *HIS4* allele was constructed to contain an AGG codon upstream and out-of-frame with the AGG codon at the initiator site. As observed for the effects of an AUG codon that is upstream and out-of-frame,[4] the UCC-initiator-tRNA directed the ribosome to the first AGG codon in the *his4* message, which precluded translation initiation at the downstream AGG. These data indicated that at least one element that is important to the ability of the ribosome to recognize the initiator region in mRNA is a cognate anticodon–codon interaction between the $tRNA_i^{met}$ and the first AUG codon in mRNA.

3.3.2. eIF-2

A genetic reversion analysis was developed by the Donahue lab to identify specific components and interactions between the preinitiation complex and mRNA which mediate the AUG initiator codon recognition process.[48] It has been discussed previously that initiator codon mutations at *HIS4* abolish translation initiation and result in a His$^-$ phenotype and when this initiator codon mutation is part of a *HIS4–lacZ* fusion construct, a white colony phenotype on X-Gal indicator plates.[4] These initiator codon mutants were reverted to His$^+$ and screened on X-Gal indicator plates for blue colonies as a sign of extragenic suppression events. The rationale was that mutations in components of the preinitiation complex that function in ribosomal recognition of a start codon would now allow the ribosome to initiate at a non-AUG codon and restore *HIS4* expression. The characterization of these encoded suppressor gene products would then tell us which components of the preinitiation complex might function in mediating the start site selection process. Three unlinked genes, *sui1*, *sui2*, and *SUI3*, were identified.

The cloning and sequencing of the *sui2* and *SUI3* genes provided important insight into their function. The wild-type *sui2* gene encodes an open reading frame of 304 amino acids that is 58% identical to the α subunit of mammalian eIF-2 and the degree of similarity is increased to 80% if conservative amino acid substitutions are considered.[49,50] The *SUI3* gene encodes the β subunit of eIF-2, a 36-kDa protein that is 42% identical to the human eIF-2β subunit.[51,52] Both the human and yeast β proteins contain two potential nucleic acid-binding domains at the same relative position. One is at the amino-terminal end and comprises three polylysine repeats and the other is at the carboxyl-terminal end and corresponds to a Zn(II) "finger" motif. The mutations in *SUI3* that confer the suppressor phenotype are located either within or slightly outside the Zn(II) finger motif. This suggests that this motif represents a nucleic acid-binding domain that mediates ribosomal recognition of the AUG initiator codon, as mutations in this region altered the start site selection process at *HIS4*.

Our genetic and molecular observations suggested that, in addition to tRNA$_i^{met}$ binding activity, eIF-2 might play a role in ribosomal recognition of a start codon. Sequence analysis of the His4 protein made in the absence of an AUG start codon appeared to confirm this interpretation. It was shown by amino acid sequence analysis that the His4 protein made *in vivo*, as a result of either *sui2* or *SUI3* suppression, began with the amino acid corresponding to the penultimate codon relative to a UUG.[49,51] Our assumption was the Met was inserted at UUG and cleaved posttranslationally by aminopeptidase. It was conceivable, however, that a mutation in either α or β might allow eIF-2 to now bind a Leu-tRNA to initiate at the complementary UUG codon and the corresponding Leu inserted at UUG was cleaved by an aminopeptidase. The significance of this distinction is that in the former case eIF-2 would be implicated to play a direct role in the recognition process allowing the tRNA$_i^{met}$ to interact with a UUG codon; in the latter case, the role of eIF-2 would be limited to tRNA binding and the cognate anticodon–codon interaction might then dictate the recognition process. In order to distinguish suppression events that allow tRNA$_i^{met}$ to initiate translation at the UUG codon from another tRNA species, we changed the penultimate amino acid codon located next to the UUG codon to a phenylalanine codon.[53] A penultimate amino acid such as phenylalanine which has a large Stokes radius will prevent the first amino acid from being cleaved by aminopeptidase. When the corresponding His4 protein was sequenced, Met was observed in the first cycle. Therefore, eIF-2 suppressor mutations afford a mismatched anticodon–codon interaction between the $tRNA_i^{met}$ and UUG codon.

The significance of these observations is that it suggests that eIF-2 is a component of the preinitiation complex that mediates recognition of the initiator region. This is consistent with the biochemical properties of translation initiation, as eIF-2 remains associated with the 40S ribosome up until the time of 80S complex formation. Our assumption is that as a result of a mutation in either the *sui2* and *SUI3* genes the preinitiation complex now has an altered specificity for a UUG codon. The mutated eIF-2, however, must maintain normal AUG recognition properties, as the suppressor genes were identified in a haploid strain and both *sui2* and *SUI3* are unique genes and essential for cell viability. We envision two possible mechanisms for the ability of these mutants to allow the $tRNA_i^{met}$ to initiate translation at a UUG codon. The first possibility is that eIF-2 is naturally involved in stabilizing the anticodon–codon interaction. A second possibility is that eIF-2 normally has an editing function that checks for a three-base-pair interaction between the $tRNA_i^{met}$ and the AUG codon. eIF-2 mutants, therefore, may have lost stringent editing function and allow $tRNA_i^{met}$ to mismatch base pair with a UUG codon. Whatever function eIF-2 might serve in the recognition process, its effect is most probably mediated directly by the $CysX_2CysX_{19}CysX_2Cys$ region of eIF-2β, as *SUI3* suppressor mutations that map to this region confer the most efficient suppression events.[48]

3.3.3. *sui1* Defines an Additional Factor Important to Start Site Selection

The third suppressor gene identified by this reversion analysis is *sui1*.[46] The *sui1* gene encodes a small protein of 12 kDa which is essential for cell viability.[51] Its protein sequence has no resemblance to any known translation initiation factor, but the characterization of *sui1* is consistent with it encoding a protein that is involved in translation initiation.[53] Conditional *sui1* mutants exhibit altered polysome profiles at the nonpermissive temperature that are consistent with a defect in translation initiation. As observed for *SUI3* suppression events, *sui1* suppression events enable the ribosome to initiate at a UUG codon in the *HIS4* coding region by promoting a mismatched base-pair interaction with the initiator-tRNA. However, the Sui1 protein is not a subunit of eIF-2 that is required for $tRNA_i^{met}$ binding. Therefore, Sui1 must define an additional factor that functions in concert with eIF-2 to enable $tRNA_i^{met}$ to establish ribosomal recognition of an AUG codon. Based on the size of the Sui1 protein and proposed functions of other initiation factors, we anticipate that *sui1* may encode the yeast equivalent of the mammalian translation initiation factor eIF-1 or eIF-1A.[15] If this turns out to be true, it would be somewhat of a surprise, as neither factor has been considered to be essential for translation initiation, whereas Sui1 clearly plays an essential role *in vivo*.

3.3.4. Other Observations on the Start Site Selection Process

The Hampsey lab has designed an alternative genetic selection scheme that is aimed at detecting changes in the AUG codon recognition process.[54] The experimental design is that an AUG codon was placed upstream and out-of-frame with the AUG start site at the *CYC1* gene. The upstream AUG precludes translation initiation at the downstream site and prevents *CYC1* expression. Revertants were selected that now allow expression of *CYC1* and eight suppressor loci (*sua1–8*; suppressor of upstream AUG) were identified.[54] Some of the suppressor genes have been characterized and they appear to fall into either of two groups. One group alters the transcription initiation site at *CYC1* to be 3′ to the upstream AUG. The yeast TFIIB gene was identified as one of these suppressors.[55] The second group appears not to alter *CYC1* transcription and therefore represent candidates for altering ribosomal recognition of the upstream AUG (M. Hampsey, personal communication). Characterization of the second group of suppressors could define other components involved in the start site process and one interesting possibility would be the identification of components that interact with sequence context; mutations in such components may enable preferential bypass of an AUG.

4. REINITIATION

The relevance of reinitiation as a mechanism of translation initiation and the factors involved were only realized through genetic studies in yeast. Reinitiation in yeast is used as a mechanism for regulating the expression of the *GCN4* gene, which encodes a positive transcriptional activator of amino acid biosynthetic genes that are subject to the general amino acid control response.[55] The expression of *GCN4* is regulated in response to amino acid starvation through four upstream AUG codons that define short open reading frames (ORFs) in its atypically long leader of 591 nucleotides. The regulatory response at *GCN4* can be simplified by considering ORF1 and ORF4 as mediating the regulation.[57] Recent studies have shown that when amino acids are plentiful the ribosomes initiate at ORF1, terminate translation, and then reinitiate at ORF4.[58] After translation of ORF4, ribosomes are proposed to fall off of the *GCN4* mRNA, precluding initiation at the *GCN4* start codon. Under amino acid starvation conditions, however, the activity of some of the ribosomes changes. Ribosomes still initiate at ORF1, but only 50% of the ribosomes reinitiate at ORF4. The other population of ribosomes bypasses ORF4 and continues to scan the leader and initiate translation at the *GCN4* start codon.

Mutations that alter this regulation have been isolated.[56] A group of genes, *GCN*s (general control nonderepressed), were identified which when mutated prevent increased levels of *GCN4* expression in response to amino acid starvation. A second group of genes, *GCD*s (general control derepressed), were identified which when mutated lead to constitutive high levels of *GCN4* expression even in the absence of an amino acid starvation signal. Gcns are considered to be positive regulators of the general control response, whereas Gcds are considered to be negative regulators of the response. The *GCD1*, *GCD2*, and *GCN3* genes in yeast have recently been implicated to be prime candidates for encoding part of the eIF-2B complex in yeast.[58,59] eIF-2B is the guanine nucleotide exchange factor that converts eIF-2–GDP to eIF-2–GTP in order to recycle eIF-2 back into the initiation process.[15] The *sui2* and *SUI3* suppressor mutants isolated in our laboratory[47] and which encode mutant forms of eIF-2α and β[49,51] have been shown to disrupt the general control response[48] by increasing *GCN4* expression in the absence of an amino acid starvation signal.[60] These mutants act as *GCD* mutants, although they were never identified by direct selection for *GCD* mutants.

The DNA sequence of the *GCN2* gene and inspection of its encoded protein have helped to put all these genetic observations into a mechanistic perspective.[61] The Gcn2 protein has at least two recognizable domains, a carboxyl-terminal domain that shows homology to histidyl-tRNA synthetase and an amino-terminal domain that is homologous to protein kinase. The Gcn2 homology to the synthetase is divergent, and therefore, as proposed,[61] this region may bind any uncharged tRNA to sense starvation of different amino acids (general control) and activate the protein kinase domain of Gcn2. The kinase domain of Gcn2 shows specific sequence identity to the mammalian protein kinases HCR[62] (heme-controlled repressor) and DAI,[63,64] the double-stranded RNA-activated inhibitor. These kinases and possibly others are believed to phosphorylate the α subunit of eIF-2 at amino acid position serine-51, under a number of different adverse physiological conditions, to downregulate translation initiation.[26] The effect of this phosphorylation event at eIF-2α is to inhibit eIF-2B activity, the guanine nucelotide exchange factor, needed to convert eIF-2–GDP to eIF-2–GTP. If the exchange reaction is inhibited, then eIF-2 is trapped in a GDP form, incapable of ternary complex formation for new rounds of translation initiation. Yeast eIF-2α (Sui2) shares considerable sequence identity with the mammalian eIF-2α subunit inclusive of the serine-51 position and the surrounding sequences,[47,48] which has suggested that a similar mechanism for controlling translation initiation rates existed in yeast via phosphorylation of eIF-2.

Recent studies have now shown that the serine-51 position of yeast eIF-2α is phosphorylated by Gcn2 *in vitro*.[65] In addition, these studies have shown that yeast cells have increased levels of a phosphorylated species of eIF-2α during amino acid starvation conditions, and mutations in serine-51 block increased phosphorylation of eIF-2α and confer a defect in the general control response by altering the regulation of *GCN4* expression through the upstream AUGs. These studies indicate that amino acid starvation, which leads to increased levels of uncharged tRNA, activates the Gcn2 kinase to phosphorylate eIF-2α. On the basis of mammalian biochemical studies, phosphorylation of eIF-2 would inhibit eIF-2B exchange activity and functional eIF-2 levels in the cell would decrease, being trapped in an inactive GDP-bound form. This impacts on reinitiation events at *GCN4*. Ribosomes containing ternary complex are capable of initiation at ORFl in the *GCN4* leader. After translation of ORF1, however, only a fraction of the ribosomes are capable of reacquiring ternary complex (eIF-2·GTP·tRNA) to reinitiate at ORF4 due to reduced levels of eIF-2–GTP. Ribosomes that cannot initiate at ORF4, therefore, continue to scan the leader region, and presumably, as a result of increased scanning time/distance, must reacquire ternary complex to initiate at the *GCN4* start site.

This mechanism of translational control at *GCN4* relies on altering the reinitiation process, which is believed to be an inefficient mechanism to arrive at translation initiation. One might also predict that this effect would extend to other genes which may have multiple upstream AUG codons in the leader. However, if eIF-2 levels are decreased in the cell, then the rate of translation initiation, in general, must also be reduced, perhaps a response needed if the growth rate of the cell is to be coordinate with rate-limiting growth conditions. Hence, Gcn2 phosphorylation of eIF-2α not only results in specific regulation of *GCN4* expression, but must have a general effect on downregulating translation initiation at all mRNAs. Recent studies have shown that expression of the human p68 kinase (DAI) in yeast leads to inhibition of the growth rate and alters polysome profiles in a fashion that is consistent with a defect in translation initiation.[66] This is the same effect that is believed to occur in mammalian cells when eIF-2α kinase activity is induced in response to adverse growth conditions.[26] Thus, genetic studies of a gene-specific translational regulatory mechanism have provided new insight into genes that encode translation initiation factors in yeast, as well as identified a common mechanism in yeast and mammals for general control of translation initiation.

5. OTHER OBSERVATIONS ON THE GENETICS OF TRANSLATION INITIATION IN YEAST

Two genes in yeast have been identified to encode the initiation factor eIF-5A which is believed to be involved in first peptide bond formation.[15] *TIF51B* and *ANB1* are the same genetic locus, the former isolated by using a human eIF-5A cDNA probe, and the latter identified for its ability to produce a transcript when cells are grown under anaerobic conditions or in the absence of heme.[67,68] *TIF51A* was isolated as part of the *TIF51B* analysis.[67] The yeast genes encode a 17-kDa protein which is 63.5% identical to the mammalian eIF-5A protein. eIF-5A undergoes a unique posttranslational modification whereby a Lys residue is converted to hypusine and mutational studies suggest that this modification is essential for its biological function. It is not clear whether eIF-5A is biologically relevant to the translation initiation process. If this proves to be the case, then the observations that its expression is regulated in response to anaerobic conditions or heme levels, and its activity is potentially regulated through hypusine modification, could provide important insights into understanding the physiology of translational control mechanisms.

6. CLOSING REMARKS

Our ability to understand the mechanism of eukaryotic translation initiation is clearly enhanced by the ability to perform genetic analysis in yeast. The genetic studies can only be interpreted, however, in light of the exhaustive biochemistry that has been performed with the mammalian system. Thus far, the ability to combine the data derived from mammalian biochemical studies and yeast genetic studies would appear to hold up quite well in making more detailed and mechanistic interpretations of the initiation process. Nevertheless, it is essential to overcome the deficiency in our understanding of the yeast factorology. More work is definitely required to understand the events that promote ribosomal binding to mRNA and the mechanism or factors that lead to ribosomal "scanning."

References

1. Steward, J. W., Sherman, F., and Schweingruber, A. M., 1971, Identification and mutational relocation of the AUG codon initiating translation of iso-1-cytochrome *c* in yeast, *J. Biol. Chem.* **246:**7429–7445.
2. Sherman, F., Steward, J. W., and Schweingruber, A. M., 1980, Mutants of yeast initiating translation of iso-1-cytocrome *c* within a region spanning 37 nucleotides, *Cell* **20:**215–222.
3. Kozak, M., 1978, How do eukaryotic ribosomes select initiation regions in messenger RNA? *Cell* **15:**1109–1123.
4. Donahue, T. F., and Cigan, A. M., 1988, Genetic selection for mutations that reduce or abolish ribosomal recognition of the *HIS4* translational initiator region, *Mol. Cell. Biol.* **8:**2955–2963.
5. Kozak, M., 1986, Point mutations define a sequence flanking the AUG initiator codon that modulates translation by eukaryotic ribosomes, *Cell* **44:**283–292.
6. Cigan, A. M., and Donahue, T. F., 1987, Sequence and structural features associated with translational initiator regions in yeast—A review, *Gene* **59:**1–18.
7. Cigan, A. M., Pabich, E. K., and Donahue, T. F., 1988, Mutational analysis of the *HIS4* translational initiator region in *Saccharomyces cerevisiae*, *Mol. Cell. Biol.* **8:**2964–2975.
8. Baim, J. B., and Sherman, F., 1988, mRNA structures influencing translation in the yeast *Saccharomyces cerevisiae*, *Mol. Cell. Biol.* **8:**1591–1601.
9. Raué, H. A., van den Heuvel, J. J., and Planta, R. J., 1990, Yeast mRNA structure and translational efficiency, in: *Posttranscriptional Control of Gene Expression* (J. E. G. McCarthy and M. F. Tuite, eds.), pp. 237–247, Springer-Verlag, Berlin.
10. Kozak, M., 1991, Structural features in eukaryotic mRNAs that modulate the initiation of translation, *J. Biol. Chem.* **266:**19867–19870.
11. Pelletier, J., and Sonenberg, N., 1985, Insertion mutagenesis to increase secondary structure within the 5′ noncoding region of a eukaryotic mRNA reduces translational efficiency, *Cell* **40:** 515–526.

12. Kozak, M., 1986, Influences of mRNA secondary structure on initiation by eukaryotic ribosomes, *Proc. Natl. Acad. Sci. USA* **83:**2850–2854.
13. Rhoads, R. E., 1988, Cap recognition and the entry of mRNA into the protein synthesis initiation cycle, *Trends Biochem. Sci.* **13:**52–56.
14. Trachsel, H. (ed.) 1990, *Translation in Eukaryotes*, CRC Press, Caldwell, New Jersey.
15. Hershey, J. W. B., 1991, Translational control in mammalian cells, *Annu. Rev. Biochem.* **60:** 717–755.
16. Sonenberg, N., 1991, Picornavirus RNA translation continues to surprise, *Trends Genet.* **7:** 105–106.
17. Meerovitch, K., Pelletier, J., and Sonenberg, N., 1989, A cellular protein that binds to the 5′-noncoding region of poliovirus RNA, *Genes Dev.* **3:**1029–1034.
18. Jang, S.-K., and Wimmer, E., 1990, Cap-independent translation of encephalmyocarditic virus RNA, *Genes Dev.* **4:**1560–1572.
19. Altman, M., Edery, I., Sonenberg, N., and Trachsel, H., 1985, Purification and characterization of protein synthesis initiation factor eIF-4E from the yeast *Saccharomyces cerevisiae*, *Biochemistry* **24:**6085–6089.
20. Altman, M., Handschin, C., and Trachsel, H., 1987, mRNA cap-binding protein: Cloning of the gene encoding protein synthesis initiation factor eIF-4E from *Saccharomyces cerevisiae*, *Mol. Cell. Biol.* **7:**998–1003.
21. Altman, M., Müller, P. P., Pelletier, J., Sonenberg, N., and Trachsel, H., 1989, A mammalian translation initiation factor can substitute for its yeast homologue *in vivo*, *J. Biol. Chem.* **264:** 12145–12147.
22. Altman, M., Sonenberg, N., and Trachsel, H., 1989, Translation in *Saccharomyces cerevisiae*: Initiation factor 4E-dependent cell-free system, *Mol. Cell. Biol.* **9:**4467–4472.
23. Brenner, C., Nakayama, N., Goebl, M., Tanaka, K., Toh, E., and Matsumoto, A., 1988, *CDC33* encodes mRNA cap-binding protein eIF-4E of *Saccharomyces cerevisiae*, *Mol. Cell. Biol.* **8:** 3556–3559.
24. Reed, S. I., 1980, The selection of *S. cerevisiae* mutants defective in the start event of cell cycle, *Genetics* **95:**561–577.
25. Altman, M., and Trachsel, H., 1989, Altered mRNA cap recognition activity of initiation factor 4E in the yeast cell cycle division mutant *cdc33*, *Nucleic Acids Res.* **17:**5923–5931.
26. Hershey, J. W. B., 1989, Protein phosphorylation controls translation rates, *J. Biol. Chem.* **264:** 20823–20826.
27. Linder, P., and Slonimski, P., 1989, An essential yeast protein encoded by duplicated genes *TIF1* and *TIF2* and homologous to the mammalian translation initiation factor eIF-4A, can suppress a mitochondrial missense mutation, *Proc. Natl. Acad. Sci. USA* **86:**2286–2290.
28. Kruszewska, A., and Szczesniak, B., 1985, Functional nuclear suppressor of mitochondrial *oxi2* mutation in yeast, *Curr. Genet.* **10:**87–93.
29. Costanzo, M. C., and Fox, T. D., 1990, Control of mitochondrial gene expression in *Saccharomyces cerevisiae*, *Annu. Rev. Genet.* **24:**91–113.
30. Linder, P., and Slonimski, P. P., 1988, Sequence of the genes *TIF1* and *TIF2* from *Saccharomyces cerevisiae* coding for a translation initiation factor, *Nucleic Acids Res.* **16:**10359.
31. Prat, A., Schmid, S. R., Buser, P., Blum, S., Trachsel, H., Nielson, P. J., and Linder, P., 1990, Expression of translational initiation factor 4A from yeast and mouse in *Saccharomyces cerevisiae*, *Biochim. Biophys. Acta* **1050:**140–145.
32. Blum, S., Mueller, M., Schmid, S. R., Linder, P., and Trachsel, H., 1989, Translation in *Saccharomyces cerevisiae* initiation factor 4A-dependent cell-free system, *Proc. Natl. Acad. Sci. USA* **86:**6043–6046.
33. Jaramillo, M., Browning, K., Dever, T. E., Blum, S., Trachsel, H., Merrick, W. C., Ravel, J. M., and Sonenberg, N., 1990, Translation initiation factors that function as RNA helicases from mammals, plants and yeast, *Biochim. Biophys. Acta* **1050:**134–139.

34. Schmid, S. R., and Linder, P., 1991, Translation initiation factor 4A from *Saccharomyces cerevisiae*: Analysis of residues conserved in the D-E-A-E family of RNA helicases, *Mol. Cell. Biol.* **11:**3463–3471.
35. Hartwell, L., 1967, Macromolecule synthesis in temperature-sensitive mutants of yeast, *J. Bacteriol.* **93:**1662–1670.
36. Moldave, K., and McLaughlin, C. S., 1988, The analysis of temperature-sensitive mutants of *Saccharomyces cerevisiae* altered in components required for protein synthesis, in: *Genetics of Translation* (M. F. Tuite ed.), pp. 271–281, Springer-Verlag, Berlin.
37. Hanis-Joyce, P. J., Singer, R. A., and Johnston, G. C., 1987, Molecular characterization of the yeast *PRT1* gene in which mutations affect translation initiation and regulation of cell proliferation, *J. Biol. Chem.* **262:**2845–2851.
38. Hanis-Joyce, P. J., 1985, Mapping *CDC* mutations in the yeast *S. cerevisiae* by *RAD52*-mediated chromosome loss, *Genetics* **110:**591–607.
39. Hanis-Joyce, P. J., Johnston, G. C., and Singer, R. A., 1987, Regulated arrest of cell proliferation mediated by yeast *prtl* mutations, *Exp. Cell Res.* **72:**134–145.
40. Altman, M., Blum, S., Wilson, T. A. W., and Trachsel, H., 1990, The 5′ leader sequence of tobacco mosaic virus RNA mediates initiation factor 4E-independent, but still initiation factor 4A-dependent, translation in yeast extracts, *Gene* **91:**127–129.
41. Altman, M., Blum, S., Pelletier, J., Sonenberg, N., Wilson, T. A. W., and Trachsel, H., 1990, Translation initiation factor-dependent extracts from *Saccharomyces cerevisiae*, *Biochim. Biophys. Acta* **1050:**155–159.
42. Rozen, F., Edery, I., Meerovitch, K., Dever, T. E., Merrick, W. C., and Sonenberg, N., 1990, Bidirectional RNA helicase activity of eukaryotic translation initiation factors 4A and 4F, *Mol. Cell. Biol.* **10:**1134–1144.
43. Yoon, H., Miller, S. P., Pabich, E. K., and Donahue, T. F., 1992, SSL1, a suppressor of a *HIS4* 5′-UTR stem-loop mutation, is essential for translation initiation initiation and affects UV resistance in yeast, *Genes Dev.* **6:**2463–2477.
44. Gulyas, K. G., and Donahue, T. F., 1992, A suppressor of a stem-loop mutation in the *HIS-4* leader encodes the yeast homolog of human ERCC-3, *Cell* **69:**1031–1042.
45. Milburn, S. C., Hershey, J. W. B., Davies, M. V., Kelleher, K., and Kaufman, R. J., 1990, Cloning and expression of eukaryotic initiation factor 4B cDNA: Sequence determination identifies a common RNA recognition motif, *EMBO J.* **9:**2783–2790.
46. Cigan, A. M., Feng, L., and Donahue, T. F., 1988, $tRNA_i^{met}$ functions in directing the scanning ribosome to the start site of translation, *Science* **242:**93–97.
47. Schulman, L. H., and Pelka, H., 1984, Recognition of tRNAs by aminoacyl-tRNA synthetases: *Escherichia coli* $tRNA^{met}$ and *E. coli* methionyl-tRNA synthetase, *Fed. Proc.* **43:**2977–2980.
48. Castilho-Valanicias, B., Yoon, H., and Donahue, T. F, 1990, Genetic characterization of the *Saccharomyces cerevisiae* translation initiation suppressors *suil*, *sui2* and *SUI3* and their effects on *His4* expression, *Genetics* **24:**483–495.
49. Cigan, A. M., Pabich, E. K., Feng, L., and Donahue, T. F., 1989, Yeast translation initiation suppressor *sui2* encodes the α subunit of eukaryotic translation initiation factor 2 and shares sequence identity with the human α subunit, *Proc. Natl. Acad. Sci. USA* **86:**2784–2788.
50. Ernst, H., Duncan, R. F., and Hershey, J. W. B., 1987, Cloning and sequencing of complementary DNAs encoding the α subunit of translational initiation factor eIF-2, *J. Biol. Chem.* **262:**1206–1212.
51. Donahue, T. F., Cigan, A. M., Pabich, E. K., and Valavicius, B., 1988, Mutations at a Zn(II) finger motif in the yeast eIF-2β gene alter ribosomal start site selection during the scanning process, *Cell* **54:**621–632.
52. Pathak, V. K., Nielson, P., Trachsel, H., and Hershey, J. W. B., 1988, Structure of the β subunit of translation initiation factor eIF-2, *Cell* **54:**633–639.
53. Yoon, H., and Donahue, T. F., 1992, The *suil* suppressor locus in *Saccharomyces cerevisiae*

encodes a translation factor that functions during tRNA$_i^{met}$ recognition of the start codon, *Mol. Cell. Biol.* **12:**248–260.

54. Hampsey, M., Na, J. G., Pinto, I., Ware, D. E., and Berroteran, R. W., 1991, Extragenic suppressors of a translation initiation defect in the *cycl* gene of *Saccharomyces cerevisiae*, *Biochimie* **73:**1445–1455.
55. Pinto, I., Ware, D. E., and Hampsey, M., 1992, The yeast *SUA7* gene encodes a homologue of human transcription factor TFIIB and is required for normal start site selection *in vivo*, *Cell* **68:** 977–988.
56. Hinnebusch, A. G., 1988, Mechanisms of gene regulation in the general control of amino acid biosynthesis in yeast *Saccharomyces cerevisiae*, *Microbiol. Rev.* **52:**248–273.
57. Hinnebusch, A. G, 1990, Involvement of an initiation factor and protein phosphorylation in translational control of *GCN4* mRNA, *Trends Biochem. Sci.* **15:**148–152.
58. Abastado, J., Miller, P. F., Jackson, B. M., and Hinnebusch, A. G., 1991, Suppression of ribosomal reinitiation at upstream open reading frames of amino acid-starved cells forms the basis for *GCN4* translational control, *Mol. Cell. Biol.* **11:**486–496.
59. Cigan, A. M., Foiani, M., Hannig, E. M., and Hinnebusch, A. G., 1991, Complex formation by positive and negative translational regulators of *GCN4*, *Mol. Cell. Biol.* **11:**3217–3228.
60. Williams, N. P., Hinnebusch, A. G., and Donahue, T. F., 1989, Mutations in the structural genes for eukaryotic translation initiation factors 2α and 2β of *Saccharomyces cerevisiae* disrupt translational control of *GCN4* mRNA, *Proc. Natl. Acad. Sci. USA* **86:**7515–7519.
61. Wek, R., and Hinnebusch, A. G., 1989, Juxtaposition of domains homologous to protein kinases and histidyl-tRNA synthetase in GCN2 protein suggests a mechanism of coupling GCN4 expression to amino acid availability, *Proc. Natl. Acad. Sci. USA* **86:**4579–4583.
62. Chen, J.-J., Throop, M. S., Gehrke, L., Kuo, I., Pal, J. K., Brodsky, M., and London, I. M., 1991, Cloning of the cDNA of the heme-regulated eukaryotic initiation factor 2α (eIF-2α) kinase of rabbit reticulocyte: Homology to yeast GCN2 protein kinase and human double-stranded RNA-dependent eIF-2α kinase, *Proc. Natl. Acad. Sci. USA* **88:**7729–7733.
63. Meur, E., Chong, K., Galabra, J., Thomas, N.S. B., Kerr, I. M., Williams, B. R. G., and Hovanessian, A. G., 1990, Molecular cloning and characterization of the human double-stranded RNA-activated protein kinase induced by interferon, *Cell* **62:**379–390.
64. Ramirez, M., Wek, R., and Hinnebusch, A. G., 1991, Ribosome-association of GCN2 protein kinase, a translational activator of the GCN4 gene of *Saccharomyces cerevisiae*, *Mol. Cell. Biol.* **11:**3027–3036.
65. Dever, T. E., Feng, L., Cigan, A. M., Wek, R., Donahue, T. F., and Hinnebusch, A. G., 1992, Phosphorylation of initiation factor 2α by protein kinase Gcn2 mediates gene specific translational control of *GCN4* in yeast, *Cell* **68:**585–596.
66. Chong, K., Feng, L., Schappent, K., Meurs, E., Donahue, T. F., Frieson, J. D., Hovanessian, A. G., and Williams, B. R. G., 1992, Human p68 kinase exhibits growth suppression in yeast and homology to the translational regulator Gcn2, *EMBO J.* **11:**1153–1562.
67. Schneir, J., Schwelberger, H. G., Smit-McBride, Z., Kang, H. A., and Hershey, J. W. B., 1991, Translation initiation factor 5A and its hypusine modification are essential for cell viability in the yeast *Saccharomyces cerevisiae*, *Mol. Cell. Biol.* **11:**3105–3114.
68. Mehta, K., Leung, D., Lafenure, L., and Smith, M., 1990, The *ANB1* locus of *Saccharomyces cerevisiae* encodes the protein synthesis initiation factor eIF-4D, *J. Biol. Chem.* **265:**8802–8807.

Chapter 5

Regulation of *GCN4* Expression in Yeast

Gene-Specific Translational Control by Phosphorylation of eIF-2α

Alan G. Hinnebusch, Ronald C. Wek, Thomas E. Dever, A. Mark Cigan, Lan Feng, and Thomas F. Donahue

1. TRANSLATIONAL CONTROL OF THE TRANSCRIPTIONAL ACTIVATOR GCN4

The GCN4 protein of *Saccharomyces cerevisiae* is a positive regulator of more than 30 unlinked genes encoding enzymes in 11 different amino acid biosynthetic pathways. GCN4 stimulates transcription from these genes in response to starvation for any one of at least ten amino acids, or when the level of an aminoacylated transfer RNA (tRNA) is reduced because of a defective aminoacyl-tRNA synthetase. This cross-pathway derepression response is known as general amino acid control. GCN4 acts directly to regulate transcription by binding to a short nucleotide sequence present upstream from each of the structural genes under its

ALAN G. HINNEBUSCH, RONALD C. WEK, THOMAS E. DEVER, AND A. MARK CIGAN • Section on Molecular Genetics of Lower Eukaryotes, Laboratory of Molecular Genetics, National Institute of Child Health and Human Development, National Institutes of Health, Bethesda, Maryland 20892. LAN FENG AND THOMAS F. DONAHUE • Department of Biology, Indiana University, Bloomington, Indiana 47405.

Translational Regulation of Gene Expression 2, edited by Joseph Ilan. Plenum Press, New York, 1993.

control. Transcriptional activation of these genes increases in response to amino acid starvation because synthesis of the GCN4 protein itself is elevated under these conditions (reviewed in refs. 1 and 2). Regulation of *GCN4* expression by amino acid availability occurs primarily at the translational level. As summarized below, this control mechanism requires short open reading frames present in the leader of *GCN4* mRNA (uORFs) (Fig. 1A) and a cascade of positive and negative *trans*-acting factors (Fig. 1C).

Increased expression of amino acid biosynthetic genes under GCN4 control in response to amino acid starvation also requires the products of the *GCN1*, *GCN2*, and *GCN3* genes. These three factors function as positive regulators by stimulating *GCN4* expression when amino acids become limiting. One piece of evidence leading to this conclusion is that mutations in *GCN2* and *GCN3* impair the derepression of a *GCN4–lacZ* fusion enzyme that occurs in wild-type cells under starvation conditions (first line in Fig. 1B). These mutations do not reduce the steady-state levels of *GCN4* or *GCN4–lacZ* messenger RNAs (mRNAs), suggesting the GCN2 and GCN3 stimulate *GCN4* expression at the translational level.[3,4]

Mutations in multiple *GCD* genes lead to high-level expression of enzymes subject to the general control in the absence of amino acid starvation.[1] It was shown that mutations in *GCD1*, *GCD2*, *GCD10*, *GCD11*, and *GCD13* all lead to constitutive derepression of the *GCN4–lacZ* fusion enzyme (Fig. 1B), and do so at levels much higher than can be explained by increased fusion mRNA abundance. Thus, GCD factors appear to be required for translational repression of *GCN4* expression when amino acids are plentiful.[4–6] Because *gcd* mutations overcome the low-level *GCN4* expression seen in *gcn1*, *gcn2*, and *gcn3* mutants, it was proposed that GCN1, GCN2, and GCN3 act indirectly as positive regulators by antagonism or repression of one or more GCD factors (Fig. 1C).[1]

Deletions of *GCD1* and *GCD2* are unconditionally lethal[7–9] and all known

Figure 1. Regulation of *GCN4* expression by upstream open reading frames in *GCN4* mRNA. **(A)** Schematic of the *GCN4* transcript (wavy arrow) shown above the protein-coding sequences (large solid rectangle) and the four short open reading frames in the mRNA leader (small solid boxes, uORFs 1–4). The sequences of the uORFs are given along with the point mutations constructed in their ATG codons.[3,15] **(B)** The effects of point mutations in the ATG codons of the four uORFs on expression of β-galactosidase activity from a *GCN4–lacZ* fusion in wild-type (wt), nonderepressible *gcn2* mutant, and constitutively derepressed *gcd1* mutant cells under nonstarvation (repressing, R) conditions and histidine starvation (derepressing, DR) conditions.[16] The uORFs are depicted as solid boxes with uORF1 on the far left; "×" indicates point mutations in the ATG codons of the uORFs. **(C)** Pathway of regulatory factors involved in translational control of *GCN4* expression. Arrows indicate stimulatory interactions; bars depict inhibition or repression. The four uORFs in the *GCN4* mRNA leader are shown as numbered boxes. uORFs 3 and 4 are the major translational barriers to *GCN4* expression under nonstarvation conditions. The inhibitory effect of these sequences is reduced under starvation conditions by uORFs 1 and 2, the greater effect being exerted by uORF1. The antagonistic interaction between the uORFs is modulated by *trans*-acting positive (*GCN*) and negative (*GCD* and *SUI*) factors in response to the abundance of aminoacylated tRNA.[10]

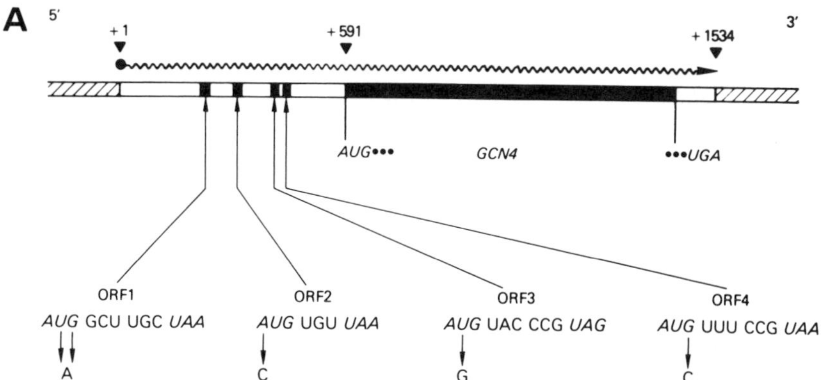

B

GCN4 UPSTREAM ORFs	*GCN4-lacZ* Enzyme Activity (U)					
	wt		*gcn2*		*gcd1*	
	R	DR	R	DR	R	DR
	10	80	5	6	280	340
	740	470	1300	1300	1400	700
	240	390	560	310	580	790
	5	12	6	11	15	16
	14	97	9	21	325	380

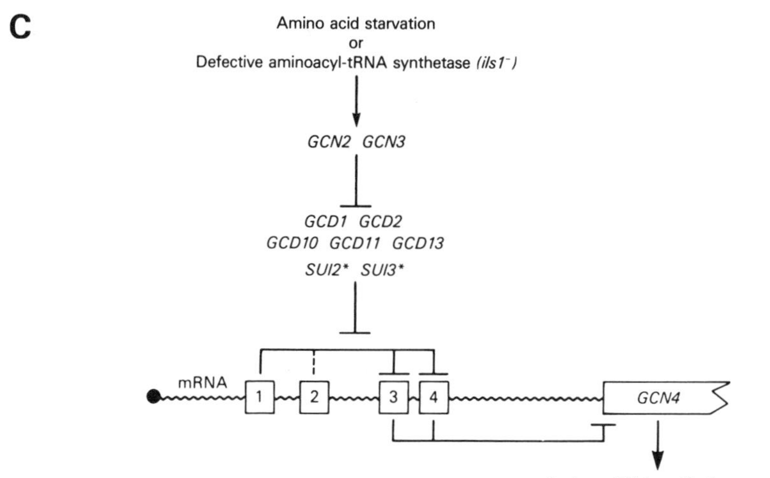

gcd point mutations lead to temperature-sensitive lethality or unconditional slow growth.[1] These secondary phenotypes indicate that GCD factors carry out essential functions in addition to their roles in regulating *GCN4* expression. In accord with this idea, certain mutations in the structural genes for the α and β subunits of yeast translation initiation factor eIF-2 (*SUI2* and *SUI3*, respectively) resemble *gcd* mutations in causing high-level unregulated expression of *GCN4* and amino acid biosynthetic genes under its control.[10] These and other findings summarized below indicate that GCD factors have essential functions in the process of translation initiation involving eIF-2. Accordingly, it was proposed that GCN1, GCN2, and GCN3 stimulate *GCN4* translation by altering the functions of one or more general translation initiation factors in response to amino acid limitation[10] (Fig. 1C).

The translational regulation of *GCN4* expression mediated by GCN and GCD factors requires the uORFs present in the long (ca. 600 nucleotides, nt) leader of *GCN4* mRNA (Fig. 1A). uORFs do not occur in most eukaryotic transcripts,[11] being present in only a few percent of *S. cerevisiae* mRNAs.[12] In addition, the introduction of a uORF into a eukaryotic transcript generally inhibits translation of the downstream protein-coding sequences.[13,14] This inhibitory effect of uORFs has been explained by proposing that ribosomes (or 40S subunits) must begin at the 5′ end of the transcript and traverse the entire mRNA leader to gain access to the AUG start codon (the scanning model). When a uORF is present, initiation occurs preferentially at that site and precludes initiation at downstream start codons, apparently because reinitiation is an inefficient process in eukaryotes.[14]

Removal of the four uORFs from the *GCN4* transcript either by deletion or point mutations in the four ATG start codons results in high-level *GCN4* expression, independent of amino acid availability and the positive regulators GCN1, GCN2, and GCN3[15,3,4,16] (see second line in Fig. 1B). In addition, insertion of the four uORFs into the leader of a heterologous yeast transcript causes expression of its protein product to be regulated in the same fashion observed for GCN4.[6] Thus, the four uORFs represent a translational control element that allows ribosomes to reach the *GCN4* start codon downstream only when cells are starved for an amino acid.

A combination of uORFs 1 and 4 (numbering from the 5′ end) is sufficient for nearly wild-type regulation of *GCN4* expression (Fig. 1B). These two uORFs have very different effects on initiation at the *GCN4* start codon. When present alone in the leader, uORF4 constitutes a very efficient translational barrier, reducing *GCN4* expression to only a few percent of that seen when no uORFs are present. By contrast, uORF1 is a leaky translational barrier when present singly, reducing *GCN4* expression only to ca. 50% of the level seen in the absence of uORFs. Moreover, when situated upstream from uORF4, uORF1 functions as a positive control element, stimulating *GCN4* expression under conditions of amino acid starvation[16,17] (Fig. 1B). Thus, recognition of uORF1 under starvation conditions allows ribosomes to traverse uORF4 sequences and initiate translation at the

GCN4 start codon. Solitary uORF3 functions as a potent translational barrier similar to uORF4, which, like the latter, can be overcome under starvation conditions when situated 3′ to uORF1.[16] Because *gcd* mutations have relatively little effect on *GCN4* expression when uORF4 (or uORF3) is present alone in the leader[16,6] (Fig. 1B), it follows that GCD factors are not required for the inhibitory effect of uORF4 on initiation at the *GCN4* start codon. Instead, these factors seem to repress *GCN4* expression primarily by antagonizing the ability of uORFl to stimulate the movement of ribosomes through uORF4 to the *GCN4* start site under nonstarvation conditions (Fig. 1C).

2. THE MECHANISM OF TRANSLATIONAL CONTROL OF *GCN4* EXPRESSION BY THE uORFs

2.1. Evidence That Ribosomes Must Translate uORF1 and Resume Scanning in Order to Initiate at *GCN4* in Amino Acid-Starved Cells

Extensive mutagenesis of the *GCN4* mRNA leader has been carried out in an effort to understand better the different functions of uORFs 1 and 4 in translational control of *GCN4* expression. Deletions of sequences surrounding the various uORFs in the *GCN4* leader can have significant quantitative effects on *GCN4* translational control; however, they do not alter the important qualitative features of the regulatory mechanism.[18] Thus, it appears that no critical regulatory sequences are present in the extended intervals between the uORFs, and that long-range secondary structure in the leader is unlikely to be important in the translational control mechanism. It was also shown that uORFs 1 and 4 could be replaced by heterologous short coding sequences without destroying regulated expression of *GCN4*.[19,20,18] The latter finding prompted the idea that uORFs 1 and 4 function in the control mechanism as translated coding sequences but that their sequences and secondary structures are not uniquely required for regulation per se. Rather, *GCN4* translational control appears to be a more general consequence of having two uORFs present in the leader. On the other hand, the heterologous uORFs function much less efficiently than authentic uORFs 1 and 4 as translational control elements. In addition, reversing the 5′–3′ order of uORFs 1 and 4 completely abolishes the derepression of *GCN4* expression,[18] showing that uORFs 1 and 4 are not functionally interchangeable. Thus, it appears that particular nucleotides either within or immediately flanking the uORFs optimize these two sequence elements for their opposing functions in translational control.

As mentioned above, solitary uORFl allows about 50% of the ribosomes scanning the leader to reach the *GCN4* start codon, whereas only 1–2% of the ribosomes can traverse solitary uORF4 and initiate at *GCN4*. The high-level *GCN4* expression seen when uORFl is present alone does not appear to result from inefficient recognition of the uORFl start codon (leaky scanning), but rather from the ability of ribosomes to reinitiate downstream following translation of uORF1. This conclusion arises in part from the observation that mutations which alter the sequence context of the uORF1 stop codon generally lead to reduced *GCN4* expression.[20,21] Numerous single amino acid substitutions made in uORF1 have no effect on expression, whereas point mutations in its stop codon that lengthen or shorten uORF1 by a single codon substantially increase the inhibitory effect of solitary uORF1 on *GCN4* expression. Insertions of various single codons between the second and third codons of uORF1 reduce *GCN4* expression, suggesting that a three-codon length is important for uORF1 function; however, these reductions are not as great as that seen when uORF1 is lengthened to four codons by a point mutation in its stop codon. Thus, it appears that sequences 3′ to the uORF1 stop codon make an important contribution to the ability of uORF1 to permit translation initiation downstream at *GCN4*.[21]

Further evidence for this last conclusion came from the fact that replacement of the ten nucleotides immediately following uORF1 with the corresponding sequence from uORF4 increased the inhibitory effect of uORF1 on *GCN4* expression. Additionally replacing the third codon of uORF1 with the rare proline codon normally found at the same position in uORF4 made solitary uORF1 indistinguishable from solitary uORF4 as a barrier to *GCN4* translation. By contrast, replacement of 25 nucleotides upstream from uORF1 with the corresponding nucleotides from uORF4 had no effect on *GCN4* expression[21] (Fig. 2). The latter suggests that uORFs 1 and 4 have very similar initiation efficiencies, a conclusion supported by measuring the synthesis rates of uORF1–*lacZ* and uORF4–*lacZ* fusion proteins when each was encoded as the 5′-proximal ORF on *GCN4* mRNA.[20,19] Thus, it appears that ribosomes initiate at uORFs 1 and 4 with similar efficiencies and that uORF1 inhibits translation downstream less than uORF4 does because of a greater probability for reinitiation following translation of uORF1 compared to uORF4. The most likely explanation for the inability of uORF4 to permit reinitiation downstream is that ribosomes dissociate from the mRNA after terminating translation at this uORF, and that this behavior is determined by sequences immediately surrounding the uORF4 stop codon. By contrast, ribosomes would remain attached to the mRNA and resume scanning following translation of wild-type uORF1.

The mutations at uORF1 mentioned above that reduce the efficiency of reinitiation at *GCN4* when uORF1 is present alone in the leader also impair the ability of uORF1 to stimulate *GCN4* expression when it is situated upstream from uORF4. The simplest explanation for this correlation is that ribosomes must translate uORF1 and resume scanning in order to traverse uORF4 sequences and initiate at *GCN4*.[18,20] One way to account for this requirement is to propose that reassembly of a competent initiation complex following termination at uORF1 occurs inefficiently in amino acid-starved cells, allowing ribosomes to scan past the start codons of uORFs 2–4 and reinitiate translation at *GCN4* instead. Under nonstarvation conditions, ribosomes would reacquire initiation factors more rap-

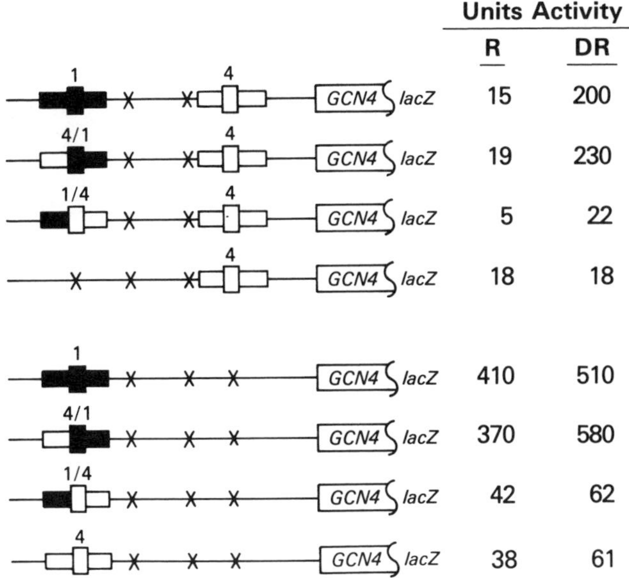

Figure 2. Sequences surrounding the stop codons of uORFs 1 and 4 distinguish the functions of these elements in translational control. *GCN4–lacZ* constructs containing point mutations in the ATG codons of uORFs 2 and 3 (X) are shown schematically with uORFs 1 and 4 and segments of 16 and 25 nucleotides located upstream and downstream from the uORFs, respectively, all indicated by boxes. Sequences at uORF1 were replaced with the corresponding sequences from uORF4 in the hybrid constructs. *GCN4–lacZ* fusion enzyme activity was measured in extracts from a constitutively repressed (R) *gcn2-1* mutant and from a constitutively derepressed (DR) *gcd1-101* mutant.[21]

idly as they scan downstream from uORF1 and reinitiate at uORF2, 3, or 4; translation of these uORFs would lead to dissociation of ribosomes from the mRNA, thereby preventing reinitiation at *GCN4*. Additional support for the idea that suppression of ribosomal reinitiation at uORFs 2–4 is responsible for increased translation of *GCN4* is presented in the following subsections.

2.2. Evidence That Ribosomes Ignore the Start Codons at uORFs 2–4 and Reinitiate at *GCN4* Following Translation of uORF1 in Amino Acid-Starved Cells

Evidence that ribosomes which translate uORF1 subsequently reach *GCN4* under starvation conditions by physically traversing uORFs 2–4, rather than binding directly to the *GCN4* start codon, is provided by the fact that insertions of sequences with the potential to form 8-base-pair (bp) stem–loop structures just upstream or downstream from uORF4 completely abolish derepression of *GCN4* translation.[22] Stem–loop structures of this size severely inhibit scanning in other yeast mRNA leaders.[23,24] The idea that ribosomes reach *GCN4* following translation of uORF1 by ignoring the start codons at uORFs 2–4 comes from the fact that mutations in the uORF4 stop codon that elongate uORF4 from 3 codons to either 46 or 93 codons have little or no effect on *GCN4* expression (Fig. 3). The 93-codon elongated-uORF4 in the latter construct actually overlaps the beginning of *GCN4* by 130 nt.[22] Thus, ribosomes that translate the elongated uORF4 in this construct would have to scan "backward" for 130 nt after terminating translation, ignoring four other potential start codons in the overlap region, in order to reach *GCN4*. (It was proven that the normal *GCN4* start codon was used exclusively for GCN4 protein synthesis in this construct; see Fig. 3 legend.) The sequence contexts of the four AUG codons in the overlap region are unlikely to preclude completely their recognition as start sites, because sequence context has a relatively small effect on initiation codon selection in yeast.[13,25,24] Therefore, it seems most likely that altering the uORF4 termination site has no effect on *GCN4* expression simply because uORF4 is not translated by ribosomes that initiate at *GCN4*.[22] These results stand in sharp contrast to the deleterious effects on *GCN4* expression associated with altering the uORF1 termination site, interpreted above

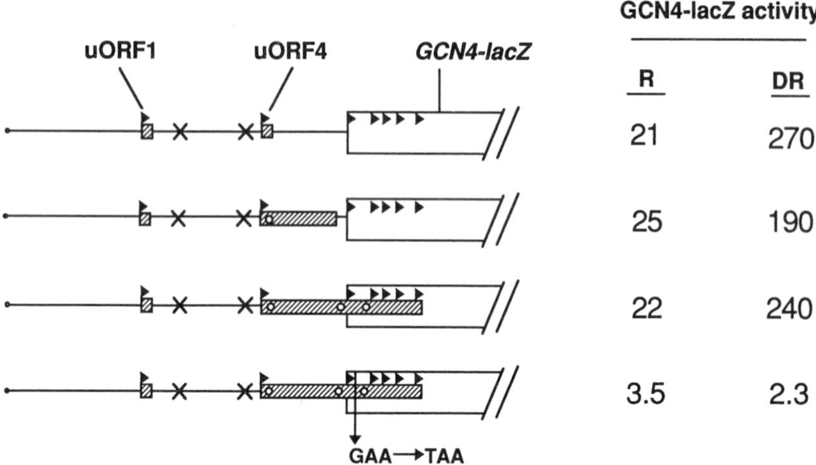

Figure 3. Elongating uORF4 to overlap *GCN4* has little or no effect on regulation of *GCN4* expression. Schematics on the left depict constructs containing uORF4 that is wild type (first line), 46 codons in length (second line), or 93 codons in length (third and fourth lines), located downstream from wild-type uORF1. An X indicates a point mutation in the ATG codons of uORFs 2 and 3. ATG codons present in all three reading frames are indicated by arrowheads. Stop codons that were mutated to elongate uORF4 are shown by open circles. The fourth construct contains a nonsense mutation in the third codon in *GCN4*. The *GCN4–lacZ* fusion enzyme activity was measured for each construct in transformants of a nonderepressible *gcn2* mutant (R) and in a constitutively derepressed *gcdl* mutant (DR). The fact that the stop codon introduced in the fourth construct abolishes *GCN4–lacZ* expression shows that the authentic *GCN4* start codon is used even when uORF4 overlaps the beginning of *GCN4* by 130 nt.[22]

to indicate that ribosomes must translate uORF1 and resume scanning to eventually reach *GCN4*.

The conclusion that ribosomes which translate *GCN4* have bypassed the uORF4 start codon is supported by biochemical measurements of translation initiation at uORF4 under repressing and derepressing conditions. By making a 1-bp insertion in the uORF4–*GCN4* overlap region of the 93-codon uORF4 construct described above, the *GCN4–lacZ* coding region was placed in-frame with uORF4. Expression of this uORF4–*lacZ* fusion was found to be regulated in a completely different fashion than the wild-type *GCN4–lacZ* fusion, being 50–70% lower in *gcdl* cells (constitutively derepressed for *GCN4*) than in *gcn2* cells (constitutively repressed for *GCN4*) (Fig. 4). This pattern of expression suggests that 50–70% of the 40S subunits scanning downstream from uORF1 under derepressing conditions fail to reinitiate at uORF4 and continue scanning to *GCN4*. The decrease in translation of the uORF4–*lacZ* fusion seen in *gcdl* cells is similar in magnitude to the increased translation of the *GCN4–lacZ* fusion that occurs in these derepressed cells (Fig. 4).[22] This relationship is expected if ribosomes are being diverted from translation of uORF4 to allow for increased initiation at *GCN4*.

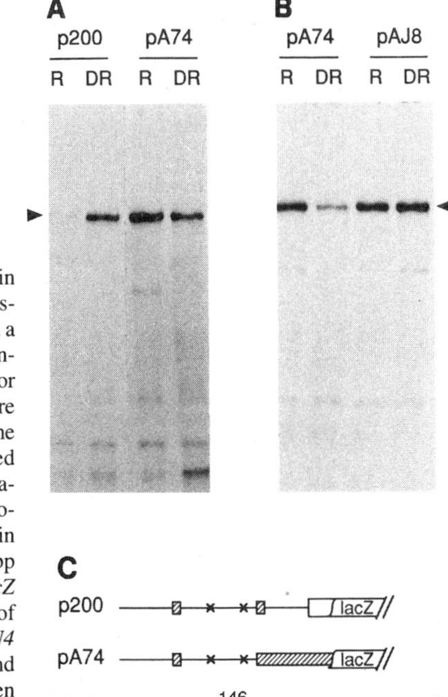

Figure 4. Synthesis of an uORF–lacZ fusion protein under repressing and derepressing conditions. Transformants of a nonderepressible *gcn2* mutant (R) and a constitutively derepressed *gcdl* mutant (DR) containing the wild-type *GCN4–lacZ* construct on p200, or the uORF4–*lacZ* constructs on pA74 or pAJ8, were pulse-labeled with [^{35}S]methionine for 5 min. The labeled fusion proteins were immunoprecipitated from samples containing equivalent amounts of labeled proteins and analyzed by SDS–PAGE and fluorography. The three constructs are shown below in schematic form. pA74 and pAJ8 contain a 1-bp insertion in the beginning of *GCN4* that fuses *lacZ* sequences in-frame to the elongated versions of uORF4 present in these constructs. (Thus, the *GCN4* start codon is out-of-frame with *lacZ* in pA74 and pAJ8.) pAJ8 contains a 146-nt insertion between uORFs 1 and 4[22] (see text for details).

It was proposed that ribosome scanning downstream from uORF1 under starvation conditions scan past uORF4 without initiating translation because the time required to scan the uORF1–uORF4 interval is insufficient for the reassembly of competent initiation complexes. The evidence for this conclusion is that progressively increasing the distance between uORF1 and uORF4 leads to a progressive decrease in *GCN4* translation under starvation conditions or in a *gcdl* mutant[22] (Fig. 5). For example, insertion of 144 nt between uORFs 1 and 4 greatly impairs derepression of *GCN4–lacZ* expression, whereas the same insertion made in the absence of uORF4, or introduced downstream from uORF4 (Fig. 5), is virtually without effect. The latter results show that the insertion impairs derepression of *GCN4* by interfering with the ability of ribosomes to traverse the uORF4 sequence and not simply because it impedes scanning. Importantly, the 144-nt insertion increases the spacing between uORF1 and uORF4 to about the same distance that normally separates uORF1 from *GCN4*. The derepression defect associated with this insertion was thus explained by proposing that 40S subunits scanning the expanded uORF1–uORF4 interval after completing translation of uORF1 now have sufficient time to reassemble the factors needed for initiation before reaching uORF4 under both starvation and nonstarvation conditions.

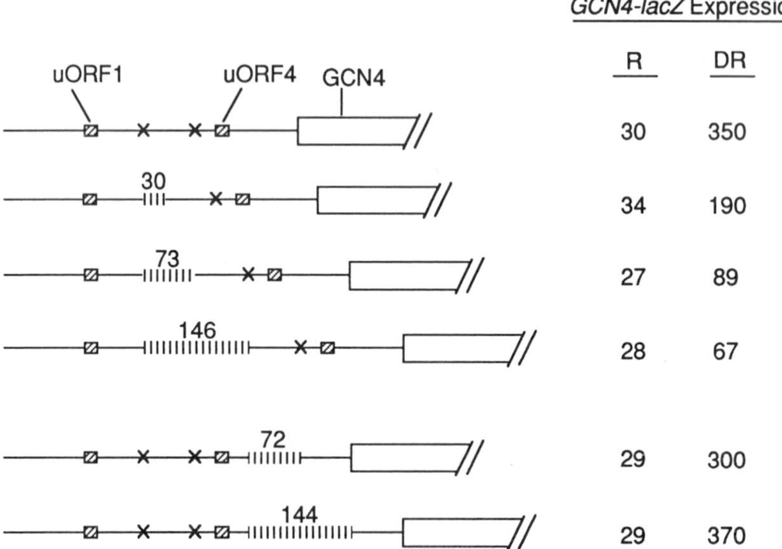

Figure 5. Increasing the distance between uORFs 1 and 4 impairs derepression of *GCN4* expression. The schematics on the left depict the constructs containing insertions that introduce 30, 73, or 146 nt between uORFs 1 and 4, or that insert 72 or 144 nt between uORF4 and *GCN4*. An X indicates a point mutation in the ATG codons of uORFs 2 and 3. Analysis of the constructs was conducted exactly as described in Fig. 2 and 3.[22]

Consequently, nearly all ribosomal subunits reinitiate at uORF4 and are thereby prevented from reaching *GCN4* (Fig. 6). Supporting this interpretation, the 144-nt insertion between uORFs 1 and 4 increases translation of the aforementioned uORF4–*lacZ* fusion in the derepressed *gcdl* mutant, eliminating the 50–70% reduction in uORF4–*lacZ* translation that occurs with the wild-type uORFl–uORF4 spacing in the *gcdl* strain[22] (Fig. 4).

Implied in the above explanation is the assumption that under nonstarvation conditions, the wild-type spacing between uORFl and uORF4 provides adequate time for nearly all ribosomes scanning downstream from uORFl to reassemble competent initiation complexes by the time they reach uORF4 (Fig. 6). Thus, the regulation of *GCN4* expression is seen to arise from the increased scanning time required for reinitiation at downstream start sites following termination at uORFl in amino acid-starved cells. As discussed below, reinitiation appears to be less

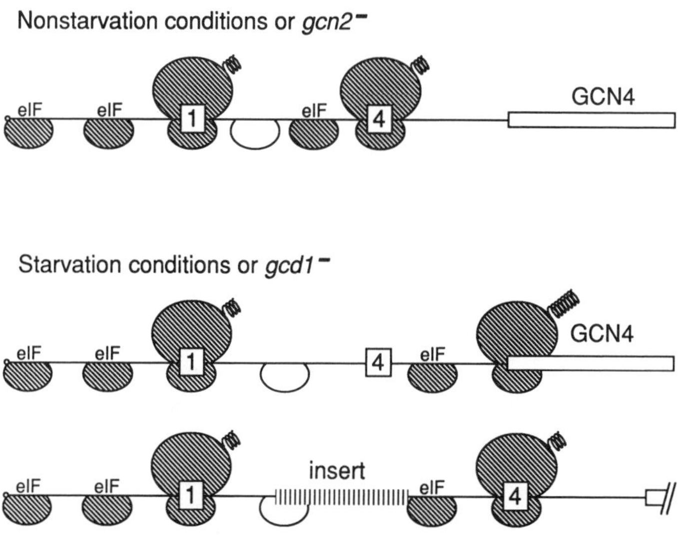

Figure 6. Model depicting suppression of reinitiation at uORF4 in amino acid-starved cells as the basis for translational derepression of *GCN4*. The *GCN4* mRNA leader is shown with uORFs 1 and 4 and the beginning of *GCN4* coding sequences indicated as boxes. 40S subunits containing the full complement of initiation factors (eIF) are shown hatched; those lacking certain factors are shown empty. 80S translating ribosomes are also hatched. Ribosomes translate uORFl and the 40S subunits resume scanning. Under repressing conditions (nonstarved wild-type or *gcn2* mutant cells), initiation factors are rapidly reassembled on the 40S subunit and reinitiation occurs efficiently at uORF4. Following uORF4 translation, no reinitiation occurs at *GCN4*, presumably because the ribosomes dissociate from the mRNA. Under derepressing conditions (starved wild-type or *gcdl* mutant cells), reassembly of initiation complexes occurs more slowly and 40S subunits are not ready to initiate when they reach uORF4; consequently, they continue scanning and reinitiate at *GCN4* instead. When the distance between uORFl and 4 is increased by an insertion to approximately the wild-type separation between uORFl and *GCN4*, most ribosomes are competent for reinitiation by the time they reach uORF4 even under derepressing conditions, thus excluding them from translation of *GCN4*.[22]

efficient under starvation conditions because the positive regulators GCN2 and GCN3 reduce the availability of certain initiation factors that must interact with 40S subunits scanning downstream from uORF1 to re-form competent initiation complexes.

3. THE ROLE OF GENERAL INITIATION FACTORS IN TRANSLATIONAL CONTROL OF *GCN4*

3.1. Mutations in Subunits of eIF-2 Impair the Regulated Expression of *GCN4*

Mutations in the structural genes for the α and β subunits of initiation factor 2 (eIF-2) in yeast, called *SUI2* and *SUI3*, respectively, have been isolated by their ability to allow an in-frame TTG codon to be used as the translational initiation site at *HIS4* in the absence of the normal ATG start codon.[26,27] This phenotype suggests that eIF-2 plays an important role in AUG recognition during the scanning process. In addition to their effects on translational start site selection, these *SUI* alleles mimic *gcd* mutations in causing high-level unregulated expression of *GCN4* and amino acid biosynthetic genes under its control (Fig. 7). Importantly, the expression of *GCN4–lacZ* constructs containing only uORF4, or no uORFs at all, is nearly insensitive to the *SUI* mutations, implying that these alterations in eIF-2 increase *GCN4* expression at the translational level by decreasing reinitiation at uORFs 2–4 irrespective of amino acid abundance. In accord with this explanation, the *SUI2* and *SUI3* mutations overcome the requirement for the positive regulators GCN2 and GCN3 for high-level *GCN4* expression.[10] The latter finding raises the possibility that eIF-2 activity is reduced by GCN2 and GCN3 in amino acid-starved cells as the means of stimulating translation initiation at *GCN4* (Fig. 1C). This idea is consistent with the fact that the *SUI2* and *SUI3* mutations that derepress *GCN4* expression appear to reduce the ability of eIF-2 to

	***GCN4-lacZ* Enzyme Activity (U)**			
uORFs present:	1 2 3 4		4 (1–3 mutated, X)	
	R	DR	R	DR
SUI+	13	75	8	21
sui2-1	200	190	23	17
SUI3-2	130	210	18	32
sui2-1 gcn2Δ	240	200	32	32

Figure 7. Mutations in eIF-2α and eIF-2β lead to constitutive derepression of *GCN4* expression. *GCN4–lacZ* fusions containing all four uORFs (columns 1 and 2) or uORF4 alone (columns 3 and 4) were assayed for β-galactosidase expression in wild-type (*SUI2*) and mutant strains containing lesions in eIF-2α (*sui2-1*) or eIF-2β (*SUI3-2*) under nonstarvation (R) or histidine starvation (DR) conditions.[10]

form ternary complexes with Met-tRNA$_i^{Met}$ and GTP.[26,27] In a related development, the binding of charged tRNA$_i^{Met}$ to the small ribosomal subunit, a reaction which involves eIF-2·GTP·Met-tRNA$_i^{Met}$ ternary complexes, was found to be diminished in a temperature-sensitive *gcd1* mutant after incubation at the nonpermissive temperature.[28] This last result could indicate that the high-level *GCN4* expression seen in *gcd1* mutants arises from reduced eIF-2 activity associated with lesions in the GCD1 protein. The findings summarized next support this interpretation.

3.2. GCD1 and GCD2 Have Essential Functions in General Translation Initiation

Recent biochemical studies on protein synthesis in mutants containing temperature-sensitive lethal *gcd* mutations strongly support the idea that GCD1 and GCD2 have general functions in translation initiation in addition to being involved in gene-specific translational control of *GCN4*. The *gcd1-101*[29,7] and *gcd2-503*[30] mutations lead to reduced rates of radiolabeled leucine incorporation into total protein following a shift to the nonpermissive temperature. Examination of total polysome profiles under the same conditions reveals a partial runoff of polysomes, a reduction in the average polysome size, and an accumulation of 80S couples (80S particles consisting of 40S and 60S subunits joined together in the absence of mRNA).[31,30] These same alterations in polysome profiles were observed in yeast strains containing the *sui2-1* mutation that affects eIF-2α,[30] or the *prt1-1* mutation[32] believed to impair the function of initiation factor-3 (eIF-3).[33] Polysome runoff is indicative of a reduced rate of translation initiation without a commensurate effect on elongation and termination rates. Additional evidence that the two *gcd* mutations impair general translation initiation is the presence of increased amounts of eIF-2α and eIF-2β subunits in 40S particles when the mutant strains are incubated at the nonpermissive temperature.[31,30] This latter phenotype could be indicative of an accumulation of 43S or 48S initiation intermediates consisting of small ribosomal subunits, eIF-3, and ternary complexes composed of eIF-2, GTP, and Met-tRNA$_i^{Met}$ (48S complexes also contain mRNA).[34]

Interestingly, the *gcd2-502* mutation leads to a different sort of polysome abnormality (Fig. 8), known as halfmers, in which polyribosomes contain extra 40S subunits that have not joined with 60S subunits to form 80S ribosomes.[30] The occurrence of halfmer polysomes suggests that a late step in the initiation pathway following the binding of 43S complexes to mRNA is impaired by the *gcd2-502* mutation, such as scanning to the AUG start codon, hydrolysis of GTP, release of eIF-2, or 40S–60S subunit joining.[34] It was suggested the *gcd1-101* and *gcd2-503* mutations might also impair one of these late steps in the initiation pathway, but because they have a more severe effect on the rate of initiation than *gcd2-502*, polysome runoff and accumulation of free 48S intermediates are observed rather than halfmer polysome formation.[30]

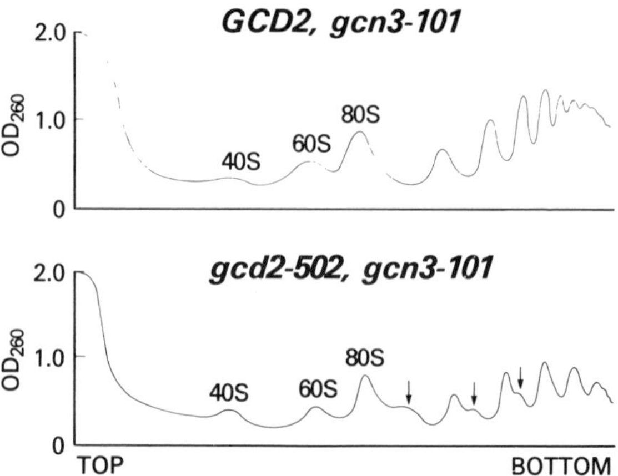

Figure 8. The *gcd2-502* mutation leads to halfmer polysome formation *in vivo*. Extracts prepared from isogenic yeast strains *gcd2-502 gcn3-101* and *GCD2 gcn3-101* grown in rich medium at 23°C were centrifuged on low-salt, 7–47% sucrose gradients to resolve polysomes and 80S ribosomes from 40S and 60S free subunits. The arrows indicate the position of halfmers, containing 40S subunits that have not joined with 60S subunits, sedimenting between the conventional polysomes containing only 80S ribosomes.[30]

The idea that 40S–60S subunit joining at the initiation codon is partially impaired by these *gcd* mutations could account for the derepression of *GCN4* translation that occurs in the mutant strains. In this view, 40S subunits scanning downstream from uORF1 containing eIF-2·GTP·Met-$tRNA_i^{Met}$ ternary complexes would have a diminished capacity to join with 60S subunits and would therefore exhibit a greater tendency to bypass uORFs 2–4 without initiating translation. An alternative possibility is that rebinding of eIF-2·GTP·Met-$tRNA_i^{Met}$ to the scanning 40S subunit is what limits reinitiation at uORFs 2–4 and that the *gcd* mutations decrease recycling of eIF-2 and the availability of ternary complexes. This latter idea, that recycling of eIF-2 is involved in *GCN4* translational control, will be explored further below.

3.3. GCD1, GCD2, and GCN3 Are Present in a High-Molecular-Weight Protein Complex That Interacts with a Fraction of eIF-2 Present in the Cell

Previous genetic studies suggested that the positive regulator GCN3 closely interacts with GCD1, GCD2, and the α subunit of eIF-2 in regulating *GCN4* translation and in carrying out the essential functions of GCD1, GCD2, and eIF-2 in the initiation of protein synthesis.[9,10,35] In agreement with this prediction, it has been shown that GCD1, GCD2, and GCN3 are integral components of a high-molecular-weight complex of about 600,000 daltons. Antibodies raised against these three proteins were used to demonstrate coelution of GCD1, GCD2, and GCN3 from a Sephacryl-300 sizing column (Fig. 9) and from a DEAE ion-exchange column, as well as comigration of these proteins in velocity sedimentation through sucrose gradients. In addition, the three proteins could be coimmunoprecipitated using antibodies against GCD1 or GCD2. Interestingly, a portion of the eIF-2 present in cell extracts also copurified with GCD1, GCD2, and GCN3 and this fraction of eIF-2 could be quantitatively coimmunoprecipitated using GCD1 antibodies. The association between eIF-2 and the GCD1-containing complex was eliminated by treatment with high concentrations of NaCl (0.5 M).[31]

There are interesting similarities between the GCD1/GCD2/GCN3 complex and the mammalian initiation factor eIF-2B that recycles eIF-2·GDP to eIF-2·GTP after each round of initiation to permit continued formation of eIF-2·GTP·Met-$tRNA_i^{Met}$ ternary complexes. Similar to what was observed in yeast,[31] eIF-2 is distributed between free and high-molecular-weight pools in mammalian cell extracts,[36] with the high-molecular-weight form being associated with the multi-subunit eIF-2B complex of about 450,000 daltons.[37–39] Moreover, the association between mammalian eIF-2 and eIF-2B is disrupted by 0.5 M KCl,[37] mimicking the salt lability observed for the interaction between yeast eIF-2 and the GCD1/GCD2/GCN3 complex.[31] In addition, there are reports of an accumulation of 43S or 48S intermediates and halfmer polysomes in mammalian cell-free transla-

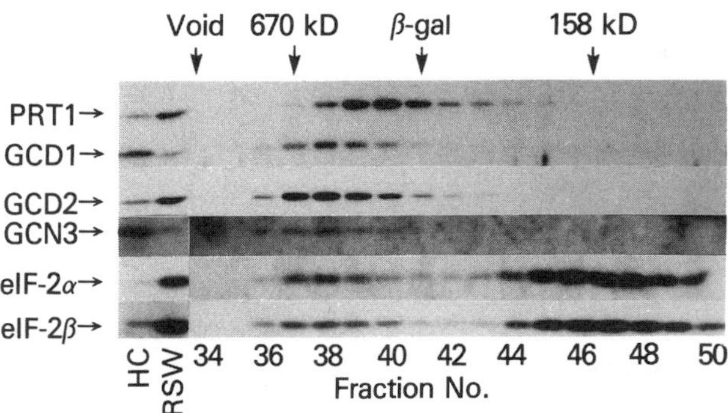

Figure 9. Evidence for complex formation by positive and negative translational regulators of *GCN4*: GCD1, GCD2, and GCN3 cofractionate in Sephacryl S-300 gel filtration chromatography and elute with a portion of the eIF-2α and eIF-2β subunits present in a ribosomal salt wash from a wild-type yeast strain. An aliquot of each column fraction was subjected to SDS–PAGE and analyzed by immunoblotting using antibodies against each of the proteins listed to the left of the panel. The results represent a composite of gels containing equivalent samples that were blotted and probed separately with the different antibodies.[31] The PRT1 protein was analyzed to determine the elution position of eIF-3, another high-molecular-weight complex that functions in translation.[33] HC, 50 μg of total cell extracts from transformants containing *GCD1*, *GCD2*, or *GCN3* on high-copy-number plasmids, which provided size markers for these proteins. RSW, 50 μg of the unfractionated ribosomal salt wash. *E. coli* β-galactosidase was added to the RSW as an internal size standard (M_r value of 420,000).

tion systems under conditions in which eIF-2B activity is inhibited,[40–42] similar to what was observed in *gcd1* and *gcd2* mutants.[31,30] Finally, the functions of both eIF-2B and the GCD1/GCD2/GCN3 complex are thought to be antagonized by protein kinases, known as DAI and HCR in the case of mammalian eIF-2B,[43] and GCN2 in the case of the GCD1/GCD2/GCN3 complex in yeast (Fig. 1C and see below).

The mammalian protein kinases DAI and HCR inhibit the activity of eIF-2B in recycling eIF-2·GDP to eIF-2·GTP by phosphorylation of the α subunit of eIF-2.[43] Phosphorylated eIF-2 forms a stable complex with eIF-2B, sequestering eIF-2B in an inactive form.[44,45,37,46] In the absence of the eIF-2-recycling activity of eIF-2B, new rounds of translation initiation are prevented from occurring.[46] This regulatory mechanism is used in mammalian cells to inhibit protein synthesis in response to various stress situations, such as virus infection (which leads to DAI activation)[47] and heme deprivation in reticulocytes (which activates HCR).[46] Amino acid limitation also leads to reduced protein synthesis by increased phosphorylation of eIF-2α in certain mammalian cells.[48,49]

The similar effects of mutations affecting eIF-2 subunits or GCD factors in derepressing *GCN4* translation, the physical similarities between the GCD1-containing complex and eIF-2B, and the opposing functions of GCD factors and the protein kinase GCN2 in regulating *GCN4* translation, all raise the interesting possibility that GCN2 stimulates *GCN4* translation by phosphorylation of eIF-2α. This would lead to inactivation of a certain fraction of eIF-2B, with consequent reduction in the recycling of eIF-2·GDP to eIF-2·GTP and diminished levels of eIF-2·GTP·Met-tRNA$_i^{Met}$ ternary complexes. As a result, rebinding of ternary complexes to scanning 40S subunits following termination at uORF1 might occur more slowly, causing many ribosomal subunits to reach uORFs 2–4 without having acquired this critical component of the initiation complex. Having bypassed the AUG codons at uORFs 2–4, most of these subunits would rebind ternary complexes before reaching the *GCN4* AUG codon and reinitiate translation there instead (see Fig. 13 below).

3.4. The Role of the Positive Regulator GCN3 in the GCD1/GCD2/GCN3 Protein Complex

Deletion of *GCN3* in an otherwise wild-type strain has no obvious phenotype besides the inability to derepress *GCN4* expression under starvation conditions.[8] This suggests that GCN3 functions primarily as a regulatory subunit in the GCD1/GCD2/GCN3 complex, mediating the inhibitory effects of the upstream positive regulators GCN1 and GCN2 on the function of the complex under conditions of amino acid starvation. For example, if GCN2 acts by phosphorylation of eIF-2α, GCN3 could mediate the stable interaction between eIF-2 and eIF-2B that causes the latter to be sequestered and inactivated. The sequence similarity noted between GCN3 and the C-terminal half of GCD2[50] might indicate that these two proteins interact directly with one another in the GCD1/GCD2/GCN3 complex; alternatively, their homologous domains might be involved in the binding of eIF-2.

The idea that GCN3 mediates the regulatory effects of GCN1 and GCN2 on the function of the GCD1/GCD2/GCN3 complex is supported by the isolation of mutations in *GCN3* that lead to high-level constitutive expression of *GCN4* (Gcd^- phenotype), completely independent of *GCN1* and *GCN2*. These mutations alter a variety of single amino acids mapping throughout GCN3 and, with one exception, do not alter the steady-state level of the protein. It is of interest that these *gcn3*^{c} alleles also resemble *gcd* mutations in causing slow growth on rich medium,[51] presumably as the result of partial inactivation of the function of the GCD1/GCD2/GCN3 complex in general translation initiation. The *gcn3*^{c} mutations could destabilize the complex, or in the context of the eIF-2·eIF-2B model mentioned above, stabilize the interaction between eIF-2 and eIF-2B which inhibits the latter, mimicking the effect of phosphorylation of eIF-2α. The fact that *gcn3*^{c} mutations derepress *GCN4* under nonstarvation conditions,[51] coupled with the observation that wild-type *GCN3* can overcome the general growth defects associated with certain *gcd* mutations,[35,9] suggests that GCN3 is present in the GCD1-containing complex under both starvation and nonstarvation conditions.

The notion that GCN3 mediates the regulatory effects of GCN2 on the GCD1/GCD2/GCN3 complex is also supported by the isolation of dominant mutations in *GCN2* that constitutively derepress *GCN4* expression and show a strong dependence on *GCN3* for this phenotype.[52,51] The *GCN2*^{c} alleles are also greatly dependent on *GCN1* for their derepressing effects on *GCN4* expression. This dependence of the *GCN2*^{c} mutations on both *GCN1* and *GCN3* is consistent with the view that GCN2 is the first step in the signal transduction pathway that couples *GCN4* expression to amino acid availability.[51] A role for GCN2 in directly sensing amino acid limitation in the cell is also suggested by the amino acid sequence of a domain flanking the protein kinase moiety of GCN2, as discussed below.

4. PHOSPHORYLATION OF eIF-2α MEDIATES TRANSLATIONAL CONTROL OF *GCN4*

4.1. Sequence Relatedness between the GCN2 Protein Kinase and the Mammalian eIF-2α Kinases DAI and HCR

GCN2 encodes a protein of 180,000 daltons containing a segment of about 425 amino acids that exhibits strong similarity with the catalytic subunit of eukaryotic protein kinases (Figs. 10 and 11).[53,54] Arginine or valine substitutions in the highly conserved lysine residue (Lys-559) that is expected to participate in the transfer of phosphate from ATP to the protein substrate abolish the *in vivo* regulatory function of GCN2 (Fig. 11).[53] In addition, these mutations destroy the ability of GCN2 to autophosphorylate *in vitro* in an immune-complex kinase

```
          .         .         .  !    * *    *    .
cAPK AKAKEDFLKKWEDPSQN...TAQ....LDQFDRIKTLGTGSFGRVMLVKH  62
GCN2 SIRRRSFNVGSRFSSIN...PATRSRYASDFEEIAVLGQGAFGQVVKARN 549
HCR  IQKIRSREVALE.........AQTSRYLNEFEELSILGKGGYGRVYKVRN 189
DAI  RKAKRSLAPRFDLPDMKETKYTVDKRFGMDFKEIELIGSGGFGQVFKAKH 285

       !      *         .        *.         .   !   ! .
     KESGNHYAMKILDKQKVVKLKQIEHTLNEKRILQAVNFPFLV........ 104
     ALDSRYYAIKKI.RHTEEKLSTI...LSEVMLLASLNHQYVVRYYAAWLE 595
     KLDGQYYAIKKILIKGATKT.DCMKVLREVKVLAGLQHPNIVGYHTAWIE 238
     RIDGKTYVIKRV.KYNNEKAE......REVKALAKLDHVNIVHYNGCW.. 326

              .         .         .         .         .
     ..................................................
     EDSM......DENVFESTDEESDLSESSSDFE.ENDLLDQSSIFKNRTNH 638
     HVHVHVQADRVPIQLPSLEVLSDQEEDRDQYGVKNDASSSSSIIFAEFSP 288
     ............DGFDYDPETSDDSLESSDYDPEN............... 349

              .         .         .         .         .
     ..................................................
     DLDNSNWDFISGSGYPDIV.............FENSSRDDENEDLDHDTS 675
     EKEKSSDECAVESQNNKLVNYTTNLVVRDTGEFESSTERQENGSIVERQL 338
     ................................SKNSSR............ 355

              .         .         .         .   !!    .
     ..............................KLEFSFKDNSNLYMVMEYVA 124
     STS..SSESQDDTDKESKSIQNVPRRRNFVKPMTAVKKKSTLFIQMEYCE 723
     LFGHNSDVEEDFTSAEESSEEDLSALRH.....TEVQYHLMLHIQMQLCE 383
     ...................................SKTKC.LFIQMEFCD 369

              .         .         .     !   .         .
     GGEMF....SHLRRIGRF.........SEPHARFYAAQI.VLTFEYLHSL 160
     NRTLYDLIHSENLNQQRD............EYWRLFRQIL.EALSYIHSQ 760
     .LSLWDWIAERNRRSRECVDESACPYVMVSVATKIF.QELVEGVFYIHNM 431
     KGTLEQWI..EKRRGEKL.........DKVLALELFEQIT.KGVDYIHSK 407

          * * .* !      .     !***.         .         .
     DLIYRDLKPENLLI...DQQGYIQVTDFGFAKRVK............... 192
     GIIHRDLKPMNIFI...DESRNVKIGDFGLAKNVHRSLDI.LKLD...SQ 803
     GIVHRDLKPRNIFLHGPDQQ..VKIGDFGLAC.....ADIIQKNAARTSR 474
     KLIHRDLKPSNIFL..VDTKQ.VKIGDFGLVTS........LKND..... 441

              .           **      .     *   .*        .
     .......GRTWTLCGTPEYLAPEIILSKG.YNKAVDWWALGVLIYEMAAG 234
     NLPGSSDNLTSAI.GTAMYVATEVLDGTGHYNEKIDMYSLGIIFFEMI.. 850
     NGE.RAPTHTSRV.GTCLYASPEQL.EGSEYDAKSDMYSVGVILLELF.. 519
     ......GKRT.RSKGTLRYMSPEQISS.QDYGKEVDLYALGLILAELL.. 481

          !  !.       ! . !       .         .         .
     YPPFF.ADQPIQIYEKIVSGKVRFPSHFSSDL....KDLLRNLL.QVDLT 278
     .YPFSTGMERVNILKKLRSVSIEFPPDFDDNKMKVEKKIIR.LLIDHDPN 898
     .QPFGTEMERAEVLTGVRAGRI...PDSLSKRCPAQAKYVQ.LLTRRNAS 564
     .HVCDTAFETSKFFTDLRDGII...SDIFDKK...EKTLLQKLLSKK.PE 523

      *       .         .
     KRFGN...LKDGVNDIKNHKWFATTDWIA 304
     KRPGARTLLNSGWLPVKHQDEVIKEALKS 927
     QRPSALQLLQSELFQNSAHVNLTLQMKII 593
     DRPNTSEILRT..LTVWKKSPEKNERHTC 550
```

Figure 10. Multiple sequence alignments of the catalytic domains of GCN2, DAI, HCR, and cAMP-dependent protein kinase (cAPK). The amino acid positions of the segments aligned in the complete sequence of each protein are indicated at the far right on each line of sequence. Sequence identities are boxed. Dots indicate the positions of gaps in the alignment; asterisks indicate nearly invariant residues present in all serine–threonine protein kinases; exclamation points indicate residues present in GCN2, DAI, and HCR that are found in no more than 5 protein kinases in the collection of 75 serine–threonine protein kinases aligned by Hanks and Quinn.[60] The figure was constructed by manually incorporating information from pairwise alignments obtained using the BestFit program and multiple sequence alignments obtained using the PileUp program.[59]

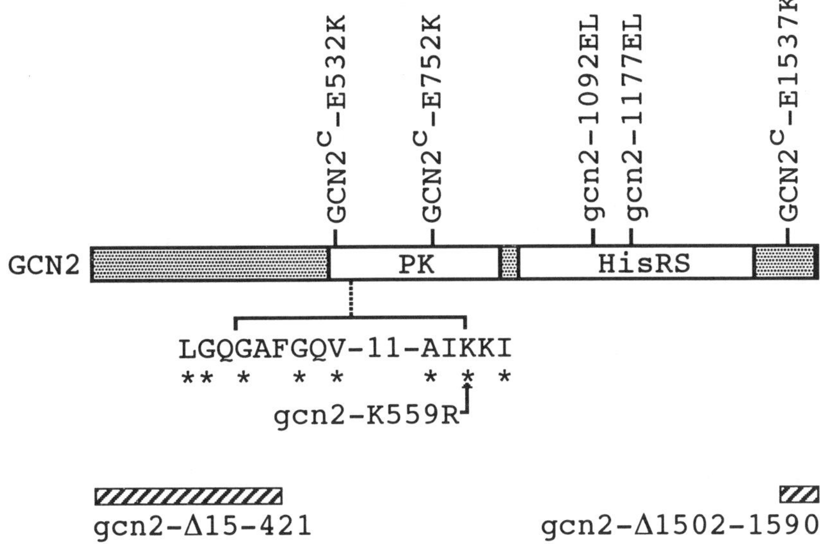

Figure 11. GCN2 contains domains homologous to protein kinases and histidyl-tRNA synthetases. The stippled box represents the GCN2 polypeptide chain, 1590 amino acids in length, with the domains similar in sequence to protein kinases (PK) and histidyl-tRNA synthetases (HisRS) indicated. Constitutively-derepressing *GCN2*ᶜ alleles are indicated above the box at the positions of the mutations, with the wild-type residue, amino acid position, and mutant residue listed in that order. Selected *gcn2* mutations that abolish derepression of *GCN4* expression are shown below the box. The sequence of the putative ATP-binding site in the protein kinase domain is shown in one-letter amino acid code, with asterisks indicating residues highly conserved among many protein kinases. The position of the *gcn2-K559R* mutations in the kinase domain is below the invariant Lys residue altered by this mutation. Two different two-codon (GluLeu) insertion alleles in the HisRS domain and two deletion alleles are indicated.[53,52]

assay.[52] These observations indicate that the protein kinase activity of GCN2 is required to stimulate *GCN4* translation.

The sequences of cloned cDNAs encoding DAI[55] and HCR[56] were reported recently and found to exhibit a relatively high degree of sequence similarity with GCN2.[57,56] Figure 10 shows a multiple sequence alignment between the catalytic domains of GCN2, HCR, DAI, and the murine cAMP-dependent protein kinase (cAPK). cAPK was chosen for these alignments because the three-dimensional structure of its catalytic domain was determined recently and is thought to provide a structural paradigm for all eukaryotic protein kinases.[58] Based on these alignments, the catalytic core of GCN2 shows 36% and 42% sequence identity with HCR and DAI, respectively, and 37% identity exists between DAI and HCR. By comparison, cAPK shows only 33%, 25%, and 30% identity with GCN2, HCR, and DAI, respectively. In addition, a variety of other protein kinase sequences that we aligned[59] exhibit, on average, 5% less sequence identity with GCN2, HCR, or DAI than exists between each of the latter three kinases and cAPK.

Another feature characteristic of GCN2, HCR, and DAI is the presence of an insert located between residues 104 and 105 in the cAPK sequence alignment. The inserts in GCN2 and HCR are quite large (136 and 116 residues, respectively) compared to that present in DAI (35 residues). The three inserts show limited sequence identity with one another (Fig. 10).

The 36–42% sequence identity seen among GCN2, DAI, and HCR is characteristic of protein kinases considered to be members of the same subfamily, e.g., cdc2 and KIN28 show this degree of sequence relatedness and have been assigned to the same branch cluster of the protein kinase phylogenetic tree.[60] Thus, it appears that GCN2, DAI, and HCR are more closely related to one another than to other protein kinases, and thus define a branch cluster on the protein kinase tree. The strong sequence similarity among GCN2, DAI, and HCR suggests that GCN2 is an eIF-2α kinase, an idea consistent with the extensive sequence identity between yeast[27] and human eIF-2α[61] in the vicinity of serine-51, the site of phosphorylation of eIF-2α by DAI and HCR.[62,63] Recent findings discussed in the next section provide strong evidence that GCN2 is an eIF-2α kinase and that it stimulates translation of *GCN4* under starvation conditions by phosphorylation of eIF-2α on serine-51. Thus, the characteristic inserts present in GCN2, DAI, and HCR may play a role in substrate recognition or in a regulatory mechanism unique to eIF-2α kinases. These functions may also involve the amino acids at 11 positions throughout their kinase domains at which the same residue occurs in GCN2, DAI, and HCR but is absent in the great majority of other protein kinases (Fig. 10).

4.2. Evidence That GCN2 Stimulates Translation of *GCN4* by Phosphorylation of eIF-2α on Serine-51

Analysis of eIF-2α phosphorylation by isoelectric focusing gel electrophoresis revealed that a hyperphosphorylated form of the protein increases in abundance when wild-type yeast cells are starved for histidine, and that this isoform is completely dependent on GCN2, being undetectable in a *gcn2*Δ strain (Fig. 12). In addition, the hyperphosphorylated form of eIF-2α is predominant

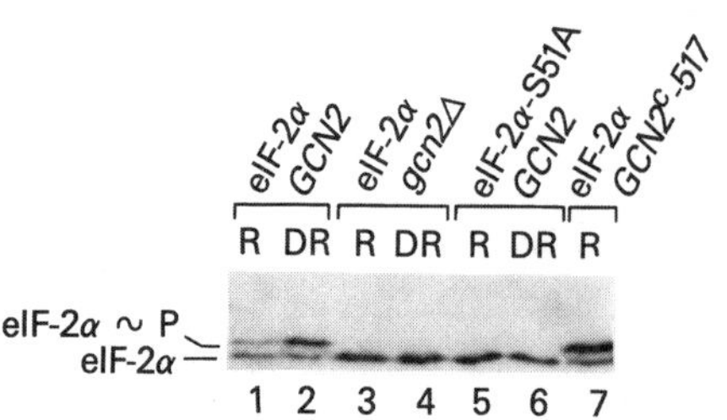

Figure 12. GCN2 protein kinase regulates the phosphorylation state of eIF-2α. Isoelectric focusing gel electrophoresis of total protein extracts followed by immunoblot analysis using eIF-2α antiserum was used to determine the phosphorylation state of eIF-2α under different conditions. Extracts were prepared from strains of the indicated genotypes grown under nonstarvation (R) or histidine starvation conditions (DR).[64]

even in the absence of starvation in strains containing a constitutively activated *GCN2*^{c} allele. Substitution of serine-51 with alanine in the eIF-2α gene (the *SUI2–S51A* allele) completely abolishes the hyperphosphorylated form of the protein in cells containing wild-type *GCN2* under starvation conditions and in derepressed *GCN2*^{c} mutants (Fig. 12). In addition to these *in vivo* results, it was shown that GCN2 protein isolated in immune complexes specifically phosphorylates the α subunit of rabbit eIF-2 and, more importantly, eIF-2α from yeast. GCN2 does not phosphorylate the mutant form of yeast eIF-2α containing the alanine-51 substitution.[64] These findings strongly suggest that GCN2 directly phosphorylates eIF-2α on serine-51 *in vivo* and that the level of eIF-2α phosphorylation increases in response to amino acid limitation.

The alanine-51 substitution in eIF-2α that prevents phosphorylation by GCN2 also impairs derepression of *GCN4* expression under starvation conditions to the same extent seen with inactivation of the GCN2 protein kinase (Table I). As expected, the alanine-51 mutation also reverses the derepressing effect of a *GCN2*^{c} mutation. Conversely, substitution of serine-51 with an aspartate residue leads to partial derepression of *GCN4* expression in the absence of amino acid starvation or the GCN2 protein kinase. These results provide strong evidence that GCN2-mediated phosphorylation of eIF-2α on serine-51 is required for increased *GCN4* expression in response to amino acid starvation. Presumably, the aspartate-51 mutation leads to only partial derepression of *GCN4* because aspartate imperfectly mimics phosphoserine at position 51 in eIF-2α. The alanine-51 and aspartate-51 substitutions and the *GCN2*^{c} mutations have no effect on the expression of a *GCN4* allele containing only uORF4 or no uORFs at all, indicating that phosphorylation of eIF-2α by GCN2 stimulates *GCN4* expression at the translational level by

Table I. *GCN4* Expression in *GCN2* and *gcn2Δ* Strains Containing *SUI2*, *SUI2–S51A*, or *SUI2–S51D* Alleles[a]

		GCN4–lacZ Expression (U)	
GCN2 allele	*SUI2* allele	R	DR
GCN2	*SUI2*	19	170
GCN2	*SUI2–S51A*	14	36
GCN2	*SUI2–S51D*	65	61
gcn2Δ	*SUI2*	14	44
gcn2Δ	*SUI2–S51A*	13	44
gcn2Δ	*SUI2–S51D*	52	50

[a]A *GCN4–lacZ* fusion was assayed in wild-type and *gcn2Δ* strains containing the indicated *SUI2* alleles. β-Galactosidase activities were measured after growing cells under nonstarvation conditions (R) or under histidine starvation conditions (DR).

reducing the inhibitory effects of uORFs 2–4 on reinitiation at the *GCN4* start codon.[64]

4.3. The Role of the GCN2 Protein Kinase in Detecting the Amino Acid Starvation Signal

Overexpression of GCN2 protein from a multicopy plasmid leads to partial derepression of *GCN4* translation,[19,52] suggesting that GCN2 protein levels limit the phosphorylation of eIF-2α under nonstarvation conditions. However, measurements of *GCN2* mRNA and protein levels have revealed little or no increase in the abundance of GCN2 in response to amino acid starvation.[52] The latter results suggest that the catalytic activity of GCN2, or access to its substrate eIF-2α, is stimulated in response to amino acid starvation. This conclusion is supported by the isolation of the *GCN2^c^* mutations mentioned above that derepress *GCN4* expression in the absence of starvation without increasing the steady-state level of GCN2 protein.[52] Two of the three best-characterized *GCN2^c^* alleles change single amino acids in the kinase domain (Fig. 11), suggesting a direct effect on GCN2 catalytic activity. A third *GCN2^c^* allele substitutes a single amino acid at the extreme C-terminus of the protein, identifying this region as a potential kinase regulatory domain. This conclusion is in accord with the fact that a deletion of these C-terminal sequences inactivates GCN2 regulatory function *in vivo* (*gcn2-Δ1502–1590*, Fig. 11) without affecting its *in vitro* autophosphorylation activity.[52] Thus, the C-terminus of GCN2 appears to be required *in vivo* to stimulate its kinase function in response to amino acid starvation.

Another strong indication that the C-terminal portion of GCN2 has a regulatory function is that 530 residues in this domain show significant similarity with the entire sequence of yeast histidyl-tRNA synthetase (HisRS), showing 22% identity and 45% similarity (Fig. 11).[53] Sequence similarity between GCN2 and the *Escherichia coli* and human HisRS sequences is also evident. Several two-codon insertions in the HisRS-like domain were shown to inactivate the *in vivo* regulatory function of GCN2 without impairing its *in vitro* autophosphorylation activity, consistent with the idea that this portion of the protein has a regulatory role in stimulating GCN2 kinase function in response to starvation.[53] Given that aminoacyl-tRNA synthetases bind uncharged tRNA as a substrate and distinguish between charged and uncharged tRNA,[65] and that accumulation of uncharged tRNA is thought to trigger the derepression of *GCN4* expression,[1] it was proposed that the HisRS-related domain of GCN2 functions to detect an increase in the level of uncharged tRNA under amino acid starvation conditions. Binding of uncharged tRNA would produce an allosteric change in the adjacent protein kinase moiety that increases the ability of GCN2 to phosphorylate eIF-2α on serine-51. Because *GCN4* expression derepresses in response to starvation for any one of at least ten amino acids, GCN2 may have diverged sufficiently from HisRS that it now lacks the ability to discriminate between different uncharged tRNA species.[53]

It was shown recently that GCN2 is a ribosome-associated protein.[57] In exponentially growing cells, essentially all of the GCN2 protein cosediments with polysomes, 80S ribosomes, or free subunits. Dissociation of the polysomes in the presence of low Mg^{2+} concentrations, or runoff of polysomes by incubation of extracts in the absence of cycloheximide, eliminates the most rapidly-sedimenting forms of GCN2, confirming a physical association with polysomes. When the polysomes are run off *in vitro*, most of the GCN2 either cosediments with free 40S and 60S ribosomal subunits or is found unassociated with ribosomes, despite a large accumulation of 80S ribosomal couples under these conditions. The same sedimentation profile for GCN2 is seen in extracts prepared from cells approaching stationary phase despite an abundance of polysomes under these conditions. These observations suggest that the ribosome association of GCN2 varies with the translation rate, showing 80S ribosome and polysome association only under conditions of rapid protein synthesis. In addition, even during exponential growth, the amount of GCN2 per ribosome is considerably greater for free 40S and 60S subunits than for 80S ribosomes or polysomes.[57] Free ribosomal subunits are associated with initiation factors that prevent subunit joining in the absence of mRNA, providing a pool of 40S and 60S subunits ready to participate in new rounds of initiation.[34] Thus, GCN2 may interact preferentially with ribosomes engaged in the initiation process. This might provide GCN2 with access to its substrate, as there is evidence that eIF-2 interacts with free ribosomal subunits.[66,41] It is also possible that ribosome binding by GCN2 allows it to monitor uncharged tRNA at the acceptor (A) site of the 80S ribosome during translation elongation. To accommodate the latter possibility with the preferential association of GCN2 with free subunits, it may be necessary to postulate that GCN2 dissociates from translating 80S ribosomes after only a few peptide bounds are formed. If activated by entry of uncharged tRNA into the A site during its limited interaction with elongating ribosomes, GCN2 would then phosphorylate eIF-2α when it subsequently binds to a free ribosomal subunit complexed with this initiation factor.[57]

5. MODEL FOR THE MOLECULAR MECHANISM OF *GCN4* TRANSLATIONAL CONTROL

Biochemical studies on mammalian systems have shown that phosphorylation of eIF-2α on serine-51 regulates translation initiation.[67,68] Genetic studies in yeast have shown that eIF-2 is directly involved in selecting the AUG start codon by the ribosome during the scanning process.[26,27] These two properties of eIF-2 are integrated into the following molecular model for *GCN4* translational control, in which phosphorylation of eIF-2α in amino acid-starved cells shifts the utilization of translation initiation sites from uORFs 2–4 to the *GCN4* AUG start codon (Fig. 13).[64]

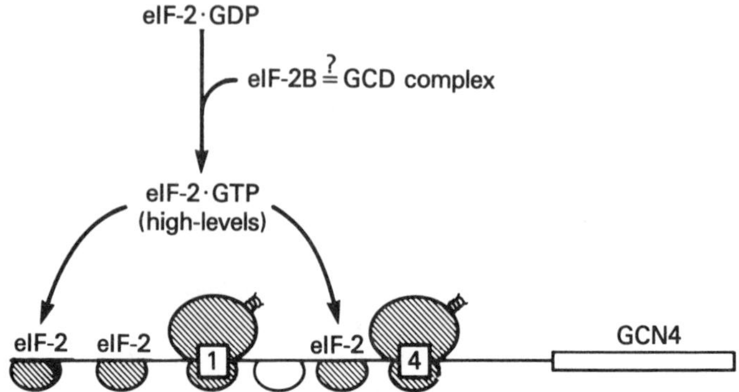
Non-Starvation Conditions
eIF-2·GDP
eIF-2B ?= GCD complex
eIF-2·GTP
(high-levels)
eIF-2
eIF-2
1
eIF-2
4
GCN4

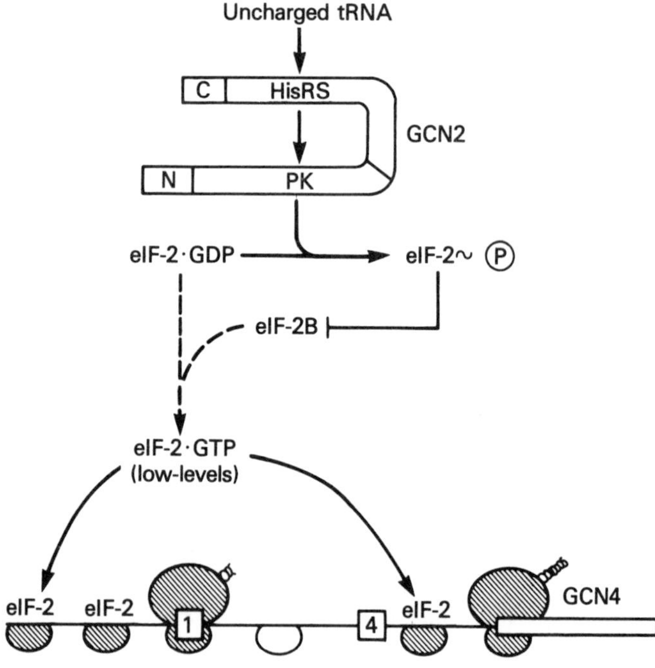
Starvation Conditions
Uncharged tRNA
C
HisRS
GCN2
N
PK
eIF-2·GDP
eIF-2~ P
eIF-2B
eIF-2·GTP
(low-levels)
eIF-2
eIF-2
1
4
eIF-2
GCN4

In the absence of amino acid starvation, scanning ribosomes initiate at the 5′-proximal uORF1 in the *GCN4* leader, terminate translation, and about 50% of the 40S subunits remain associated with the mRNA and resume scanning downstream. The majority of these ribosomal subunits reacquire the ternary complex consisting of eIF-2, GTP, and Met-$tRNA_i^{Met}$ before reaching the start codons for uORF2, 3, or 4 and reinitiate translation at one of these uORFs. These ribosomes dissociate from the mRNA following termination at uORFs 2–4, such that essentially none reaches the *GCN4* start codon under these conditions.

Under amino acid starvation conditions, uncharged tRNA accumulates and stimulates GCN2 protein kinase activity by interacting with the domain of GCN2 that is similar in sequence to histidyl-tRNA synthetases. GCN2 phosphorylates the α subunit of eIF-2. Phosphorylation of eIF-2α reduces the amount of active eIF-2 available for ternary complex formation by partially inhibiting eIF-2B, the guanine nucleotide exchange factor that replaces GDP with GTP on eIF-2 following each round of initiation. The high-molecular-weight, GCD1-containing complex may be the yeast equivalent of eIF-2B, explaining why mutations in eIF-2 subunits and GCD factors all have the same derepressing effect on *GCN4* translation. Because of the diminished amounts of ternary complexes present under starvation conditions, about 50% of the 40S subunits scanning downstream after terminating translation at uORF1 fail to rebind ternary complexes containing initiator tRNA before reaching uORF4 and thus ignore the start codons at uORFs 2–4. Most of these subunits bind ternary complexes before they reach the *GCN4* AUG codon and reinitiate translation there instead. Because only a few percent of the ribosomes bypass uORFs 2–4 under nonstarvation conditions, the fact that under starvation conditions 50% of the subunits ignore these sites and initiate at *GCN4* accounts for the large derepression ratio observed for *GCN4* expression.

←

Figure 13. Model for the role of eIF-2 in *GCN4* translational control.[64] *GCN4* mRNA is shown with uORFs 1 and 4 and the beginning of the *GCN4* coding sequences all indicated as boxes (uORFs 2 and 3 were omitted for simplicity, as they are dispensable for regulation). 40S ribosomal subunits are shown hatched when they are associated with eIF-2 and are thus competent to initiate translation (the eIF-2 bound to 40S subunits is meant to indicate a ternary complex composed of eIF-2, GTP, and Met-$tRNA_i^{Met}$); unshaded 40S subunits are not competent to initiate. 80S translating ribosomes are hatched. Under nonstarvation conditions, eIF-2·GDP is readily recycled to eIF-2·GTP by eIF-2B, leading to high levels of eIF-2·GTP and ternary complex formation. The ternary complexes thus formed reassemble with 40S ribosomes scanning downstream following termination at uORF1, causing reinitiation at uORF4. Following uORF4 translation, no reinitiation occurs at *GCN4*, possibly because the ribosomes dissociate from the mRNA. Under starvation conditions, uncharged tRNA accumulates and stimulates the GCN2 protein kinase activity. GCN2 phosphorylates eIF-2α and the phosphorylated eIF-2 sequesters eIF-2B, inhibiting the recycling of eIF-2·GDP to eIF-2·GTP. The resulting low level of eIF-2·GTP and ternary complex formation diminishes the rate at which competent initiation complexes are reassembled on 40S subunits following translation of uORF1. Thus, many 40S subunits scanning downstream from uORF1 are not competent to initiate at uORF4; these subunits acquire ternary complex while scanning the uORF4–*GCN4* interval, and initiate translation at *GCN4* instead. As indicated in the upper panel, biochemical studies[31] suggest that the products of the *GCD* genes may encode subunits of yeast eIF-2B.

Our model assumes that reduced ternary complex formation under starvation conditions has a much greater impact on reinitiation events at uORFs 2–4 and at *GCN4* compared to "primary" initiation events at 5′-proximal AUG codons, such as the uORF1 start site. This difference could exist because reinitiation is inherently less efficient than primary initiation,[69] making it more sensitive to modest reductions in eIF-2 activity. Alternatively, since 40S subunits are expected to contain ternary complexes when they interact with the 5′ end of mRNA,[34] reducing the availability of ternary complexes may diminish somewhat the rate of conventional initiation at 5′-proximal AUGs, but it should not lead to extensive leaky scanning at these sites in the manner suggested for reinitiating ribosomes at uORFs 2–4. Thus, by imposing a reinitiation mechanism on the translation of *GCN4*, the presence of uORF1 makes *GCN4* expression exquisitely sensitive to small reductions in eIF-2 activity that have little impact on conventional initiation events on the majority of cellular mRNAs.

6. IMPLICATIONS FOR TRANSLATIONAL CONTROL OF OTHER EUKARYOTIC mRNAs

Nearly 10% of the 699 different vertebrate mRNAs considered in a recent study,[11] including two-thirds of the mRNA transcripts from oncogenes, have one or more ATG codons present upstream from the start site for protein synthesis. In several instances, it has been shown that multiple uORFs found in extended mRNA leaders similar to that of *GCN4* are recognized as translational start sites and inhibit initiation at the downstream protein coding sequences: cytomegalovirus β-gene transcript,[70] Rous sarcoma virus RNA,[71,72] cauliflower mosaic virus RNA,[73,74] and SV40 "late-early" transcripts.[75] Translation of one of these mRNAs could involve reinitiation at the downstream protein-coding sequences following recognition of the uORFs in the leader, provided that the 5′-proximal uORF resembles *GCN4* uORF1 in allowing ribosomal scanning to resume following its own translation. As discussed above, the efficiency of reforming an initiation complex following termination at a uORF is thought to be reduced in yeast cells under amino acid starvation conditions because the activity of eIF-2 is partially impaired under these circumstances. This results in an increased scanning distance required for efficient reinitiation. Conditions known to reduce eIF-2 function in mammalian cells include hemin deprivation,[46] virus infection,[47] treatment with heavy metals,[76] and reduced levels of an amino acid[77] or tRNA-aminoacylation activity.[78] It is possible that under these or other stress situations in which protein synthesis is downregulated at the initiation step, an interplay between multiple uORFs could modulate the translation of particular mRNAs in a fashion similar to that elucidated for *GCN4*. We can also envision a regulatory mechanism in which a single uORF in the mRNA leader directs the utilization of alternative downstream start sites for protein synthesis as the result of variable scanning-distance requirements for reinitiation under different growth conditions.

References

1. Hinnebusch, A. G., 1988, *Microbiol. Rev.* **52:**248–273.
2. Hinnebusch, A. G., 1990, *Prog. Nucleic Acids Res. Mol. Biol.* **38:**195–240.
3. Hinnebusch, A. G., 1984, *Proc. Natl. Acad. Sci. USA* **81:**6442–6446.
4. Hinnebusch, A. G., 1985, *Mol. Cell. Biol.* **5:**2349–2360.
5. Harashima, S., and Hinnebusch, A. G., 1986, *Mol. Cell. Biol.* **6:**3990–3998.
6. Mueller, P. P., Harashima, S., and Hinnebusch, A. G., 1987, *Proc. Natl. Acad. Sci. USA* **84:**2863–2867.
7. Hill, D. E., and Struhl, K., 1988, *Nucleic Acids. Res.* **16:**9253–9265.
8. Hannig, E. M., and Hinnebusch, A. G., 1988, *Mol. Cell. Biol.* **8:**4808–4820.
9. Paddon, C. J., and Hinnebusch, A. G., 1989, *Genetics* **122:**543–550.
10. Williams, N. P., Hinnebusch, A. G., and Donahue, T. F., 1989, *Proc. Natl. Acad. Sci. USA* **86:** 7515–7519.
11. Kozak, M., 1987, *Nucleic Acids Res.* **15:**8125–8149.
12. Cigan, A. M., and Donahue, T. F., 1987, *Gene* **59:**1–13.
13. Sherman, F., and Stewart, J. W., 1982, In: *The Molecular Biology of the Yeast Saccharomyces, Metabolism and Gene Expression* (J. N. Strathern, E. W. Jones and J. R. Broach, eds.), pp. 301–334, Cold Spring Harbor Laboratory, Cold Spring Harbor, New York.
14. Kozak, M., 1984, *Nucleic Acids Res.* **12:**3873–3893.
15. Thireos, G., Driscoll-Penn, M., and Greer, H., 1984, *Proc. Natl. Acad. Sci. USA* **81:**5096–5100.
16. Mueller, P. P., and Hinnebusch, A. G., 1986, *Cell* **45:**201–207.
17. Tzamarias, D., Alexandraki, D., and Thireos, G., 1986, *Proc. Natl. Acad. Sci. USA* **83:**4849–4853.
18. Williams, N. P., Mueller, P. P., and Hinnebusch, A. G., 1988, *Mol. Cell. Biol.* **8:**3827–3836.
19. Tzamarias, D., and Thireos, G., 1988, *EMBO J.* **7:**3547–3551.
20. Mueller, P. P., Jackson, B. M., Miller, P. F., and Hinnebusch, A. G., 1988, *Mol. Cell. Biol.* **8:** 5439–5447.
21. Miller, P. F., and Hinnebusch, A. G., 1989, *Genes Dev.* **3:**1217–1225.
22. Abastado, J. P., Miller, P. F., Jackson, B. M., and Hinnebusch, A. G., 1991, *Mol. Cell. Biol.* **11:** 486–496.
23. Baim, S. B., and Sherman, F., 1988, *Mol. Cell. Biol.* **8:**1591–1601.
24. Cigan, A. M., Pabich, E. K., and Donahue, T. F., 1988, *Mol. Cell. Biol.* **8:**2964–2975.
25. Donahue, T. F., and Cigan, A. M., 1988, *Mol. Cell. Biol.* **8:**2955–2963.
26. Donahue, T. F., Cigan, A. M., Pabich, E. K., and Castilho-Valavicius, B., 1988, *Cell* **54:**621–632.
27. Cigan, A. M., Pabich, E. K., Feng, L., and Donahue, T. F., 1989, *Proc. Natl. Acad. Sci. USA* **86:** 2784–2788.
28. Tzamarias, D., Roussou, I., and Thireos, G., 1989, *Cell* **57:**947–954.
29. Wolfner, M., Yep, D., Messenguy, F., and Fink, G. R., 1975, *J. Mol. Biol.* **96:**273–290.
30. Foiani, M., Cigan, A. M., Paddon, C. J., Harashima, S., and Hinnebusch, A. G., 1991, *Mol. Cell. Biol.* **11:**3203–3216.
31. Cigan, A. M., Foiani, M., Hannig, E. M., and Hinnebusch, A. G., 1991, *Mol. Cell. Biol.* **11:** 3217–3228.
32. Hartwell, L. H., and McLaughlin, C. S., 1969, *Proc. Natl. Acad. Sci. USA* **62:**468–474.
33. Moldave, K., and McLaughlin, C. S., 1988, in: *Genetics of Translation* (M. F. Tuite, ed.), pp. 271–281, Springer-Verlag, Berlin.

34. Moldave, K., 1985, *Annu. Rev. Biochem.* **54:**1109–1149.
35. Harashima, S., Hanning, E. M., and Hinnebusch, A. G., 1987, *Genetics* **117:**409–419.
36. Rowlands, A. G., Montine, K. S., Henshaw, E. C., and Panniers, R., 1988, *Eur. J. Biochem.* **175:** 93–99.
37. Konieczny, A., and Safer, B., 1983, *J. Biol. Chem.* **258:**3402–3408.
38. Salimans, M., Goumans, H., Amesz, H., Beene, R., and Voorma, H., 1984, *Eur. J. Biochem.* **145:** 91–98.
39. Thomas, N. S. B., Matts, R., Petryshyn, R., and London, I., 1984, *Proc. Natl. Acad. Sci. USA* **81:** 6998–7002.
40. De Benedetti, A., and Baglioni, C., 1983, *J. Biol. Chem.* **258:**14556–14562.
41. Gross, M., Redman, R., and Kaplansky, D. A., 1985, *J. Biol. Chem.* **260:**9491–9500.
42. Gross, M., Wing, M., Rundquist, C., and Rubino, M. S., 1987, *J. Biol. Chem.* **262:**6899–6907.
43. Pain, V. M., 1986, *Biochem. J.* **235:**625–637.
44. Matts, R., Levin, D., and London, I., 1983, *Proc. Natl. Acad. Sci. USA* **80:**2559–2563.
45. Siekierka, J., Manne, V., and Ochoa, S., 1984, *Proc. Natl. Acad. Sci. USA* **81:**352–356.
46. Safer, B., 1983, *Cell* **33:**7–8.
47. Schneider, R. J., and Shenk, T., 1987, In: *Translational Regulation of Gene Expression* (J. Ilan, ed.), pp. 431–445, Plenum Press, New York.
48. Clemens, M. J., Galpine, A. Austin, S. A., Panniers, R., Henshaw, E. C., Duncan, R., Hershey, J. W., and Pollard, J. W., 1987, *J. Biol. Chem.* **262:**767–771.
49. Scorsone, K. A., Panniers, R., Rowlands, A. G., and Henshaw, E. C., 1987, *J. Biol. Chem.* **262:** 14538–14543.
50. Paddon, C. J., Hannig, E. M., and Hinnebusch, A. G., 1989, *Genetics* **122:**551–559.
51. Hannig, E. H., Williams, N. P., Wek, R. C., and Hinnebusch, A. G., 1990, *Genetics* **126:** 549–562.
52. Wek, R. C., Ramirez, M., Jackson, B. M., and Hinnebusch, A. G., 1990, *Mol. Cell. Biol.* **10:** 2820–2831.
53. Wek, R. C., Jackson, B. M., and Hinnebusch, A. G., 1989, *Proc. Natl. Acad. Sci. USA* **86:**4579–4583.
54. Roussou, I., Thireos, G., and Hauge, B. M., 1988, *Mol. Cell. Biol.* **8:**2132–2139.
55. Meurs, E., Chong, K., Galabru, J., Thomas, N. S. B., Kerr, I. M., Williams, B. R. G., and Hovanessian, A. G., 1990, *Cell* **62:**379–390.
56. Chen, J.-J., Throop, M. S., Gehrke, L., Kuo, I., Pal, J. K., Brodsky, M., and London, I. M., 1991, *Proc. Natl. Acad. Sci. USA* **88:**7729–7733.
57. Ramirez, M., Wek, R. C., and Hinnebusch, A. G., 1991, *Mol. Cell. Biol.* **11:**3027–3036.
58. Knighton, D. R., Zheng, J., Ten Eyck, L. F., Xuong, N.-H., Taylor, S. S., and Sowadski, J. M., 1991, *Science* **253:**407–414.
59. Devereux, J., Haeberli, P., and Smithies, O., 1984, *Nucleic Acids Res.* **12:**387–395.
60. Hanks, S. K., and Quinn, A. M., 1991, *Meth. Enzymol.* **200:**38–62.
61. Ernst, H., Duncan, R. F., and Hershey, J. W. B., 1987, *J. Biol. Chem.* **262:**1206–1212.
62. Colthurst, D. R., Campbell, D. G., and Proud, C. G., 1987, *Eur. J. Biochem.* **166:**357–363.
63. Pathak, V. K., Schindler, D., and Hershey, J. W. B., 1988, *Mol. Cell. Biol.* **8:**993–995.
64. Dever, T. E., Feng, L., Wek, R. C., Cigan, A. M., Donahue, T. D., and Hinnebusch, A. G.,1992, *Cell* **68:**585–596.
65. Schimmel, P. R., and Soll, D., 1979, *Annu. Rev. Biochem.* **48:**601–648.
66. Thomas, N. S. B., Matts, R. L., Levin, D. H., and London, I. M., 1985, *J. Biol. Chem.* **260:**9860–9866.
67. London, I. M., Levin, D. H., Matts, R. L., Thomas, N. S. B., Petryshyn, R., and Chen, J. J., 1987, In: *The Enzymes*, Vol. 18 (P. D. Boyer and E. G. Krebs, eds.), pp. 359–380, Academic Press, New York.
68. Hershey, J. W. B., 1991, *Annu. Rev. Biochem.* **60:**717–755.

69. Kozak, M., 1989, *J. Cell Biol.* **108:**229–241.
70. Geballe, A. P., and Mocarski, E. S., 1988, *J. Virol.* **62:**3334–3340.
71. Hensel, C. H., Petersen, R. B., and Hackett, P. B., 1989, *J. Virol.* **63:**4986–4990.
72. Petersen, R. B., Moustakas, A., and Hackett, P. B., 1989, *J. Virol.* **63:**4787–4796.
73. Futterer, J., Gordon, K., Bonneville, J. M., Sanfacon, H., Pisan, B., Penswick, J., and Hohn, T., 1988, *Nucleic Acids Res.* **16:**8377–8391.
74. Fuetterer, J., Gordon, K., Sanfacon, H., Bonneville, J.-M., and Hohn, T., 1990, *EMBO J.* **9:**1697–1707.
75. Khalili, K., Brady, J., and Khoury, G., 1987, *Cell* **48:**639–645.
76. Hurst, R., Schatz, J. R., and Matts, R. L., 1987, *J. Biol. Chem.* **262:**15939–15945.
77. Kimball, S. R., Antonetti, D. A., Brawley, R. M., and Jefferson, L. S., 1991, *J. Biol. Chem.* **266:** 1969–1976.
78. Pollard, J. W., Galpine, A. R., and Clemens, M. J., 1989, *Eur. J. Biochem.* **182:**1–9.

Chapter 6

Co- and Posttranslational Processes and Mitochondrial Import of Yeast Cytochrome *c*

Fred Sherman, Richard P. Moerschell, Susumu Tsunasawa, Rolf Sternglanz, and Mark E. Dumont

1. INTRODUCTION

The two nuclear genes, *CYC1* and *CYC7*, encoding the mitochondrial proteins iso-1-cytochrome *c* and iso-2-cytochrome *c*, respectively, in the yeast *Saccharomyces cerevisiae* comprise one of the most thoroughly studied gene–protein systems of eukaryotes. As summarized in Fig. 1, all steps of *CYC1* gene expression have been systematically examined, including transcription, translation, cotranslational and posttranslational modification, mitochondrial import, heme attachment, and enzymatic function. This system has been used to address questions

---

FRED SHERMAN • Departments of Biochemistry and Biophysics, School of Medicine and Dentistry, Rochester, New York 14642. RICHARD P. MOERSCHELL • Department of Biochemistry, School of Medicine and Dentistry, Rochester, New York 14642; *present address*: Institute for Protein Research, Osaka University, Suita, Osaka 565, Japan. SUSUMU TSUNASAWA • Institute for Protein Research, Osaka University, Suita, Osaka 565, Japan. ROLF STERNGLANZ • Department of Biochemistry, State University of New York at Stony Brook, Stony Brook, New York 11794. MARK E. DUMONT • Department of Biochemistry, School of Medicine and Dentistry, Rochester, New York 14642.

Translational Regulation of Gene Expression 2, edited by Joseph Ilan. Plenum Press, New York, 1993.

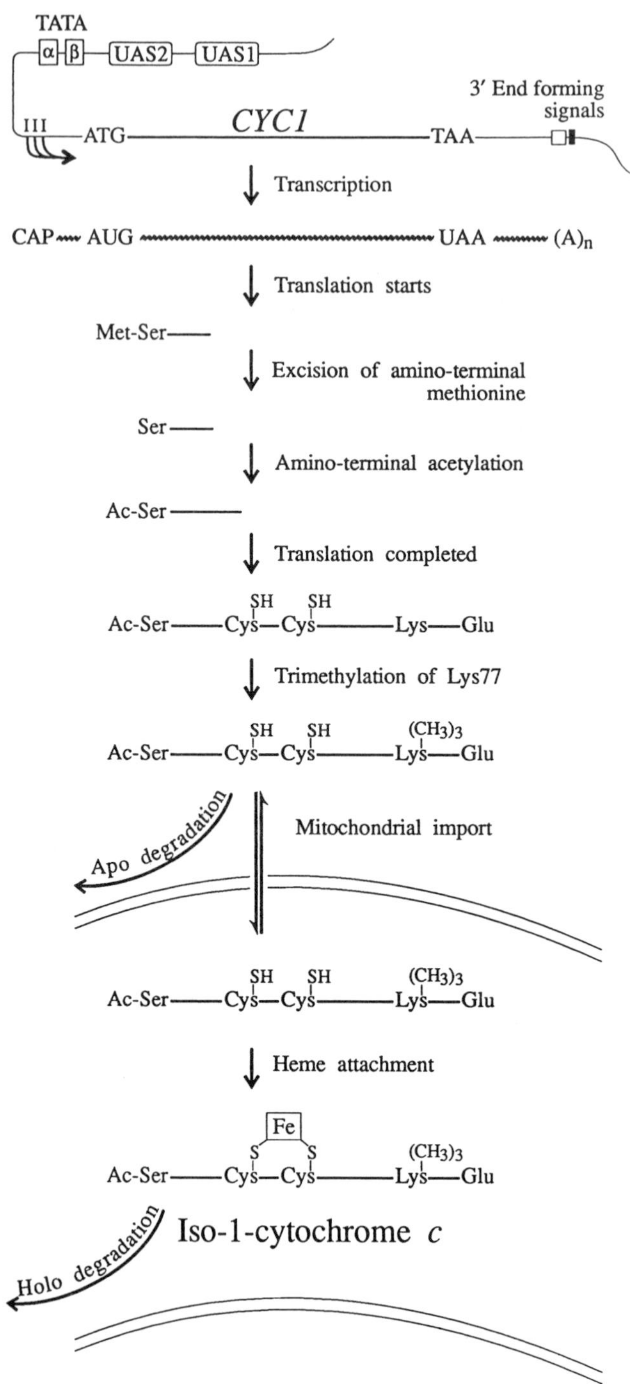
TATA
α
β
UAS2
UAS1
3′ End forming signals
III
ATG
CYC1
TAA
Transcription
CAP AUG UAA (A)n
Translation starts
Met-Ser
Excision of amino-terminal methionine
Ser
Amino-terminal acetylation
Ac-Ser
Translation completed
SH SH
Ac-Ser Cys Cys Lys Glu
Trimethylation of Lys77
SH SH (CH3)3
Ac-Ser Cys Cys Lys Glu
Apo degradation
Mitochondrial import
SH SH (CH3)3
Ac-Ser Cys Cys Lys Glu
Heme attachment
Fe
S S (CH3)3
Ac-Ser Cys Cys Lys Glu
Holo degradation
Iso-1-cytochrome c

concerning general principles involved in all of these processes, as well as serving as a model system for investigation in other diverse areas such as mutagenesis, recombination, protein folding, and protein stability. Also, early studies of yeast iso-1-cytochrome *c* played a major role in identification of chain-terminating and chain-initiating codons and nonsense suppressors.[13,14]

The two forms of cytochrome *c*, iso-1-cytochrome *c* and iso-2-cytochrome *c*, constitute, respectively, 95% and 5% of the total cytochrome *c* complement in derepressed cells of *S. cerevisiae*.[15] Mutations that directly or indirectly affect the structure or levels of these iso-cytochromes *c* have been used in studies of the expression and maturation of these proteins. In this chapter, we review the co- and posttranslational maturation and mitochondrial import of yeast iso-cytochromes *c*, and we discuss the relevance of these findings to processes generally occurring in eukaryotes.

2. *CYC* MUTANTS

Cytochrome *c* deficiencies can result from a variety of distinct classes of mutations that act generally or specifically. The *hem* mutants, defective in porphyrin or heme synthesis, lack cytochrome *c* and all other heme proteins.[16] Similarly, mutants particularly sensitive to catabolite repression have low levels of cytochrome *c* and other mitochondrial cytochromes.[17] The *hap* mutations, which generally alter transcription of certain heme proteins, either have little effect on the levels of iso-1-cytochrome *c* (*hap1*),[18] or are pleiotropic, reducing the levels of various cytochromes (*hap2* and *hap3*).[19] There are genes, however, that specifically and solely affect iso-1-cytochrome *c* and iso-2-cytochrome *c*. The genetic symbol *cyc* is reserved for such genes, and these major loci are listed in Table I.

Figure 1. The steps leading from the *CYC1* gene to the formation of an altered form of iso-1-cytochrome *c*, containing an abnormal Ac-Ser- terminus to illustrate amino-terminal acetylation. (Wild-type iso-1-cytochrome *c* contains an amino-terminal threonine residue which is unacetylated.) All of these processes have been investigated, including the requirements of the UAS1 and UAS2 elements[1,2] and two TATA elements (α and β)[3] for initiation of transcription at I sites; 3′-end formation;[4] and translation initiation[5,6] and termination.[7] During translation of the nascent apo-iso-1-cytochrome *c*, the amino-terminal methionine is excised by methionine aminopeptidase, and the penultimate residue is acetylated.[8] After completion of translation, lysine-77 of apo-iso-1-cytochrome *c* is trimethylated by the specific *S*-adenosylmethionine:cytochrome *c*-lysine *N*-methyltransferase.[9] This apo-iso-1-cytochrome *c* is degraded by a specific protease, especially if import is impaired.[10] The apo-iso-1-cytochrome *c* is believed to diffuse reversibly across the outer mitochondrial membrane and then bind to cytochrome *c* heme lyase.[11] Cytochrome *c* heme lyase, which is encoded by the *CYC3* gene, covalently attaches the heme moiety to the apo-cytochrome *c* within mitochondria.[12] Attachment of heme produces a conformational change, trapping holo-iso-1-cytochrome *c* in the intermembrane space and promoting binding to the outside of the inner membrane, where it carries out its physiological functions.[11] Recent experiments suggested that holo-iso-1-cytochrome *c* is degraded, presumably by a general mitochondrial protease. D. Pearse and F. Sherman (unpublished results).

Table I. Loci Specifically and Solely Affecting Cytochrome *c*

Locus	Gene product	Deficiencies in mutants		Ref.
		Iso-1	Iso-2	
CYC1	Iso-1-cytochrome *c*	0	+	20
CYC7	Iso-2-cytochrome *c*	+	0	21
CYC3	Heme lyase	0	0	12
CYC2	Import factor	±	±	22

3. *CYC1* MUTANTS AND ALTERED ISO-1-CYTOCHROMES *c*

The first methods to detect cytochrome *c*-deficient mutants were developed over 25 years ago.[13,14] A large number of altered iso-1-cytochromes *c* were uncovered in early studies by analyzing the series of mutations of the type $CYC1^+$ → *cyc1-x* → *CYC1-x-y*, where $CYC1^+$ denotes the wild-type gene that encodes iso-1-cytochrome *c*, *cyc1-x* denotes mutations that cause deficiency or nonfunction of iso-1-cytochrome *c*, and *CYC1-x-y* denotes intragenic reversions that restore at least partial activity and that give rise to either the normal or altered iso-1-cytochrome *c*. Over 500 *cyc1-x* mutants have been isolated and characterized and over 100 different iso-1-cytochrome *c* sequences have been obtained from *CYC1-x-y* revertants.[23]

Numerous altered iso-1-cytochromes *c* have also been generated using standard methods of site-directed mutagenesis, which relies on single-stranded *Escherichia coli* vectors containing the target sequence and a short synthetic oligonucleotide containing the desired alterations. Of more importance, we have described a more convenient procedure for producing specific alteration of genomic DNA by transforming yeast directly with synthetic oligonucleotides.[24,25] This procedure is easily carried out by transforming a defective *cyc1* mutant and selecting for revertants that are at least partially functional. The oligonucleotide used for transformation contains a sequence that corrects the defect and produces additional alterations at nearby sites. This technique is ideally suited for producing a large number of specific alterations that change a completely nonfunctional allele to at least a partially functional form. This selection procedure used with *cyc1* mutants allows recovery of altered iso-1-cytochromes *c* with less than 1% of the normal activity. This procedure is illustrated with the *cyc1-31* mutant (Fig. 2) that has been used to construct various single and multiple amino acid alterations at the amino-terminal region of iso-1-cytochrome *c*.[8]

Met cleavage	Acetylation	-1 1 2 3 4 5 6 (Met)**Thr**-Glu-Phe-Lys-Ala-Gly-	
		ATA ATG ACT GAA TTC AAG GCC GGT	*CYC1*$^{+}$
		ATA ATG ACT GAA TA- AAG GCC GGT	*cyc1-31*
0	0	ATA ATA ATG TTG TTC TTG GCC GGT Met-**Leu**-Phe-Leu-Ala-Gly-	*CYC1-850*
+	0	ATA ATA ATG GGT TTC TTG GCC GGT (Met)**Gly**-Phe-Leu-Ala-Gly-	*CYC1-841*
0	+	ATA ATA ATG GAA TTC TTG GCC GGT Ac-Met-**Glu**-Phe-Leu-Ala-Gly-	*CYC1-853*
+	+	ATA ATG TCT GAA TTC TTG GCC GGT (Met)**Ser**-Glu-Phe-Leu-Ala-Gly- Ac	*CYC1-793*

Figure 2. Examples of the creation of altered iso-1-cytochromes *c* by transforming the *cyc1-31* strain with synthetic oligonucleotides and selecting for functional transformants. Amino acid sequences of the amino-terminal region of the iso-1-cytochromes *c* are presented along with the corresponding DNA sequences of *CYC1* alleles. Nucleotides of the transformants that differ from the *cyc1-31* sequence are underlined. The penultimate residues are denoted in boldface. The *cyc1-31* mutant completely lacks iso-1-cytochrome *c* because of the frameshift and TAA nonsense mutations. Altered iso-1-cytochromes *c* with four types of amino termini are illustrated, without (0) and with (+) cleavage of the amino-terminal methione and without (0) and with (+) amino-terminal acetylation. Adapted from ref. 8.

4. CLEAVAGE OF AMINO-TERMINAL METHIONINE

The two cotranslational processes, cleavage of amino-terminal methionine and amino-terminal acylation, are by far the most common modification events, affecting nearly all proteins. Protein synthesis initiates with formylmethionine in prokaryotes, mitochondria, and chloroplasts, whereas proteins initiate with methionine in the cytosol of eukaryotes. The formyl group is removed from prokaryotic proteins by a deformylase, resulting in methionine at amino termini. The amino-terminal methionine is cleaved from the nascent chains of most prokaryotic and eukaryotic proteins by methionine aminopeptidase. Subsequently, amino-terminal acetylation can occur on certain proteins, either containing or lacking the initiator methionine. This amino-terminal acetylation occurs on over one-half of eukaryotic proteins, buy only rarely on prokaryotic proteins (see refs. 26 and 27 for reviews). Another, but rare, cotranslational acylation is *N*-myristoylation, in which myristic acid is linked via an amide bond to the amino-terminal glycine residues of a variety of eukaryotic proteins.[28]

Advances in our knowledge of the enzymes involved in eukaryotic amino-terminal processing have been made with yeast, and the determinations of which specific amino acid sequences are processed have been made with the iso-1-

cytochrome *c* system. In addition, the biological significance can be assessed with yeast mutants deficient in these protein modifications.

Early work with cell-free systems revealed enzymatic activities that remove amino-terminal residues of methionine from certain proteins in both prokaryotes and eukaryotes (see refs. 26 and 27 for reviews). This excision takes place before completion of nascent polypeptide chains and before other amino-terminal processing events such as amino-terminal acetylation. Further amino-terminal maturation occurs if the protein is secreted, usually by removal of signal sequences which are generally 15–30 residues in length. Methionine aminopeptidases from both prokaryotes[29] and yeast[30] have been purified, characterized, and shown to have the same or similar specificities. DNA sequence analysis of the cloned genes indicated that the deduced amino acid sequences of the two proteins are 60% similar and 42% identical.[30a]

We wish to emphasize that, in contrast to the import of most if not all other mitochondrial proteins, the import of cytochrome *c* does not involve an amino-terminal signal and does not require a membrane potential as an energy source. Because the amino-terminal region of iso-1-cytochrome *c* is dispensable for its biosynthesis and function, amino-terminal processing can be freely investigated with essentially any alteration. Thus, altered forms of iso-1-cytochrome *c* proved to be ideally suited for investigating the specificity of methionine aminopeptidase and amino-terminal acetylation. Cleavage of amino-terminal methionine was shown to be dependent on adjacent sequences in a study reported 20 years ago, in which protein sequencing of altered iso-1-cytochromes *c* from revertants of the initiation mutants revealed that methionine was cleaved from penultimate residues of threonine and alanine but not valine, leucine, isoleucine, or arginine. Studies of other altered iso-1-cytochromes *c*[31] were the basis for our hypothesis that methionine is completely removed from penultimate residues having radii of gyration of 1.29 Å or less (glycine, alanine, serine, cysteine, threonine, proline, and valine).[32] This hypothesis was confirmed from the results of transforming the *cyc1-31* mutant with a complete set of synthetic oligonucleotides encoding all possible amino acid residues (Figs. 2 and 3).[8] Furthermore, the degree of methionine cleavage was quantitatively estimated to an accuracy of 3–4% by high-performance liquid chromatography (HPLC) analysis of amino-terminal peptides. A similar pattern of cleavage also was observed in the following systems: prokaryotic methionine aminopeptidase *in vitro*;[29,33,34] mutational derivatives of methionyl-tRNA synthetase;[35] human growth hormone *in vivo* in *E. coli*;[36] mutational derivatives of thaumatin protein *in vivo* in yeast;[37] mutational derivatives of hemoglobin in a mammalian *in vitro* system;[38] and with yeast methionine aminopeptidase *in vitro*.[30] Only minor differences were observed between the quantitative results obtained *in vivo* with yeast iso-1-cytochrome *c* and with the two proteins from *E. coli*.[8,35,36]

The study by Moerschell *et al.*[8] also revealed that antepenultimate (the third residue) proline residues can inhibit methionine cleavage from certain residues

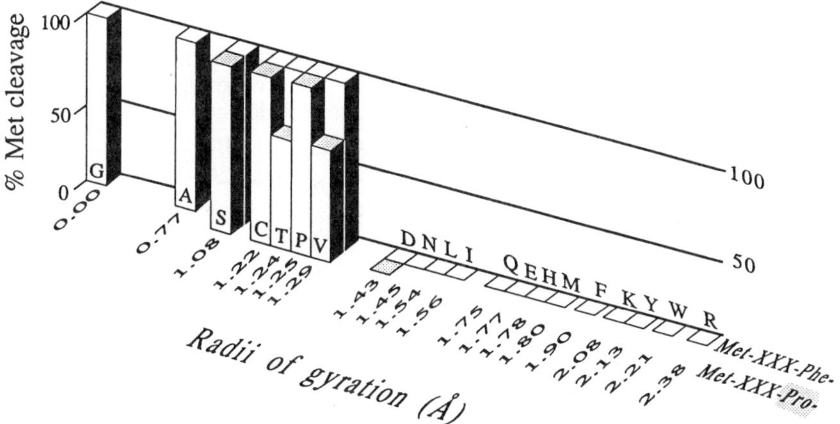

Figure 3. The percent cleavage of the amino-terminal methonine residue and radii of gyration of the penultimate residues. The complete series of altered iso-1-cytochromes *c* had the sequences Met-XXX-Phe-Leu-, whereas representatives had the sequences Met-XXX-Pro-Leu-, with antepenultimate proline residues.[8]

with intermediate sizes of side chains. In contrast to sequences having other antepenultimate residues, methionine was only partially cleaved from the sequences Met-Thr-Pro- and Met-Val-Pro- (Fig. 3). In other studies with *E. coli*, antepenultimate proline residues partially inhibited cleavage from Met-Ala-Pro-[35,39] and Met-Thr-Pro-.[40] Methionine cleavage was completely inhibited from the Met-Val-Pro- sequence of a mutant human hemoglobin.[41,42] These findings could be explained by a tendency of antepenultimate proline residues to distort topologically the amino terminus and interfere with the action of methionine aminopeptidase.

5. STABILITY AND AMINO TERMINI

Because all newly synthesized proteins begin with methionine, it might be expected that methionine aminopeptidase would allow the formation of a range of termini. What is striking, however, is that the specificity of methionine aminopeptidase has been evolutionarily conserved between prokaryotes and eukaroytes, resulting in the formation of only a subset of possible termini. Also remarkable, this specificity is almost a perfect mirror image of a subset of ubiquitin-dependent degradation signals that is composed, in part, of amino termini (Fig. 4).

Degradation of eukaryotic proteins by the ubiquitin system involves several discrete steps, including the binding of E3 (the ubiquitin-protein ligase) to specific

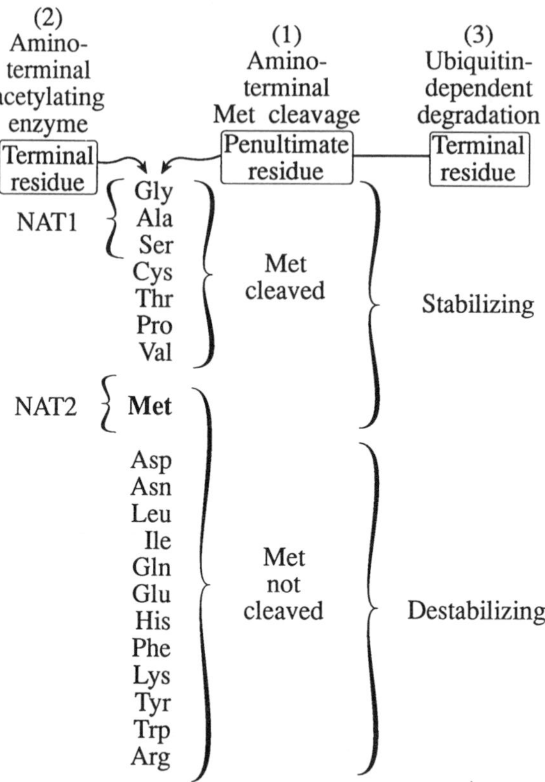

Figure 4. Properties of penultimate and amino-terminal residues and the relationships among (1) the penultimate residues preventing cleavage of amino-terminal methionine residues (see Fig. 3), (2) the amino-terminal acetylating enzymes, and (3) the amino-terminal residues destablizing proteins. (1) This pattern of amino-terminal methionine cleavage appears equivalent in both prokaryotes and eukaryotes. (2) Eukaryotic amino-terminal acetylation depends on at least two enzyme systems, NAT1 and NAT2, which act cotranslationally on at least a subset of proteins with the indicated amino termini. Other NAT enzymes acting on other subsets may also exist. Acetylation occurs more rarely in prokaryotes, and does not follow the same pattern. Amino-terminal acetylation occurring posttranslationally involves different acetyltransferases than those shown, apparently with different specificities. (3) Ubiquitin-dependent degradation requires a specific internal lysine residue as well as the indicated amino-terminal residues. Proteins having other signals not at amino-termini also can be degraded by the ubiquitin system. Prokaryotes presumably have a degradation system also dependent on amino-terminal residues.

sites on the protein. One group of signals, denoted types I–III[43] or N-degron,[44] is comprised, in part, of a destabilizing amino-terminal residue. At least in yeast, the amino-terminal residues serving as destabilizing signals include arginine, lysine, histidine (type I basic site), leucine, phenylalanine, tyrosine, isoleucine, and tryptophan (type II bulky-hydrophobic site). Aspartic acid and glutamic acid residues are converted to destabilizing residues only after conjugation to arginine;

and asparagine and glutamine are likewise converted after deamidation and arginine conjugation.[45] These destabilizing residues all have side chains with large radii of gyration. Thus, with the exception of methionine, potential destabilizing residues correspond to the group of penultimate residues which prevent cleavage of the adjacent initiator methionine (Fig. 4).

It is also surprising not only that yeast and *E. coli* have retained the same specificity for methionine aminopeptidase, but that both can degrade proteins based on a similar "N-end rule,"[45a] even though prokaryotes lack ubiquitin-dependent degradation systems. Apparently, the *E. coli* Clp protease degrades proteins with arginine, lysine, histidine, leucine, phenylalanine, tyrosine, and tryptophan termini, either directly or after the action of certain amino acid transferases. Thus, *E. coli* proteins are normally protected from the action of the Clp protease because of the lack of cleavage of methionine from these destabilizing residues. The similarities in the structure and specificity of yeast and *E. coli* methionine aminopeptidase imply their evolutionary precedence over the diverse Clp and ubiquitin degradation systems defining the two "N-end rules." If destabilizing amino-terminal residues comprise the "N-end rule," the penultimate residues determining methionine cleavage could be denoted as an "N-end ruler."

Although this relationship between the degradation and methonine cleavage appears to have been evolutionarily maintained, we wish to stress that eukaryotic degradation requires structural determinants in addition to the amino-terminal residue,[46,47] and that other signals not involving amino termini may be more prominent for the ubiquitin ligation system. Thus, not all eukaryotic proteins with large amino-terminal residues are necessarily unstable, and some proteins with small amino-terminal residues may be degraded by the ubiquitin system. Also, there is a paucity of known natural proteins that are degraded by the "N-end rule" pathway. One cited example of endoproteolytic cleavage that generates a destabilizing residue is the short-lived bacteriophage λ cII protein.[45] It also has been suggested that damaged or abnormal proteins may be degraded by the "N-end rule" pathway.[45,46] This relationships between "N-end rule" and methionine cleavage may be preserved in eukaryotes and prokaryotes to maintain some other, yet unexplained regulation of a subset of proteins.

6. AMINO-TERMINAL ACETYLATION

Amino-terminal acetylation of proteins is catalyzed by amino-terminal acetyltransferases which transfer acetyl groups from acetyl-CoA to termini of α-amino groups of most eukaryotic proteins and, more rarely, to prokaryotic proteins. Similar to amino-terminal methionine cleavage, amino-terminal acetylation is one of the most common protein modifications in eukaryotes, representing approximately 85% of the different protein species. Amino-terminal acetylation occurs cotranslationally in *in vitro* translation systems, usually when there are between 20 and 50 residues extruding from the ribosome (see refs. 26 and 27 for reviews). Acetylation may occur on the initiator methionine residue, but more often on penultimate residues after methionine cleavage. Acetylation also can occur after translation is completed on amino termini of special proteins, such as actins from some but not all organisms;[48,49] or at internal sites after specific proteolytic processing, as in cases of peptide hormones.[50–53] Apparently, cotranslational and posttranslational acetylation involve different sets of acetyltransferases having different specificities. Also, posttranslational acetylation of each protein or group of peptides involves a different acetylation system. Amino-terminal acetyltransferase that presumably acts cotranslationally has been purified from yeast[54] and rat liver.[55,56] The enzymes from the two species appear to have different structures, as indicated by differences in amino acid compositions and isoelectric points.

6.1. Structural Requirements for Amino-Terminal Acetylation

Proteins susceptible to amino-terminal acetylation have a variety of different amino-terminal sequences, with no simple consensus motif, although serine, alanine, and methionine are the most common amino-terminal residues.[26] Attempts to decipher the rules by which eukaryotic proteins are cotranslationally acetylated have involved examination of published amino acid sequences, systematic studies of altered proteins *in vivo*, especially iso-1-cytochrome *c*, and systematic studies of synthetic peptides *in vitro*, especially with the rat liver enzyme.[56] The action of the rat liver amino-terminal acetyltransferase on 65 synthetic peptides *in vitro* revealed specificities that differed from the specificities observed with yeast NAT1 enzyme *in vivo* (see below).[56] It remains to be seen if this difference is due to inherent differences between the two enzymes or to differences of acetylation occurring *in vivo* and *in vitro*. Nevertheless, specific alterations of iso-1-cytochrome *c* can be used to define systematically possible consensus sequences *in vivo*, thereby obtaining information that could not be accessible from just examining protein data bases.

In normal yeast strains, the amino-terminal methionine of iso-1-cytochrome *c* is cleaved and the newly exposed threonine is not acetylated (Fig. 2). Early studies revealed, however, that certain mutant proteins having alterations in the amino-terminal region were acetylated, including Ac-Met-Ile-Arg-, Ac-Met-Ile-Lys-, Ac-Met-Met-Asn-, Ac-Met-Asn-Asn-, Ac-Met-Glu-Phe-, Ac-Met-Asp-Phe-, and Ac-Ser-Glu-Phe-.[31] Interestingly, several of these acetylated iso-1-cytochrome *c* sequences did not have any counterparts in the protein data base. In some sequences, only two or three specific residues appear to specify acetylation. In spite of the diverse eukarotic sequences that are acetylated, one simple pattern emerged in the first systematic study by Moerschell *et al.*,[8] which included all 20 residues in the Met-*All*-Phe-Leu- amino terminus (Table II). In this series, only

Table II. Examples of Amino-Terminal Acetylation of Altered Iso-1-cytochromes *c* with Amino-Terminal Methionine[a]

% Acetylation	Sequence	Allele
100	Ac-Met-Glu-Phe-Leu-	*CYC1-853*
100	Ac-Met-Asp-Phe-Leu-	*CYC1-848*
79	Ac-Met-Asn-Phe-Leu-	*CYC1-849*
0	Met-Etc-Phe-Leu-	—
0	Etc-Phe-Leu-	—
55	Ac-Met-Glu-Phe-Lys-	*CYC1-838*
67	Ac-Met-Asp-Pro-Leu-	*CYC1-878*
~100	Ac-Met-Met-Asn-Ser-	*CYC1-183-T*, etc.
~100	Ac-Met-Met-Asn-Met-	*CYC1-242-N*
~100	Ac-Met-Asn-Asn-Asn-	*CYC1-345-H*
~100	Ac-Met-Ile-Arg-Ile-	*CYC1-31-Y*
~100	Ac-Met-Ile-Lys-Phe-	*CYC1-9-BU*

[a]Results of the series Met-*All*-Phe-Leu-, having all 20 residues, are summarized at the top of the table (etc. denotes other residues). These and other results established that Ac-Met-Glu- and Ac-Met-Asp- are sufficient, but not necessary, for amino-terminal acetylation in eukaryotes, and that the degree of acetylation is influenced by other nearby residues. Adapted from refs. 8 and 31 and R. P. Moerschell, S. Tsunasawa, and F. Sherman (unpublished results).

Met-Glu- (*CYC1-853*) and Met-Asp- (*CYC1-848*) were completely acetylated, whereas Met-Asn- (*CYC1-849*) was partially acetylated (top of Table II). We believe that Met-Glu- and Met-Asp- are members of the class of Met-Acidic-proteins, whereas Met-Asn-Phe- (*CYC1-849*) may be a member of a different class that may include Ac-Met-Asn-Ans- (*CYC1-345-H*) and possibly other iso-1-cytochromes *c* (bottom of Table II).[31] Thus, Met-acidic- termini are sufficient, but not necessary to direct methionine acetylation because some Met-Ile- and Met-Met- iso-1-cytochromes *c* are also acetylated (bottom of Table II).[31] Furthermore, the diminished acetylation of Met-Asp-Pro- (*CYC1-878*) (middle of Table II) suggests that antepenultimate proline residues inhibit acetylation in addition to methionine cleavage. Acetylation of Met-acidic- termini also may be diminished by basic residues in the terminal region, as exemplified by *CYC1-838* (middle of Table II).

Corroborative evidence for the acetylation of all eukaryotic proteins beginning with Met-Glu- and Met-Asp- proteins was obtained from examination of the literature and of a protein data base. Only proteins whose sequences were determined by both protein and nucleic acid sequencing were considered in the analysis, because the presence of an AUG initiator codon in the DNA sequence established that the protein was not created by internal cleavage, whereas the protein sequence revealed amino-terminal modifications. Only 13 examples of

Met-Glu- and Met-Asp- were uncovered after excluding proteins translated *in vitro* and proteins acetylated posttranslationally. All seven of the eukarotic proteins from this group were acetylated, whereas all six of the prokaryotic proteins were not acetylated.[8] Furthermore, Met-Glu- and Met-Asp- were among the acetylated termini in an *in vivo* study of 20 forms of thaumatin proteins in yeast[37] and in an *in vitro* study of 20 forms of human β-globin translated with reticulocyte and wheat germ systems.[38]

Ongoing studies with altered forms of iso-1-cytochromes *c* are aimed at deciphering other consensus sequences for acetylation. These studies are hampered, however, by the complexity of amino-terminal acetylation, which we believe is due in part to the existence of multiple amino-terminal acetyltransferases, each acting on different patterns of amino-terminal sequences.[57,58]

6.2. Mutants Deficient in Amino-Terminal Acetylation, *nat1⁻*

We have reported that the *NAT1* and *ARD1* genes encode an amino-terminal acetyltransferase in the yeast *S. cerevisiae*. Both *nat1⁻* and *ard1⁻* mutants are unable to carry out amino-terminal acetylation of a subset of normally acetylated proteins.[57] The *nat1⁻* mutant was originally uncovered by screening a collection of lethal temperature-sensitive mutants for protein acetyltransferase activity *in vitro*. Furthermore, regions of the *NAT1* gene and the chloramphenicol acetyltransferase genes of bacteria have limited but significant homology. The previously identified *ard1⁻* mutant was first suspected to be related to *nat1⁻* because of certain similar phenotypes. In addition to lacking acetyltransferase activity, strains with mutations in either *NAT1* or *ARD1* exhibit slower growth, derepression of the silent mating type locus *HMLα*, and failure to enter G_0. Also, a 20-fold increase of acetyltransferase activity occurred when both *NAT1* and *ARD1* were concomitantly overexpressed, but not when either gene was individually overexpressed. Thus, either the *NAT1* and *ARD1* proteins function together as an amino-terminal acetyltransferase, or *ARD1* is required for the function of *NAT1*. Some of these findings were confirmed by Lee *et al.*,[59,60] who denoted *nat1* and *aaa1*.

Analysis of two-dimensional gels revealed that at least 20 proteins, and possibly many more,[57] or approximately 20% of the different species of soluble proteins,[61] were affected in *nat1⁻* strains. Also, amino acid sequencing revealed that an altered iso-1-cytochrome *c* normally having an Ac-Ser-Glu-Phe- terminus was unacetylated in *nat1⁻* strain. Also, histones H2B1 and H2B2 were not amino-terminally acetylated in *nat1⁻* strains, whereas the normally acetylated histones H2A and H4 remained acetylated. The apparent acetylation of histones H2A and H4 in *nat1⁻* strains indicated that there must be at least one other amino-terminal acetyltransferase in yeast, presumably acting on a different class of terminal sequences.[57] Furthermore, Lee *et al.*[58] have demonstrated amino-terminal acetyltransferase activity of *nat1⁻* (denoted *aaa⁻*) extracts *in vitro* on certain peptides. We have investigated the specificity of the NAT1 acetyltransferase *in vivo* by analyzing the percent acetylation of variant iso-1-cytochrome c. The results, summarized in Table III, clearly distinguished two groups, one containing termini with methionine that are acetylated in both normal *NAT1*$^+$ and mutant *natl*$^-$ strains, and the other group with serine or glycine termini that are not acetylated in *natl*$^-$ strains. Further information comes from the lack of acetylation of 14 different proteins in *natl*$^-$ strains (Table IV). These results suggests that the NAT1 acetyltransferase acts on at least a subset of proteins with serine, glycine, and alanine termini, but not on at least some proteins with methionine amino-termini. Furthermore, proteins with a wide range of different sequences adjacent to serine termini are substrates of the NAT1 acetyltransferase (Table IV).

6.3. Biological Importance of Amino-Terminal Acetylation

There are a surprisingly few examples demonstrating the importance of amino-terminal acetylation for stability or function of proteins. Unfortunately, most of these conclusions are based on comparisons of acetylated and unacetylated forms that had additional differences in amino acid sequence. Alterations at amino termini, including loss of acetylation, decreased thermal stabilities of NADP-specific glutamate dehydrogenase from *Neurospora crassa*.[64] Similarly, the unacetylated form of the *Escherichia coli* ribosomal 5S protein is more thermosensitive than the acetylated form.[65] A decrease in half-life was observed for a form of murine hypoxanthine phosphoribosyltransferase from mouse, which had an amino-terminal replacement of proline for acetylalanine.[66] Hershko *et al.*[67]

Table III. Percent Amino-Terminal Acetylation of Altered Iso-1-cytochromes *c* from *NAT1*$^+$ and *natl-3* Strains[a]

		% Acetylation	
		NAT1$^+$	*natl-3*
CYC1-793	ATA ATG TCT GAA TTC TTG GCC Ac- (Met) Ser - Glu - Phe - Leu - Ala -	100	0
CYC1-RAT11	AAA ATG GGT GAT GTT GAA AAA Ac- (Met) Gly - Asp - Val - Glu - Lys -	70	0
CYC1-872	ATA ATA ATG AAC AAC TTG GCC Ac - Met - Asn - Asn - Leu - Ala -	100	~95
CYC1-848	ATA ATA ATG GAC TTC TTG GCC Ac - Met - Asp - Phe - Leu - Ala -	100	100
CYC1-853	ATA ATA ATG GAA TTC TTG GCC Ac - Met - Glu - Phe - Leu - Ala -	100	100

[a]Methionine residues cleaved before acetylation are denoted by (Met). Underlined amino acid residues and nucleotides denote those that differ from normal (see Fig. 2), except *CYC1-RAT11*, which encodes rat cytochrome *c*,[62] and which therefore greatly differs from *CYC1*$^+$. From R. P. Moerschell, S. Tsunasawa, J. R. Mullen, R. Sternglanz, and F. Sherman (unpublished results).

Table IV. Deficiency of Amino-Terminal Acetylation of Certain Proteins in *nat1*⁻ Strains

Protein	Amino-terminal sequence
Yeast ribosomal protein[a]	
22	Ser-Val-Glu-Pro-Val-Val-Val-Ile-Asp-Gly-
23	Ser-Gln-Pro-Val-Val-Val-Ile-Asp-Ala-Lys-
30	Ser-Ala-Pro-Gln-Ala-Lys-Ile-Leu-Ser-Gln-
40	Ser-Ala-Pro-Gln-Ala-Lys-Ile-Leu-Ser-Gln-
39 (L16)	Ser-Ala-Lys-Ala-Gln-Asn-Pro-Met-Arg-Asp-
41	Ser-Thr-Glu-Leu-Thr-Val-Gln-Ser-Glu-Arg-
50	Ser-Asp-Ala-Val-Thr-Ile-Arg-Thr-Arg-Lys-
61R	Ser-Ala-Val-Pro-Ser-Val-Gln-Ala-Phe-Lys-
Yeast nonhistone protein 4 (*NHP4*)[b]	Ser-Asp-Trp-Asp-Thr-Asn-Thr-Ile-Ile-Gly-
Human ACTH peptide[c]	Ser-Tyr-Ser-Met-Glu-His-Phe-Arg-Trp-Gly-
Yeast alcohol dehydrogenase I peptide[c]	Ser-Ile-Pro-Glu-Thr-Gln-Lys-Gly-Val-Ile-
Human superoxide dismutase protein[d] and peptide[c]	Ala-Thr-Lys-Ala-Val-Cys-Val-Leu-Lys-Gly-
Human basic fibroblast growth factor[d]	Ala-Ala-Gly-Ser-Ile-Thr-Thr-Leu-Pro-Ala-
Human acidic fibroblast growth factor[d]	Ala-Glu-Gly-Glu-Ile-Thr-Thr-Phe-Thr-Ala-

[a] *In vivo*[63]
[b] *In vivo* (David Kolodrubetz, unpublished results).
[c] *In vitro*[58]
[d] *In vivo* (Lawrence Cousens, unpublished results).

observed that amino-terminal acetylated cytochrome *c* and enolase from mammalian sources were not degraded *in vitro* by what was presumed to be a ubiquitin-dependent system, in contrast to the nonacetylated correspondents from yeast, which were good substrates. Also Matsuura *et al.*[68] suggested that amino-terminal acetylation protected apo-cytochrome *c* from degradation *in vitro*. In contrast, Sokolik and Cohen[69] used *NAT1*⁺ and *nat1*⁻ yeast strains to prepare acetylated and unacetylated pairs of rat and yeast cytochrome *c*, respectively (Table II), and observed equal extents of ubiquitin conjugation within each pair, although both yeast forms were more highly ubiquitinated than both of the rat forms. Thus, the difference in ubiquitination of mammalian and yeast cytochrome *c* is due to differences other than amino-terminal acetylation. Furthermore, Mayer *et al.*[70] observed ubiquitin-dependent degradation of amino-terminal acetylated proteins in a crude reticulocyte lysate. The presence of most of the unacetylated derivatives at approximately normal levels in two-dimensional gels demonstrated that acetylation is not generally required for stability;[57,61] because certain proteins were diminished, enhanced, lost, and gained in *nat1*⁻ strains, however, some of the alterations on two-dimensional gels could be indirect, perhaps due to the effects of unacetylated regulatory proteins.[61]

A few examples of acetylation affecting the function of various proteins have been reported. Amino-terminal acetylation, occurring posttranslationally, causes increased melanotropic effects of α-melanocyte-stimulating hormone, while it reduces the analgesic action of β-endorphin.[50,71] Nonacetylated cytoplasmic actin from cultured *Drosophila* cells is less efficient in the assembly of microfilaments than the acetylated form.[72] The oxygen affinity of nonacetylated forms of feline and possibly other β-hemoglobin chains becomes insensitive to the modifying effects of organic phosphates than the acetylated forms.[73] It should be noted that all of the effects observed *in vivo* and involving proteins modified cotranslationally had differences in amino acid sequences in addition to the lack of acetylation.

The most significant means for assessing the general importance of amino-terminal acetylation comes from the defects and, more surprising, the lack of defects in *natl*⁻ and *ardl*⁻ mutants, which lack the major amino-terminal acetyltransferase. As described above, the silent mating loci, particularly *HML*α, are partially derepressed in *natl*⁻ and *ardl*− mutants, leading to a partial mating defect in *MAT***a** strains. In addition, *natl*− and *ardl*− mutants exhibit defects of slow growth, inability of homozygous diploid strains to sporulate, and the failure to enter G_0 when limited for nutrients.[57] Presumably, these multiple defects are due to the lack of amino-terminal acetylation of one or more specific proteins requiring acetylation for function. Diminished function caused by the lack of acetylation of the SIR3 protein, for example, can explain the partial derepression of *HML*,[74] whereas diminished function of any one of a number proteins in the cAMP pathway can explain the failure to enter G_0 and the inability of homozygous diploids to sporulate.[57] Although viability of *natl*− and *ardl*− mutants certainly was not anticipated, especially in light of the lack of acetylation of many ribosomal proteins in the mutant strains (Table IV), many investigators have been shocked to find viable mutants with few or no obvious detrimental phenotypes after disrupting other "essential" yeast genes.

In summary, studies with yeast mutants have indicated that there are at least two acetyltransferases that act cotranslationally on many eukaryotic proteins. NAT1, the major amino-terminal acetyltransferase, acts on at least a subset of proteins with serine, glycine, and alanine amino-termini, but not on any of the tested methionine termini (Tables III and IV). A second acetyltransferase, NAT2, may act cotranslationally on all or a subset of proteins with methionine termini (Table III).[58] Each of the acetyltransferases may recognize several groups of sequences at amino termini, and acetylation may be inhibited by amino acid changes within or adjacent to these consensus sequences. Although amino-terminal acetylation may be required for stability and function of certain proteins, viability of *natl*− mutants lacking the major acetyltransferases suggest that the role of acetylation may be subtle and not absolute for most proteins. Although there is an intriguing relationship between the pattern of methionine cleavage and a component of a ubiquitin-dependent degradation signal, the acetylation of certain proteins with mainly serine, alanine, and methionine termini (Fig. 4) does not explain its role for prevention of degradation. Possibly only a subset of proteins actually require this modification for activity or stability, whereas the remainder are acetylated only because their termini fortuitously correspond to consensus sequences.

7. CYTOCHROME *c* HEME LYASE AND MITOCHONDRIAL IMPORT

Cytochrome *c* is a mitochondrial protein located on the outer surface of the inner mitochondrial membrane, where it participates in electron transport. Like most mitochondrial proteins, cytochrome *c* is encoded in the nucleus, synthesized in the cytoplasm, and then translocated into mitochondria. Like all known proteins that are imported into mitochondria, it is capable of being imported posttranslationally in *in vitro* assays. In contrast to most other imported mitochondrial proteins, however, cytochrome *c* does not require a membrane potential across the inner membrane,[75] the precursor is not cleaved during import (see ref. 76 for review) and the amino-terminal region is not required for proper localization.[77] Nevertheless, cytochrome *c* does undergo a major posttranslational modification: attachment of heme to the apoprotein to form holo-cytochrome *c*. This modification is catalyzed by the enzyme cytochrome *c* heme lyase, which attaches the cysteine residues 19 and 22 in the apoprotein to the propionic side chains.

Dumont *et al.*[12] demonstrated that cytochrome *c* heme lyase was encoded in yeast by the nuclear gene *CYC3*. Strains carrying *cyc3*$^-$ null mutations completely lacked both iso-1-cytochrome *c* and iso-2-cytochrome *c* (Table I). Enzyme assays of attachment of heme to apo-cytochrome *c* by mitochondrial extracts from yeast bearing different copy numbers of the functional *CYC3* gene suggested that *CYC3* encodes cytochrome *c* heme lyase. Mitochondrial extracts prepared from *cyc3*$^-$ strains exhibited greatly reduced heme lyase activity in the assay, while extracts from strains bearing multiple copies of *CYC3* were greatly enriched in the activity.[78] Furthermore, heme lyase activity was detected in an *E. coli* expression system containing the yeast *CYC3* gene.[78] A similar gene has been cloned from *Neurospora*.[79]

Because of the important role of the heme in cytochrome *c* folding and evidence that an analog of protohemin could block uptake of apocytochrome *c* into mitochondria isolated from *Neurospora*,[80] Dumont *et al.*[11,78] examined the relationship between heme attachment and import of cytochrome *c* into mitochondria in *in vitro* and *in vivo* systems. Apo-cytochrome *c* transcribed and translated *in vitro* could be imported with high efficiency into mitochondria isolated from normal yeast strains. Little import of apo-cytochrome *c* occurred with mitochondria isolated from *cyc3*$^-$ strains, however, which lack cytochrome *c* heme lyase. Similar results have been obtained with *Neurospora* mitochondria.[81] In addition, amino acid substitutions in apo-cytochrome *c* at either of the two cysteine residues that are the sites of the thioether linkages to heme, or at an immediately adjacent histidine that serves as a ligand of the heme iron, resulted in a substantial reduction in the ability of the precursor to be translocated into mitochondria *in vitro*. In contrast, replacement of the methionine serving as the other heme ligand had no detectable effect on import of apo-cytochrome *c* in this system.[78] The requirements of these amino acid residues *in vitro* are consistent with results obtained *in vivo*. Of the 30 single amino acid substitutions in iso-1-

cytochrome *c* examined *in vivo* by Hampsey *et al.*,[23] only replacements at residues Cys-19, Cys-22, and His-23 were found to result in a complete deficiency of holo-iso-1-cytochrome *c*. The two cysteine residues are the sites of the thioether bond linkages in holo-cytochrome *c*. The histidine immediately adjacent to these cysteines is one of the ligands of the heme iron and may play a role in recognition by the heme lyase. The lack of cytochrome *c in vivo* apparently results from a failure to have heme attached, causing a blockage in import.

The role of covalent heme attachment for import of cytochrome *c* into mitochondria observed *in vitro* was investigated *in vivo* by determining the distribution of apo- and holo-cytochrome *c* in the cytoplasm and mitochondria from normal and mutant strains of yeast.[11] The study was carried out with apo-iso-2-cytochrome *c* because of the instability of apo-iso-1-cytochrome *c* (see next section). Consistent with the results obtained *in vitro*, *cyc3*⁻ yeast strains, lacking heme lyase, accumulate apo-iso-2-cytochrome *c* in the cytoplasm. Cytoplasmic accumulation was also observed with an altered form of iso-2-cytochrome *c* containing two serine residues substituted for the normal cysteine residues that are the sites of heme attachment, even in *CYC3*⁺ strains. A low but significant level of 5% of this altered apo-iso-2-cytochrome *c*, however, was located in mitochondria. The percentage of imported apo-iso-2-cytochrome *c* could be greatly increased by expressing high levels of heme lyase with a multicopy plasmid, suggesting that the apo-iso-2-cytochrome *c* was binding to heme lyase.

Immunoblots probed with antibodies recognizing cytochrome *c* heme lyase demonstrated that this enzyme is inaccessible from the outer mitochondrial surface but is exposed to the intermembrane space, and that it remains tightly associated with the inner mitochondrial membrane upon subfractionation of mitochondria. Thus, although apocytochrome *c* and heme lyase must interact for cytochrome *c* to accumulate inside the outer mitochondrial membrane, heme lyase is found predominantly in a location that would be expected to be inaccessible to apocytochrome *c*. We suggest that apocytochrome *c* is reversibly imported to the intermembrane space as part of the pathway leading to holo-cytochrome *c* formation. Accumulation of apo-cytochrome *c* in mitochondria is apparently dependent on heme lyase, presumably through a trapping mechanism, but does not require heme attachment.

8. APO-CYTOCHROME *c* DEGRADATION AND REGULATION

Because heme attachment is required for efficient mitochondrial import of cytochrome *c*, blockage of heme attachment *in vivo* would be expected to lead to an accumulation of apo-cytochrome *c* in the cytoplasm. In fact, while *cyc3*⁻ mutants lack both holo-iso-1-cytochrome *c* and holo-iso-2-cytochrome *c*, apo-iso-1-cytochrome *c* is also absent in *cyc3*⁻ strains, although apo-iso-2-cytochrome *c* is present at approximately the same level at which holo-iso-2-cytochrome *c* is found in related *CYC3*$^{+}$ strains.[10] The lack of apo-iso-1-cytochrome *c* is not due to a deficiency of either transcription or translation, but to rapid degradation of the protein. Apo-cytochromes *c*, encoded by composite cytochrome *c* genes composed of the central portion of iso-2-cytochrome *c* flanked by amino and carboxyl regions of iso-1-cytochrome *c*, exhibit increased stability compared with apo-iso-1-cytochrome *c*. A region encompassing no more than four amino acid differences between iso-1-cytochrome *c* and iso-2-cytochromes *c* is sufficient to partially stabilize the protein. In contrast to what is observed *in vivo* with the apo forms, the holo forms of the composite iso-cytochromes *c* were even less stable to thermal denaturation than iso-1-cytochrome *c* or iso-2-cytochrome *c*. Apparently certain regions or structures of apo-iso-1-cytochrome *c* are signals that make the protein a substrate for a cytosolic degradation system.[10]

Degradation of apo-iso-1-cytochrome *c* may play a role in regulating levels of the various forms of cytochromes *c* in cells. Blocking heme attachment to apo-cytochrome *c* substantially inhibits import into mitochondria *in vitro* and *in vivo*. If this occurs under certain physiological conditions, the relative stability of apo-iso-2-cytochrome *c* could lead to its accumulation, and eventually to a relative increase in holo-iso-2-cytochrome *c* when heme levels increase. Thus, the differential stability of the apo-iso-cytochromes *c* may be part of a regulatory process that increases the relative amount of iso-2-cytochrome *c* compared with iso-1-cytochrome *c* under certain physiological conditions.

Although holo-iso-1-cytochrome *c* and holo-iso-2-cytochrome *c* are normally present at approximately 95% and 5% relative levels, respectively, in derepressed cells, the absolute amount and relative proportions of the two iso-cytochromes *c* are strongly dependent on the response to heat shock[82] and growth conditions such as catabolite (glucose) repression and the degree of aerations.[13] Under anaerobic conditions, or under aerobic conditions in the presence of high glucose levels, there is generalized repression of the synthesis of a number of mitochondrial and other enzymes, including those involved in the utilization of nonfermentable carbon sources. During the initial phase of catabolite derepression, in cultures that have exhausted fermentable substrates and are just inititating synthesis of these enzymes, holo-iso-2-cytochrome *c* can constitute as much as 25% of the total amount of the iso-cytochromes *c*.[13] Similarly, when anaerobically grown yeast are induced with oxygen, the two iso-cytochromes *c* are produced at different rates, depending on the initial physiological state of the cultures, with holo-iso-2-cytochrome *c* sometimes comprising more than half the total complement of cytochrome *c*. Yeast grown under partially catabolite-repressed or partially anaerobic conditions generally contain high proportions of holo-iso-2-cytochrome *c*. On the other hand, yeast grown under certain conditions of extreme aeration can have less than 1% holo-iso-2-cytochrome *c*. Even though mutants lacking iso-2-cytochrome *c* do not exhibit any obvious defect, and the structure and function of the two cytochromes *c* are very similar, higher levels of holo-iso-2-cytochrome *c* may confer some slight evolutionary advantage for cells that are partially repressed.

Partially repressed yeast may use both differential transcription and differential stability of the apo-proteins as regulatory mechanisms for maintaining elevated proportions of iso-2-cytochrome *c*. Laz *et al.*[83] demonstrated that *CYC7* messenger RNA (mRNA) is present at elevated levels compared with *CYC1* mRNA at the onset of derepression when glucose is exhausted from growth media. Furthermore, *CYC7* mRNA levels do not appear to be regulated by heme[83] in spite of the fact that heme strongly affects *CYC1* mRNA levels[84] and appears to be an important general mediator of cellular responses to growth conditions. The complexity of transcriptional regulation of *CYC1* and *CYC7* is revealed by differences of HAP1 binding sites *in vitro*.[85]

Catabolite repression or anaerobic growth conditions may diminish synthesis of heme lyase or other components required for import, including heme, causing a reduction in the efficiency of import. Apo-iso-2-cytochrome *c*, the more stable of the two apo forms, would then be expected to be enriched in the cytoplasm and would thus constitute a higher fraction of the material that ultimately is imported and converted to holo-cytochrome *c*.

Recently, Pillar and Bradshaw[82] demonstrated that transcription of iso-2-cytochrome *c* was not only induced by the stationary-growth phase, but also by heat shock and low cAMP levels. In contrast, *CYC1* transcription was repressed in the stationary phase and was unaffected by heat shock and low cAMP levels. Furthermore, heat-shock-induced transcription of *CYC7* mRNA occurred both aerobically and anaerobically.

A striking parallel exists between the regulation of *CYC1* and *COX5a* on one hand and *CYC7* and *COX5b* on the other. The *COX5a* and *COX5b* genes, which encode two iso forms of yeast cytochrome *c* oxidase subunit V, respond differently to the degree of aeration and to heme concentration.[86,87] Furthermore, cytochrome *c* oxidase with either subunits Va or Vb have different kinetic properties *in vivo*.[88] It is tempting to speculate that cytochrome *c* oxidase having either subunits Va and Vb interact respectively with iso-1-cytochrome *c* and iso-2-cytochrome *c* to produce kinetic properties that are favorable for the corresponding environmental conditions.

9. STABILIZATION OF HOLO-CYTOCHROME *c*

Individual amino acid residues each contribute to the overall stability of the protein. Evolutionary selection presumably optimizes protein structure with respect to function and stability so that "random" amino acid replacements generally are detrimental or at least neutral. Consistent with this view are numerous replacements in iso-1-cytochrome *c* that resulted in a full range of altered stabilities, from little or no effect to greatly diminished stabilities. Das *et al.*,[89] however, combined genetic selection with oligonucleotide-directed mutagenesis to produce an altered iso-1-cytochrome *c* with an increased stability *in vitro* and *in vivo*. Reversion of two missense mutants resulted in second-site replacements of Asn-57 to Ile-57. Introduction of the Ile-57 replacement in an otherwise normal sequence caused a 17°C increase in the transition temperature T_m of unfolding, corresponding to a greater than twofold increase in the free energy change of thermal unfolding; at 25°C, $\Delta G°$ for the normal iso-1-cytochrome *c* was 5.4 kcal/mole, while that for the Ile-57 mutant protein was 12.5 kcal/mole. Thermodynamic stabilities of the Ile-57 mutant and iso-1-cytochromes *c* with other replacements at position 52 suggested that hydrophobic interactions were the main factor for enhancing stability.[90]

Surprisingly, the Ile-57 iso-1-cytochrome *c* was found to be present at a higher level *in vivo* than the normal form, and appeared to be degraded at a lower rate than the normal protein (D. Pearse and F. Sherman, unpublished results). A yeast homologue of the proteolytic activities uncovered in mammalian mitochondria[91,92] is a likely candidate for the enzyme causing degradation of holocytochrome *c*. It is indeed puzzling why a stabile cytochrome *c* similar to the Ile-57 protein was not naturally selected and maintained by evolutionary pressures. It is tempting to speculate that the wild-type protein is evolutionarily maintained partially labile because of a regulatory requirement based on its degradation rate.

ACKNOWLEDGMENTS. We wish to thank Dr. Peter Rubenstein (University of Iowa) and Dr. Robert E. Cohen (University of California, Los Angeles) for useful comments. This investigation was supported by U.S. Public Health Service Research Grants R01 GM12702 and GM28220 and by the Ministry of Education, Science, and Culture of Japan Grants-in-Aid for International Research Program, Joint Research Grant 63044090.

REFERENCES

1. Guarente, L., 1987, Regulatory proteins in yeast, *Annu. Rev. Genet.* **21:**425–452.
2. Forsburg, S. L., and Guarente, L., 1989, Communication between mitochondria and the nucleus in regulation of cytochrome genes in the yeast *Saccharomyces cerevisiae*, *Annu. Rev. Cell Biol.* **5:** 153–180.
3. Li, W.-Z., and Sherman, F., 1991, Two types of TATA elements for the *CYC1* gene of the yeast *Saccharomyces cerevisiae*, *Mol. Cell. Biol.* **11:**666–676.
4. Russo, P., Li, W.-Z., Hampsey, D. M., Zaret, K. S., and Sherman, F., 1991, Distinct *cis*-acting signals enhance 3′ endpoint formation of *CYC1* mRNA in the yeast *Saccharomyces cerevisiae*, *EMBO J.* **10:**563–571.
5. Sherman, F., and Stewart, J. W., 1982, Mutations altering initiation of translation of yeast iso-1-cytochrome *c*; Contrast between the eukaryotic and prokaryotic process, in: *Molecular Biology of Yeast Saccharomyces: Metabolism and Gene Expression* (J. N. Strathern, E. W. Jones, and J. R. Broach, eds.), pp. 301–333, Cold Spring Harbor Laboratory, Cold Spring Harbor, New York.
6. Clements, J., Laz, T., and Sherman, F., 1989, The role of yeast mRNA sequences and structures in translation, in: *Yeast Genetic Engineering* (P. J. Barr, A. J. Brake, and P. Valenzuela, eds.), pp. 65–82, Butterworth, Boston.

7. Sherman, F., Ono, B., and Stewart, J. W., 1979, Use of the iso-1-cytochrome *c* system for investigating nonsense mutants and suppressor in yeast, in: *Nonsense Mutations and tRNA Suppressors* (J. E. Cellis and J. D. Smith, eds.), pp. 133–153, Academic Press, New York.
8. Moerschell, R. P., Hosokawa, Y., Tsunasawa, S., and Sherman, F., 1990, The specificities of yeast methionine aminopeptidase and acetylation of amino-terminal methionine *in vivo*: Processing of altered iso-1-cytochromes *c* created by oligonucleotide transformation, *J. Biol. Chem.* **265:**19638–19643.
9. Frost, B., and Paik, W. K., 1990, Cytochrome *c* methylation, in: *Protein Methylation* (W. K. Paik and S. Kim, eds.), pp. 60–77, CRC Press, Boca Raton, Florida.
10. Dumont, M. E., Mathews, A. J., Nall, B. T., Baim, S. B., Eustice, D. C., and Sherman, F., 1990, Differential stability of two apo-isocytochromes *c* in the yeast *Saccharomyces cerevisiae*, *J. Biol. Chem.* **265:**2733–2739.
11. Dumont, M. E., Cardillo, T. S., Hayes, M. K., and Sherman, F., 1991, Role of cytochrome *c* heme lyase in mitochondria import and accumulation of cytochrome *c* in yeast, *Mol. Cell. Biol.* **11:** 5487–5496.
12. Dumont, M. E., Ernst, J. F., and Sherman, F., 1987, Identification and sequence of the gene encoding cytochrome *c* heme lyase in the yeast *Saccharomyces cerevisiae*, *EMBO J.* **6:**235–241.
13. Sherman, F., and Stewart, J. W., 1971, Genetics and biosynthesis of cytochrome *c*, *Annu. Rev. Genet.* **5:**257–296.
14. Sherman, F., 1990, Studies of yeast cytochrome *c*: How and why they started and why they continued, *Genetics* **125:**9–12.
15. Sherman, F., Taber, H., and Campbell, W., 1965, Genetic determination of iso-cytochromes *c* in yeast, *J. Mol. Biol.* **13:**21–39.
16. Gollub, E. G., Liu, K., Dayan, J., Adlersberg, M., and Sprinson, D. B., 1977, Yeast mutants deficient in heme biosynthesis and a heme mutant additionally blocked in cyclization of 2,3-oxidosqualene, *J. Biol. Chem.* **252:**2846–2854.
17. Parker, J. H., and Mattoon, J. R., 1969, Mutants of yeast with altered oxidative energy metabolism: Selection and genetic characterization, *J. Bacteriol.* **100:**647–657.
18. Verdière, J., Creusot, F., Guarente, L., and Slonimski, P. P., 1986, The overproducing *CYP1* mutation and the underproducing *hap1* mutations are alleles of the same gene which regulates in *trans* the expression of the structural genes encoding iso-cytochromes *c*, *Curr. Genet.* **10:** 339–342.
19. Pinkham, J. L., and Guarente, L., 1985, Cloning and molecular analysis of the *HAP2* locus: A global regulator of respiratory genes in *Saccharomyces cerevisiae*, *Mol. Cell. Biol.* **5:**3410–3416.
20. Sherman, F., Stewart, J. W., Margoliash, E., Parker, J., and Campbell, W., 1966, The structural gene for yeast cytochrome *c*, *Proc. Natl. Acad. Sci. USA* **55:**1498–1504.
21. Downie, J. A., Stewart, J. W., Brockman, N., Schweingruber, A. M., and Sherman, F., 1977, Structural gene for yeast iso-2-cytochrome *c*, *J. Mol. Biol.* **113:**369–384.
22. Dumont, M. E., Schlichter, J. B., Cardillo, T. S., Hayes, M. K., Bethlendy, G., and Sherman, F., 1993, *CYC2* encodes a factor involved in mitochondrial import of yeast cytochrome *c*, *Mol. Cell. Biol.*, in press.
23. Hampsey, D. M., Das, G., and Sherman, F., 1988, Yeast iso-1-cytochrome *c*: Genetic analysis of structure–function relationships, *FEBS Lett.* **231:**275–283.
24. Moerschell, R. P., Tsunasawa, S., and Sherman, F., 1988, Transformation of yeast with synthetic oligonucleotides, *Proc. Natl. Acad. Sci. USA* **85:**524–528.
25. Moerschell, R. P., Das, G., and Sherman, F., 1991, Transformation of yeast directly with synthetic oligonucleotides, *Meth. Enzymol.* **194:**362–369.
26. Driessen, H. P. C., deJong, W. W. Tesser, G. I., and Bloemendal, H., 1985, The mechanism of N-terminal acetylation of proteins, *CRC Crit. Rev. Biochem.* **18:**281–325.
27. Kendall, R. L., Yamada, R., and Bradshaw, R. A., 1990, Cotranslational amino-terminal process, *Meth. Enzymol.* **185:**398–407.

28. Gordon, J. I., Duronio, R. J., Rudnick, D. A., Adams, S. P., and Gokel, G. W., 1991, Protein N-myristoylation, *J. Biol. Chem.* **266:**8647–8650.
29. Ben-Bassat, A., Bauer, K., Chang, S. Y., Myambo, K., Boosman, A., and Chang, S., 1987, Processing of the initiation methionine from proteins: Properties of the *Escherichia coli* methionine aminopeptidase and its gene structure, *J. Bacteriol.* **169:**751–757.
30. Chang, Y.-H., Teichert, U., and Smith, J. A., 1990, Purification and characterization of a methionine aminopeptidase from *Saccharomyces cerevisiae*, *J. Biol. Chem.* **265:**19892–19897.

30a. Chang, Y.-H., Teichert, U., and Smith, J. A., 1992, Molecular cloning, sequencing, deletion, and overexpression of a methionine aminopeptidase gene from *Saccharomyces cerevisiae*, *J. Biol. Chem.* **267**:8007–8011.

31. Tsunasawa, S., Stewart, J. W., and Sherman, F., 1985, Amino-terminal processing of mutant forms of yeast iso-1-cytochrome *c*: The specificities of methionine aminopeptides and acetyltransferase, *J. Biol. Chem.* **260:**5382–5391.
32. Sherman, F., Stewart, J. W., and Tsunasawa, S., 1985, Methionine or not methionine at the beginning or a protein, *BioEssays* **3:**27–31.
33. Miller, C. G., Strauch, K. L., Kukral, A. M., Miller, J. L., Wingfield, P. T., Mazzei, G. J., Werlen, R. C., Graber, P., and Movva, N. R., 1987, N-terminal methionine-specific peptidase in *Salmonella typhimurium*, *Proc. Natl. Acad. Sci. USA* **84:**2718–2722.
34. Wingfield, P., Graber, P., Turcatti, G., Movva, N. R., Pelletier, M., Craig, S., Rose, K., and Miller, C. G., 1989, Purification and characterization of a methionine-specific aminopeptidase from *Salmonella typhimurium*, *Eur. J. Biochem.* **180:**23–32.
35. Hirel, H.-P., Schmitter, J.-M., Dessen, P., Fayat, G., and Blanquet, S., 1989, Extent of N-terminal methionine excision from *Escherichia coli* proteins is governed by the side-chain length of the penultimate amino acid, *Proc. Natl. Acad. Sci. USA* **86:**8247–8251.
36. Dalbøge, H., Bayen, S., and Pedersen, J., 1990, *In vivo* processing of N-terminal methionine in *E. coli*, *FEBS* **266:**1–3.
37. Huang, S., Elliott, R. C., Liu, P. S., Koduri, R. K., Weickmann, J. L., Lee, J. H., Blair, L. C., Gosh-Dastidar, P., Bradshaw, R. A., Bryan, K. M., Einarson, B., Kendall, R. L., Kolacs, K. H., and Saito, K., 1987, Specificity of cotranslational amino-terminal processing of proteins in yeast, *Biochemistry* **26:**8242–8246.
38. Boissel, J. P., Kasper, T. J., and Bunn, H. F., 1988, Cotranslational amino-terminal processing of cytosolic proteins: Cell-free expression of site-directed mutants of human hemoglobin, *J. Biol. Chem.* **263:**8443–8449.
39. Yamada, T., Kato, K., Kawahara, K., and Nishimura, O., 1986, Separation of recombinant human interleukin-2 and methionyl interleukin-2 produced in *Escherichia coli*, *Biochem. Biophy. Res. Commun.* **135:**837–843.
40. Devlin, P. E., Drummond, R. J., Toy, P., Mark D. F., Watt, K. W., and Devlin, J. J., 1988, Alteration of amino-terminal codons of human granulocyte-colony-stimulating factor increases expression levels and allow efficient processing by methionine aminopeptidase in *Escherichia coli*, *Gene* **65:** 13–22.
41. Prchal, J. T., Cashman, D. P., and Kan, Y. W., 1986, Hemoglobin Long Island is caused by a single mutation (adenine to cytosine) resulting in a failure to cleave amino-terminal methionine, *Proc. Natl. Acad. Sci. USA* **83:**24–27.
42. Barwick, R. C., Jones, R. T., Head, C. G., Shih, M. F.-C., Prchal, J. T., and Shih, D. T.-B., 1985, Hb Long Island: A hemoglobin variant with a methionyl extension at the NH_2 terminus and a prolyl substitution for the normal histidyl residue 2 of the β chain, *Proc. Natl. Acad. Sci. USA* **82:** 4602–4605.
43. Ciechanover, A., and Schwartz, A. L., 1989, How are substrates recognized by the ubiquitin-mediated proteolytic system? *TIBS* **14:**483–488.
44. Varshavsky, A., 1991, Naming a target signal, *Cell* **64:**13–15.
45. Gonda, D. K., Bachmair, A., Wünning, J. W., Tobias, W. S., and Varshavsky, A., 1989, Universality and structure of the N-end rule, *J. Biol. Chem.* **264:**16700–16712.

45a. Tobias, J. W., Shrader, T. E., Rocap, G., and Varshavsky, A., 1991, The N-end rule in bacteria, *Science* **254**:1374–1377.
46. Bachmair, A., and Varshavsky, A., 1989, The degradation signal in a short-lived protein, *Cell* **56:** 1019–1032.
47. Dunten, R. L., Cohen, R. E., Gregori, L., and Chau, V., 1991, Specific disulfide cleavage is required for ubiquitin conjugation and degradation of lysozyme, *J. Biol. Chem.* **266:**3260–3267.
48. Sheff, D. R., and Rubenstein, P. A., 1989, Identification of N-acetylmethionine as the product released during the NH2-terminal processing of a pseudo-class I actin, *J. Biol. Chem.* **264:**11491–11496.
49. Cook, R. K., Sheff, D. R., and Rubenstein, P. A., 1991, Unusual metabolism of the yeast actin amino terminus, *J. Biol. Chem.* **266:**16825–16833.
50. Symth, D. G., and Zakarian, S., 1980, Selective processing of β-endorphin in regions of porcine pituitary, *Nature* **288:**613–615.
51. Glembotski, C. C., 1982, Acetylation of α-melanotropin and β-endorphin in the rat intermediate pituitary, *J. Biol. Chem.* **257:**10493–10500.
52. Glembotski, C. C., 1982, Characterization of the peptide acetyltransferase activity in bovine and rat intermediate pituitaries responsible for the acetylation of β-endorphin and α-melanotropin, *J. Biol. Chem.* **257:**10501–10509.
53. Chappell, M. C., O'Donohue, T. L., Millington, W. R., and Kempner, E. S., 1986, The size of enzymes acetylating α-melanocyte-stimulating hormone and β-endorphin, *J. Biol. Chem.* **261:** 1088–1091.
54. Lee, F.-J. S., Lin, L.-W., and Smith, J. A., 1988, Purification and characterization of an N^{α}-acetyltransferase from *Saccharomyces cerevisiae*, *J. Biol. Chem.* **263:**14948–14955.
55. Yamada, R., and Bradshaw, R. A., 1991, Rat liver polysome N^{α}-acetyltransferase: Isolation and characterization, *Biochemistry* **30:**1010–1016.
56. Yamada, R., and Bradshaw, R. A., 1991, Rat liver polysome N^{α}-acetyltransferase: Substrate specificity, *Biochemistry* **30:**1017–1021.
57. Mullen, J. R., Kayne, P. S., Moerschell, R. P., Tsunasawa, S., Gribskov, M., Colavito-Shepanski, M., Grunstein, M., Sherman, F., and Sternglanz, R., 1989, Identification and characterization genes and mutants for an N-terminal acetyltransferase from yeast, *EMBO J.* **8:**2067–2075.
58. Lee, F.-J. S., Lin, L.-W., and Smith, J. A., 1990, Identification of methionine N^{α}-acetyltransferase from *Saccharomyces cerevisiae*, *J. Biol. Chem.* **265:**3603–3606.
59. Lee, F.-J. S., Lin, L.-W., and Smith, J. A., 1989 N^{α} acetylation is required for normal growth and mating of *Saccharomyces cerevisiae*, *J. Bacteriol.* **171:**5795–5802.
60. Lee, F.-J. S., Lin, L.-W., and Smith, J. A., 1989, Molecular cloning and sequencing of a cDNA encoding N^{α}-acetyltransferase from *Saccharomyces cerevisiae*, *J. Biol. Chem.* **264:**12339–12343.
61. Lee, F.-J. S., Lin, L.-W., and Smith, J. A., 1989, N^{α}-acetyltransferase deficiency alters protein synthesis in *Saccharomyces cerevisiae*, *FEBS Lett.* **256:**139–142.
62. Clements, J. M., O'Connell, L. I., Tsunasawa, S., and Sherman, F., 1989, Expression and activity of a gene encoding rat cytochrome *c* in the yeast *Saccharomyces cerevisiae*, *Gene* **83:**1–14.
63. Takakura, H., Tsunasawa, S., Miyagi, M., and Warner, J. R., 1992, N-terminal acetylation of ribosomal proteins of *Saccharomyces cerevisiae*, *J. Biol. Chem.* **267:**5442–5445.
64. Siddig, M. A. M., Kinsey, J. A., Fincham, J. R. S., and Keighren, M., 1980, Frameshift mutations affecting the N-terminal sequence of *Neurospora* NADP-specific glutamate dehydragenase, *J. Mol. Biol.* **137:**125–135.
65. Cumberlidge, A. G., and Isono, K., 1979, Ribosomal protein modification in *Escherichia coli*. I. A mutant lacking the N-terminal acetylation of proteins S5 exhibits thermosensitivity, *J. Mol. Biol.* **131:**169–189.
66. Johnson, G. G., Kronert, W. A., Bernstein, S. I., Chapman, V. M., and Smith, K. D., 1988, Altered turnover of allelic variants of hypoxanthine phosphoribosyltransferase is associated with N-terminal amino acid sequence variation, *J. Biol. Chem.* **263:**9079–9082.

67. Hershko, A., Heller, H., Eytan, E., Kaklij, G., and Rose, I. A., 1984, Role of the α-amino group of protein in ubiquitin-mediated protein breakdown, *Proc. Natl. Acad. Sci. USA* **81:**7051–7025.
68. Matsuura, S., Arpin, M., Hannum, C., Margoliash, E., Sabatini, D. D., and Morimoto, T., 1981, *In vitro* synthesis and posttranslational uptake of cytochrome *c* into isolated mitochondria: Role of a specific addressing signal in the apocytochrome, *Proc. Natl. Acad. Sci. USA* **78:**4368–4372.
69. Sokolik, C. W., and Cohen, R. E., 1992, Ubiquitin conjugation to cytochrome *c*. Structure of the yeast iso-1 conjugate and possible recognition determinants, *J. Biol. Chem.* **267:**1067–1071.
70. Mayer, A., Siegel, N. R., Schwartz, A. L., and Ciechanover, A., 1989, Degradation of proteins with acetylated amino termini by the ubiquitin system, *Science* **244:**1480–1483.
71. Symth, D. G., Massey, D. E., Zakarian, S., and Finnie, M. D. A., 1979, Endorphins are stored in biologically active and inactive forms: Isolation of α-N-acetyl peptides, *Nature* **279:**252–254.
72. Berger, E. M., Cox, G. Weber, L., and Kenney, J. L., 1981, Actin acetylation in *Drosophia* tissue culture cells, *Biochem. Genet.* **19:**321–331.
73. Moo-Penn, W. F., Bechter, K. C., Schmidt, R. M., Johnson, M. H., Jue, D. L., Schmidt, D. E., Jr., Dunlap, W. M., Opella, S. J., Bonaventura, J., and Bonaventura, C., 1977, Hemoglobin Raleigh (β_1 Valine → Acetylalanine). Structural and functional characterization, *Biochemistry* **16:**4872–4879.
74. Stone, E. M., Swanson, M. J., Romeo, A. M., Hicks, J. B., and Sternglanz, R., 1991, The *SIR1* gene of *Saccharomyces cerevisiae* and its role as an extragenic suppressor of several mating-defective mutants, *Mol. Cell. Biol.* **11:**2253–2262.
75. Zimmerman, R., Hennig, B., and Neupert, W., 1981, Different transport pathways of individual precursor proteins into mitochondria, *Eur. J. Biochem.* **116:**455–460.
76. Hartl, F.-U., Pfanner, N., Nicholson, D. W., and Neupert, W., 1989, Mitochondrial protein import, *Biochim Biophys. Acta* **998:**1–45.
77. Sherman, F., and Stewart, J. W., 1973, Mutations at the end of the iso-1-cytochrome *c* gene of yeast, in: *The Biochemistry of Gene Expression in Higher Organisms* (J. K. Pollak and J. W. Lee, eds.), pp. 56–86, Australian and New Zealand Book Co., Sydney.
78. Dumont, M., Ernst, J. F., Hampsey, D. M., and Sherman, F., 1988, Coupling of heme attachment to import of cytochrome *c* into yeast mitochondria: Studies with heme lyase deficient mitochondria and altered apocytochromes *c*, *J. Biol. Chem.* **263:**15928–15937.
79. Drygas, M. E., Lambowitz, A. M., and Nargang, F. E., 1989, Cloning and analysis of the *Neurospora* gene for cytochrome *c* heme lyase, *J. Biol. Chem.* **264:**17897–17906.
80. Hennig, B., and Neupert, W., 1981, Assembly of cytochrome *c*. Apocytochrome *c* is bound to specific sites on mitochondria before its conversion to holocytochrome *c*, *Eur. J. Biochem.* **121:** 302–312.
81. Nargang, F. E., Drygas, M. E., Kwong, P. L., Nicholson, D. W., and Neupert, W., 1988, A mutant of *Neurospora crassa* deficient in cytochrome *c* heme lyase activity can not import cytochrome *c* into mitochondria, *J. Biol. Chem.* **263:**9388–9394.
82. Pillar, T. M., and Bradshaw, R. E., 1991, Heat Shock and stationary phase induce transcription of the *Saccharomyces cerevisiae* iso-2 cytochrome *c* gene, *Curr. Genet.* **20:**185–188.
83. Laz, T. M., Pietras, D. F., and Sherman, F., 1984, Differential regulation of the duplicated iso-cytochrome *c* genes in yeast, *Proc. Natl. Acad. Sci. USA* **81:**4475–4479.
84. Guarente, L., and Mason, T., 1983, Heme regulates transcription of the *CYC1* gene of *S. cerevisiae* via an upstream activation site, *Cell* **32:**1279–1286.
85. Kim, K. S., Pfeifer, K., Powell, L., and Guarente, L., 1990, Internal deletions in the yeast transcription activator HAP1 have opposite effects at two sequence elements, *Proc. Natl. Acad. Sci. USA* **87:**4524–4528.
86. Poyton, R. O., Trueblood, C. E., Wright, R. W., and Farrell, L. E., 1988, Expression and function cytochrome *c* oxidase subunit isologues, *Ann. N. Y. Acad. Sci.* **55:**289–307.
87. Hodge, M. R., Kim, G., Singh, K., and Cumsky, M. G., 1989, Inverse relationship of the yeast *COX5A* genes by oxygen and heme, *Mol. Cell. Biol.* **5:**1958–1964.

88. Waterland, R. A., Basu, A., Chance, B., and Poyton, R. O., 1991, The isoforms of yeast cytochrome *c* oxidase subunit V alter the *in vivo* kinetic properties of the holoenzyme, *J. Biol. Chem.* **266:**4180–4186.
89. Das, G., Hickey, D. R., McLendon, D., McLendon, G., and Sherman, F., 1989, Dramatic thermostabilization of yeast iso-1-cytochrome *c* by an Asn57 → Ile57 replacement, *Proc. Natl. Acad. Sci. USA* **86:**496–499.
90. Hickey, D. R., Berghuis, A. M., Lafond, G., Jaeger, J. A., Cardillo, T. S., McLendon, D., McLendon, Das, G., Sherman, F., Brayer, G. D., and McLendon, G., 1991, Enhanced thermodynamic stabilities of yeast iso-1-cytochromes *c* with amino acid replacements at positions 52 and 102, *J. Biol. Chem.* **226:**11686–11694.
91. Duque-Magalhães, M. C., and Ferreira, M. M. M., 1980, Cytochrome *c* degrading activity in rat liver mitochondria, *Biochem. Biophys. Res. Commun.* **93:**106–112.
92. Desautels, M., and Goldberg, A. L., 1982, Demonstration of an ATP-dependent, vanadate-sensitive endoprotease in the matrix of rat liver mitochondria, *J. Biol. Chem.* **257:**11673–11679.

Chapter 7

eIF-4E Phosphorylation and the Regulation of Protein Synthesis

Robert M. Frederickson and Nahum Sonenberg

1. INTRODUCTION

Increasing evidence has accumulated over the past few years indicating an important role for translational control in the regulation of gene expression and cell growth in eukaryotic cells. An overall increase in the rate of protein synthesis is a fundamental component of the response of eukaryotic cells to growth stimulatory agents.[1,2] Furthermore, there exists a set of growth-regulated proteins whose translation is selectively stimulated in response to mitogenic agents.[3] Increased translation is necessary for both entry into and transit through the cell cycle,[4,5] and during mitosis the cap-dependent translation of eukaryotic cellular mRNAs is largely inhibited.[6,7] The recent striking finding that overexpression of the mRNA 5′ cap-binding protein, eukaryotic translation initiation factor-4E (eIF-4E), results in malignant transformation of fibroblast cells,[8,9] implicates this factor as a limiting component in growth control pathways, and provides a basis for understanding the role translation plays in cellular proliferation. Indeed, this result suggests the existence of a set of translationally attenuated growth-regulatory

Robert M. Frederickson and Nahum Sonenberg • Department of Biochemistry and McGill Cancer Centre, McGill University, Montréal, Québec, Canada H3G 1Y6.

Translational Regulation of Gene Expression 2, edited by Joseph Ilan. Plenum Press, New York, 1993.

messenger RNAs (mRNAs), whose synthesis is specifically enhanced relative to the bulk of cellular proteins upon eIF-4E overexpression. The specific stimulation of the expression of these important proteins may then feed into and further activate signaling pathways controlling cellular proliferation.

Several viruses have evolved mechanisms to commandeer the cellular protein synthetic machinery to translate selectively their own viral mRNAs, to the detriment of host protein synthesis.[10] One well-studied example is the cap-independent mechanism of poliovirus mRNA translation.[11] Recently, it has become clear that this process has its origins in a cellular cap-independent mechanism of protein synthesis initiation. Macejak and Sarnow have shown that the immunoglobin heavy-chain binding protein, BiP, can be translated by an internal initiation mechanism.[12] This cap-independent system of cellular translation may be analogous to constitutively active transcription promoters, often associated with "housekeeping" genes, to allow certain messages to remain refractory to cell-cycle (mitotic)[7] or stress-induced[13] signals effected at the level of translation.

2. REGULATION OF RATES OF PROTEIN SYNTHESIS

2.1. Introduction

Many variables contribute to the regulation of general translation rates in eukaryotic cells. Two important determinants are the abundance and activity of the very large number of translational components, such as ribosomal proteins and translation factors. In eukaryotic tissues, long-term adjustments in the rate of protein synthesis under different nutritional conditions may be effected by changes in the cellular ribosome content.[14] Indeed, it has been suggested that modulation of ribosomal protein abundance may play a role in either the development or progression of certain malignancies.[15–18] Changes in the pool size of translation initiation factors also occur, such as the two- to threefold increase in eIF-2α, eIF-2β, and eIF-4E proteins and an eight- to ten-fold increase in eIF-4A proteins upon mitogenic stimulation of T cells.[19–20] These rapid increases in the pool size of limiting translation factors such as eIF-4E and eIF-2α are modest, roughly parallel to those of both cellular mRNA and ribosome content, and cannot fully account for the enhanced rate of translation per ribosome. Other rapid changes in protein synthesis, such as the four- to sixfold reduction apparent upon serum deprivation of tissue culture cells, can equally not be explained by the only minor changes (10–40% decrease) in levels of both translation factors and ribosomes, which are coordinate with similar decreases in total cellular protein.[21] The pool size and activity of other translation factors, such as those controlling the elongation step, may also contribute to adjustments in translation rates.[22]

2.2. Pnosphorylation Regulates Initiation Factor Activity

Rapid changes in protein synthesis are largely due to increased or decreased activity of the initiation process.[14] These effects arise from changes in the phosphorylation state, and thus activity of the various initiation factors.[23,24] Reversible serine/threonine phosphorylation has long been known to play a fundamental role in the regulation of a wide variety of cellular processes,[25] and is likely the primary means of adjusting initiation factor activity to effect changes in translation rates. Phosphorylation of the α subunit of eIF-2, for example, is an important means of downregulating translation rates. Phosphorylation of this initiation factor by an eIF-2 kinase[26] traps the guanine exchange factor (GEF) in an inactive eIF-2–GDP complex. This effectively depletes the cell of the limiting amounts of GEF (responsible for eIF-2 recycling, through exchange of GTP for GDP) and leads to a global downregulation of translation through inhibition of ternary complex formation (see Fig. 1). This occurs both upon hemin deprivation in reticulocyte lysate and as part of the host antiviral response. More recently, a large body of evidence has accumulated associating phosphorylation of translation initiation factor eIF-4E with a rapid stimulation of the overall translation rate, and specifically enhanced synthesis of growth-related proteins.[24,27]

Mitogen- and growth-factor-induced signals are initiated by binding of factors to specific cell surface receptors.[28] These receptors exhibit, are associated with, or activate downstream tyrosine kinase activity, and tyrosine phosphorylation is implicated as an important initial event in signal transduction. Subsequent phosphorylation of other proteins on serines, threonines, and tyrosines is believed to further transmit signals to key effector molecules.[29,30] The identity of the relevant downstream targets remains to be determined, however. Since an overall enhancement in the rate of protein synthesis is necessary for entry into and transit through the cell cycle,[4,5] some of these key effector molecules of the mitogenic response must include those that determine translation rates. Results from the authors' and other labs have demonstrated that the mRNA 5′ cap-binding protein eIF-4E is rapidly phosphorylated upon exposure of cultured cells to a wide array of mitogens and growth factors.[24] This is significant insofar as eIF-4E is believed to be the limiting factor in the binding of mRNA to ribosomes,[31] which itself is the rate-limiting step of protein synthesis in normal cycling cells.[14] Moreover, eIF-4E which has had the major site of serine phosphorylation mutated to an alanyl residue does not demonstrate the mitogenic and oncogenic potential of wild-type eIF-4E when overexpressed in or injected into cultured cells.[8,9,32] At this point, it is pertinent to note that changes in eIF-4E phosphorylation occur in conjunction with similar modifications of other initiation factors and translational components, such as eIF-4B, the p220 subunit of eIF-4F, and the 40S ribosomal subunit protein S6,[23,24] suggesting that eIF-4E phosphorylation may not be the sole determinant for regulation of translation rates in response to growth-promoting agents (see

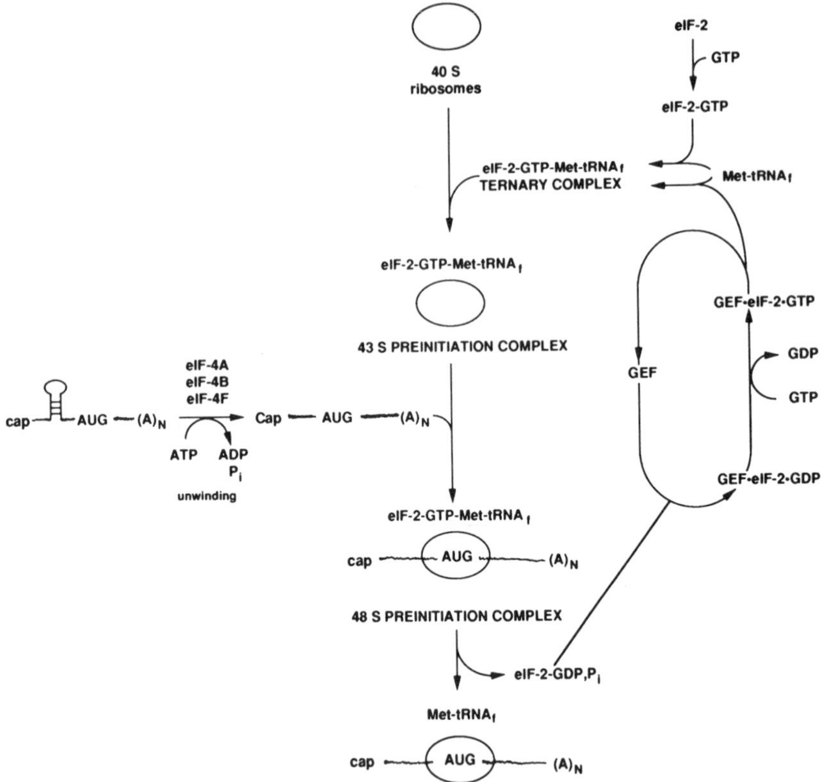

Figure 1. Formation of mRNA ribosome complexes. The model depicts only a small number of initiation factors which participate in initiation complex formation. RNA binding occurs to 43S preinitiation complexes that consist of the 40S small ribosomal subunit in association with the ternary complex (eIF-2–GTP–Met-tRNA) and two other initiation factors, eIF-3 and eIF-4C (not shown in the figure). Three initiation factors participate in mRNA binding to the 43S preinitiation complex: eIF-4A, eIF-4B, and eIF-4F. Their function is described in refs. 34 and 35. In addition, ATP is required as a cofactor of the generation of energy for the unwinding of mRNA secondary structure. For more details the reader is referred to a recent review by Merrick[61]; the figure is reproduced from a recent review.[24]

Section 7). Nevertheless, current evidence points to a central role for this translation factor in both the global and specific changes in translation rates that occur upon modulation of cell growth. The scope of this report is to survey recent advances in our understanding of the role of eIF-4E and its phosphorylation in the control of cell growth. More detailed information on the biochemistry of the cap-binding proteins and the process of translation initiation can be found in earlier reviews from this,[33,34] and other[35] laboratories.

3. eIF-4E IS A PHOSPHOPROTEIN

3.1. Introduction

The possibility of posttranslational modification of eIF-4E was suggested early on with the finding that the factor could be resolved into at least two isoelectric variants. Sonenberg *et al.*[36] reported two isoelectric variants of uncomplexed eIF-4E in rabbit reticulocytes, and additional minor forms have been detected in human erythrocytes[37] and HeLa cells.[31,38] Rychlik *et al.*[37] demonstrated two major isoelectric variants of eIF-4E in reticulocytes; the pIs were determined to be roughly 5.9 and 6.3, and the pI 5.9 form was shown to be phosphorylated *in vivo*. The phosphorylated residue has been identified as serine, and in a subsequent study the phosphorylated residue was reported to be serine codon 53.[39] The existence of two major isoelectric variants has more recently been confirmed by immunoblotting of crude cell lysates from a variety of sources fractionated by two-dimensional isoelectric focusing (IEF)–polyacrylamide gel electrophoresis.[40–43] The existence of much less abundant isoelectric variants, present below the detection limit of the immunological assay, cannot be formally ruled out. The detection of other minor phosphorylated forms by IEF indicates the possibility of other phosphorylations on eIF-4E.[44] eIF-4E synthesized in a reticulocyte lysate becomes phosphorylated, resulting in both major and minor forms separable by IEF. Surprisingly, *in vitro* synthesis of a mutant eIF-4E protein containing an alanine substituted for serine-53 also becomes phosphorylated in reticulocyte lysate, albeit with different kinetics than the wild-type protein.[44] As serine-53 is not present in this protein, the phosphorylation must occur at a second site, the identity of which remains undetermined.

3.2. eIF-4E Is a Component of the Mitogen-Activated Cellular Phosphorylation Cascade

Notwithstanding the existence of a putative second site of phosphorylation, the evidence is quite convincing that eIF-4E activity is tightly correlated with phosphorylation at serine-53. Several studies with cells metabolically labeled with ^{32}P-orthophosphate have shown that the rate of eIF-4E phosphorylation increases in response to a wide variety of growth factors, mitogenic agents, and phosphatase inhibitors.[41,42,45–52] A compilation of the agents which can induce eIF-4E phosphorylation, along with the corresponding reference, is shown in Table I. In all cases tested so far, this increase occurred at the single primary site of *in vivo* phosphorylation, according to tryptic peptide analysis of the phosphoprotein. Agents which induce eIF-4E phosphorylation include various purified peptide growth factors, such as EGF, PDGF, and TNF-α, and a differentiation-promoting factor, NGF. As stated earlier, these factors exert their effects on cell growth and differentiation through binding and activation of specific cell surface receptor

Table I. Effectors Promoting eIF-4E Phosphorylation[a]

Effector	Cells	Ref.
Serum	NIH 3T3	41
	Swiss 3T3	45
Epidermal growth factor	Human mammary epithelial (184A1N4)	46
Insulin	3T3-L1	47
Nerve growth factor	PC12	42
Platelet-derived growth factor	NIH 3T3	41
	WI-38 human lung fibroblasts	48
Tumor necrosis factor-α	HeLa	49
	Bovine aortic endothelial cells	49
	ME-180	49
	Human foreskin fibroblasts	49
	U937	49
Lipopolysaccharide	B lymphocytes	50
Okadaic acid	Human mammary epithelial (184A1N4)	46
Phorbol ester	Reticulocytes	47
	3T3-L1	47
	NIH 3T3	41
pp60$^{\text{v-}src}$	NIH 3T3	41
	Rat 3Y1	41
p21$^{\text{Ha-v-}ras}$	Rat 1	52

[a]Adapted from ref. 22.

phosphotyrosine kinases (PTKs). Enhanced phosphorylation and concomitant activation of downstream serine/threonine/tyrosine protein kinases are thought to further transmit both mitogenic and differentiation-inducing signals to relevant effector molecules. Thus, oncogenic variants of components of these signaling pathways would also be predicted to maintain elevated rates of eIF-4E phosphorylation in the absence of serum or growth factors. Indeed, expression of pp60$^{\text{v-}src}$, or a transforming variant of cellular pp60^{src}, in fibroblasts results in a marked increase in the rate of phosphorylation eIF-4E in serum-deprived cells.[41] This effect is not restricted to Src, as expression of other tyrosine kinase oncoproteins, such as Lck or Fps, has a similar effect on 4E phosphorylation,[8,41] indicating that eIF-4E phosphorylation is an important downstream outcome of tyrosine kinase activation. Moreover, these results predict that activated variants of other proto-oncogene members of the cellular signaling cascade, such as Ras (see Section 6), as well as certain second messengers would promote eIF-4E phosphorylation.

Response to second messenger analogs varies. Phorbol esters, which bind to and activate various protein kinase C (PKC) subtypes[53] induce eIF-4E phosphorylation in several cell lines.[41,42,47,50,51] This is consistent with various reports demonstrating that out of a variety of purified kinases tested, only PKC phosphorylates eIF-4E *in vitro* with high efficiency.[54–57] The relevance of these data for eIF-4E phosphorylation by PKC in intact cells is difficult to assess. Coinjection of eIF-4E and PKC potentiates the mitogenic activity of the former in quiescent 3T3 cells.[56] There is also evidence, however, for PKC-independent eIF-4E phosphorylation: while downregulation of PKC through long-term exposure of cells to phorbol ester has revealed that the insulin-induced phosphorylation of eIF-4E in 3T3-L1 cells is highly dependent upon activation of this kinase,[47] the NGF-induced response in PC12 cells is unaffected by PKC down-regulation.[42] The latter result is consistent with the fact that neuritogenesis and differentiation of PC12 cells by NGF is also unaffected by downregulation of PKC by PMA.[58] A similar result was found for B lymphocytes stimulated by lipopolysaccharide (LPS), where the eIF-4E phosphorylation rate was shown to be insensitive to H7 and HA1004, two inhibitors of PKC and cyclic nucleotide-dependent kinases.[50] Thus, it seems likely that other kinases exist *in vivo* which mediate eIF-4E phosphorylation depending on the cell and the effector in question. Another second messenger analog, dibutyryl cyclic AMP (dbcAMP), an activator of protein kinase A[30] (cyclic nucleotide-dependent protein kinase, PKA) family members, does not induce eIF-4E phosphorylation in 3T3 or PC12 cells (R. M. Frederickson and N. Sonenberg, unpublished results); moreover, purified PKA does not phosphorylate the p25 subunit of eIF-4F *in vitro*.[54]

There is also a net increase in the steady-state levels of the phosphorylated variant of eIF-4E in certain systems. Two-dimensional IEF analyses of cell lysates[40–43] or purified eIF-4E[50,55] from a variety of sources and under a variety of growth conditions has revealed that unphosphorylated and monophosphorylated eIF-4E exist at roughly a 1:1 ratio in normal cycling cells. This ratio is altered, however, upon stimulation of cell growth or differentiation in favor of the phosphorylated variant. Upon treatment of PC12 cells with NGF,[42] LPS activation of B cells,[50] or expression of v-*src* in 3T3 fibroblasts,[41] the ratio shifts from equivalent levels of the two isoelectric variants in unstimulated cells to roughly 3:1 in favor of the phosphorylated variant. Such a shift represents only a 30% increase in the level of phosphorylated eIF-4E and obviously cannot account for the large increases in phosphorylation determined from the labelling studies.

Where there is phosphorylation there is dephosphorylation, however, and the regulation of eIF-4E phosphorylation clearly involves modulation of both kinases and phosphatases. Treatment of cells in culture with okadaic acid, an inhibitor of protein phosphatases pp1 and pp2A, further enhances EGF-induced phosphorylation of eIF-4E.[46] Thus, eIF-4E protein phosphatase activities may contribute to the regulation of eIF-4E phosphorylation *in vivo* by EGF. The action of both regulated kinases and phosphatases on eIF-4E leads to alterations in the rate of turnover of the phosphate moiety in stimulated cells. Rychlik *et al.*[50] have shown an increased rate of phosphorylation/dephosphorylation in B cells stimulated to proliferate with LPS or phorbol ester, as the change in the ratio of the two forms as measured by IEF analysis cannot fully account for the enhanced labeling (up to 50-fold). There are indications for changes in phosphate turnover rates in NGF-stimulated PC12 cells,[42] Src-transformed 3T3 cells,[41] and Ras-transformed cloned rat embryo fibroblasts (CREFs).[52] Such an increased turnover of the phosphate moiety could be explained by a model wherein the phosphate was consumed upon completion of or at some point during each initiation cycle. Such an increased turnover could equally merely reflect increases in activity of both eIF-4E kinase and phosphatase activities in stimulated cells. Elucidation of this question requires a better understanding of how phosphorylation affects eIF-4E activity.

4. PHOSPHORYLATION AND eIF-4E ACTIVITY

4.1. Introduction

The enhancement of eIF-4E phosphorylation by such a wide variety of growth stimulatory agents and the dramatic results of its overexpression in fibroblast cells indicate a key role for this translation factor in the mitogenic response and the concomitant increased efficiency of the initiation of protein synthesis. The role that eIF-4E phosphorylation plays in modulating its activity is still elusive. As has been shown with ribosomal protein S6, phosphorylation of eIF-4E has been shown to enhance its activity upon addition to an *in vitro* translation system.[57] These findings still provide little insight as to how phosphorylated eIF-4E is able to stimulate translation.

The most readily measured and well-defined activity of eIF-4E is its binding to the 5′ cap structure (m^7GpppX, where X is any nucleotide) present on all eukaryotic cellular (except organellar) mRNAs (reviewed in refs. 34 and 35). Both phosphorylated and unphosphorylated eIF-4E isoforms bind to the cap structure. Moreover, so does recombinant protein containing a substitution of alanine for serine-53.[44] There has been no detailed comparison, however, of the dissociation constants of cap binding to unphosphorylated, monophosphorylated, or mutant eIF-4E proteins. A small change in the binding constant of eIF-4E to mRNA may have a significant effect on the initiation of translation, especially if it is accompanied by similar changes in the affinities of other translation factors for mRNA and/or eIF-4E.

Whereas eIF-4E is thought to be the first protein with which the mRNA comes into contact in the initiation process, mRNA binding to ribosomes and entry into the initiation cycle is mediated by a considerable number of translation initiation factors, and is a complex regulated process[34,35,59,60] (see Fig. 1 for a schematic representation). Subtle changes in the strengths of any of these interactions could have a significant effect on the formation of higher order complexes. In light of this, it is interesting that Joshe-Barve *et al.* have shown that replacement of the major site of phosphorylation with an alanyl residue prevents association of eIF-4E with the 48S initiation complexes.[44] In contrast, eIF-4E^{ser} isolated from these complexes was found to be highly phosphorylated when analyzed by tube gel IEF. This finding is consistent with a role for eIF-4E phosphorylation in promoting any of the various interactions leading to the entry of an mRNA into the 48S complex. One possibility is eIF-4F complex formation itself. It is not known whether eIF-4E first binds to the cap structure *in vivo* as a monomer, as part of the trimeric eIF-4F complex, or in the context of an eIF-4E–p220 dimer.[61] eIF-4F is a heterotrimeric complex consisting of (1) eIF-4E, which contains the cap-binding domain,[36] (2) eIF-4A$_c$, the founding member of the DEAD box family of RNA helicases,[62] and (3) a 220-kDa polypeptide whose function is not known. eIF-4E can also be purified in a functional complex with p220 alone, however, as eIF-4A seems to be more loosely associated than the other two subunits.[61,62]

eIF-4F has been reported to bind more strongly to the cap structure than eIF-4E.[63] This fact suggests the possibility that eIF-4E phosphorylation, perhaps in conjunction with p220 phosphorylation (see section 7), could affect the stability of the complex and/or its interaction with mRNA, and provides a rationale for how eIF-4E phosphorylation could affect cap binding indirectly. Indeed, heat shock has been shown to result in both reduced phosphorylation of eIF-4E and reduced recovery of the eIF-4E–p220 complex by cap-column chromatography.[61,64] These changes are concomitant with the shutoff of cap-dependent host protein translation.[13] It has recently been shown by IEF analysis, however, that there is only a small difference in the extent of phosphorylation of uncomplexed eIF-4E and eIF-4E in the 4F complex, suggesting another determinant contributes to the lesion in eIF-4F activity in lysates from heat-shocked cells.[64] As phosphorylation of eIF-4E is enhanced in response to growth factors and mitogens, it would be of interest to determine whether eIF-4F complex formation is concomitantly stimulated by these agents.

Another potential role for eIF-4F phosphorylation could be nucleocytoplasmic transport of mRNAs. Indeed, the 5′ cap itself is a multifunctional component of mRNA, involved not only in efficient translation, but also nucleocytoplasmic transport, splicing, and stabilization of mRNA against 5′ exonucleolytic degradation.[33,65] Recently, Lejbkowicz *et al.* have shown that a significant proportion of cellular eIF-4E localizes to the nucleus.[66] A nuclear locale has also been reported for a minor fraction of the p220 pool.[67] Thus we cannot rule out the possibility that phosphorylation of eIF-4E may affect the transport of capped messages, and possibly other bound initiation factors such as p220, out of the nucleus for translation.

4.2. Oncogenic Potential of eIF-4E Is Dependent on Serine-53

Various lines of evidence have clearly demonstrated that the mitogenic and oncogenic potential of eIF-4E is linked to the phosphorylated isoform. Lazaris-Karatzas *et al.* have shown that overexpression of eIF-4E by as little as fivefold can result in malignant transformation of primary or established fibroblast cells.[8] eIF-4E-overexpressing cells were able to grow in soft agar, and formed tumors when injected into nude mice. This effect is dependent on the presence of serine codon 53; replacement by an alanyl residue abrogated the transforming capability. This links the phosphorylated form with oncogenic potential. De Benedetti and Rhoads overexpressed mutant and wild-type eIF-4E in HeLa cells.[68] Overexpression of the wild-type, but not mutant, protein resulted in more rapid cell growth, culminating in the development of multinucleation and cell death within a month, thus demonstrating that eIF-4E affects cell growth. A similar link between the mitogenic activity of eIF-4E and phosphorylation at serine codon 53 was observed by microinjection studies in quiescent NIH 3T3 cells.[32] Injection of wild-type eIF-4E resulted in a rapid induction of DNA synthesis and morphological transformation; mutant eIF-4E, however, as well as several other translation factors, had no effect. The importance of the phosphorylation site was further highlighted in subsequent studies where it was shown that coinjected protein kinase C was able to potentiate the effects of injected wild-type, but not the alanyl mutant eIF-4E.[56] PKC was shown to phosphorylate eIF-4E directly and specifically in this study, at least *in vitro*, suggesting that this effect could be direct *in vivo*. It is difficult to know, however, whether the lack of an effect upon coinjection of the mutant protein eIF-4E[Ala-53] and protein kinase C could not be explained by a block in DNA synthesis induction by PKC due to a *trans*-dominant effect of the mutant eIF-4E. Such a *trans*-dominant negative effect of eIF-4E[Ala-53] was suggested by the slowed growth of NIH 3T3 cells overexpressing the mutant protein.[8] In most of the aforementioned functional studies, mutation of the phosphorylation site completely abolished the mitogenic and oncogenic potential of eIF-4E, suggesting that phosphorylation of eIF-4E does not merely enhance its activity, but is indispensable for its activity. This suggestion assumes no gross defects in eIF-4E conformation in the mutant protein due to the conservative substitution of an alanyl residue for the serine, which might render the protein inactive, independent of the loss of the phosphorylation site.

5. REGULATION OF EXPRESSION OF SPECIFIC GENES BY eIF-4F

5.1. Introduction

The inhibitory effects of long 5′ noncoding regions containing stable secondary structures and/or upstream AUGs on translation prompted the idea that the 5′ untranslated region (UTR) modulates the translation of mRNAs encoding proteins involved in the regulation of cell growth and differentiation.[8,69,70] Translationally regulated genes encode growth factors (c-*sis*, ref. 71 human fibroblast growth factor, FGF, ref. 72), tyrosine kinases (*lck*, ref. 73), and transcription factors such as c-*myc*.[74] An attractive possibility is that overexpression of eIF-4E results in greater amounts of active eIF-4F complex, and hence increased unwinding activity and mitigation of translational repression of such growth-related mRNAs. In normal cells, a similar alleviation of translational repression may occur upon stimulation of cell growth or induction of differentiation, by increasing the activity of eIF-4F through phosphorylation of eIF-4E and other components of the initiation process. Recent results of Manzella *et al.* are consistent with this hypothesis.[75] In the latter report, the authors demonstrate the specific effect of insulin on the translation of hybrid mRNAs containing either the entire 5′ UTR, or only the 115 5′-most bases, of ornithine decarboxylase (ODC). This effect correlated with the simultaneous induction of the phosphorylation of initiation factors eIF-4B and eIF-4E upon such treatment. In a different approach, Fagan *et al.* have shown a direct and specific effect of transfected eIF-4E on translation initiation of a growth-regulated cellular transcript encoding ornithine aminotransferase (OAT) in a retinoblastoma cell line expressing low levels of endogenous eIF-4E.[76] No induction of ODC protein was observed upon transfection of the cDNA containing the alanine substitution for serine codon 53, suggesting that the effect was dependent upon phosphorylation of the exogenous eIF-4E.

5.2. Viral Infection and eIF-4F Activity

More insight into the significance of eIF-4E phosphorylation can be gained from recognition that dephosphorylation of this factor accompanies the shutoff of host protein synthesis which occurs during adenovirus infection of cells.[40] This suggests that modification or inactivation of eIF-4F may be a common viral mechanism to shut down host protein synthesis: poliovirus accomplishes this through cleavage of the p220 subunit of eIF-4F,[77] whereas adenovirus achieves the same result by interfering with phosphorylation of the eIF-4E component of the complex. Whereas both poliovirus and adenovirus mRNA translation initiate by cap-independent mechanisms, however, RNA helicase activity may still be required for this process. This raises the question as to whether dephosphorylation of eIF-4E (or cleavage of p220) actually inactivates the eIF-4F complex, or merely modifies it to favor the viral mechanisms of internal initiation of translation. Indeed, there is some evidence that suggests a role for eIF-4F and another factor involved in unwinding of RNA secondary structure, eIF-4B, in internal initiation.[78]

Interestingly, the adenovirus-mediated hypophosphorylation of eIF-4E is prevented by the base analog 2-aminopurine (2-AP), which has previously been shown to interfere with adenovirus suppression of host translation.[79] This suppression is tightly correlated with the presence of activatable interferon-induced, double-stranded RNA-dependent kinase (DAI), which is also sensitive to 2-AP.[80] These facts have led Huang and Schneider to suggest that activation of this kinase and underphosphorylation of eIF-4E may be controlled by similar pathways induced by interferon and dsRNA, or perhaps a viral gene product.[40] Identification of the proteins that control phosphorylation of eIF-4E should provide some insight into this possibility.

6. Ras MEDIATES eIF-4E PHOSPHORYLATION AND MITOGENIC ACTIVITY

A crucial question to the understanding of the mechanism of signal transduction and its link with the translational apparatus is how the growth factor- and mitogen-mediated signals are transmitted from their respective receptors to eIF-4E and other translational components. If the mechanism of signal transduction to ribosomal protein S6 is a representative model, then the process could be quite complex, highly regulated, and may be under control of redundant pathways (reviewed in refs. 81 and 82). Tyrosine phosphorylation mediated by binding of growth factors to their receptors or via activated tyrosine kinase oncoproteins such as Src is implicated as an important initial event in signal transduction.[28] Ras functions downstream of Src in a common signaling pathway.[83,84] Phosphotyrosine kinases (PTKs) are believed to transmit their signals through the GTPase-activating protein (GAP), which is a major player in the regulation of Ras function.[85] This indicates that growth factor[40,41,45–51] and pp60^{src}-induced[41] increases in eIF-4E phosphorylation might be Ras dependent. This idea has received support through use of a *trans*-dominant inhibitory Ras mutant containing a Ser-17 to Asn mutation which results in a preference for Ras binding to GDP over GTP.[86] Expression of the Ras dominant negative mutant in PC12 cells inhibits both cellular differentiation[87] and NGF-induced stimulation of eIF-4E phosphorylation.[42] In fact, there is a time- and dose-dependent reduction in eIF-4E phosphorylation in response to NGF in PC12 cells expressing this Ras mutant. These findings indicate that NGF induces both differentiation and eIF-4E phosphorylation in PC12 cells through activation of a Ras-dependent signaling pathways(s), and places Ras upstream of eIF-4E in this pathway. Consistent with this result, transformation of CREF cells with oncogenic Ras results in elevated rates of eIF-4E phosphorylation.[52]

Other lines of investigation have shown that Ras functions downstream of eIF-4E. Lazaris-Karatzas *et al.* have recently reported that Ras is a key factor in the eIF-4E-mediated oncogenic transformation of fibroblast cells.[88] Cells overexpressing this factor have elevated levels of Ras activity, as evidenced by the increased proportion of GTP-bound Ras. Overexpression of GAP, the negative regulator of cellular Ras,[85] in eIF-4E-overexpressing cells causes reversion of the transformed phenotype. Furthermore, the authors show that neutralizing antibodies to Ras or the dominant negative mutant of Ras inhibit the mitogenic activity of eIF-4E injected into fibroblasts. eIF-4E phosphorylation was not greatly affected in GAP-overexpressing cells, however, indicating that reversion of the transformed phenotype was not due to inhibition of eIF-4E phosphorylation.

Thus, a separate Ras-dependent reaction, perhaps more sensitive to the GDP/GTP ratio, has been rendered limiting upon Ras downregulation via GAP overexpression. Taken together, these data constitute a strong and convincing argument that eIF-4E exerts its mitogenic and oncogenic activities via the activation of Ras.

One interesting possibility is that the mutant Ras, or the neutralizing antibodies, may inhibit the activity of (a) polypeptide(s) whose translation is specifically stimulated upon injection of eIF-4E. This idea does have precedence, as the specific effect of overexpression of eIF-4E on the translation of a cellular message (OAT) has been demonstrated[76] (see Section 5). In Fig. 2 we present a model adapted from ref. 88 which presents this idea in schematic form. Consistent with this hypothesis, recent results from our laboratory demonstrate that mRNAs containing extensive secondary structure in their 5′ UTR could be translated efficiently in cells overexpressing eIF-4E.[89]

The link between Ras and eIF-4E has been further established through the use of the transformation complementation assay in rat embryo fibroblasts (REFs). As is the case with *ras*, eIF-4E can transform REFs only when complemented by a

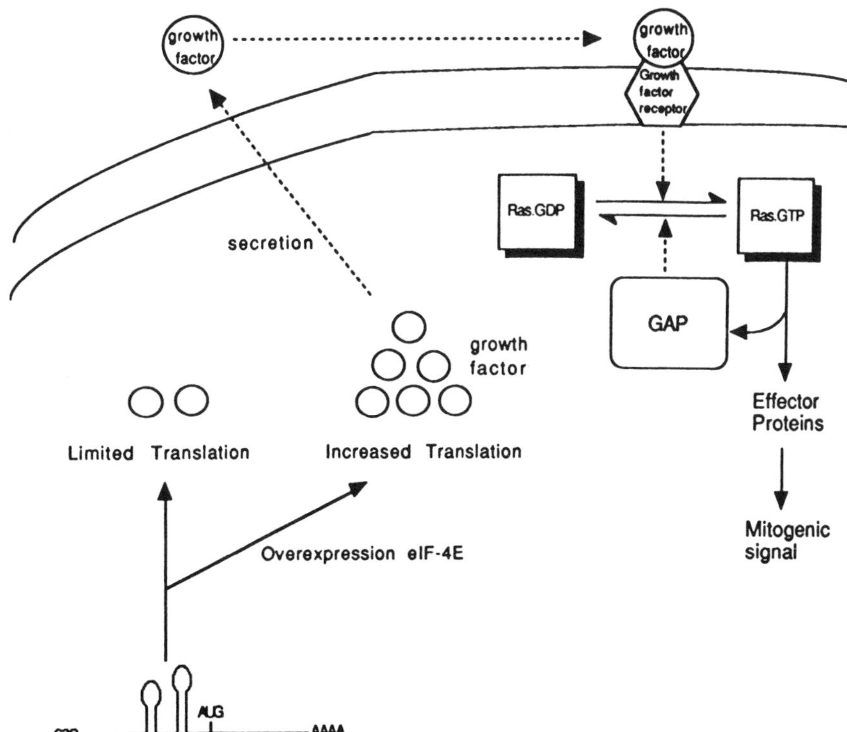

Figure 2. Schematic model of malignant transformation by eIF-4E. Adapted from ref. 88.

nuclear oncogene such as *myc* or E1A.[9] Taken together, these results strongly suggest that eIF-4E is an intracellular transducer of extracellular signals, functioning in diverse signal transduction pathways, and that it is dependent on a Ras pathway for its phosphorylation and mitogenic activity. These findings provide a strong link between the Ras signaling system and the regulation of a rate-limiting step in translation.

7. eIF-4B AND p220 ARE ALSO PHOSPHORYLATED UPON MITOGENIC STIMULATION OF CELLS

Initiation factors other than eIF-4E are also phosphorylated in response to growth factors and mitogens.[23] The p220 polypeptide,[90] a component of the cap-binding complex eIF-4F, is multiply phosphorylated in response to treatment of quiescent cells with phorbol-12-myristate-13-acetate,[51] insulin,[47] and epidermal growth factor (EGF).[48] Initiation factor 4B, which cooperates with eIF-4F or eIF-4A in the unwinding of helical RNA,[91] is also a phosphoprotein. Multiple phosphorylations on eIF-4B occur during the early G_1 phase of the cell cycle,[92] and in response to treatment of 3T3-L1 cells with PMA and insulin.[47,51] Interestingly, phosphorylation of both eIF-4B and p220 occurs in response to insulin in PKC-downregulated 3T3-L1 cells.[47] This contrasts with the PKC dependence of eIF-4E phosphorylation in response to insulin treatment. Although there is no direct evidence that the phosphorylation state of eIF-4B or p220 affects the activity of these factors *in vitro* or *in vivo*, it is conceivable that phosphorylation of these factors is important for their interaction with phosphorylated eIF-4E in the activation of translation initiation.

8. CONCLUDING REMARKS

The dramatic effect of eIF-4E overexpression on cellular proliferation not only documents the limiting nature of eIF-4E activity in the cell, but engenders the provocative conclusion that a translation factor can be a protooncogene product. It may be that eIF-4E is only the first of a novel class of translation-related protooncogene products, which may eventually include factors which regulate its activity. These results also show that eIF-4F is an intermediary in growth control pathways, and thus interacts with components of signal transduction pathways, such as p21 Ras.

Indeed, translation initiation factors are targets of mitogen- and growth factor-modulated kinases and/or phosphatases. In the case of eIF-4E, phosphorylation is induced in response to a wide variety of agents, including growth factors and tyrosine kinase oncogenes. Thus, tyrosine kinase-mediated pathways control translation factor phosphorylation. Figure 3 is a simple working model

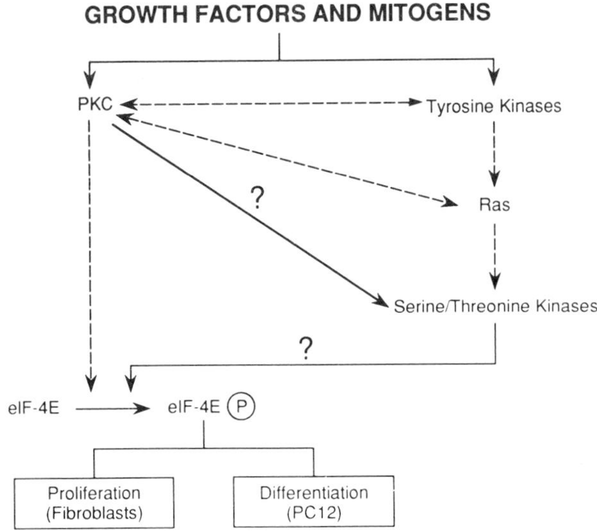

Figure 3. Model for signal transduction to eIF-4E. The figure illustrates our current understanding of the relay of extracellular signals to eIF-4E, resulting in its phosphorylation. Both PKC-dependent and PKC-independent pathways are implicated. The reader is referred to the text for details. Broken lines depict possible indirect signaling sequences. Solid lines accompanied by a quesion mark indicate pathways for which there exists no definitive evidence, but which are nonetheless possible. Adapted from ref. 24.

which incorporates our current understanding of the pathways controlling eIF-4E phosphorylation, and should serve to complement the model of eIF-4E-mediated malignant transformation presented in Fig. 2. Phosphorylation of eIF-4E is also induced by PMA, implicating PKC-mediated pathways. Different factors can effect PKC-dependent and PKC-independent eIF-4E phosphorylation pathways. The Ras pathway is a major player in relaying signals to eIF-4E. It is critical for the mitogenic effect of eIF-4E in fibroblasts (Fig. 2) and it is implicated in NGF-induction of eIF-4E phosphorylation in PC12 cells. At least one serine/threonine kinase must link Ras and eIF-4E in this scheme, as Ras itself does not exhibit kinase activity. PKC may function distally to Ras and mediate eIF-4E phosphorylation either directly or indirectly, depending on the cell and agent, or it may function in a separate pathway. The existence of both PKC-dependent and PKC-independent pathways controlling phosphorylation of eIF-4F and eIF-4B suggests the possibility of multiple, perhaps redundant signaling pathways which regulate their activity.

It is striking that phosphorylation of eIF-4E and other components of the initiation process occurs in response to diverse factors which effect either proliferation or differentiation, depending on the cell and agent in question. In particular, phosphorylation of eIF-4E may be a necessary and common event in signal transmission in the cell. The ultimate manifestation of the activation of signal transduction pathways is likely to depend on the genetic background of the cell in question. The particular repertoire of translationally repressed mRNAs could differ among different cell types; hence, the phosphorylation of eIF-4E would lead to enhanced translation of either growth- or differentiation-related mRNAs in a cell-type-specific manner.

ACKNOWLEDGMENTS. Work in the authors' laboratory was funded by grants to N.S. from the Medical Research Council of Canada and the National Cancer Institute of Canada. During the course of this work R.M.F. was a recipient of Predoctoral Studentships from the Medical Research Council of Canada and the National Cancer Institute of Canada.

REFERENCES

1. Hershko, A., Mammont, P., Shields, R., and Tomkins, G. M., 1971, "Pleiotropic response," *Nature New Biol.* **232:**206–211.
2. Rudland, P. S., and Jiminez de Asua, L., 1979, Action of growth factors in the cell cycle, *Biochem. Biophys. Acta* **560:**91–133.
3. Baserga, R., 1990, The cell cycle: Myths and realities, *Cancer Res.* **50:**6769–6771.
4. Brooks, R. F., 1976, Regulation of the fibroblast cell cycle by serum, *Nature* **260:**248–250.
5. Brooks, R. F., 1977, Continuous protein synthesis is required to maintain the probability of entry into S phase, *Cell* **12:**311–317.
6. Fan, H., and Penman, S., 1970, Regulation of protein synthesis in mammalian cells, *J. Mol. Biol.* **50:**665–670.
7. Bonneau, A.-M., and Sonenberg, N., 1987, Involvement of the 24 kDa cap binding protein in regulation of protein synthesis in mitosis, *J. Biol. Chem.* **262:**11134–11139.
8. Lazaris-Karatzas, A., Montine, K. S., and Sonenberg, N., 1990, Malignant transformation by a eukaryotic initiation factor subunit that binds to the mRNA 5′ cap, *Nature* **345:**544–547.
9. Lazaris-Karatzas, A., and Sonenberg, N., 1991, The mRNA 5′ cap-binding protein, eIF-4E, cooperates with v-*myc* or E1A in the transformation of primary rodent fibroblasts, *Mol. Cell. Biol.* **12:**1234–1238.
10. Sonenberg, N., 1990, Measures and countermeasures in the modulation of initiation factor activities by viruses, *New Biol.* **2:**402–409.
11. Sonenberg, N., 1990, Poliovirus translation, *Curr. Top. Microbiol. Immunol.* **161:**23–47.
12. Macejak, D. G., and Sarnow, P., 1991, Internal initiation of translation mediated by the 5′ leader of a cellular mRNA, *Nature* **353:**90–94.
13. Duncan, R., and Hershey, J. W. B., 1984, Heat shock-induced translational alterations in HeLa cells, *J. Biol. Chem.* **259:**11882–11889.
14. Jagus, R., Anderson, W. F., and Safer, B., 1981, Initiation of mammalian protein biosynthesis, *Prog. Nucleic Acid Res. Mol. Biol.* **25:**127–185.
15. Ou, J. H., Yen, T. S., Wang, Y. F., Kam, W. K., and Rutter, W. J., 1987, Cloning and characterization of a human ribosomal protein gene with enhanced expression in fetal and neoplastic cells, *Nucleic Acids Res.* **15:**8919–8934.
16. Chester, K. A., Robson, L., Begent, R. H. J., Talbot, I. C., Pringle, J. H., Primrose, R. J. H., Macpherson, G., Boxer, P., Southall, P., and Malcom, A. D. B., 1989, Identification of a human

ribosomal protein mRNA with increased expression in colorectal tumours, *Biochim, Biophys. Acta* **1009:**297–300.
17. Ferrari, S., Manfredi, R., Tagliafico, E., Rossi, E., Donelli, A., Tortelli, G., and Tortelli, U., 1990, Noncoordinated expression of S6, S11, and S14 ribosomal protein genes in leukemic blast cells, *Cancer Res.* **50:**5825–5828.
18. Sharp, M. G. F., Adams, S. M., Elvin, P., Walker, R. A., Brammar, W. J., and Varley, J. M., 1990, A sequence previously identified as metastasis-related encodes an acidic ribosomal phosphoprotein, P2, *Br. J. Cancer* **61:**83–88.
19. Mao, X., Green, J., Safer, B., Frederickson, R. M., Miyamoto, S., Sonenberg, N., and Thompson, C. B., 1992, Regulation of gene expression of translation initiation factors upon human T-cell activation, *J. Biol. Chem.* **267:**20444–20450.
20. Boal, T., Chiorini, J., Cohen, R., Miyamoto, S., Frederickson, R. M., Sonenberg, N., and Safer, B., 1993, Regulation of eukaryotic initiation factor expression during T-cell activation, *Biochem. Biophys. Acta*, in press.
21. Duncan, R., and Hershey, J. W. B., 1985, Regulation of initiation factors during translational repression caused by serum depletion, *J. Biol. Chem.* **260:**5486–5492.
22. Ryazonov, A. G., Rudkin, B. B., and Spirin, A. S., 1991, Regulation of protein synthesis at the elongation step, *FEBS Lett.* **285:**170–175.
23. Hershey, J. W. B., 1989, Protein phosphorylation controls translation rates, *J. Biol. Chem.* **264:** 20823–20826.
24. Frederickson, R. M., and Sonenberg, N., 1992, Signal transduction and regulation of translation, *Semin. Cell Biol.* **3**(2):105–113.
25. Krebs, E. G., and Beavo, J. A., 1979, Phosphorylation and dephosphorylation of enzymes, *Annu. Rev. Biochem.* **48:**923–959.
26. Proud, C. G., 1986, Guanine nucleotides, protein phosphorylation and the control of translation, *Trends Biochem. Sci.* **11:**73–77.
27. Rhoads, R. E., 1991, Protein synthesis, cell growth, and oncogenesis, *Curr. Opin. Cell Biol.* **3:** 1019–1024.
28. Ullrich, A., and Schlesinger, J., 1990, Signal transduction by receptors with tyrosine kinase activity, *Cell* **61:**203–212.
29. Hunter, T., 1987, A thousand and one protein kinases, *Cell* **50:**823–829.
30. Edelman, A. M., Blumenthal, D. K., and Krebs, E. G., 1987, Protein serine/threonine kinases, *Annu. Rev. Biochem.* **56:**567–613.
31. Duncan, R. F., Milburn, S. C., and Hershey, J. W. B., 1987, Regulated phosphorylation and low abundance of HeLa cell initiation factor eIF-4F suggests a role in translational control, *J. Biol. Chem.* **262:**380–388.
32. Smith, M. R., Jaramillo, M. L., Liu, Y., Dever, T. E., Merrick, W. C., Kung, H., and Sonenberg, N., 1990, Translation initiation factors induce DNA synthesis and transform NIH 3T3, *New Biol.* **2:**648–654.
33. Edery, I., Pelletier, J., and Sonenberg, N., 1987, Role of eukaryotic messenger RNA cap-binding protein in regulation of translation in: *Translational Regulation of Gene Expression* (J. Ilan, ed.), pp. 335–365, Plenum Press, New York.
34. Sonenberg, N., 1988, Cap-binding proteins of eukaryotic messenger RNA: Function in initiation and control of translation, *Prog. Nucleic Acid Res. Mol. Biol.* **35:**173–207.
35. Rhoads, R. E., 1988, Cap recognition and the entry of mRNA into the protein synthesis initiation cycle, *Trends Biochem. Sci.* **13:**52–56.
36. Sonenberg, N., Rupprecht, K. M., Hecht, S. M., and Shatkin, A. J., 1979, Eukaryotic mRNA cap-binding protein: Purification by affinity chromatography on sepharose coupled m^{7}GDP, *Proc. Natl. Acad. Sci. USA* **76:**4345–4349.
37. Rychlik, W., Gardner, P. R., Vanaman, T. C., and Rhoads, R. E., 1986, Structural analysis of the messenger RNA cap-binding protein, *J. Biol. Chem.* **261:**71–75.

38. Buckley, B., and Ehrenfeld, E., 1986, Two-dimensional gel analysis of the 24-kDa cap binding protein from poliovirus-infected and uninfected HeLa cells, *Virology* **152:**497–501.
39. Rychlik, W., Russ, M. A., Rhoads, R. E., 1987, Phosphorylation site of eukaryotic initiation factor 4E, *J. Biol. Chem.* **262:**10434–10437.
40. Huang, J., and Schneider, R. J., 1991, Adenovirus inhibition of cellular protein synthesis involves inactivation of cap-binding protein, *Cell* **65:**271–280.
41. Frederickson, R. M., Montine, K. S., and Sonenberg, N., 1991, Phosphorylation of eukaryotic translation initiation factor 4E is increased in *src*-transformed cell lines, *Mol. Cell. Biol.* **11:**2896–2900.
42. Frederickson, R. M., Mushynski, W. E., and Sonenberg, N., 1992, Phosphorylation of translation initiation factor eIF-4E is induced in a *ras*-dependent manner during nerve growth factor-mediated PC12 cell differentiation, *Mol. Cell. Biol.* **12:**1239–1247.
43. Gierman, T. M., Frederickson, R. M., Sonenberg, N., and Pickup, D. J., 1992, The eukaryotic translation initiation factor 4E is not modified during the course of vaccinia virus replication, *Virology* **188:**934–937.
44. Joshi-Barve, S., Rychlik, W., and Rhoads, R. E., 1990, Alteration of the major phosphorylation site of eukaryotic initiation factor 4E prevents its association with the 48S initiation complex, *J. Biol. Chem.* **265:**2979–2983.
45. Kaspar, R. L., Rychlik, W., White, M. W., Rhoads, R. E., and Morris, D. R., 1990, Simultaneous cytoplasmic redistribution of ribosomal protein L32 mRNA and phosphorylation of eukaryotic initiation factor 4E after mitogenic stimulation of Swiss 3T3 cells, *J. Biol. Chem.* **265:**3619–3622.
46. Donaldson, R. W., Hagedorn, C. H., and Cohen, S., 1991, Epidermal growth factor or okadaic acid stimulates phosphorylation of eukaryotic initiation factor 4F, *J. Biol. Chem.* **266:**3162–3166.
47. Morley, S. J., and Traugh, J. A., 1990, Differential stimulation of phosphorylation of initiation factors eIF-4F, eIF-4B, eIF-3, and ribosomal protein S6 by insulin and phorbol esters, *J. Biol. Chem.* **265:**10611–10616.
48. Bu, X., and Hagedorn, C. H., 1991, Platelet-derived growth factor stimulates phosphorylation of the 25 kDa mRNA cap binding protein (eIF-4E) in human lung fibroblasts, *FEBS Lett.* **283:**219–222.
49. Marino, M., Pfeffer, L. M., Guidon, D. T., and Donner, D. B., 1989, Tumor necrosis factor induces phosphorylation of a 28 kDa mRNA cap-binding protein in human cervical carcinoma cells, *Proc. Natl. Acad. Sci. USA* **86:**8417–8421.
50. Rychlik, W., Rush, J. S., Rhoads, R. E., and Waechter, C. J., 1990, Increased rate of phosphorylation/dephosphorylation of the translation initiation factor eIF-4E correlates with the induction of protein and glycoprotein biosynthesis in activated B-lymphocytes, *J. Biol. Chem.* **265:**19467–19471.
51. Morley, S. J., and Traugh, J. A., 1989, Phorbol esters stimulate phosphorylation of eukaryotic initiation factors 3, 4B, and 4F, *J. Biol. Chem.* **264:**2401–2404.
52. Rinker-Schaeffer, C. W., Austin, V., Zimmer, S., and Rhoads, R. E., 1992, *ras*-transformation of cloned rat embryo fibroblasts results in increased rates of protein synthesis and phosphorylation of eukaryotic initiation factor 4E, *J. Biol. Chem.* **267:**10659–10664.
53. Nishizuka, Y., 1988, The molecular heterogeneity of protein kinase C and its implications for cellular regulation, *Science* **334:**661–665.
54. Tuazon, P. T., Merrick, W. C., and Traugh, J. A., 1989, Comparative analysis of phosphorylation of translation initiation and elongation factors by seven protein kinases, *J. Biol. Chem.* **264:**2772–2777.
55. Tuazon, P. T., Morley, S. J., Merrick, W. C., Rhoads, R. E., Traugh, J. A., 1990, Association of initiation factor eIF-4E in a cap binding protein complex (eIF-4F) is critical for and enhances phosphorylation by protein kinase C, *J. Biol. Chem.* **265:**10617–10621.
56. Smith, M. R., Jaramillo, M., Tuazon, P. T., Traugh, J. A., Liu, Y., Sonenberg, N., and Kung, H., 1991, Modulation of the mitogenic activity of eukaryotic initiation factor 4E by protein kinase C, *New Biol.* **3:**601–607.

57. Morley, S. J., Dever, T. E., Etchison, D., and Traugh, J. A., 1991, Phosphorylation of eIF-4F by protein kinase C or multipotential S6 kinase stimulates protein synthesis at initiation, *J. Biol. Chem.* **266:**4669–4672.
58. Reinhold, D. S., and Neet, K. E., 1989, The lack of a role for protein kinase C in neurite extension and in the induction of ornithine decarboxylase by nerve growth factor in PC12 cells, *J. Biol. Chem.* **264:**3538–3544.
59. Pain, V. M., 1986, Initiation of protein synthesis in mammalian cells, *Biochem. J.* **235:**624–637.
60. Merrick, W. C., 1992, Mechanism and regulation of eukaryotic protein synthesis, *Microbiol. Rev.* **56:**291–315.
61. Lamphear, B. J., and Panniers, R., 1990, Cap binding protein complex that restores protein synthesis in heat-shocked Ehrlich cell lysates contains highly phosphorylated eIF-4E, *J. Biol. Chem.* **265:**5333–5336.
62. Nielsen, P. J., and Trachsel, H., 1988, The mouse protein synthesis initiation factor 4A gene includes two related functional genes which are differentially expressed, *EMBO J.* **7:**2097–2105.
63. Lee, K. A. W., Edery, I., and Sonenberg, N., 1985, Isolation and structural characterization of cap-binding proteins from poliovirus-infected cells, *J. Virol.* **54:**515–524.
64. Lamphear, B. J., and Panniers, R., 1991, Heat shock impairs the interaction of cap-binding protein complex with 5′ mRNA cap, *J. Biol. Chem.* **266:**2789–2794.
65. Shatkin, A. J., 1976, Capping of eukaryotic mRNAs, *Cell* **9:**645–653.
66. Lejbkowicz, F., Goyer, C., Darveau, A., Neron, S., Lemieux, R., and Sonenberg, N., 1992, A fraction of the mRNA 5′ cap-binding protein, eIF-4E, localizes to the nucleus, *Proc. Natl. Acad. Sci. USA* **89:**9612–9616.
67. Etchison, D., and Etchison, R. J., 1987, Monoclonal antibody-aided characterization of cellular p220 in uninfected and poliovirus-infected HeLa cells: Subcellular distribution and identification of conformers, *J. Virol.* **61:**2702–2710.
68. De Benedetti, A., and Rhoads, R. E., 1990, Overexpression of eukaryotic protein synthesis factor 4E in HeLa cells results in aberrant growth and morphology, *Proc. Natl. Acad. Sci. USA* **87:**8212–8216.
69. Pelletier, J., and Sonenberg, N., 1985, Insertional mutagenesis to increase secondary structure within the 5′ noncoding region of a eukaryotic mRNA reduces translational efficiency, *Cell* **40:** 515–526.
70. Frederickson, R. M., Lazaris-Karatzas, A., and Sonenberg, N., 1990, The eukaryotic mRNA cap-binding protein (eIF-4E): Phosphorylation and regulation of cell growth, in: *Post-Transcriptional Control of Gene Expression* (J. E. G. McCarthy and M. F. Tuite, eds.), pp. 497–509, Springer-Verlag, Berlin.
71. Rao, C. D., Pech, M., Robbins, K. C., and Aaronson, S. A., 1986, The 5′ untranslated sequence of the c-*sis*/platelet-derived growth factor-2 transcript is a potent translation inhibitor, *Mol. Cell. Biol.* **8:**284–292.
72. Bates, B., Hardin, J., Zhan, X., Drickamer, K., and Goldfarb, M., 1991, Biosynthesis of human fibroblast growth factor-5, *Mol. Cell. Biol.* **11:**1840–1845.
73. Marth, J. D., Overell, R. W., Meier, K. E., Krebs, E. G., and Perlmutter, R. M., 1988, Translational activation of the *lck* proto-oncogene, *Nature* **332:**171–173.
74. Darveau, A., Pelletier, J., and Sonenberg, N., 1985, Differential efficiencies of *in vitro* translation of mouse c-*myc* transcripts differing in the 5′ untranslated region, *Proc. Natl. Acad. Sci. USA* **82:** 2315–2319.
75. Manzella, J. M., Rychlik, W., Rhoads, R., Hershey, J. W. B., and Blackshear, P. J., 1991, Insulin induction of ornithine decarboxylase: Importance of mRNA secondary structure and phosphorylation of initiation factors eIF-4B and eIF-4F, *J. Biol. Chem.* **266:**2383–2389.
76. Fagan, R. J., Lazaris-Karatzas, A., Sonenberg, N., and Rozen, R., 1991, Translational control of ornithine aminotransferase: Modulation by initiation factor eIF-4E, *J. Biol. Chem.* **266:**16518–16523.

77. Etchison, D., Milburn, S. C., Edery, I., Sonenberg, N., and Hershey, J. W. B., 1982, Inhibition of HeLa cell protein synthesis following poliovirus infection correlates with the proteolysis of a 220,000-dalton polypeptide associated with eukaryotic initiation factor 3 and a cap binding protein complex, *J. Biol. Chem.* **257**:14806–14810.
78. Scheper, G. C., Voorma, H. O., and Thomas, A. A. M., 1992, Eukaryotic initiation factors-4E and -4F stimulate 5′ cap dependent as well as internal initiation of protein synthesis, *J. Biol. Chem.* **267**:7269–7274.
79. Huang, J., and Schneider, R. J., 1989, Adenovirus inhibition of cellular protein synthesis is prevented by the drug 2-aminopurine, *Proc. Natl. Acad. Sci. USA* **87**:7115–7119.
80. O'Malley, R. P., Duncan, R. F., Hershey, J. W. B., and Mathews, M. B., 1989, Modification of protein synthesis initiation factors and the shut-off of host protein synthesis in adenovirus-infected cells, *Virology* **168**:112–118.
81. Kozma, S. C., Ferrari, S., and Thomas, G., 1989, Unmasking a growth factor/oncogene activated S6 phosphorylation cascade, *Cell Signalling* **1**:219–225.
82. Erikson, R. L., 1990, Structure, expression, and regulation of protein kinases in the phosphorylation of ribosomal protein S6, *J. Biol. Chem.* **266**:6007–6010.
83. Mulcahy, L. S., Smith, M. R., and Stacey, D. W., 1985, Requirement of *ras* proto-oncogene function during serum-stimulated growth of NIH 3T3 cells, *Nature* **313**:241–243.
84. Smith, M. R., DeGudicibus, S. J., and Stacey, D. W., 1986, Requirement for c-*ras* proteins during viral oncogene transformation, *Nature* **320**:540–543.
85. Trahey, M., and McCormick, F., 1987, A cytoplasmic protein stimulates normal N-*ras* p21 GTPase but does not affect oncogenic mutants, *Science* **238**:542–545.
86. Feig, L. A., and Cooper, G. M., 1988, Inhibition of NIH 3T3 cell proliferation by a mutant *ras* protein with preferential affinity for GDP, *Mol. Cell. Biol.* **8**:3235–3243.
87. Szeberényi, J., Cai, H., and Cooper, G. M., 1990, Effect of a dominant inhibitory Ha-*ras* mutation on neuronal differential of PC12 cells, *Mol. Cell. Biol.* **10**:5324–5332.
88. Lazaris-Karatzas, L., Smith, M. R., Frederickson, R. M., Jaramillo, M. L., Liu, Y., Kung, H., and Sonenberg, N., 1992, Ras mediates translation initiation factor 4E-induced malignant transformation, *Genes Dev.* **6**:1631–1642.
89. Koromilas, A. E., Lazaris-Karatzas, A., and Sonenberg, N., 1992, mRNAs containing extensive secondary structure in their 5′ non-coding region translate efficiently in cells overexpressing initiation factor eIF-4E, *EMBO J.* **11**:4153–4158.
90. Yan, R., Rychlik, W., Etchison, D., and Rhoads, R. E., 1992, Amino acid sequence of the human protein synthesis initiation factor eIF-4γ, *J. Biol. Chem.* **267**:23226–23231.
91. Rozen, F., Edery, I., Meerovitch, K., Dever, T. E., Merrick, W. C., and Sonenberg, N., 1990, Bidirectional RNA helicase activity of eukaryotic translation initiation factors 4A and 4F, *Mol. Cell. Biol.* **10**:1134–1144.
92. Duncan, R. F., and Hershey, J. W. B., 1985, Regulation of initiation factors during translational repression caused by serum-depletion, *J. Biol. Chem.* **260**:5493–5497.

Chapter 8

Interferon-Induced and Double-Stranded RNA-Activated Proteins as Key Enzymes Regulating Protein Synthesis

Ara G. Hovanessian

1. INTRODUCTION

In 1971, Hunt and Ehrenfeld[1,2] demonstrated for the first time the inhibition of protein synthesis in rabbit reticulocytes by double-stranded RNA (dsRNA) extracted from poliovirus-infected HeLa cells. This inhibition was confirmed by a wide variety of synthetic and naturally occurring dsRNAs.[3] During this time, Ian Kerr and collaborates[4–6] demonstrated a similar type of inhibition of protein synthesis by dsRNA in cell-free systems prepared from interferon-treated cells. In addition to this, they showed that when cell-free systems were prepared from interferon-treated and virus-infected cells, protein synthesis was inhibited. Thus, the requirement for dsRNA was abolished by virus infection. This was due to the production of viral dsRNA-like molecules during the infection. These observations pointed out two requirements for the inhibition of protein synthesis in cell-free systems: first, interferon treatment of cells, and second, addition of dsRNA in the

Ara G. Hovanessian • Unité de Virologie et Immunologie Cellulaire (UA CNRS 1157), Institut Pasteur, 75015 Paris, France.

Translational Regulation of Gene Expression 2, edited by Joseph Ilan. Plenum Press, New York, 1993.

cell-free system. Accordingly, the initial objectives were then to specify the components that were induced by interferon and which interacted with dsRNA. Eventually, it became clear that an enzyme responsible for the synthesis of a low-molecular-weight oligonucleotide inhibitor (later referred to as 2-5A) and a protein kinase were likely to be involved.[7–14]

The 2-5A synthetase is responsible for the synthesis of 2′-5′ linked oligomers of adenosine, ppp(A2′p5′A)$_n$, where n is 2 or more. For convenience, these mixtures of oligonucleotides are referred to as 2-5A, and the enzyme which synthesizes them is referred to as 2-5A synthetase. The function of 2-5A is to activate a latent endoribonuclease responsible for degradation of viral or cellular RNAs.[15–21] The other interferon-induced enzyme, the protein kinase, is responsible for the phosphorylation of the protein synthesis initiation factor eIF-2, thus mediating inhibition of protein synthesis.[7,–10,14,22–24] Although the phosphorylated eIF-2 could initiate protein synthesis, it cannot be recycled. For this reason the protein synthesis becomes inhibited due to the lack of availability of active eIF-2.[25,26] Accordingly, the 2-5A synthetase and the protein kinase could be functional in two independent pathways by which protein synthesis becomes inhibited (Fig. 1). Natural or synthetic dsRNAs are efficient activators of these enzymes. A dsRNA of 30–50 base pairs is the minimal requirement for the activation.[27,28] It should be emphasized that single-stranded RNAs (ssRNA) might as well activate these enzymes due to their secondary structure which forms stable duplex regions. For example, the human immunodeficiency virus (HIV)-1 leader RNA, which contains a double-stranded inverted repeat, is able to activate these enzymes.[29–31] The levels of the protein kinase and the 2-5A synthetase are significantly enhanced in interferon-treated cells (reviewed in ref. 32). These enzymes are activated during virus infections, probably due to the presence of viral RNA: genomic dsRNA or ssRNA with stable duplex structure or dsRNA in the form of viral RNA replicative intermediates or dsRNA generated from symmetrical transcription of the viral DNA template.[33] Consequently, protein synthesis becomes blocked in interferon-treated, virus-infected cells. Accordingly, both enzyme systems have been implicated in the mechanism of the antiviral action of interferon.[33,34]

Besides interferon-treated cultured cells, analogous dsRNA-activated enzymes are also described in rabbit reticulocytes, thus demonstrating the reason dsRNA inhibits protein synthesis in reticulocyte lysates.[22,35,36] The presence of these enzymes in rabbit reticulocytes also indicates that they might be implicated in other functions besides the mechanism of the antiviral action of interferon. In mice, the 2-5A synthetase and the protein kinase are widely distributed in different tissues.[37–39] These enzymes are constantly induced due to the production of interferon, which itself is induced under different physiological conditions.[40] In humans both of these enzymes are detectable in peripheral blood mononuclear cells.[41–44] Their levels might be used as convenient markers to monitor the presence of circulating interferon in patients with virus infections and also to

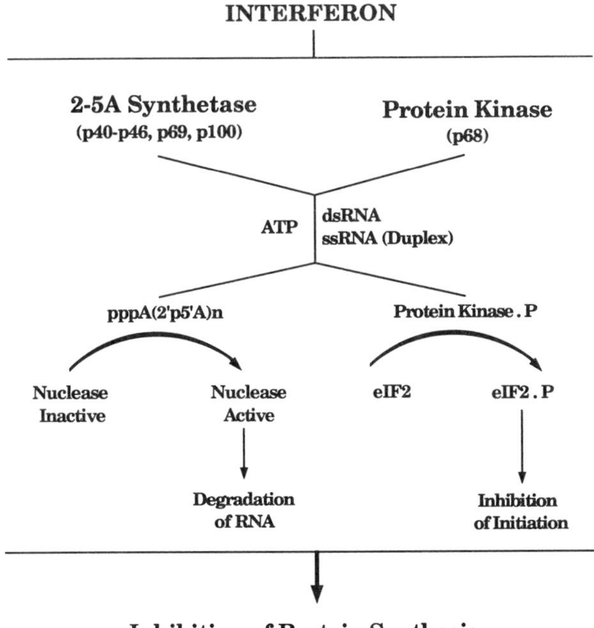

Figure 1. Two independent dsRNA-activated inhibitory pathways leading to inhibition of protein synthesis. Interferon treatment of cells results in the induction of the two dsRNA-activated enzymes, the 2-5A synthetase and the protein kinase. In human cells at least three major forms of 2-5A synthetase, p40–p46, p69, and p100, have been described. The protein kinase is a 68-kDa protein (p68). Both enzymes are activated by dsRNA or ssRNA containing stable duplex structures. 2-5A synthesized by 2-5A synthetase activates a nuclease responsible for RNA degradation. The phosphorylated protein kinase catalyzes phosphorylation of eIF-2alpha, leading to the inhibition of initiation of protein synthesis.

estimate the immediate response of patients to treatment with interferon or to inducers of interferon.

2. THE 2-5A SYSTEM

2.1. Enzymes in the 2-5A System

Three types of enzymes are involved in the 2-5A system: (1) the 2-5A synthetase which synthesize 2′-5′ linked oligoadenylates,[11–13] (2) the 2-5A-activated endonuclease which degrades RNA,[15,16] and (3) a phosphodiesterase which converts 2-5A molecules into AMP and ATP.[45] The level of 2-5A synthetase is usually increased 10- to 100-fold in response to interferon, whereas the levels of the nuclease and the phosphodiesterase are generally not significantly modified. Besides the synthesis of 2-5A oligomers, *in vitro* the 2-5A synthetase also has the capacity to catalyze the transfer of a nucleotide monophosphate moiety (AMP, GMP, UMP, CMP, TMP) to the 2′-OH end of a preformed 2-5A molecule or to a nucleotide with the structure RpA such as NAD^+ and tRNA.[46–48] Whether these latter reactions are really functional *in vivo* in cells, however, remains uncertain. The activated nuclease degrades both cellular and viral mRNAs as well as ribosomal RNA.[15–17,19,20,49] This enzyme cleaves RNAs on the 3′ side of UN sequences to yield UpNp-terminated products. The most frequent cleavages occur after UA, UG, and UU, with much less frequent cleavages after CA and AC.[50,51] The activation of the endonuclease is fully reversible. Removal of 2-5A from the activated enzyme results in the reversion of the endonuclease to its latent state, but it remains sensitive to activation.[52] The endonuclease has an apparent molecular weight of 80 kDa,[53] whereas the phosphodiesterase might be a 40-kDa protein.[45]

2.2. Different Forms of 2-5A Synthetase

Human cells contain at least three major forms of 2-5A synthetase: the small, 40- and 46-kDa isoforms, and two large forms, 69 and 100 kDa.[54,55] The existence of 40- and 46-kDa (p40 and p46) forms of 2-5A synthetase has been confirmed by cloning of cDNAs corresponding to these proteins.[56–59] These enzymes are coded by 1.6- and 1.8-kilobase (kb) messenger RNAs (mRNAs), which are derived from the same gene by differential splicing. Accordingly, p40 and p46 are identical in their first 346 amino acid residues, but are different at their carboxyl-terminal ends.

The existence of 69- and 100-kDa forms (p69 and p100) of 2-5A synthetase was confirmed by monoclonal and polyclonal antibodies.[54,60,61] These large forms of 2-5A synthetase are probably encoded by another gene or genes different from that encoding the p40–p46. Polyclonal antibodies raised against purified preparations of p69 and p100 were found to be specific to either form and cross-reacted only slightly with the other forms of 2-5A synthetase.[61] Thus, there might be some minor epitopes in common between p100, p69, and p40–p46. Figure 2 shows an immunoblot assay using anti-p100 polyclonal antibodies and crude extracts from control and interferon-treated Daudi cells. These antibodies react mainly with p100, but also cross-react with p69 and p46 at a much lower affinity. The identity of the 56-kDa protein induced by interferon is not known. It might be another 2-5A synthetase or a different protein which shares a common peptide sequence with p46, p69, and p100 2-5A synthetases.

2.3. Specific Features of p69 and p100

The p69 and p100 are differentially expressed in different human cells treated with interferon.[54,55,62] The three types of interferon are capable of inducing these

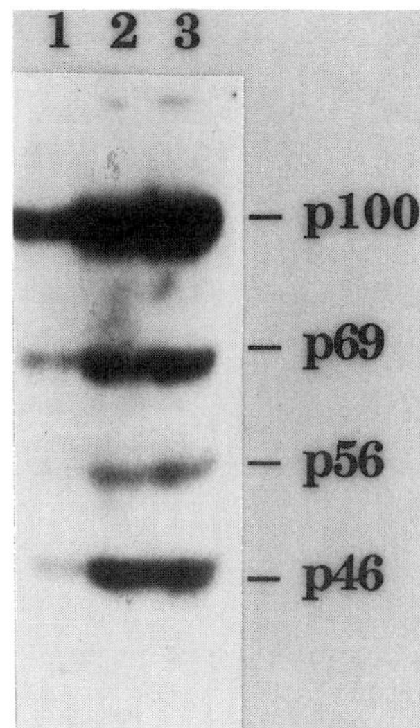

Figure 2. The three major forms of 2-5A synthetase. Extracts from control (lane 1) or interferon-treated Daudi cells at 50 (lane 2) and 500 (lane 3) units/ml of human alpha-interferon were analyzed by immunoblotting using polyclonal antibodies raised against p100. An autoradiograph is shown.

enzymes as long as the cells respond to the interferon type. The p100 is tightly associated with the ribosomes,[54,63] whereas p69 is found mostly associated with cell membranes.[54] Accordingly, p69 but not p100 is found to be myristilated.[62] Immunoenzymatic staining analysis of interferon-treated cells indicated that a proportion of p69 is concentrated around the nuclei and the rest is distributed in a specific pattern in the cytoplasm, whereas p100 is found in a diffused state in the cytoplasm.[62] Experiments on the induction and synthesis of p69 and p100 in interferon-treated cells suggested different mechanisms for the induction of p69 and p100 in a given cell line. Furthermore, they indicated the presence of cell-specific factors regulating the expression of these enzymes in response to interferon treatment. The differential expression and differences for the induction patterns of p69 and p100 favor the presence of distinct genes encoding these enzymes.

One of the most significant differences between p69 and p100 is in the pattern of 2-5A oligomers synthesized by each enzyme (Fig. 3). At optimum concentrations of dsRNA, p69 has the capacity to synthesize higher oligomers, whereas p100 has the tendency to synthesize more of the dimeric form.[60] As the dimeric form has no function, one might argue whether the function of p100 is to catalyze other reactions than making dimers, i.e., *in vitro* one might be pushing the purified p100 to synthesize 2-5A. These p69 and p100 forms of 2-5A synthetase have different pH and ATP concentration requirements and show different affinities to dsRNA for activation. In view of all the differences between these enzymes, it is possible to suggest that they might be activated under different physiological conditions and might have different functions.

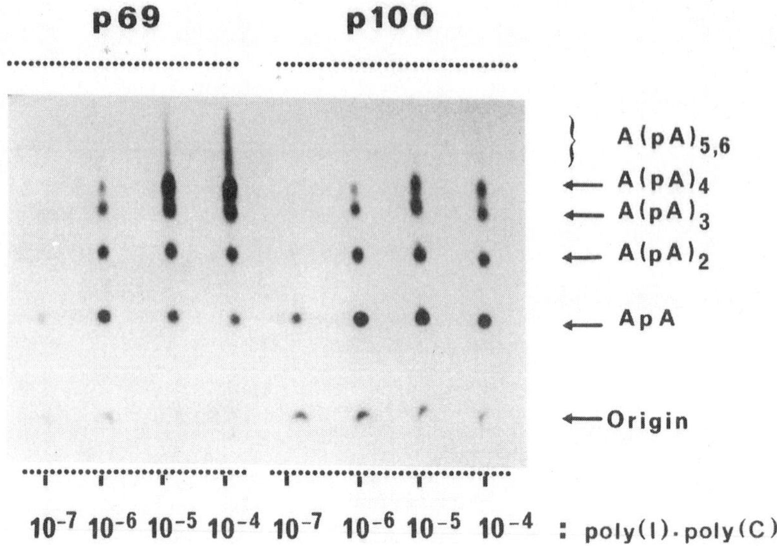

Figure 3. Synthesis of 2′,5′-linked oligoadenylates by p69 and p100. Purified p69 and p100 were assayed for 2-5A synthetase activity in the presence of [α-^{32}P]ATP and different concentrations of poly(I)·poly(C), 10^{-7}, 10^{-6}, 10^{-5}, and 10^{-4} g/ml. After bacterial alkaline phosphatase treatment, 2-5A "cores" were analyzed by electrophoresis on Whatman 3 MM paper.[60] An autoradiograph is shown.

2.4. cDNA Sequence of p69 Reveals Two Homologous Functional Domains

By the use of polyclonal antibodies specific to p69, we have recently reported the molecular cloning of three full-length cDNAs which identify four interferon-induced RNAs of 5.7, 4.5, 3.7, and 3.2 kilobases (kb).[64] Analysis of the nucleotide sequence of three full-length cDNAs (5.6, 3.1, and 2.9 kb) revealed that they have a common open reading frame of 683 amino acids with different 3′ termini: cDNAs 5.6 and 3.1 have an extension of four amino acids, whereas cDNA 2.9 has an extension of 44 amino acids. *In vitro* transcription–translation of cDNAs 3.1 and 2.9 kb generated proteins of 69 and 71 kDa, respectively. Both proteins bind a monoclonal antibody specific for p69 and can synthesize 2-5A.

The deduced amino acid sequence of p69 revealed that it can be divided into two homologous and adjacent domains of 324 and 340 amino acids at its NH_2- and COOH-termini. These two domains have 46% identical amino acids and 18% conservative amino acid homologies. Each domain is also highly homologous to the first 346 amino acids common to the two isoforms of the small 2-5A synthetase, p40 and p46. The catalytic domain of the small 2-5A synthetase should be located within these 346 amino acids since this domain is highly conserved (82% strict amino acid homology) between murine and human small 2-5A synthetase.[65] Gel filtration experiments under native conditions have suggested that in Daudi cells p69 exists as a dimer whereas p46 exists as a tetramer. This and the presence of two homologous domains in p69 and their homology with p40 suggest that the functioning of 2-5A synthetase activity might require the presence of four catalytic domains which can be provided by four molecules of p40 or two molecules of p69.

2.5. The Function of the 2-5A System

Introduction of 2-5A into cells by the calcium phosphate coprecipitation technique results in the activation of the endonuclease, leading to degradation of RNA and consequently inhibition of protein synthesis.[18,20] 2-5A introduced in virus-infected cells mediates degradation of viral RNA in addition to cellular RNA, leading to inhibition of virus production.[21] Therefore 2-5A might exert both antiproliferative and antiviral effects. The inhibitory effects of 2-5A are transient since 2-5A becomes degraded by the 2′,5′-phosphodiesterase. In general, the 2-5A synthetase in intact cells is not active. Activation of this enzyme, however, occurs during the course of virus infections such as encephalomyocarditis virus (EMC), reovirus, SV40, herpes virus, and vaccinia virus.[19,66–70] The evidence for this is the accumulation of 2-5A molecules in such virus-infected cells. It should be emphasized that besides EMC virus, the other virus infections do not appear to be affected by the 2-5A system. This might be in part due to the synthesis of "modified" 2-5A molecules. For example, in the case of herpes virus-infected cells, two 2-5A related products have been isolated which function as analog inhibitors, i.e., they can prevent the activation of the endonuclease.[67]

In interferon-treated EMC virus-infected cells, the endonuclease is active and could be demonstrated by the specific pattern of RNA degradation.[49,70] In this latter virus infection, not only is 2-5A present and the endonuclease active, but it also plays a part in the inhibition of virus growth. The introduction of a 2-5A analog inhibitor (which binds but does not activate the endonuclease) into EMC virus-infected cells abrogates the antiviral action of interferon to inhibit virus growth, thus demonstrating the role of the 2-5A system in the anti-EMC virus response. Finally, expression of cDNA clones encoding the small 2-5A synthetase in transfected cells provides evidence that the constitutive expression of the synthetase could mediate an antiviral effect against picornaviruses.[71–73]

3. THE INTERFERON-INDUCED dsRNA-ACTIVATED PROTEIN KINASE

3.1. Features of the Protein Kinase (p68 kinase)

This protein kinase (see ref. 74 for a recent review) is induced by interferon and it is manifested by autophosphorylation of a 68-kDa protein (p68) in human cells or a 65-kDa protein (p65) in murine cells.[8–10,14,75–81] In both p68 and p65 the phosphorylation occurs on serine and threonine residues.[23,76] The protein kinase is associated with the rough microsomal pellet, but only a small proportion of it is found bound tightly to ribosomal subunits.[63] By the use of monoclonal antibodies[77] we were able to purify the protein kinase from human cells (p68 kinase) to homogeneity and to characterize its enzymatic activity.[76–81] The p68 kinase is characterized by two distinct protein kinase activities (Fig. 1). The first one is functional for its autophosphorylation, whereas the second one is responsible for the phosphorylation of the exogenous substrate, the alpha subunit of eIF-2. When activated by dsRNA in the presence of divalent cations and ATP, p68 is autophosphorylated by a mechanism which we believe is intramolecular,[81] although others have suggested that it is intermolecular.[82] Whatever is the case, the phosphorylated p68 is then capable of catalyzing phosphorylation of eIF-2. Phosphorylation of the exogenous substrate eIF-2 is not dependent on dsRNA. It might occur as long as p68 is phosphorylated. It should be noted that p68 has several phosphate-binding sites.[76] Consequently, there is a strong correlation between the degree of phosphate content of p68 and its kinase activity on the exogenous substrate.[78] Dephosphorylation of the p68 kinase can be catalyzed by a manganese ion-dependent class I phosphatase.[83] Histones are alternative substrates for the kinase, but the significance of this is not known.[14]

Another protein kinase associated with the heme-regulated inhibitor (HCR) can also phosphorylate eIF-2.[22] Thus, phosphorylation of the alpha subunit of eIF-2 can be catalyzed by two different protein kinases, both of which, however, phosphorylated the same peptide on eIF-2.[24] In order to rule out any misunderstanding, the dsRNA-activated protein kinase should not be referred to as eIF-2 kinase, as has been done previously. The kinase has also been referred to as dsRNA-activated inhibitor (DAI). Once again this is misleading, since the kinase itself is not the inhibitory agent. A more suitable abbreviation, for example, could be DAK, for dsRNA-activated kinase.

3.2. The Protein Kinase–eIF-2 System

When activated, the protein kinase becomes autophosphorylated and catalyzes the phosphorylation of the alpha subunit of protein synthesis initiation factor eIF-2. Phosphorylation of eIF-2 then blocks the eIF-2B mediated exchange of GDP for GTP required for catalytic utilization of eIF-2. These events lead to limitations in functional eIF-2 leading to inhibition of initiation of protein synthesis.[25] The significance of the dsRNA-activated protein kinase–eIF-2 system has been demonstrated in adenovirus-infected human cells.[84,85] In such infected cells, adenovirus-coded VAI RNA binds to p68 kinase and prevents its activation during virus infection.[86] The translation of viral mRNAs, therefore, occur efficiently under the inhibition conditions of the p68 kinase–eIF-2 system. On the other hand, in cells infected with an adenovirus deletion mutant (*dl*331) which fails to synthesize VAI RNA, p68 kinase becomes activated—leading to its autophosphorylation and phosphorylation of eIF-2—and as a consequence, protein synthesis becomes drastically inhibited. Activation of p68 kinase in cells infected with adenovirus *dl*331 is probably due to the production of viral dsRNA in response to a symmetrical transcription of both viral DNA strands.[87] Besides the adenovirus *dl*331 infection, enhanced phosphorylation of p68/p65 (human/murine kinase) and eIF-2 has been reported during infection of interferon-treated cells with different types of viruses.[88–93] In view of this, it has been suggested that the protein kinase–eIF-2 system might play a role, at least in part, in the mechanism of the antiviral action of interferon.[74,94]

3.3. Regulation of the Protein Kinase Activity

Since the discovery of the kinase, several studies have emphasized the specificity of dsRNA in activating the protein kinase activity. The dsRNA can be synthetic poly(I)·poly(C) or natural (reovirus dsRNA) (reviewed in ref. 95). In contrast to this, we have shown that the kinase can also be activated by polyanions such as heparin, dextran sulfate, chondroitin sulfate, and poly(L-glutamine).[80] The only common feature between these compounds and dsRNA is their polyanionic nature, thus indicating that activation of the protein kinase is dependent on the polyanionic nature of the activator. These results emphasize the possibility that various activators might exist in different types of cells. Nevertheless, it is most likely that in intact cells the activators of the protein kinase are RNA molecules, either dsRNAs or single-stranded RNAs with secondary structures containing double-stranded regions (see Section 1). Accordingly, the 5′ untranslated region (leader region or TAR sequence) of all HIV mRNAs can activate the kinase due to its stem–loop structure.[30,96]

One of the important features of the p68 kinase is that it is activated by low concentrations of dsRNA, but its activation is inhibited by high concentrations of dsRNA (Fig. 4). A possible explanation for this effect might be the presence of two dsRNA binding sites on the p68 kinase: low- and high-affinity sites.[81] At low concentrations, dsRNA becomes bound to the high-affinity binding site and in the presence of Mn, the p68 kinase can bind ATP and be autophosphorylated. At high concentrations of dsRNA, both binding sites become occupied, leading to an irreversibly inactivated p68 kinase. Probably by a similar mechanism, VAI RNA blocks the activity of p68 kinase. In fact, VAI RNA can activate or inhibit activation, depending on its concentration,[55] and since the concentration of VAI is high in adenovirus-infected cells, the activation of the kinase is blocked.

3.4. Control of the Protein Kinase System by Viruses

The kinase becomes activated during different virus infections, leading to the inhibition of protein synthesis. As a defence mechanism, different viruses have developed specific strategies to regulate the functioning of p68 kinase (Fig. 5).[97]

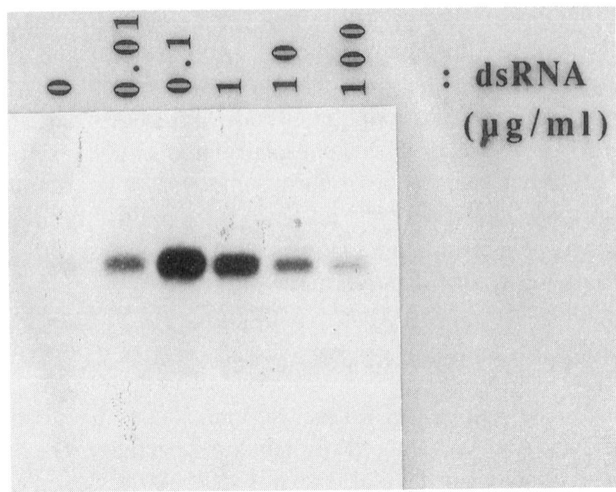

Figure 4. The dsRNA-activated protein kinase is activated by low concentrations of dsRNA, whereas high concentrations of dsRNA block its activation. Purified protein kinase from human cells (p68 kinase) was assayed for autophosphorylation, i.e., for protein kinase activity, by incubation with [α-^{32}P]ATP and different concentrations of poly(I)·poly(C) as indicated. Samples were analyzed by polyacrylamide gel electrophoresis.[81] An autoradiograph is shown.

For example, adenovirus-encoded VAI RNA complexes with and inactivates p68 kinase.[86] Poliovirus infection induces the kinase degradation,[93] while infection by another picornavirus, encephalomyocarditis virus, possibly causes its sequestration[63] (also see Section 4.1). Human immunodeficiency virus may mediate the downregulation of the kinase via action of the Tat regulatory protein,[98] whereas influenza virus blocks kinase activity by activation of a cellular inhibitor of p68 kinase.[92,99] Finally, reovirus and vaccinia virus appear to downregulate the kinase by encoding gene products that bind to and sequester the activator of p68 kinase.[100,101] It should be noted that in spite of the action of all such control mechanisms on the kinase, under certain circumstances the kinase remains functional. For example, in poliovirus-infected cells, the kinase remaining not degraded is found to be highly phosphorylated.[93]

3.5. Features of the Protein Kinase Deduced from Its cDNA

We have recently reported the molecular cloning of a cDNA encoding the human protein kinase, p68 kinase.[102] The deduced amino acid sequence of this cDNA predicts a protein of 550 amino acids containing all of the conserved domains specific for members of the protein kinase family, including the catalytic domain characteristic of serine/threonine kinases. *In vitro* translation of the full-length kinase cDNA yields a protein of 68 kDa that binds dsRNA, is recognized

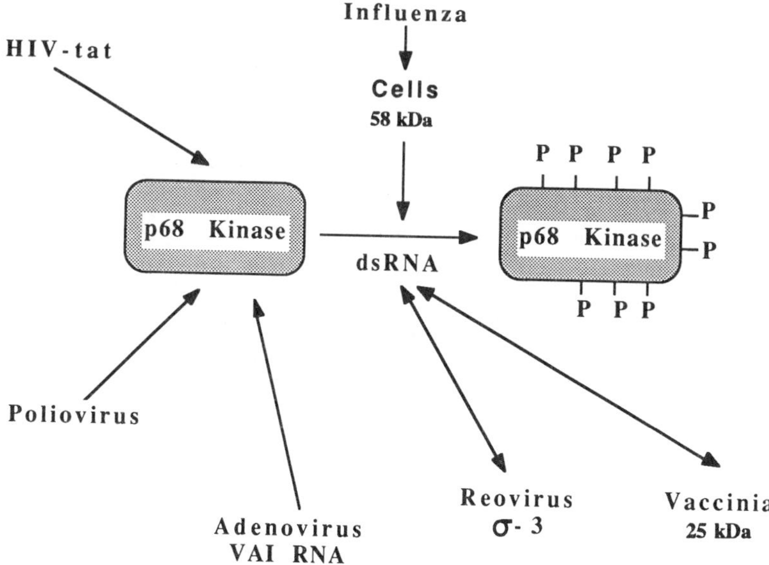

Figure 5. Control of the dsRNA-activated protein kinase by viruses. HIV-tat by an unknown mechanism reduces the amount of the kinase present in a cell line. Poliovirus infection results in degradation of the kinase. A cellular factor of 58 kDa becomes activated during influenza infection and causes inhibition of the autophosphorylation reaction and also phosphorylation of eIF-2. Direct interaction of VAI RNA with the kinase irreversibly inactivates the protein kinase. Reovirus protein σ-3 is a major viral protein that binds specifically to dsRNA. Vaccinia virus synthesizes a 25-kDa protein that binds dsRNA and can prevent *in vitro* activation of the kinase.

by a monoclonal antibody raised against the native p68 kinase, and is autophosphorylated. The deduced amino acid sequence of the kinase indicates that it is hydrophilic, a feature that may account for its ability to bind RNA. The kinase sequence, however, does not contain previously described consensus sequences characteristic of RNA and DNA binding proteins. The protein kinase sequence manifests several potential phosphorylation sites. In particular, there is a phosphorylation site homologous to that of cAMP-dependent protein kinase, and another site homologous to that of the histone H1 kinase.[103,104]

The mRNA of p68 kinase is a single 2.5-kb RNA strongly induced by interferon. In Daudi cells, the steady-state levels of this mRNA are increased few hours after treatment with interferon, with a maximal level at 16 hr. The degree of the enhancement is correlated with the dose of interferon. The induction of kinase mRNA was confirmed by measuring the rate of transcription in isolated nuclei. This is a very rapid process with a maximal level at 2 hr.[102]

A computer search for protein homology did not reveal previously described proteins identical to p68 kinase. As expected, however, there were homologies with protein kinases and kinase-related proteins.[105] The best homology was with a cGMP-dependent protein kinase,[106] despite the fact that p68 kinase is not activated by cGMP. The next best candidates were tyrosine protein kinases: for instance, p68 kinase shares 9.5–10.6% strict homology (without any gaps) with the kinase-related transforming protein, murine HCK, and human Yes-1.[107,108] These observations suggest that p68 kinase belongs to a subfamily of protein serine/threonine kinases that is related to protein tyrosine kinases. Normally, these latter kinases are cellular homologues of retroviral oncogene products. The p68 kinase sequence also revealed the presence of a 32-amino acid insert located within the catalytic domain upstream of conserved subdomain VI.[102] This insert resembles a variable region found in a class of receptor tyrosine kinases, including the platelet-derived growth factor receptor, the colony-stimulating factor 1 receptor, and the c-*kit* protooncogene product.[105] This kinase insert domain in the platelet-derived growth factor receptor interacts with a phosphatidylinositol 3-kinase and may also be involved in the recognition of other cellular targets.[109,110] Accordingly, the insert in p68 kinase might be involved in specific substrate recognition.

3.6. Function of p68 Kinase

The target of the phosphorylated p68 kinase is the alpha subunit of eIF-2. Phosphorylation of eIF-2alpha at the serine residue 51 then causes an inhibition of protein synthesis at the initiation step of translation.[111–113] This latter could be demonstrated by the growth of adenovirus lacking VAI RNA in cells expressing wild-type or mutant eIF-2. Adenovirus *dl*720 mutants grows very poorly in human 293 cells because of inhibition of protein synthesis due to phosphorylation of eIF-2. In 293 cells expressing the serine-to-alanine mutant of eIF-2alpha, however, the growth of adenovirus *dl*720 is restored since protein synthesis under these conditions is not inhibited.[114] These results confirm the critical role of eIF-2 phosphorylation in the mechanism of inhibition of protein synthesis.

Using the information given by the amino acid sequence of p68 kinase, two significant observations were deduced on the activity and on the dsRNA binding site.[115,116] (1) The mutations of the Lys residue in the protein kinase catalytic subdomain II (amino acid residue 295) to Arg or to Pro resulted in the generation of a protein devoid of activity. This is in total agreement with previously published results on other protein kinases, since this invariable Lys is directly involved in ATP binding and phosphotransfer reaction.[105] (2) The preparation of truncated proteins suggested the presence of a dsRNA binding site at the NH_2 terminal of p68 kinase located between amino acid residues 155 and 242. It should be emphasized that this latter does not rule out the possibility for the presence of another dsRNA binding site at the COOH terminal of p68 kinase. These two observations are employed below to demonstrate the functioning of p68 kinase in mechanisms such as antiviral action[117] and cell growth inhibition.[116]

3.6.1. Expression of p68 Kinase in Murine NIH 3T3 Cells

The cDNA encoding p68 kinase was transfected in murine NIH 3T3 cells using the pcDNA1/*neo* expression vector. Several stable clones were selected expressing the wild-type kinase.[117] The transfected kinase showed properties identical to the natural kinase, such as subcellular localization in the ribosomal fraction and dsRNA dependence for autophosphorylation. Preliminary results indicated that constitutive expression of the wild-type p68 kinase renders cells resistant to EMC virus infection in contrast to cells expressing the corresponding Lys-to-Arg mutant.

In parallel to these studies, we isolated some cells expressing wild-type kinase but not manifesting an apparent effect on the yield of EMC virus. For example, a clone referred to as 13.3 which expressed high levels of p68 kinase did not manifest any resistance to EMC virus infection. Upon treatment with interferon, this clone responded by the induction of the murine homologue of the dsRNA-activated kinase and the 2-5A synthetase. However, EMC virus yield was reduced at most by 1 log compared to 3–4 logs in the corresponding infected control cells. This latter result suggests that constitutive expression of a functional kinase *in vivo* might lead to counterselection of cells which manifest certain alterations, affecting the functioning of the kinase system and consequently resulting in a reduced antiviral protection.

3.6.2. Expression of p68 Kinase in Yeast, *Saccharomyces cerevisiae*

The α subunit of eIF-2 in the yeast shares 80% amino acid homology with its human equivalent when conservative substitutions are considered.[118] Because of this strong homology, Chong *et al.*[116] investigated the action of p68 kinase expression in yeast. Expression of the wild type but not the Lys-to-Arg mutant resulted in a dramatic growth-suppressing phenotype. The wild-type kinase was found to be activated in yeast probably by the dsRNA genome of fungal viruses (ScVs) which contaminate most laboratory strains of *Saccharomyces cerevisiae*.[119,120] Two sets of experiments were then carried out to demonstrate that inhibition of yeast growth is due to activation of p68 kinase and its functioning, i.e., phosphorylation of yeast eIF-2. Coexpression of the wild-type kinase with either the NH_2 terminus (fragment 1–256 amino acids) of p68 kinase or the yeast eIF-2alpha mutant Ser-51 to Ala resulted in the recovery of yeast growth. The NH_2-terminus domain probably antagonized activation of the wild-type p68 kinase by binding to dsRNA *in vivo*. On the other hand, overexpression of the eIF-2 mutant resulted in an increase in the eIF-2 pool which was not sensitive to the action of p68 kinase.

These results provide the first direct evidence that p68 kinase can mediate inhibition of growth through the phosphorylation of eIF-2. Previously inhibition of growth in yeast has been reported by the overexpression of the yeast protein kinase GCN2, which is involved in translational control of amino acid biosynthesis in yeast.[121,122] It will be interesting to determine whether p68 kinase is the human functional homologue of GCN2.

4. CONCLUSIONS

The role played by dsRNA as an interferon inducer and also as an activator of the interferon-induced enzymes 2-5A synthetase and p68 kinase remains an intriguing phenomenon.[33] Consequently, it remains tempting to speculate whether these enzymes are functional in the mechanism of interferon induction by dsRNA.

4.1. Localized Activation for the dsRNA-Activated Enzymes?

Baglioni[123] has suggested a mechanism for discrimination between viral and cellular mRNA based on observations in which mRNA covalently linked to dsRNA is degraded preferentially in extracts from interferon-treated cells compared to mRNA not linked to dsRNA.[124,125] These results suggest that in interferon-treated, virus-infected cells, localized activation of the nuclease in the vicinity of the replicative intermediate may be responsible for the mechanism of discrimination between viral and cellular mRNA. Such a mechanism might be operational in interferon-treated cells depending on the multiplicity of virus used during infection. At low multiplicity of virus, the small amounts of 2-5A synthesized in the vicinity of the replicative intermediate might be functional for degradation of viral RNA. At high multiplicities of virus, however, both viral and cellular RNA become degraded due to the production of large amounts of 2-5A. Other *in vitro* results on the functioning of the kinase system are in accord with a similar localized activation model.[26,126] Consistent with this, it has been demonstrated that inhibition of p68 kinase by VAI RNA *in vivo* results in a strong and selective increase in the translational efficiency of mRNA transcribed from a transfected plasmid.[127,128] Most probably, dsRNA generated from symmetric transcription of the transfected plasmid is the cause for the localized activation of the kinase.

We have previously reported that the level of p68 kinase is dramatically reduced in nonionic-detergent NP40 extracts obtained from interferon-treated cells during infection with EMC virus.[129] More recently, however, we were able to explain the cause of this reduction simply by the reduced NP40 solubility of p68 kinase occurring during EMC virus infection.[63] Consequently, p68 kinase can be recovered by extraction with an ionic detergent. Reduced NP40 extractibility of p68 kinase is dependent on the multiplicity of virus infection and seems to be specific, since other cellular proteins as well as the 100-kDa 2-5A synthetase (p100) are not modified. Immunofluorescence studies using specific antibodies demonstrated that p68 kinase which is distributed evenly in the cytoplasm of HeLa cells becomes concentrated around the nuclei after EMC virus infection. As a consequence of aggregating around the nuclei, p68 kinase might then resist extraction by NP-40. The aggregated kinase is found to be already activated probably due to binding to the replicative form and/or to replicative intermediates of EMC virus RNA. Through this process, the functioning of p68 kinase might be guaranteed by a localized activation in the replication complexes of EMC virus.

4.2. Wider Significance of the dsRNA-Activated Enzymes

The protein kinase and the 2-5A synthetase discovered as interferon-induced enzymes play key roles in inhibitory mechanisms leading to inhibition of protein synthesis. Constitutively expressed low levels of these enzymes are also present in most cells not treated with interferon. The presence of these enzymes in control cells could also be due to serum stimulation (unpublished results). Besides their indirect antiviral properties, the protein kinase and the 2-5A synthetase might be implied in cell growth and differentiation.[130–135]

The demonstration that different forms of 2-5A synthetase exist and manifest different properties suggests alternative roles for the different forms of 2-5A synthetase and their product molecules.[136] The p100 2-5A synthetase has been suggested to play a role in the mechanisms of splicing of precursor RNA transcripts, but no convincing evidence has been reported. The function of the alternative products of 2-5A synthetase has not been demonstrated. Certainly, addition of an AMP in 2′,5′ linkage to such important metabolites as NAD^+, ADP-ribose, and ApppppA should create significant physiological modifications.[46] Similarly, addition of an AMP to a tRNA molecule should cause its inactivation.[48]

The expression of p68 kinase in yeast demonstrated that at least one of the functions of this kinase is to phosphorylate eIF-2. *In vitro*, the protein kinase can also phosphorylate other substrates such as histones. *In vivo*, however, phosphorylation of histones is not increased in interferon-treated cells. Nevertheless, the question remains open whether the protein kinase causes the phosphorylation of nuclear or cytoplasmic proteins other than eIF-2. Recent reports have suggested that the induction of interferon by virus or dsRNA and the induction of certain genes by interferon might require the functioning of the protein kinase.[137–141] The expression of a cDNA encoding p68 kinase in eukaryotic cells will be invaluable in the further analysis of the role(s) of this enzyme.

References

1. Ehrenfeld, E., and Hunt, T., 1971, Double-stranded poliovirus RNA inhibits initiation of protein synthesis by reticulocyte lysates, *Proc. Natl. Acad. Sci. USA* **68:**1075–1078.

2. Hunt, T., Ehrenfeld, E., 1971, Cytoplasm of poliovirus-infected HeLa cells inhibits cell-free hemoglobin synthesis, *Nature* **230:**91–94.
3. Hunter, T., Hunt, T., and Jackson, R. G., 1975, The characteristics of inhibition of protein synthesis by double-stranded ribonucleic acid in reticulocyte lysate, *J. Biol. Chem.* **250:**409–417.
4. Kerr, I. M., Brown, R. E., and Ball, L. A., 1974, Increased sensitivity of cell-free protein synthesis of dsRNA after interferon treatment, *Nature* **250:**57–59.
5. Friedman, R. M., Metz, D. H., Esteban, R. M., Towell, D. R., Ball, L. A., and Kerr, I. M., 1972, Mechanism of interferon action: Inhibition of viral messenger ribonucleic acid translation in L-cell extracts, *J. Virol.* **10:**1184–1198.
6. Kerr, I. M., Friedman, R. E., Brown, R. E., Ball, A. L., and Brown, J. C., 1974, Inhibition of protein synthesis in cell-free systems from interferon-treated infected cells: Further characterization and effect of formylmethionyl-tRNA, *J. Virol.* **13:**9–21.
7. Roberts, W. K., Clemens, M. J., and Kerr, I. M., 1976, Interferon induced inhibition of protein synthesis in L cell extracts: An ATP dependent step in the activation of an inhibition by dsRNA, *Proc. Natl. Acad. Sci. USA* **73:**3136–3140.
8. Roberts, W. K., Hovanessian, A. G., Brown, R. E., Clemens, M. J., and Kerr, I. M., 1976, Interferon-mediated protein kinase and low molecular weight inhibitor of protein synthesis, *Nature* **264:**477–480.
9. Lebleu, B., Sen, G. C., Shaila, S., Carer, B., and Lengyel, P., 1976, Interferon, dsRNA and protein phosphorylation, *Proc. Natl. Acad. Sci. USA* **73:**335–341.
10. Zilberstein, A., Federman, P., Shulman, L., and Revel, M., 1976, Specific phosphorylation *in vitro* of a protein associated with ribosomes of interferon-treated mouse L cells, *FEBS Lett.* **68:** 119–124.
11. Kerr, I. M., Brown, R. E., and Hovanessian, A. G., 1977, Nature of the inhibitor of cell-free protein synthesis formed in response to interferon and dsRNA, *Nature* **268:**540–542.
12. Hovanessian, A. G., Brown, R. E., and Kerr, I. M., 1977. Synthesis of low molecular weight inhibitor of protein synthesis with enzyme from interferon-treated cells, *Nature* **268:**537–540.
13. Kerr, I. M., and Brown, R. E., 1978, pppA2′p5′A3′p5′A: An inhibitor of protein synthesis synthesized with an enzyme fraction from interferon-treated cells, *Proc. Natl. Acad. Sci. USA* **75:** 256–260.
14. Hovanessian, A. G., and Kerr, I. M., 1979, The (2′5′) oligoadenylate pppA2′p5′A2′p5′A synthetase and protein kinase(s) from interferon-treated cells, *Eur. J. Biochem.* **93:**515–526.
15. Clemens, M. J., and Williams, B. R. G., 1978, Inhibition of protein synthesis by pppA2′p5′A2′p5′A: A novel oligonucleotide synthesized by interferon-treated L cell extracts, *Cell* **13:**565–572.
16. Baglioni, C., Minks, M. A., and Maroney, P. A., 1978, Interferon action may be mediated by activation of a nuclease by pppA2′p5′A2′p5A, *Nature* **279:**684–687.
17. Brown, G. E., Lebleu, B., Kawakita, M., Shaila, S., Sen, G. C., and Lengyel, P., 1976, Increased endonuclease activity in an extract from mouse Ehrlich ascites tumor cells which had been treated with a partially purified interferon preparation: Dependence on dsRNA, *Biochem. Biophys. Res. Commun.* **69:**114–122.
18. Williams, B. R. G., Golgher, R. R., Brown, R. E., Gilbert, C. S., and Kerr, I. M., 1979, Natural occurrence of 2-5A in interferon-treated EMC virus infected L cells, *Nature* **282:**581–586.
19. Williams, B. R. G., and Kerr, I. M., 1978, Inhibition of protein synthesis by 2′5′ linked adenine oligonucleotides in intact cells, *Nature* **276:**88–89.
20. Hovanessian, A. G., Wood, J. N., Meurs, E., and Montagnier, L., 1979, Increased nuclease activity in cells treated with pppA2′p5′A2′p5′A, *Proc. Natl. Acad. Sci. USA* **76:**3261–3265.
21. Hovanessian, A. G., and Wood, J. N., 1980, Anticellular and antiviral effects of pppA (2′p5′A)n, *Virology* **101:**81–90.
22. Farrel, P. J., Balkow, K., Hunt, T., Jackson, J., and Trachsel, H., 1977, Phosphorylation of initiation factor IF-E2 and the control of reticulocyte protein synthesis, *Cell* **11:**187–200.

23. Samuel, C. E., Farris, D. A., and Eppstein, D. A., 1977, Mechanism of interferon action, kinetics of interferon action in mouse L929 cells: Translation inhibition, protein phosphorylation and messenger RNA methylation and degradation, *Virology* **83:**56–68.
24. Samuel, C., 1979, Mechanism of interferon action: Phosphorylation of protein synthesis initiation factor eIF2 in interferon-treated human cells by a ribosome-associated kinase processing site specificity similar to hemin-regulated rabbit reticulocyte kinase, *Proc. Natl. Acad. Sci. USA* **76:**515–526.
25. Safer, B., 1983, 2B or not 2B: Regulation of the catalytic utilization of eIF2, *Cells* **33:**7–8.
26. De Benedett, A., and Baglioni, C., 1985, Kinetics of dephosphorylation of eIF2 (P) and reutilization of mRNA, *J. Biol. Chem.* **260:**3135–3139.
27. Johnston, M. I., and Torrence, P. F., 1984, The role of interferon-induced proteins, double-stranded RNA and 2′,5′ oligoadenylate in the interferon-mediated inhibition of viral translation, in *Interferon*, Volume 3: *Mechanism of Production and Action* (R. M. Friedman, ed.), pp. 189–298, Elsevier, Amsterdam.
28. Minks, M. A., West, D. K., Benvin, S., and Baglioni, C., 1979, Structural requirements of double-stranded RNA for the activation of 2′5′ oligo(A) polymerase and protein kinase of interferon-treated cells, *J. Biol. Chem.* **254:**10180–10183.
29. Edery, I., Pretryshyn, R., and Sonenberg, N., 1989, Activation of double-stranded RNA-dependent kinase (dsl) by the TAR region of HIV-1 mRNA: A novel translational control mechanism, *Cell* **56:**303–312.
30. Sen Gupta, D. N., and Silverman, R. H., 1989, Activation of interferon regulated dsRNA dependent enzymes by human immunodeficiency virus 1 leader RNA, *Nucleic Acids Res.* **17:** 969–978.
31. Schröder, H. C., Ugarkovic, D., Wenger, R., Reuter, P., Okamoto, T., and Müller, W. E. G., 1990, Binding of Tat protein to TAR region of human immunodeficiency virus type 1 blocks TAR-mediated activation of (2′–5′) oligoadenylate synthetase, *AIDS Res. Hum. Retroviruses* **6:**659–672.
32. Lebleu, B., and Content, J., 1982, Mechanisms of interferon action: Biochemical and genetic approaches, in: *Interferon 1982*, Volume 4 (I. Gresser, ed.), pp. 47–94, Academic Press, New York.
33. Lengyel, P., 1987, Double-stranded RNA and interferon action, *J. Interferon Res.* **7:**511–519.
34. Kerr, I. M., 1987, The 2-5A system: A personal view, *J. Interferon Res.* **7:**505–510.
35. Hovanessian, A. G., 1980, Double-stranded RNA dependent protein kinase(s) in rabbit reticulocyte lysates analogous to that from interferon-treated cells, *Biochimie* **62:**775–778.
36. Hovanessian, A. G., and Kerr, I. M., 1978, Synthesis of an oligonucleotide inhibitor of protein synthesis in rabbit reticulocyte lysates analogous to that formed in extracts from interferon-treated cells, *Eur. J. Biochem.* **84:**149–159.
37. Hovanessian, A. G., and Riviere, Y., 1980, Interferon-mediated induction of 2-5A synthetase and protein kinase in the liver and spleen of mice infected with NDV or injected with poly(I)·poly(C), *Ann. Virol. Inst. Pasteur* **131E:**501–602.
38. Krishnan, I., and Baglioni, C., 1980, 2′5′-oligo(A) polymerase activity in serum of mice infected with EMC virus or treated with interferon, *Nature* **285:**485–488.
39. Krust, B., Rivière, Y., and Hovanessian, A. G., 1982, p67K kinase in different tissues and plasma of control and interferon-treated mice, *Virology* **120:**240–246.
40. Galabru, J., Robert, N., Buffet-Janvresse, C., Rivière, Y., and Hovanessian, A. G., 1985, Continuous production of interferon in normal mice: The effect of anti-interferon globulin, sex, age, strain and environment on the levels of 2-5A synthetase and p67K kinase, *J. Gen. Virol.* **66:**711–718.
41. Schattner, A., Merlin, G., Wallach, D., Rosenberg, H., Bino, T., Hahn, I., Levin, S., and Revel, M. J, 1981, Monitoring of interferon therapy by assay of (2′-5′) oligo-isoadenylate synthetase in human peripheral white blood cells, *J. Interferon Res.* **1:**587–594.

42. Schattner, A., Wallach, D., Merlin, G., Hahn, T., Levin, S., Ramot, B., and Revel, M. J., 1982, Variation of (2′-5′) oligo synthetase level in lymphocytes and granulocytes of patients with viral infections and leukemia, *J. Interferon Res.* **2:**355–363.
43. Buffet-Janvresse, C., Magard, H., Robert, N., and Hovanessian, A. G., 1983, Assay and the levels of 2-5A synthetase in lymphocytes of patients with viral, bacterial and autoimmune diseases, *Ann. Immunol. Inst. Pasteur* **134D:**247–258.
44. Buffet-Janvresse, C., Vannier, J. P., Laurent, A. G., Robert, N., and Hovanessian, A. G., 1986, Enhanced level of double-stranded RNA-dependent protein kinase in peripheral blood mononuclear cells of patients with viral infections. *J. Interferon Res.* **6:**85–96.
45. Schmidt, A., Chernajovsky, Y., Shulman, L., Federman, P., Berissi, H., and Revel, M., 1979, An interferon induced phosphodiesterase degrading (2′-5′) oligoisoadenylate and CCA terminus of tRNA, *Proc. Natl. Acad. Sci. USA* **76:**4788–4792.
46. Ball, L. A., and White, C. N., 1979, Induction, purification, and properties of 2′,5′ oligoadenylate synthetase, in: *Regulation of Macromolecular II Synthesis by Low Molecular Weight Mediators* (H. Koch and D. Ritters, eds.), pp. 303–318, Academic Press, New York.
47. Justesen, J., Ferbus, D., and Thang, M. N., 1980, Elongation mechanism and substrate specificity of 2′,5′-oligoadenylate synthetase, *Proc. Natl. Acad. Sci. USA* **77:**4618–4622.
48. Justesen, J., Worm-Leonhard, H., Ferbus, D., and Petersen, H. U., 1985, The interferon-induced enzyme 2-5A synthetase adenylates tRNA, *Biochimie* **67:**651–655.
49. Wreschner, D. H., James, T. C., Silverman, R. H., and Kerr, I. M., 1981, Ribosomal RNA cleavage nuclease activation and 2-5A (ppp(A2′p)$_n$A) in interferon-treated cells, *Nucleic Acids Res.* **9:**1571–1581.
50. Wreschner, D. H., McCauley, J. W., Skehel, J. J., and Kerr, I. M.,1981, Interferon action sequence specificity of the ppp(A2′p)$_n$A-dependent ribonuclease, *Nature* **289:**414–417.
51. Floyd-Smith, G., Slattery, E., and Lengyel, P., 1981, Interferon action: RNA cleavage pattern of a (2′-5′) oligo-adenylate-dependent endonuclease, *Science* **219:**1030–1032.
52. Slattery, E., Ghosh, N., Samanta, H., and Lengyel, P., 1979, Interferon, double-stranded RNA and RNA degradation: Activation of an endonuclease by (2′-5′)$_A$n, *Proc. Natl. Acad. Sci. USA* **76:**4778–4782.
53. Jacobsen, H., Czarnieki, W., Krause, D., Friedman, R. M., and Silverman, R. H., 1983, Interferon-induced synthesis of 2-5A dependent RNase in mouse JLS V9R cells, *Virology* **125:** 496–501.
54. Hovanessian, A. G., Laurent, A. G., Chebath, J., Galabru, J., Robert, N., and Svab, J., 1987, Identification of 69 kd and 100 kd forms of 2-5A synthetase in interferon-treated human cells by specific monoclonal antibodies, *EMBO J.* **6:**1273–1280.
55. Chebath, J., Benech, P., Hovanessian, A. G., Galabru, J., and Revel, M., 1987, Four different forms of interferon-induced 2′-5′ oligo (A) synthetase identified by immunoblotting in human cells, *J. Biol. Chem.* **262:**3852–3857.
56. Benech, P., Merlin, G., Revel, M., and Chebath, J., 1985, 3′ end structure of the human (2′-5′) oligo A synthetase gene: Prediction of two distinct proteins with cell-type-specific expression, *Nucleic Acids Res.* **13:**1267–1281.
57. Benech, P., Mory, Y., Revel, M., and Chebath, J., 1985, Structure of two forms of the interferon-induced (2′-5′) oligo A synthetase of human cells based on cDNAs and gene sequences, *EMBO J.* **4:**2249–2256.
58. Saunders, M. E., Gewert, D. R., Tugwell, M. E., McMahon, M., and Williams, B. R. G., 1985, Human 2-5A synthetase: Characterization of a novel cDNA and corresponding gene structure, *EMBO J.* **4:**1761–1768.
59. Wathelet, M., Moutschen, S., Cravador, A., Dewit, L., Defilippi, P., Huez, G., and Content, J., 1986, Full-length sequence and expression of the 42 kDa 2-5A synthetase induced by human interferon, *FEBS Lett.* **196:**113–117.
60. Hovanessian, A. G., Svab, J., Marié, I., Robert, N., Chamaret, S., and Laurent, A. G., 1988,

Characterization of 69–100 Kda forms of 2-5A synthetase from interferon-treated human cells, *J. Biol. Chem.* **263:**4945–4949.

61. Marié, I., Galabru, J., Svab, J., and Hovanessian, A. G., 1989, Preparation and characterization of polyclonal antibodies specific for the 69 and 100 k-Dalton forms of human 2-5A synthetase. Biochem, *Biophys. Res. Commun.* **160:**580–587.
62. Marié, I., Svab, J., Robert, N., Galabru, J., and Hovanessian, A. G., 1990, Differential expression and distinct structure of 69 and 100 kDa forms of 2-5A synthetase in human cells treated with interferon, *J. Biol. Chem.* **265:**18601–18607.
63. Dubois, M. F., and Hovanessian, A. G., 1990, Modified subcellular localization of interferon-induced p68 kinase during encephalomyocarditis virus infection, *Virology* **179:**591–598.
64. Marié, I., and Hovanessian, A. G., 1992, The 69-kd 2-5A synthetase is composed of two homologous and adjacent functional domains, *J. Biol. Chem.* **267:**9933–9939.
65. Ichii, Y., Fukunaga, R., Shiojiri, S., and Sokawa, Y., 1986, Mouse 2-5A synthetase cDNA: Nucleotide sequence and comparison to human 2-5A synthetase, *Nucleic Acids Res.* **14:**10117.
66. Nilsen, T. W., Maroney, P. A., and Baglioni, C., 1982, Synthesis of (2′-5′)oligoadenylate and activation of an endoribonuclease in interferon-treated HeLa cells infected with reovirus, *J. Virol.* **42:**1039–1045.
67. Hersh, C. L., Brown, R. E., Roberts, W. K., Suyryd, E. A., Kerr, I. M., and Stark, G. R., 1984, Simian virus 40-infected, interferon-treated cells contain 2′,5′oligoadenylates which do not activate cleavage of RNA, *J. Biol. Chem.* **259:**1731–1737.
68. Cayley, P. J., Davies, J. A., McCullagh, K. G., and Kerr, I. M., 1984, Activation of the ppp(A2′p)nA system in interferon-treated, herpes simplex virus-infected cells and evidence for novel inhibitors of the ppp(A2′)nA-dependent RNase, *Eur. J. Biochem.* **143:**165–174.
69. Paez, E., and Esteban, M., 1984, Resistance of vaccinia virus to interferon is related to an interference phenomenon between the virus and the interferon system, *Virology* **134:**12–28.
70. Silverman, R. H., Cayley, J. P., Knight, M., Gilbert, C. S., and Kerr, I. M., 1982, Control of the ppp(A2′p)$_n$A system in HeLa cells: Effects of interferon and virus infection, *Eur. J. Biochem.* **124:**131–138.
71. Chebath, J., Benech, P., Revel, M., and Vigneron, M., 1987, Constitutive expression of (2′-5′) oligo A synthetase confers resistance to picornavirus infection, *Nature* **330:**587–588.
72. Rysiecki, G., Gewert, D. R., and Williams, B. R. G., 1989, Constitutive expression of a 2′-5′-oligoadenylate synthetase cDNA results in increased antiviral activity and growth suppression, *J. Interferon Res.* **9:**649–657.
73. Coccia, E. M., Romeo, G., Nissim, A., Marziali, G., Albertini, R., Affabris, E., Battistini, A., Fiorucci, G., Orsatti, R., Rossi, G. B., and Chebath, J., 1990, A full-length murine 2-5A synthetase cDNA transfected in NIH-3T3 cells impairs EMCV but not VSV replication, *Virology* **179:**228–233.
74. Hovanessian, A. G., 1989, The double-stranded RNA-activated protein kinase induced by interferon: dsRNA-PK, *J. Interferon Res.* **9:**641–647.
75. Berry, M. J., Knutson, G. S., Lasky, S. R., Muneemitsu, S. M., and Samuel, C. F., 1985, Purification and substrate specificities of the double-stranded RNA-dependent protein kinase from untreated and interferon-treated mouse fibroblasts, *J. Biol. Chem.* **260:**11240–11247.
76. Krust, B., Galabru, J., and Hovanessian, A. G., 1984, Further characterization of the protein kinase activity mediated by interferon in mouse and human cells, *J. Biol. Chem.* **259:**8494–8498.
77. Laurent, A. G., Krust, A. G., Galabru, J., Svab, J., and Hovanessian, A. G., 1985, Monoclonal antibodies to interferon induced 68,000-Mr protein and their use for the detection of double-stranded RNA dependent protein kinase in human cells, *Proc. Natl. Acad. Sci. USA* **82:**4341–4345.
78. Galabru, J., and Hovanessian, A. G., 1985, Two interferon-induced proteins are involved in the protein kinase complex dependent on double-stranded RNA, *Cell* **43:**685–694.

79. Galabru, J., and Hovanessian, A. G., 1987, Autophosphorylation of the protein kinase dependent on double-stranded RNA, *J. Bio. Chem.* **262:**15538–15544.
80. Hovanessian, A. G., and Galabru, J., 1987, The double-stranded RNA-dependent protein kinase is also activated by heparin, *Eur. J. Biochem.* **167:**467–473.
81. Galabru, J., Katze, M. G., Robert, N., and Hovanessian, A. G., 1989, The binding of double-stranded RNA and adenovirus VAI RNA to interferon-induced protein kinase, *Eur. J. Biochem.* **178:**581–589.
82. Kostura, M., and Mathews, M. B., 1989, Purification and activation of the double-stranded RNA-dependent eIF2 kinase DAI, *Mol. Cell. Biol.* **9:**195–200.
83. Szyska, R., Kudlicki, W., Kramer, G., Hardesty, B., Galabru, J., and Hovanessian, A. G., 1989, A type 1 phosphoprotein phosphatase active with phosphorylated Mr 68,000 initiation factor 2 kinase, *J. Biol. Chem.* **264:**3827–3831.
84. O'Malley, R. P., Mariano, T. M., Siekierka, J., and Mathews, M. B., 1986, A mechanism for the control of protein synthesis by adenovirus VAI RNA, *Cell* **44:**391–400.
85. Kitajewski, J., Schneider, R. J., Safer, B., Munemitsu, S. M., Samuel, C. E., Thimmapaya, B., and Shenk, T., 1986, Adenovirus VAI RNA antagonizes the antiviral action of interferon by preventing activation of the interferon induced eIF2 kinase, *Cell* **45:**195–200.
86. Katze, M. G., De Corato, D., Safer, B., Galabru, J., and Hovanessian, A. G., 1987, Adenovirus VAI RNA complexes with the 68,000-Mr protein kinase to regulate its autophosphorylation and activity, *EMBO J.* **6:**689–697.
87. Maran, A., and Mathews, M. B., 1988, Characterization of the double-stranded RNA implicated in the inhibition of protein synthesis in cells infected with a mutant adenovirus defective for VA RNA, *Virology* **264:**106–113.
88. Gupta, S. L., Holmes, S. L., and Mehra, L. L., 1982, Interferon action against reovirus: Activation of interferon-induced protein kinase in mouse L-929 cells upon reovirus infection, *Virology* **120:**495–499.
89. Nilsen, T. W., Maroney, P. A., and Baglioni, C., 1982, Inhibition of protein synthesis in reovirus-infected HeLa cells with elevated levels of interferon-induced protein kinase activity, *J. Biol. Chem.* **257:**14,593–14,596.
90. Samuel, C. E., Duncan, G. S., Knutson, G. S., and Hershey, J. W. B., 1984, Mechanism of interferon action. Increased phosphorylation of protein synthesis initiation factor eIF-2 in interferon-treated, reovirus-infected L929 fibroblasts *in vitro* and *in vivo*, *J. Biol. Chem.* **259:** 13,451–13,457.
91. Rice, A. P., Duncan, R., Hershey, J. W. B., and Kerr, I. M., 1985, Double-stranded RNA-dependent protein kinase and 2-5A system are both activated in interferon-treated, encephalomyocarditis virus-infected HeLa cells, *J. Virol.* **54:**894–898.
92. Katze, M. G., Tomita, J., Black, T., Krug, R. M., Safer, B., and Hovanessian, A. G., 1988, Influenza virus regulates protein synthesis during infection by repressing autophosphorylation and activity of the cellular 68,000 Mr protein kinase, *J. Virol.* **62:**3710–3717.
93. Black, T. L., Safer, B., Hovanessian, A. G., and Katze, M. G., 1989, The cellular 68,000 Mr protein kinase is highly autophosphorylated and activated yet significantly degraded during poliovirus infection: Implications for translational regulation, *J. Virol.* **63:**2244–2251.
94. Lengyel, P., 1981, Enzymology of interferon action—A short survey, *Meth. Enzymol.* **79:** 135–148.
95. Johnston, M. I., and Torrence, P. F., 1984, The role of interferon-induced proteins, double-stranded RNA and 2′,5′ oligoadenylate in the interferon-mediated inhibition of viral translation, in: *Interferon*, Volume 3: *Mechanism of Production and Action* (R. M. Friedman, ed.), pp. 189–298, Elsevier, Amsterdam.
96. Roy, S., Agy, M., Hovanessian, A. G., Sonenberg, N., and Katze, M. G., 1991, The integrity of the stem structure of human immunodeficiency virus type 1 Tat-responsive sequence RNA is

required for interaction with the interferon-induced 68,000 Mr protein kinase, *J. Virol.* **65:** 632–640.

97. Sonenberg, N., 1990, Measures and countermeasures in the modulation of initiation factor activities by viruses, *New Biol.* **2:**402–409.
98. Roy, S., Katze, M. G., Edery, I., Hovanessian, A. G., and Sonenberg, N., 1990, Control of the interferon-induced 68,000 Mr protein kinase by the HIV-1 TAT gene product, *Science* **247:**1216–1219.
99. Lee, T. G., Tomita, J., Hovanessian, A. G., and Katze, M. G., 1990, Purification and partial characterization of a cellular inhibitor of the interferon-induced protein kinase of Mr 68,000 from influenza virus-infected cells, *Proc. Natl. Acad. Sci. USA* **87:**6208–6212.
100. Imani, F., and Jacobs, B. L., 1988, Inhibitory activity for the interferon induced protein kinase is associated with the reovirus serotype 1 σ 3 protein, *Proc. Natl. Acad. Sci. USA* **85:**7887–7891.
101. Akkaraju, G. R., Whitaker-Dowling, P., Younger, J. S., and Jagus, R., 1989, Vaccinia specific kinase inhibitory factor prevents translational inhibition by double-stranded RNA in rabbit reticulocyte lysate, *J. Biol. Chem.* **264:**10321–10325.
102. Meurs, E., Chong, K., Galabru, J., Thomas, S. B., Kerr, I. M., Williams, B. R. G., and Hovanessian, A. G., 1990, Molecular cloning and characterization of cDNA encoding human double-stranded RNA activated protein kinase induced by interferon, *Cell* **62:**379–390.
103. Edelman, A. M., Blumenthal, D. K., and Krebs, E. G., 1987, Protein serine/threonine kinases, *Annu. Rev. Biochem.* **56:**567–613.
104. Langan, T. A., 1978, Methods for the assessment of site-specific histone phosphorylation, *Meth. Cell Biol.* **19:**127–142.
105. Hanks, S. K., Quinn, A. M., and Hunter, T., 1988, The protein kinase family: Conserved features and deduced phylogeny of the catalytic domains, *Science* **241:**42–52.
106. Takio, K., Blumenthal, D. K., Walsh, K. A., Titani, K., and Krebs, E. G., 1986, Amino acids sequence of rabbit skeletal muscle myosin light chain kinase, *Biochemistry* **25:**8049–8057.
107. Klemsz, M. J., McKercher, S. R., and Maki, R. A., 1987, Nucleotide sequence of the mouse hck gene, *Nucleic Acids Res.* **15:**9600.
108. Sukegawa, J., Semba, K., Yamanashi, Y., Nishizawa, M., Miyajima, N., Yamamoto, T., and Toyoshima, K., 1987, Characterization of cDNA clones for the human c-*yes* gene, *Mol. Cell. Biol.* **7:**41–47.
109. Kazlauskas, A., and Cooper, J. A., 1989, Autophosphorylation of the PDGF receptor in the kinase insert region regulates interactions with cell proteins, *Cell* **58:**1121–1133.
110. Taylor, G. R., Reedijk, M. R., Rothwell, V., Rohrschneider, L., and Pawson, T., 1989, The unique insert of cellular and viral fms protein tyrosine kinase domains is dispensible for enzymatic and transforming activities, *EMBO J.* **8:**2029–2037.
111. Colthurst, D. R., Campbell, D. G., and Proud, C. G., 1987, Structure and regulation of eukaryotic initiation factor eIF2 sequence of the α subunit phosphorylated by the haem-controlled repressor by the double-stranded RNA-activated inhibitor, *Eur. J. Biochem.* **166:** 357–363.
112. Pathak, V. K., Schindler, D., and Hershey, J. W. B., 1988, Generation of a mutant form of protein synthesis initiation factor eIF2 lacking the site of phosphorylation by eIF2 kinase, *Mol. Cell. Biol.* **8:**993–995.
113. Pain, V. M., 1986, Initiation of protein synthesis in mammalian cells, *Biochem. J.* **235:**625–637.
114. Davies, M. V., Furtado, M., Hershey, J. W. B., Thimmapaya, B., and Kaufman, R. J., 1989, Complementation of adenovirus associated RNA I gene deletion by expression of a mutant eukaryotic translation initiation factor, *Proc. Natl. Acad. Sci. USA* **86:**9163–9167.
115. Katze, M. G., Wambach, M., Wong, M. L., Garfinkel, M., Meurs, E., Chong, K., Williams, B. R. G., Hovanessian, A. G., and Barber, G. N., 1991, Functional expression and RNA binding

analysis of the interferon-induced, dsRNA activated 68,000 Mr protein kinase in a cell-free system, *Mol. Cell. Biol.* **11:**5497–5505.

116. Chong, E., Feng, L., Donahue, T. F., Friesen, J. D., Hovanessian, A. G., and Williams, B. R. G., 1991, Human p68 kinase exhibits growth suppression in yeast and homology to the translational regulator GCN2, *EMBO J.* **11:**1553–1562.
117. Meurs, E., Watanabe, Y., Barber, G. N., Katze, M., and Hovanessian, A. G., 1992, The expression of the interferon-induced dsRNA activated p68 kinase in transfected mammalian cells, *J. Virol.* **66:**5805–5814.
118. Cigan, A. M., Pabich, E. K., Feng, L., and Donahue, T. F., 1989, Yeast translation initiation suppressor sui2 encodes the α subunit of eukaryotic initiation factor 2 and shares sequence identity with the human α subunit, *Proc. Natl. Acad. Sci. USA* **86:**2784–2788.
119. Schnitt, J. J., and Tipper, D. J., 1990, K_{28}, a unique double-stranded RNA killer virus of *Saccharomyces cerevisiae*, *Mol. Cell. Biol.* **10:**4807–4815.
120. Wicker, R. A., 1986, Double-stranded RNA replication in yeast: The killer system, *Ann. Rev. Biochem.* **55:**373–395.
121. Tzamarias, D., and Thireos, G., 1988, Evidence that the GCN2 protein kinase regulates reinitiation by yeast ribosomes, *EMBO J.* **7:**3547–3551.
122. Hinnebusch, A. G., 1988, Mechanisms of gene regulation in the general control of amino acid biosynthesis in *Saccharomyces cerevisiae*, *Microbiol. Rev.* **52:**248–273.
123. Baglioni, C., 1979, Interferon induced enzymatic activities and their role in the antiviral state, *Cell* **17:**255–264.
124. Nilsen, T. W., and Baglioni, C., 1979, Mechanism for discrimination between viral and host mRNA in interferon-treated cells, *Proc. Natl. Acad. Sci. USA* **76:**2600–2604.
125. Baglioni, C., Nilsen, T. W., Maroney, P. A., and Ferra, F., 1982, Molecular mechanisms of action of interferon, *Tex. Rep. Biol. Med.* **41:**471–478.
126. De Benedetti, A., and Baglioni, C., 1984, Inhibition of mRNA binding to ribosomes by localized activation of dsRNA-dependent protein kinase, *Nature* **311:**79–81.
127. Kaufman, R. J., and Murtha, P., 1987, Translational control mediated by eucaryotic initiation factor-2 is restricted to specific mRNAs in transfected cells, *Mol. Cell. Bio.* **7:**1568–1571.
128. Akusjarvi, G., Svensson, C., and Nygard, O., 1987, A mechanism by which adenovirus-associated RNA_I controls translation in a transient expression assay, *Mol. Cell. Biol.* **7:**549–551.
129. Hovanessian, A. G., Galabru, J., Meurs, E., Buffet-Janvresse, C., Svab, J., and Robert, N., 1987, Rapid decrease in the levels of the double-stranded RNA-dependent protein kinase during virus infections, *Virology* **159:**126–136.
130. Pretryshyn, R., Chen, J. J., and London, I. M., 1988, Detection of activated double-stranded RNA-dependent protein kinase in 3T3-F442A cells, *Proc. Natl. Acad. Sci. USA* **85:**1427–1431.
131. Wells, V., and Mallucci, L., 1985, Expression of the 2-5A system during the cell cycle, *Exp. Cell Res.* **159:**27–36.
132. Mechti, N., Affabri, E., Romeo, G., Lebleu, B., and Rossi, G. B., 1984, Role of interferon and 2′,5′oligoadenylate synthetase in erythroid differentiation of Friend leukemia cells, *J. Biol. Chem.* **259:**3261–3265.
133. Sokawa, Y., Nagata, K., and Ichikawa, Y., 1981, Induction and function of (2′-5′) oligoadenylate synthetase in differentiation of mouse myeloid leukemia cells, *Exp. Cell Res.* **135:**191–197.
134. Ferbus, D., Testa, U., Titeux, M., Louache, F., and Thang, M. N., 1985, Induction of (2′-5′) oligoadenylate synthetase activity during granulocyte and monocyte differentiation, *Mol. Cell. Biochem.* **67:**125–133.
135. Kumar, R., and Mendelsohn, J., 1990, Growth regulation of A431 cells, *J. Biol. Chem.* **265:** 4578–4582.
136. Hovanessian, A. G., 1991, Interferon-induced and double-stranded RNA-activated enzymes: A specific protein kinase and 2′,5′-oligoadenylate synthetase, *J. Interferon Res.* **11:**199–205.
137. Marcus, P. I., and Sekellick, M. J., 1988, Interferon induction by viruses XVI. 2-Aminopurine

blocks selectively and reversibly an early stage in interferon induction, *J. Gen. Virol.* **69:**1637–1645.

138. Zinn, K., Keller, A., Whittemore, L. A., and Maniatis, R., 1988, 2-Aminopurine selectively inhibits the induction of β-interferon, c-*fos* and c-*myc* gene expression, *Science* **240:**210–213.
139. Tiwari, R. K., Kusari, J., Kumar, R., and Sen, G. C., 1988, Gene induction by interferons and double-stranded RNA: Selective inhibition by 2-aminopurine, *Mol. Cell. Biol.* **8:**4289–4294.
140. Lew, D. J., Decker, T., and Darnel, J. E., Jr., 1989, Alpha interferon and gamma interferon stimulate transcription of a single gene through different signal transduction pathways, *Mol. Cell. Biol.* **9:**5404–5411.
141. Williams, B. R. G., 1991, Signal transduction and transcriptional regulation of interferon-α-stimulated genes, *J. Interferon Res.* **11:**207–213.

Chapter 9

Translational Regulation by Vaccinia Virus

Rostom Bablanian

1. POXVIRUS GROUP

Poxviruses are DNA-containing viruses that infect both vertebrate and invertebrate hosts. Smallpox, caused by an orthopoxvirus, was a major scourge to humanity for centuries. Vaccination against smallpox was introduced by Jenner in 1798 and by means of this method it was eventually possible to eradicate this disease over a decade ago.[1] Vaccinia viruses (VV) are the most frequently studied members of the orthopoxvirus group. Recently, the entire genome of the Copenhagen strain of VV was sequenced because it is recognized that these viruses have the potential of being used as live recombinant expression vectors for molecular studies as well as for developing new vaccines.[2] Studies with VV have resulted in several novel findings which not only have contributed to the field of virology, but have also enriched our understanding of important eukaryotic processes. The RNA polymerase,[3,4] the poly(A) polymerase,[5,6] and the 5′ cap of messenger RNAs (mRNAs)[7] were all reported in studies with VV about the same time or before they were recognized in eukaryotic cells.

Rostom Bablanian • Department of Microbiology and Immunology, SUNY, Health Science Center at Brooklyn, Brooklyn, New York 11203.

Translational Regulation of Gene Expression 2, edited by Joseph Ilan. Plenum Press, New York, 1993.

2. POXVIRUS-ASSOCIATED ENZYMES

Among all the animal viruses, poxviruses are probably the least dependent on the host cell for their multiplication. A unique feature of these DNA-containing viruses is their mode of replication, which takes place almost exclusively in the cytoplasm of the infected cell. Poxviruses have evolved mechanisms to conduct replication without the involvement of the host nucleus. Consequently, they possess a large number of enzymes which are packaged in the infecting virion and are involved in a multitude of functions culminating in the replication of the progeny virus in the cytoplasm of the host cell. Enzymes which have been isolated from purified VV include the RNA polymerase, the poly(A) polymerase, the capping enzyme complex, the RNA methyl transferase, two nucleotide triphosphate phosphohydrolases, a nicking-closing enzyme, a DNA topoisomerase, and a protein kinase (for a review, see Moss[8]).

3. VACCINIA VIRUS-INDUCED CYTOPATHOLOGY

Vaccinia viruses are highly cytocidal. Infection of cells by these viruses gives rise to both morphological and biochemical lesions. When cells are infected at high multiplicities a morphological lesion in the form of cell rounding appears within 1 hr after infection. This cell rounding persists through the infection until late, when infected cells begin to fuse, forming polykaryocytes. Early studies suggested that virus-associated component(s) may be involved in early virus-induced morphological changes. It was shown later, however, that in order for these changes to occur, virus-induced polypeptide synthesis was required. The nature and the mechanism of action of this putative polypeptide(s) remains unknown. Cell fusion usually occurs late in infection. If cells are infected at high multiplicities in the presence of inhibitors of RNA synthesis, however, "fusion from without" may also be seen with certain strains of VV (for review see Bablanian[9,10] and Schrom and Bablanian[11]). Infection of cells at high multiplicities also causes a dramatic reduction in DNA, RNA, and protein synthesis (for review see Bablanian[10]). It was suggested that the inhibition of host DNA synthesis was caused by a component of the virus[12] and it was later demonstrated that a virus-associated DNase enters the nucleus and acts on single-stranded DNA.[12,13] The synthesis of RNA in infected cells was inhibited by 3 hr after infection[14,15] and about 50% of some host mRNAs was degraded in infected cells at about this time.[16] By far the most studied biochemical lesion of VV has been inhibition of host-cell protein synthesis. Depending on the cell type and multiplicity of infection used, this event becomes apparent as early as 20 min after infection.[17] (for review see Bablanian[10]).

This chapter will be concerned with the mechanisms of inhibition of host-cell protein synthesis (shutoff) by VV. The mechanisms proposed for shutoff by vaccinia virus infections fall into four general categories: (1) alterations in intracellular ionic environment, (2) degradation of host-cell mRNAs, (3) inhibition caused by virus components, and (4) inhibition caused by virus-induced or virus-directed RNAs. It is important to emphasize that VV-induced shutoff is selective: host-cell protein synthesis is inhibited, whereas viral protein synthesis persists. Therefore, for any hypothesis on shutoff to be viable, it must take selectivity into consideration.

4. ROLE OF INTRACELLULAR IONIC ENVIRONMENT IN VACCINIA VIRUS–INDUCED SHUTOFF

Many cytocidal viruses alter the permeability of the membrane of infected cells causing an imbalance in the sodium/potassium ion concentration gradient.[18] It has been demonstrated that VV infection causes an early and transient alteration in membrane permeability followed by a gradual resealing of the membrane by 2 hr after infection.[19] A role for this transient leakiness of VV-infected cell membranes in shutoff has not been advanced.

5. ROLE OF DEGRADATION OF HOST-CELL mRNAs IN SHUTOFF

Degradation of only host-cell mRNAs after VV infection would certainly appear to be a viable hypothesis in causing selective shutoff as was defined above. The only study on this phenomenon, however, shows a 40–60% degradation of two prominent host-cell mRNAs at 3 hr after infection.[16] Shutoff in VV-infected cells is greater than 80% within 20 min after infection, using the same type of cells at comparable multiplicities of infection,[17] suggesting that degradation per se does not contribute significantly toward this selective inhibition. During the advanced stages of infection, however, selective degradation would undoubtedly favor the translation of viral mRNAs.

6. ROLE OF VACCINIA VIRUS COMPONENTS IN SHUTOFF

Early observations concerning VV-induced shutoff suggested that some constituent of the VV particle may be responsible for causing this selective inhibition.[20] This conclusion was based on the assumption that 5 μg/ml of actinomycin D is sufficient to prevent virus-induced RNA synthesis. Subsequent work, however, demonstrated that under these conditions of infection, small size virus-induced RNA synthesis does occur.[21,22] Additional evidence in support of the involvement of a virion component in shutoff came from *in vitro* studies where VV cores were added directly to cell-free protein-synthesizing systems. In one study, a large amount of viral cores was added to a reticulocyte lysate system, causing inhibition of endogenous protein synthesis.[23] On this basis, it was concluded that a structural component of the core was responsible for this inhibition. In a second study, where a coupled transcription–translation reticulocyte lysate system was used, it was demonstrated that the addition of 1.2×10^9 VV cores/ml resulted in transcription followed by translation. If the amount of cores was increased tenfold, however, transcription increased significantly but translation was inhibited.[24] These authors suggested that this inhibition of polypeptide synthesis may be related to the increased amount or viral cores used. In a third study a basic phosphoprotein component of the cores was isolated which was shown to inhibit protein synthesis in a cell-free system. This inhibition of protein synthesis, however, was not selective, since this component from the cores also inhibited VV mRNA translation.[25] A fourth study implicated VV surface tubules in shutoff.[26] These authors showed that surface tubules obtained by nonionic detergent Nonidet P40 (NP40) inhibited globin translation in a messenger-dependent, reticulocyte lysate cell-free system, but did not exclude the possibility that NP40 may have stayed bound to these hydrophobic proteins[27,28] and could have been the cause of this inhibition. Since vaccinia virus mRNA translation in the presence of these surfaces tubules was not tested, it is not known whether these tubules are selective inhibitors. In other studies using surface tubules, it was shown that this virion component did not inhibit protein synthesis either in HeLa cells or in the reticulocyte lysate cell-free system, but rather, acted from within the infected cells to cause degeneration of cells at late times after infection.[29,30] In a coupled transcription–translation reticulocyte lysate cell-free system rendered message dependent, an attempt was made to define the *in vitro* inhibition caused by VV cores.[31] These authors demonstrated that when they boiled cores and then added them to the cell-free system, these were just as inhibitory for protein synthesis as the untreated cores, and thus concluded that the inhibition could not be caused by a protein component of the cores. It is known that VV particles, in addition to possessing DNA as their genome, also contain RNA.[32,33] In view of the above reports that high concentrations of cores inhibit protein synthesis in cell-free systems, and that this inhibition was not due to a protein component of the virion,[31] it was of interest to determine whether RNA extracted from VV cores had an inhibitory effect on translation in the reticulocyte lysate system. The results from this study showed that indeed core-associated RNA was able to inhibit protein synthesis in a messenger-dependent reticulocyte lysate system. This inhibition of translation was selective; host-cell mRNA translation was inhibited, whereas VV mRNA translation was not.[34] Thus, an RNA component of the virion cores does fit the criterion established above, namely, that it inhibits protein synthesis in a selective manner.

7. ROLE OF RNA IN SHUTOFF

Data supporting the view that VV-induced RNA is involved in shutoff come from studies using inhibitors of RNA and protein synthesis. The observation was made that in infected HeLa cell suspensions treated with 5–10 μg/ml of actinomycin D (ACD) there was an enrichment of a small polyadenylated class of RNAs which sedimented at 5S,[35] as opposed to normal VV transcripts, which sediment at 10–14S.[14,15] On the basis of these results, it was suggested that the small poly(A)-rich RNAs may compete with host mRNAs and contribute to shutoff.[35] In other studies, a correlation between the rate of RNA synthesis and the extent of shutoff was established in several VV-infected cell types.[36–38] It was demonstrated that in the presence of high doses of ACD and also in the presence of cordycepin, shutoff did not occur in VV-infected cells.[37] Using VV irradiated with various doses of ultraviolet light to infect cells in the presence of 20 μg/ml of ACD, it was demonstrated that small viral RNA species were associated with shutoff.[39] Earlier work employing an *in vitro* system had suggested that small polyadenylated molecules like those found in ACD-treated infected cells were also present when cells were infected with UV-irradiated virus.[40]

7.1. Early Vaccinia Virus Transcripts and Shutoff

On the basis of the above results, it was speculated that a component(s) of early VV transcripts was a primary candidate for the inhibition of host-cell protein synthesis. Therefore, the effect of VV *in vitro*-transcribed RNAs was tested on translation of exogenous mRNAs in nuclease-treated reticulocyte lysates. It was demonstrated that the translation of several cellular mRNAs and encephalomyocarditis (EMC) viral mRNAs was inhibited by the *in vitro*-transcribed VV RNA products in the cell-free system. The translation of VV mRNAs obtained from infected cells, however, was only minimally affected by these *in vitro* transcripts.[35]

7.2. Poly(A) and Its Role in Shutoff

Additional studies indicated that the selective inhibitory property of the *in vitro* RNAs resided only in small, nontranslated polyadenylated RNAs.[41] This selective inhibitory property of the small poly(A)-containing RNAs was abrogated only when the RNA was deadenylated, an observation implicating poly(A) in this inhibition.[42] Furthermore, when the effect of synthetic poly(A) on the translation of exogenous mRNAs in reticulocyte lysates was assayed, it was found that poly(A) inhibited the translation of HeLa-cell mRNAs more effectively than the translation of vaccinia virus mRNAs, again demonstrating selectivity in translation.[42] It has been shown that in the presence of ATP, virus cores synthesize only poly(A)[43] utilizing the core-associated poly(A) polymerase.[44–46] Using this procedure, it was demonstrated that VV cores synthesized only poly(A) and that this poly(A) inhibited protein synthesis in the cell-free system in a similar selective manner as synthetic poly(A) or the poly(A) obtained after the removal of the 5′-end leader from short polyadenylated RNAs of normal VV *in vitro* transcripts.[47]

Thus, the data obtained with all these small, nontranslated polyadenylated RNAs from *in vitro* transcripts (designated as POLADS) indicate that POLADS are selective inhibitors of protein synthesis and that their active moiety is the 3′-end poly(A).

7.3. POLADS Found in Vaccinia Virus-Infected Cells

At this juncture, it was of great importance to determine if POLADS were also present in infected cells. Indeed, POLADS were found in three different cell types infected with VV and had similar properties as the small polyadenylated RNAs synthesized *in vitro* from viral cores. POLADS obtained from these infected cells inhibited the translation of HeLa-cell mRNAs and not that of VV mRNAs when tested in the nuclease-treated reticulocyte lysate system. The active moiety of these *in vivo* POLADS was also poly(A), similar to that found in *in vitro* POLADS.[48]

7.4. Role of Poly(A) in Translation

Because the active moiety of POLADS was shown to be the 3′-poly(A) of this molecule, it was important to determine what role poly(A) plays in the translation of cellular and VV mRNAs. The poly(A) found at the 3′ terminus of most eukaryotic mRNAs has been implicated in a number of cellular functions, but the precise role of poly(A) in translation has not yet been determined.[49–51] An initial observation showed that several polynucleotides, including poly(A), inhibited translation in reticulocyte lysates at the level of initiation.[52] In a later study, however, it was suggested that poly(A), as opposed to other homopolymers, was a specific inhibitor of protein synthesis.[53] These authors demonstrated that free poly(A) preferentially affected the rate of translation of poly(A)$^+$ rather than poly(A)$^-$ mRNAs, leading to the suggestion that poly(A)$^+$ mRNAs required a component found in the lysate which was being sequestered by free poly(A). A candidate for this component was speculated to be the poly(A)-binding protein (PAB).[53] These results were substantiated by the observation that the translation of poly(A)$^+$ globin mRNA was sensitive, but poly(A)$^-$ tobacco mosaic virus (TMV) was resistant to exogenously added free poly(A) in the reticulocyte lysate system.[54] Employing reovirus mRNAs, which are known to be naturally poly(A)$^-$,[55] it was demonstrated that the overall translation of these mRNAs was also resistant to free poly(A) added to an L-cell lysate system.[56] More recently, it was shown that inhibition of translation by poly(A) could be reversed by addition of purified PAB,[54] and it has been proposed that the PAB bound to the 3′ terminus of mRNAs may facilitate translation through an interaction with the 60S ribosomal subunit.[49]

Vaccinia virus mRNAs, whether transcribed *in vitro* or *in vivo*, possess poly(A) at their 3′ terminus like most eukaryotic mRNAs.[57,44,58] This poly(A) also resembles eukaryotic poly(A) in that it is added to viral mRNA posttranscriptionally.[59] A terminal riboadenylate transferase has been found in vaccinia virus cores which is capable of adding sequences of (A)s to various primers.[45,60] The length of the 3′-poly(A) of VV mRNAs transcribed either *in vitro* or *in vivo* was found to be similar in size for the majority of the chains, i.e., around 100–170 residues long.[58,59] Not much is known about the 5′ ends of VV mRNAs except that the late mRNAs predominantly contain poly(A) of variable lengths at their 5′ ends[61–63] with an undetermined role.

7.5. Exogenously Added Poly(A) and Its Role in Translation of mRNAs Devoid of Poly(A) Tails

Since both host-cell and VV mRNAs are known to be capped and polyadenylated, reasons were sought to determine why VV mRNAs are more efficient than HeLa-cell mRNAs in initiating their translation in the presence of POLADS or poly(A). Therefore, we sought the special feature(s) that provide VV mRNAs with a leading edge in translation over HeLa mRNAs. It was previously demonstrated that between 5 and 10% of the mRNAs synthesized both *in vitro* and *in vivo* by vaccinia virus are devoid of poly(A) tails and that the polypeptides synthesized in a Krebs II ascites cell-free extract either by poly(A)$^+$ or poly(A)$^-$ mRNAs are both quantitatively and qualitatively similar.[59] As was mentioned above, in VV-infected cells, a population of nontranslated polyadenylated sequences (POLADS) is produced with poly(A) chain lengths from 50 to over 200 nucleotides. These POLADS have a similar effect on translation as free poly(A): they inhibit the translation of HeLa mRNAs, but have little or no effect on the translation of VV mRNAs in the reticulocyte lysate system.[48] In infected cells, shutoff or selective inhibition occurs in the presence of these POLADS; therefore, *a priori* we have to assume that VV mRNAs must be less sensitive to the effect of poly(A) and thus are behaving in a similar manner as mRNAs without a poly(A) tail, which have been shown to be more resistant to the effect of free poly(A).[53,54,56] When VV mRNAs [from the total infected cell extract or after oligo(dT)–cellulose selection] were translated in the nucleased-treated reticulocyte lysates they were found to be more resistant to exogenously added poly(A) than were similarly obtained HeLa mRNAs. However, in comparison, the VV mRNAs obtained from the total cytoplasmic RNA were much more resistant to the effect of exogenously added poly(A) than the VV mRNAs selected by oligo(dT)–cellulose chromatography. This result suggested that in the total RNA pool from VV-infected cells some of the mRNAs may be without poly(A) tails, giving rise to a population of mRNAs which was more resistant to the inhibitory effect of poly(A) (Fig. 1).[64] Another observation supporting this view is seen in Fig. 2, where it is shown that poly(A)$^+$ VV mRNAs are more resistant than poly(A)$^+$ HeLa mRNAs to inhibition by poly(A). Even more dramatic is the almost complete lack of inhibition of translation of poly(A)$^-$ VV mRNA by poly(A).[64] Thus, while the presence of a poly(A) tail clearly yields an mRNA that is

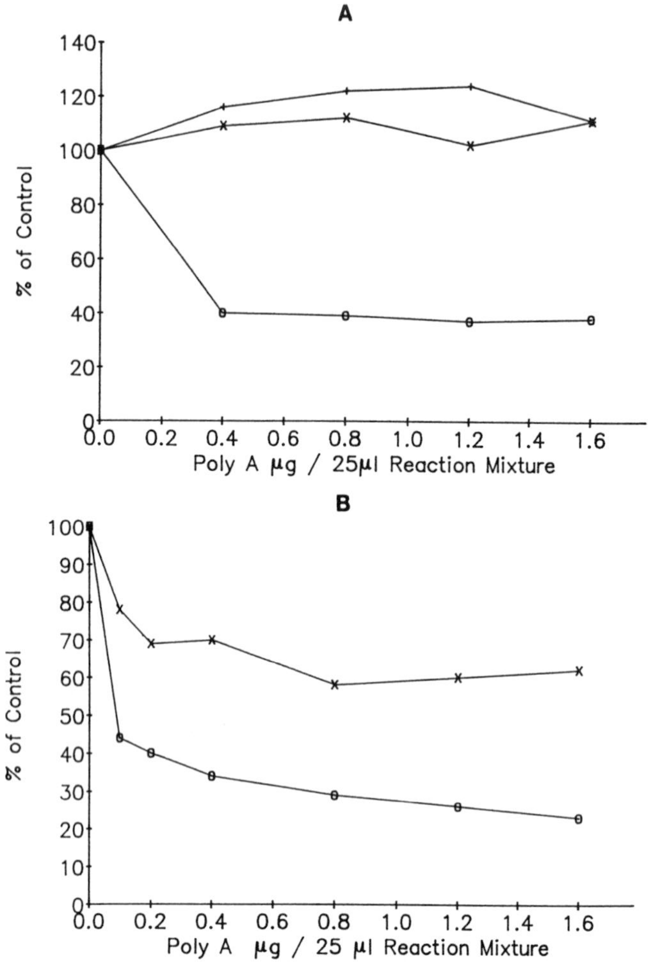

Figure 1. Translation of HeLa and VV mRNAs obtained from total cytoplasmic RNA or after oligo(dT)–cellulose chromatography. **(A)** (□) HeLa (15 μg) (×) early VV (5 μg), and (+) late VV (7.5 μg) total RNA obtained from uninfected or infected cells were added to the nucleased reticulocyte lysates in the presence of various concentrations of poly$(A)_{150\text{-}300}$. Twenty microliters of the reaction mixtures was subjected to 5–15% polyacrylamide gel electrophoresis (PAGE) and exposed to an X-ray film. The extent of translation was determined by laser densitometry of the autoradiograms and plotted as percent of control. **(B)** (□) HeLa (0.6 μg) and (×) early VV (0.2 μg) mRNAs were added to the nucleased reticulocyte lysates in the presence of various concentrations of poly$(A)_{150\text{-}300}$. Twenty microliters of the reaction mixtures was subjected to 5–15% PAGE and exposed to X-ray film. The extent of translation was determined by laser densitometry of the autoradiogram and plotted as percent of control. From Bablanian *et al.*[64]

sensitive to translation inhibition by poly(A), VV mRNA contains some unique feature that is not present in HeLa mRNAs that minimizes this inhibition.

7.6. Do Vaccinia Virus mRNAs Initiate Internally?

It is possible that this unique element is also responsible for the lack of sensitivity of VV mRNAs to inhibition by m^7GTP. A scanning model proposed by Kozak[65] has shown that the 43S ribosomal complex binds to the 5′ terminus of an mRNA and scans the 5′ noncoding region until it reaches the initiation codon. In recent years a number of atypical picornavirus mRNAs have been found which lack a cap structure and initiate translation internally upstream in the 5′ noncoding region without scanning.[66–68] Vaccinia virus mRNAs are known to be capped, monocistronic structures and one would assume that these mRNAs would be cap dependent. It was therefore an unexpected finding that the cap analog m^7GTP was significantly less inhibitory for the translation of VV mRNAs compared to HeLa-cell mRNAs (Fig. 3).[64] These data suggest that the cap structure is not absolutely required for the translation of VV early and late mRNAs as it is for HeLa-cell mRNAs and may also indicate that VV mRNAs can initiate internally. It is of interest to note that the 26S mRNA of Semliki Forest virus, which is known to be capped, is also not sensitive to a cap analog.[69] Both Semliki Forest 26S and VV mRNAs must possess additional features which make them capable of initiating under conditions of decreased cap-specific initiation factors. One of these features could be the 5′ secondary structure with respect to the activity of eIF-4F in regulating translation. Considerable data have been accumulated which indicate that translational efficiency is inversely correlated with secondary structure.[54,70–75]

7.7. Inhibition of Translation by Poly(A) Is Partially Reversed by eIF-4A

To test the effect of initiation factors on the poly(A)-mediated inhibition of translation, we used several purified initiation factors, including eIF-4A, eIF-4F, and eIF-4B. Only purified eIF-4A was capable of partially reversing the inhibition of HeLa-cell mRNA translation caused by a poly(A) chain of 150–300 nucleotides, but not the inhibition cause by longer chains of poly(A). On the other hand, the poly(A)-mediated inhibition of early VV mRNAs caused by the longer chains of poly(A) was completely reversed by eIF-4A.[64] Early in infection, VV mRNAs which are more resistant to the effect of poly(A) might escape its inhibitory effects by having a reduced requirement for unwinding of the mRNA by eIF-4A to allow ribosome binding. While this may be part of the poly(A) inhibition scheme, the major portion of the reversal of poly(A)-mediated inhibition of translation is probably mediated by PAB. The reversal by eIF-4A is in addition to the reversal brought about by PAB, so that eIF-4A reversal becomes more evident when PAB is above limiting quantities.[64]

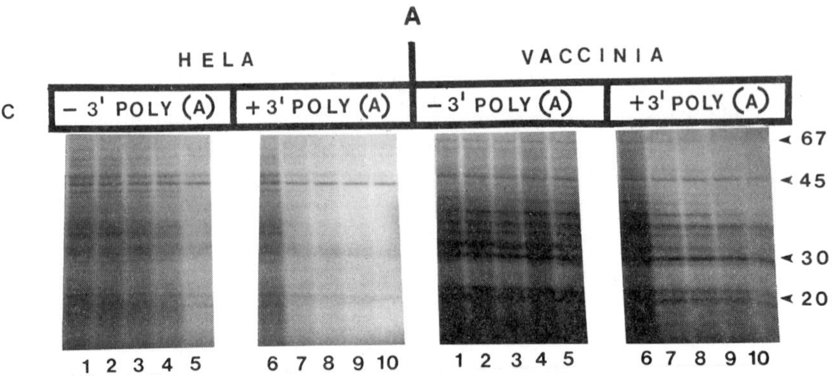
A
HELA
VACCINIA
C
− 3' POLY (A)
+ 3' POLY (A)
− 3' POLY (A)
+3' POLY (A)
67
45
30
20
1 2 3 4 5
6 7 8 9 10
1 2 3 4 5
6 7 8 9 10

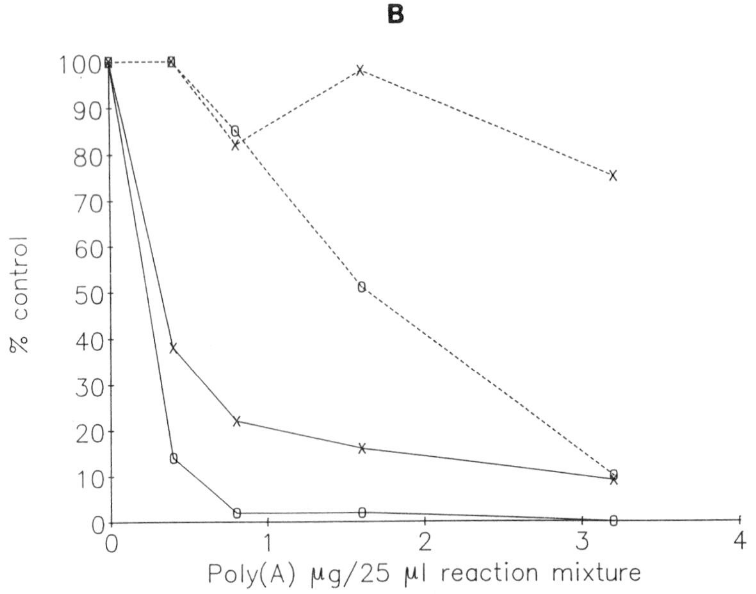
B
% control
Poly(A) µg/25 µl reaction mixture
0
10
20
30
40
50
60
70
80
90
100
1
2
3
4

7.8. Inhibition of Translation by Poly(A) or POLADS Is Completely Reversed by PAB

It has been demonstrated that *in vitro* inhibition of translation by poly(A) can be essentially overcome by the exogenous addition of PAB, which indicates that this inhibition may be due to the depletion of endogenous PAB.[54] These observations in the *in vitro* system have been confirmed by *in vivo* data which demonstrate that PAB is essential for translation, because depletion of PAB results in the inhibition of both initiation and poly(A) tail shortening which is necessary for initiation of translation.[76] Exogenously added poly(A) probably acts as a sink to soak up free PAB, thus depriving mRNA poly(A) tails of the ability to interact with PAB. The inhibitory effect of POLADS with different poly(A) lengths on the translation *in vitro* of HeLa and viral mRNAs can be reversed in the presence of PAB. Reversal of inhibition in the presence of PAB underscores the critical role of the poly(A) moiety of POLADS in the induction of inhibition. Although POLADS with different poly(A) lengths differ in their efficacy as inhibitors, they most likely inhibit protein synthesis in the cell-free system in a similar manner as free poly(A), because earlier work had demonstrated that the inhibition by poly(A) can be reversed by PAB.[54] The binding of PAB to the poly(A) of mRNA transcripts is believed to be necessary for the optimal conformational folding of the transcript prior to its initial association with the ribosome.[77] It is therefore possible that POLADS possessing longer poly(A) tails (50–200 nucleotides) sequester more PAB and thereby decrease its availability to bind to the poly(A) tail of HeLa mRNA transcripts, which possess shorter poly(A) tails (50–70 nucleotides).[78,79] This, in turn, renders their efficient initiation of translation more unlikely. If PAB associates stoichiometrically with poly(A), its ability to reverse a translationally inhibited system is directly commensurate to the amount of inhibitor present, be it in the form of POLADS or free poly(A). The addition of exogenous PAB to an inhibited *in vitro* translation system can counteract the depletion of this factor and thus reverse the inhibition. Both HeLa-cell and VV poly(A)$^+$ mRNAs seem to

←

Figure 2. The effect of deadenylation of HeLa and VV mRNAs on the poly(A)-mediated inhibition of translation in the reticulocyte lysate system. **(A)**HeLa-cell and early VV mRNAs were deadenylated. Nucleased-treated reticulocyte lysates were programmed with deadenylated or mock-treated HeLa (0.6 μg/25 μl assay) or VV (0.6 μg/25 μl assay) mRNAs in the presence of various concentrations of poly(A)$_{150\text{-}300}$. Lanes 1–5, deadenylated mRNAs; lanes 6–10, mock-treated mRNAs. Lanes 1 and 6, no poly(A); lanes 2 and 7, poly(A), 0.4 μg; lanes 3 and 8, poly(A), 0.8 μg; lanes 4 and 9, poly(A), 1.6 μg; and lanes 5 and 10, poly(A), 3.2 μg. Twenty microliters of the reaction mixture was subjected to 5–15% PAGE and exposed to an X-ray film. The numbers on the right of the autoradiograms represent molecular mass (kilodaltons) of marker proteins. **(B)** The relative translational efficiency of deadenylated compared to adenylated HeLa and early VV mRNAs in the presence of various concentrations of poly(A). The percent translation of deadenylated HeLa and VV mRNAs was compared to that of the mRNAs with poly(A)s obtained by laser densitometry. The translation of poly(A)$^-$ and poly(A)$^+$ mRNAs in the presence of various concentrations of poly(A) is plotted. (□—□) HeLa mRNA with poly(A); (×—×) VV mRNA with poly(A); (□- - -□) HeLa mRNA without poly(A); (×- - -×) VV mRNA without poly(A).

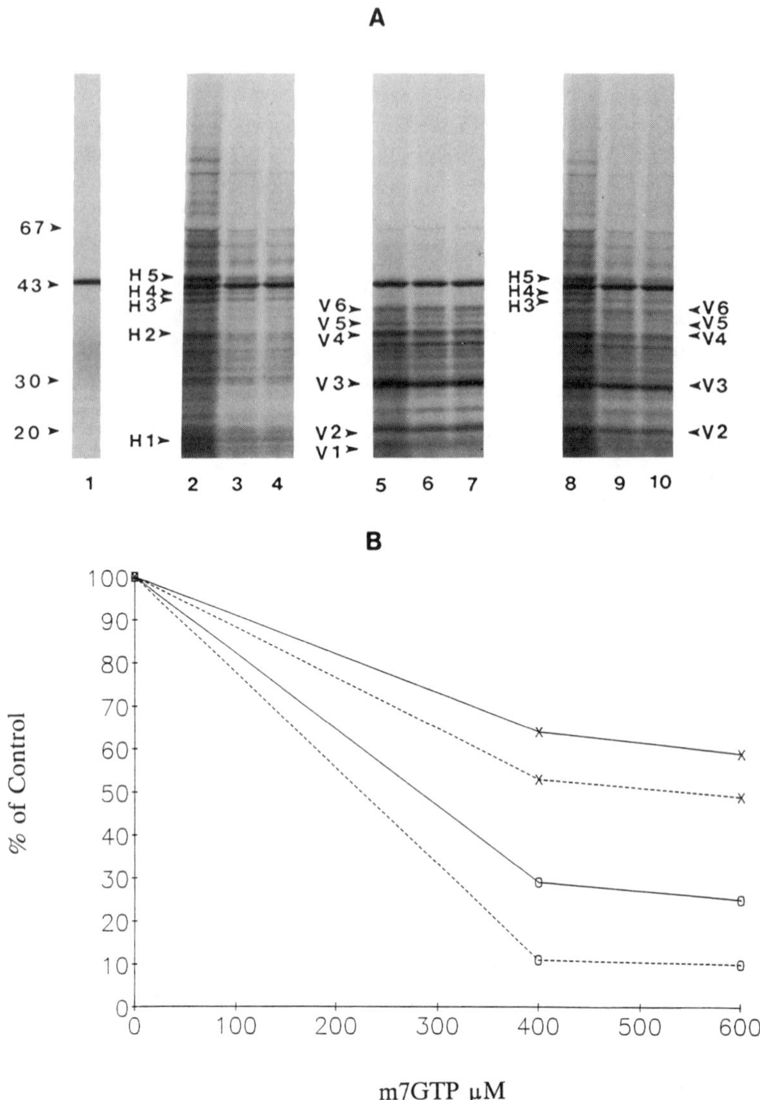

Figure 3. The translational efficiency of the mixed HeLa and VV mRNAs in the presence of various concentrations of m^{7}GTP. **(A)** HeLa and early VV mRNAs, either separately or mixed together, were translated in the reticulocyte lysate system in the presence of various concentrations of m^{7}GTP. The arrows specify the individual polypeptides, HeLa (H) or VV (V). Lane 2, HeLa mRNA; lanes 3 and 4, HeLa mRNA with 400 and 600 μM of m^{7}GTP, respectively; lane 5, early VV mRNA; lanes 6 and 7, early VV mRNA with 400 and 600 μM of m^{7}GTP, respectively; lane 8, HeLa + early VV mRNA; lanes 9 and 10, HeLa + early VV mRNA with 400 and 600 μM of m^{7}GTP, respectively. Lane 1, no added RNA. The numbers on the left represent the molecular mass (kilodaltons) of marker proteins. **(B)** Translation (percent of control) of various polypeptides in the presence of 400 and 600 μM of m^7 GTP. (□—□) HeLa polypeptides; (□- - -□) HeLa polypeptides from the mixture; (×—×) VV polypeptides; (×- - -×) VV polypeptides from the mixture. From Bablanian *et al.*[64]

require PAB for initiation of their translation. The disparity in poly(A) tail lengths between HeLa-cell mRNAs (50–70 residues) and VV mRNAs (50–170)[59] may be advantageous for VV transcripts, as it would enhance their likelihood of binding to PAB and their subsequent translation in the presence of POLADS. Therefore, a general paucity of PAB would compromise the translation of HeLa mRNAs rather than that of VV mRNAs.

7.9. Inhibition of Host-Cell and Vaccinia Virus Protein Synthesis by Aberrant POLADS

An observation made many years ago by several investigators showed that when cells were infected with UV-irradiated VV or infected in the presence of ACD both host and viral protein synthesis were inhibited. It was therefore of interest to determine if POLADS were produced in these infected cells and were related to this type of inhibition. We isolated and tested the effect of POLADS produced when cells were infected under normal conditions (V-POLADS), with UV-irradiated VV (UV-POLADS), or infected in the presence of actinomycin D (ACD) (ACD-POLADS). Under these aberrant conditions of infection many more POLADS were produced, and, on average, possessed longer poly(A) tails than those found in normal infections.[80] In order to explain differences in the inhibitory activity of the various types of POLADS, we sought to correlate putative variations in the length of the poly(A) tails of the various types of POLADS with their inhibitory activity, since it was demonstrated that longer poly(A) chains (>200 residues) are more inhibitory on the translation *in vitro* of host and viral transcripts than are smaller chains.[47,53] Thus, POLADS of the same size class were subjected to enzymatic treatment to remove the leader sequence at the 5′ terminus of the RNA; the remaining poly(A) was then terminally labeled. By comparing the 3′-poly(A) of POLADS of a particular class, we were able to correlate the range of poly(A) tail lengths of a certain type of POLADS with its expected inhibitory potential. When equal amounts of each type of POLADS were assayed for their inhibitory activity on HeLa mRNA translation, UV-POLADS and ACD-POLADS were found to be more inhibitory, most probably because they had a greater percentage of longer poly(A) tails than did V-POLADS. To determine the effect of POLADS on the translation of both host and viral mRNAs, we measured the translation *in vitro* of HeLa-cell and vaccinia viral mRNAs in the presence of incremental concentrations of a single class of POLADS (up to 320 nucleotides). V-POLADS compromised both HeLa-cell and VV mRNA translation in a dose-dependent manner. Their inhibitory activity on the translation of HeLa mRNAs, however, was comparatively greater. UV-POLADS at identical concentrations, however, displayed greater inhibitory activity on HeLa and VV protein synthesis than did V-POLADS. Similarly, ACD-POLADS were equally potent as UV-POLADS in inducing a dose-dependent inhibition of translation of both HeLa and VV mRNA transcripts. As was shown to be the case with HeLa-cell protein synthesis, the comparatively greater inhibitory potency of UV-POLADS and

ACD-POLADS on vaccinia virus protein synthesis most likely resulted from a greater proportion of lengthy poly(A) tails in the aliquots containing these POLADS.[80] It is unclear what regulates the addition of adenosine residues to leader sequences to achieve poly(A) tails of various lengths in VV-infected cells; a polyadenylation termination factor or signal has not been identified. It is conceivable, however, that expression of the viral gene encoding a putative regulatory factor is affected by ACD or UV irradiation. Thus, the extension of oligonucleotide primers in cells infected with UV-irradiated virus or virus in the presence of ACD proceeds unregulated, leading to the formation of UV-POLADS and ACD-POLADS with longer poly(A) tails than those synthesized under normal conditions of infection. The synthesis of POLADS in VV-infected cells during the early phase of infection favors the selective translation of VV versus host transcripts. It is likely that the virus evolved this mechanism to establish selectivity early in infection.[48] It appears, however, that during infection with UV-irradiated virus or virus grown in the presence of ACD, cells produce POLADS capable of debilitating the synthesis of the virus' own proteins. These POLADS have aberrantly long poly(A) tails and are over produced under these conditions. Whatever the precise cause or mechanism of this phenomenon, it is obvious that the deregulated production of POLADS has a deleterious effect on host as well as on virus protein synthesis, which suggests that, most likely, their synthesis is normally controlled.

8. CONCLUDING REMARKS

Vaccinia virus is a large and complex animal virus and probably possesses several strategies for bringing about selective inhibition of host protein synthesis. The strategy of using POLADS for this purpose may be one of the many tactics employed by VV that we are just beginning to understand. There remains a great deal of work to be accomplished before we can definitely claim that POLADS indeed are involved in shutoff. Future work in this area of research in our laboratory will be concerned with the study of the regulation of POLADS in intact cells. We will study the effect of expressing POLADS *in vivo* under inducible conditions, to determine if inhibition of host-cell protein synthesis will occur. We will also express the PAB gene in a recombinant VV in infected cells, to determine if shutoff will be prevented under these conditions. It is hoped that these approaches will lead to a better understanding of POLADS in relation to the phenomenon of shutoff.

ACKNOWLEDGMENTS. I wish to thank Santo Scribani and Nike Cacoullos for their help in the preparation of the manuscript. I also wish to express my gratitude to Dr. Mariano Esteban for valuable comments and his critical reading of the manuscript.

References

1. Fenner, F., Henderson, D. A., Arita, I., Jerzek, Z., and Ladnyi, I. D., Smallpox and its eradication, 1988, WHO, Geneva.
2. Goabel, S. C., Johnson, G. P., Perkus, M. E., Davis, S. W., Winslow, J. P., and Paoeletti, E., 1990, *Virology* **179:**247–266.
3. Kates, J. R., and McAuslan, B. R., 1967, *Proc. Natl. Acad. Sci. USA* **58:**134–141.
4. Munyon, W. E., Paoletti, E., and Grace, J. T., 1967, *Proc. Natl. Acad. Sci. USA* **58:**2280–2288.
5. Kates, J. R., and Beeson, J., 1970, *J. Mol. Bio.* **50:**19–23.
6. Brakel, C., and Kates, J. R., 1974, *J. Virol.* **14:**715–723.
7. Wei, C. M., and Moss, B., 1975, *Proc. Natl. Acad. Sci. USA* **72:**318–322.
8. Moss, B., 1990, In *Virology*, 2nd ed. (B. N. Fields and D. M. Knipe, eds.), pp. 2079–2111, Raven Press, New York.
9. Bablanian, R., 1975, *Prog. Med. Virol.* **19:**40–83.
10. Bablanian, R., 1984, In *Comprehensive Virology* (H. Frankel-Conrat and R. R. Wagner, eds.), pp. 391–429, Plenum Press, New York.
11. Schrom, M., and Bablanian, R., 1981, *Arch. Virol.* **70:**173–187.
12. Pogo, B. G. T., and Dales, S., 1973, *Proc. Natl. Acad. Sci. USA* **70:**1726–1729.
13. Pogo, B. G. T., and Dales, S., 1974, *Virology* **58:**377–386.
14. Becker, Y., and Joklik, W. K., 1964, *Proc. Natl. Acad. Sci. USA* **51:**577–585.
15. Salzman, N. P., Shatkin, A. J., and Sebring, E. D., 1964, *J. Mol. Biol.* **8:**405–416.
16. Rice, A. P., and Roberts, B. E., 1983, *J. Virol.* **47:**529–539.
17. Metz, D. H., and Esteban, M., 1972, *Nature* **238:**385–388.
18. Carrasco, L., and Lacal, J. C., 1983, *Pharmacol. Ther.* **23:**109.
19. Carrasco, L., and Esteban, M., 1982, *Virology* **117:**62–69.
20. Moss, B., 1968, *J. Virol.* **2:**1028–1037.
21. Rosemond-Hornbeak, H., and Moss, B., 1975, *J. Virol.* **16:**34–42.
22. Schrom, M., and Bablanian, R., 1979, *J. Gen. Virol.* **44:**625–638.
23. Ben-Hamida, F., and Beaud, G., 1978, *Proc. Natl. Acad. Sci. USA* **75:**175–179.
24. Cooper, J. A., and Moss, B., 1978, *Virology* **88:**149–165.
25. Ben-Hamida, F., Person, A., and Beaud, G., 1983, *J. Virol.* **45:**452–455.
26. Mbuy, G. N., Morris, R. E., and Bubel, H. C., 1982, *Virology* **116:**137–147.
27. Holloway, P. W., 1973, *Anal. Biochem.* **53:**304–308.
28. Helenius, A., and Simons, K., 1975, *Biochim. Biophys. Acta* **415:**29–79.
29. Wolstenholme, J., Woodward, C. G., Burgoyne, R. D., and Stephen, J., 1977, *Arch. Virol.* **53:** 25–37.
30. Burgoyne, R. D., and Stephen, J., 1979, *Arch. Virol.* **59:**107–119.
31. Pelham, H. R. B., Sykes, J. M. M., and Hunt, T., 1978, *Eur. J. Biochem.* **82:**199–209.
32. Planterose, D. N., Nishimura, C., and Salzman, N. P., 1962, *Virology* **18:**294–301.
33. Roening, G., and Holowczak, J. A., 1974, *J. Virol.* **47:**704–708.
34. Coppola, G., and Bablanian, R., 1983, *Proc. Natl. Acad. Sci. USA* **80:**75–79.
35. Rosemond-Hornbeak, H., and Moss, B., 1975, *J. Virol.* **16:**34–42.
36. Bablanian, R., 1975, *Prog. Med. Virol.* **19:**40–83.
37. Bablanian, R., Baxt, B., Sonnabend, J., and Esteban, M., 1978, *J. Gen. Virol.* **39:**403–413.
38. Schrom, M., and Bablanian, R., 1979, *Virology* **99:**319–328.
39. Bablanian, R., Coppola, G., Scribani, S., and Esteban, M., 1981, *Virology* **112:**1–12.
40. Gershowitz, A., and Moss, B., 1979, *J. Virol.* **31:**849–853.
41. Bablanian, R., Coppola, G., Masters, P., and Banerjee, A. K., 1986, *Virology* **148:**375–380.
42. Bablanian, R., and Banerjee, A. K., 1986, *Proc. Natl. Acad. Sci. USA* **83:**1290–1294.
43. Kates, J., 1970, *Cold Spring Harbor Symp. Quant. Biol.* **35:**743–752.
44. Kates, J., and Beeson, J., 1970, *J. Mol. Bio.* **50:**1–18.

45. Brown, M., Dorson, J. W., and Bollum F. J., 1973, *Virology* **12:**203–208.
46. Moss, B., Rosenblum, E. N., and Garon, C. F., 1973, *Virology* **55:**143–156.
47. Bablanian, R., Goswami, S. K., Esteban, M., and Banerjee, A. K., 1987, *Virology* **161:**366–373.
48. Su, M. J., and Bablanian, R., 1990, *Virology* **179:**679–693.
49. Sachs, A. B., and Davis, R. W., 1989, *Cell* **58:**857–867.
50. Brawerman, G., 1981, *Crit. Rev. Biochem.* **10:**1–38.
51. Voorma, H. O., Goumnas, H., Amersz, N., and Benne, R., 1983, *Curr. Top. Cell Regul.* **22:**8883–8887.
52. Lodish, H. F., and Nathan, D. G., 1972, *J. Biol. Chem.* **247:**7822–7829.
53. Jacobson, A., and Favreau, M., 1983, *Nucleic Acids Res.* **11:**6353–6368.
54. de Sa, M. F. G., Strandart, N., de Sa, C. M., Akhayat, O., Husca, M., and Scherrer, K., 1988, *Eur. J. Biochem.* **176:**521–526.
55. Stoltzfus, C. M., Shatkin, A. J., and Banerjee, A. K., 1973, *J. Biol. Chem.* **182:**7993–7998.
56. Lemay, G., and Millward, S., 1986, *Arch. Biochem. Biophys.* **249:**7822–7829.
57. Kates, J., 1970, *Cold Spring Harbor Symp. Quant. Biol.* **35:**743–752.
58. Sheldon, R., Jurale, C., and Kates, J., 1972, *J. Virol.* **14:**214–224.
59. Nevins, J. A., and Joklik, W. K., 1975, *Virology* **63:**1–14.
60. Moss, B., and Rosenblum, E. N., 1974, *J. Virol.* **14:**86–98.
61. Bertholet, C., Van Meir, E., ter Heggeler-Bordier, B., and Wittek, R., 1987, *Cell* **50:**153–162.
62. Schwer, B., Visca, P., Vos, J. C., and Stunnenberg, H. G., 1987, *Cell* **50:**163–169.
63. Ahn, B. Y., and Moss, B., 1989, *J. Virol.* **63:**226–232.
64. Bablanian, R., Goswami, S. K., Esteban, M., Banerjee, A. K., and Merrick, W. C., 1991, *J. Virol.* **65:**4449–4460.
65. Kozak, M., 1978, *Cell* **15:**1109–1123.
66. Jang, S. S., Krausslich, H. G., Nicklin, M. J. H., Duke, G. M., Palmenberg, A. C., and Wimmer, E., 1988, *J. Virol.* **62:**2636–2643.
67. Pelletier, J., and Sonenberg, N., 1989, *J. Virol.* **63:**441–444.
68. Rueckert, R., 1990, In: *Virology*, Vol. 1, 2nd ed. (B. N. Fields, D. M. Knipe, R. B. Chanock, M. S. Hirsh, J. L. Melnick, T. P. Monath, and B. Roizman, eds.), pp. 507–548, Raven Press, New York.
69. van Steeg, H., Thomas, A., Verbeek, S., Kasparaitis, M., Voorma, H. O., and Benne, R., 1981, *J. Virol.* **38:**728–736.
70. Guan, K. L., and Weiner, H., 1989, *J. Biol. Chem.* **264:**17764–17769.
71. Kozak, M., 1980, *Cell* **19:**79–80.
72. Kozak, M., 1988, *Mol. Cell. Biol.* **8:**2737–2744.
73. Kozak, M., 1989, *Mol. Cell. Biol.* **9:**5134–5142.
74. Lowson, T. G., Ray, B. K., Dodds, J. T., Grifo, J. A., Abramson, R. D., Merrick, W. C., Betsch, D. F., Weith, H. L., and Thach, R. E., 1986, *J. Biol. Chem.* **261:**13979–13989.
75. Pelletier, J., and Sonenberg, N., 1988, *Nature* **334:**320–325.
76. Schreier, M. H., and Staehelin, T., 1973, *J. Mol. Biol.* **73:**329–349.
77. Baer, B. W., and Kornberg, R. D., 1980, *Proc. Natl. Acad. Sci. USA* **77:**1890–1892.
78. Brawermann, G., and Diez, J., 1975, *Cell* **5:**271–280.
79. Sheiness, D., and Darnell, J. E., 1973, *Nature New Biol.* **241:**265–268.
80. Cacoullos, N., and Bablanian, R., 1991, *Virology* **184:**747–751.

Chapter 10

Translation Control by Adenovirus Virus-Associated RNA I

Bayar Thimmapaya, Ghanashyam D. Ghadge, Prithi Rajan, and Sathyamangalam Swaminathan

1. ADENOVIRUS GROUP

Adenoviruses were first isolated by Rowe and co-workers from human adenoid tissues—hence the name adenoviruses. These are nonenveloped DNA viruses, composed of only DNA and proteins, and are widespread in nature. Forty-one of the 93 strains identified so far are of human origin, while the rest have been isolated from monkeys, dogs, cattle, rodents, and birds.[1–3] Clinical studies have shown that these viruses can cause a variety of diseases in humans, most of which involve the respiratory tract, the eye, and the gastrointestinal tract. For the most part, these infections lead to self-limited illnesses or latent persistent infections and are usually followed by complete recovery. Adenoviruses can maintain inapparent infections within host animals for months or even years. Lymphoid cells may serve as reservoirs for these persistent infections. Several members of this group can also induce malignant tumors in newborn rodents and therefore

BAYAR THIMMAPAYA, GHANASHYAM D. GHADGE, PRITHI RAJAN, AND SATHYAMANGALAM SWAMINATHAN • Department of Microbiology and Immunology, and Robert H. Lurie Cancer Center, Northwestern University Medical School, Chicago, Illinois 60611.

Translational Regulation of Gene Expression 2, edited by Joseph Ilan. Plenum Press, New York, 1993.

adenoviruses are classified as members of the DNA tumor virus group. All the members of this group tested so far have been shown to transform rodent cells in culture.[4] Despite their tumorigenic potential in experimental models, these viruses have not been shown to be tumorigenic in humans. Adenoviruses can be grown to high titers in a variety of cultured mammalian cells. Human cells of epithelial origin are best suited for growing human adenoviruses. Because of the small size of their genome, the ease with which they can be grown in rapidly growing cultured cells, and their ability to transform rodent cells *in vitro*, adenoviruses have served as important model systems to study the regulation of eukaryotic gene expression and virus–host-cell interactions. Such studies have led to important discoveries pertaining to the transcriptional and translational regulatory mechanisms of the cellular and viral systems, and virus-mediated neoplastic transformation. In this regard, adenovirus type 2 (Ad2) and 5 (Ad5) are the best understood. This review summarizes the role of the two adenovirus-encoded small RNAs in the regulation of translation in human cells infected with Ad2 or Ad5.

2. GENOME STRUCTURE AND TRANSCRIPTIONAL MAP OF HUMAN ADENOVIRUSES

The viral genome consists of a linear double-stranded DNA of about 36,000 base pairs (bp) which is divided into 100 map units (m.u.) with the left end representing 0 m.u. and the right end 100 m.u. Nucleotide sequences of the Ad2[5,6] and most of the Ad5 genome[7] and the detailed transcriptional map of both Ad2 and Ad5 have been published and the protein products encoded by the different spliced transcripts of the viral genome have been identified (reviewed in Levine[8]). Figure 1 shows an oversimplified transcriptional map of human Ad2. Both strands of the viral genome are transcribed at early and late times after infection. Early after infection and before DNA synthesis, six discrete regions of the viral genome are transcribed. These are early transcription region 1A or E1A, E1B, E2A, E2B, E3,

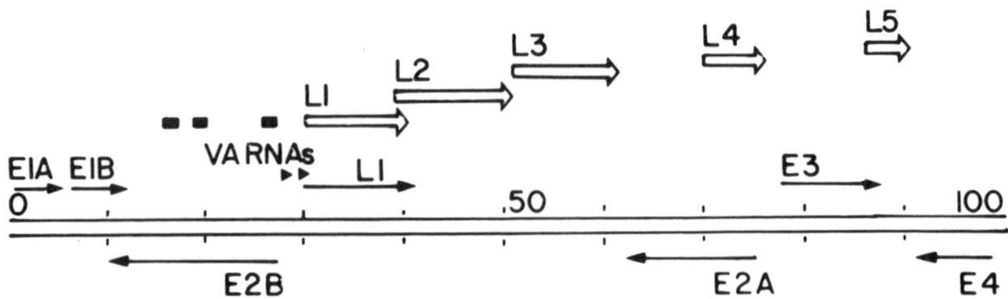

Figure 1. Transcriptional map of adenovirus type 2 or 5. Early transcriptional regions E1A, E1B, E2A, E2B, E3, E4, and L1 are shown by arrows with single lines. Late mRNAs L1–L5 are shown by open arrows. The solid boxes are tripartite leader segments. The position of the two VA RNA genes is indicated.

and E4. Each of these regions produces a family of spliced messenger RNAs (mRNAs) and therefore each generates different polypeptides (reviewed in Ziff[9]). The functions of most of the proteins encoded by these messages have been identified. For example, complete transformation of rodent cells in culture requires polypeptides encoded by both E1A and E1B.[10] One of the polypeptides encoded by E1A is also required for efficient transcription of other viral early promoters (reviewed in Flint and Shenk[11]). E2A and E2B together encode three polypeptides, all of which are vital for viral DNA synthesis.[12] E3 encodes several polypeptides, one of which allows the virus-infected cells to escape immuno surveillance (reviewed in Wold and Gooding[13]). The polypeptides encoded by the E4 region contain multiple functions, including a transactivation function.[14,15] The two polypeptides encoded by the L1 region are essential for virus encapsidation.[16] Late after infection, that is, after DNA synthesis, there is a large increase in the transcriptional activity of the major late promoter located at 16.5 m.u. Transcription initiated at 16.5 m.u. continues to the right end of the viral chromosome. This long primary transcript undergoes extensive splicing and polyadenylation giving rise to five families of late messages (L1–L5) (reviewed in Ziff[9]). Most of the polypeptides encoded by these families of mRNAs are capsid proteins. One of the late proteins, the 100-kDa protein, is not found in capsids and is believed to play at least two roles, one in the assembly of the capsid proteins[17] and the other in the translation of viral mRNAs.[18] All the mRNAs of the major late transcription unit are coterminal at their 5′ end and contain a 200-nucleotide segment designated as the tripartite leader. The tripartite leader consists of three segments which are derived from three different regions of the viral chromosome.[19] This 5′ terminal noncoding segment plays a major role in the translational regulation of mRNAs that contain these sequences.[20,21] Two other genes that are transcribed at late times after virus infection are the protein IX and the IVa2 genes. In the lytic cycle, the viral gene expression is therefore controlled at a number of levels which include transcription, splicing, transport of mRNAs from the nucleus to the cytoplasm, translation, and polyadenylation. Many of these control mechanisms, most likely, are similar to those that operate in a normal eukaryotic cell.

3. VIRUS-ASSOCIATED RNAs

In 1966, Weissman and co-workers discovered large amounts of a small RNA in KB cells infected with Ad2 and named it virus-associated RNA (VA RNA).[22] The nucleotide sequence of this RNA was then determined and the RNA was found to be of about 160 nucleotides.[23] It was soon discovered that the gene for this RNA was virus-encoded and transcribed by RNA polymerase III.[24,25] In 1975, another small RNA (VAII RNA) was detected in Ad-infected cells. This was similar in size to the major species (VAI RNA), but was present in much smaller amounts when compared to the former.[26] Both VA genes are embedded in the major late transcription unit, around 30 m.u. of the viral chromosome, and are transcribed independently from their own respective promoters in a rightward direction.[27–29] VAI RNA is heterogenous at the 5′ end with transcription starting at two positions, three bases apart.[30–32] The longer species starts with an A residue, whereas the shorter species starts with a G residue.[30] Both the VAI and VAII RNAs are heterogenous at their 3′ ends because transcription terminates at one of several U residues.[33] The promoters for both genes are intragenic, as is typical of RNA polymerase III genes.[33,34] The promoter of the VAI RNA gene has been extensively analyzed. It maps between nucleotides +10 and +69 with respect to the A start.[34,35] Mutational analyses show that the promoter sequence consists of two distinct and separable sequence blocks, A and B (approximately nucleotides +10 to +18 and +54 to +69, respectively).[36,37] The distance between the two blocks is somewhat flexible and can be increased to 75 bp from its normal distance of 35 bp without seriously affecting transcription. Even a small deletion, however, between the two sequence blocks severely reduces transcription.[36,37] In this respect, the promoter resembles the promoter of a eukaryotic transfer RNA (tRNA) gene, raising the possibility that the VA RNA genes might have evolved from a eukaryotic tRNA gene.[34–37] The detailed map of the two VA RNA genes on the Ad chromosome, the location of the two sequence blocks critical for transcription, and the levels of the two VA RNAs synthesized at late times in Ad infections are shown in Fig. 2.

Transcription of the two VA genes begins at early times after infection, giving rise to roughly equal quantities of the RNAs. During the early-to-late transition, however, the transcription of the VAI gene increases dramatically, leading to its accumulation to very high levels, approximately 10^7 molecules per cell.[26] Transcription of the VAII gene is also increased at late times, but the RNA accumulates to a much reduced level compared to the VAI species. The ratio between the VAI and VAII species in a productively infected cell is roughly 40:1. The two genes apparently compete for a limiting transcription factor in virus-infected cells. Since the VAI gene is a superior competitor, it is able to transcribe to a much higher level.[34] There is genetic evidence in support of this hypothesis. The accumulation of the VAII RNA is increased by more than tenfold in cells infected with VAI$^-$/VAII$^+$ Ad mutants compared to those infected with the wild-type (VAI$^+$/VAII$^+$)

→

Figure 2. Enlarged diagram of the region of the Ad5 chromosome which codes for the two VA RNA genes, transcriptional control elements of the VAI gene, and the VA RNAs synthesized by Ad5. **(A)** The location of the two VA genes on the viral chromosome and the intragenic promoter. The VA transcripts are shown by arrows. The promoter region of the VAI RNA gene is shown below the viral chromosome. Boxes A and B are the two transcriptional control regions of the intragenic promoter. **(B)** The intragenic promoter sequences and the location of transcriptional control elements of the VAI gene promoter. The single-strand sequence here reads as the sense strand of the VAI RNA. **(C)** The VA RNAs synthesized in cells infected by Ad5. The cells were infected with wild-type (WT) Ad5 at 25 PFU per cell for 20 hr. During the last 4 hr of the infection, cells were labeled with ^{32}P inorganic phosphate. Total RNA was extracted and analyzed on an 8% DNA sequencing gel. Only the relevant portion of the gel is shown.

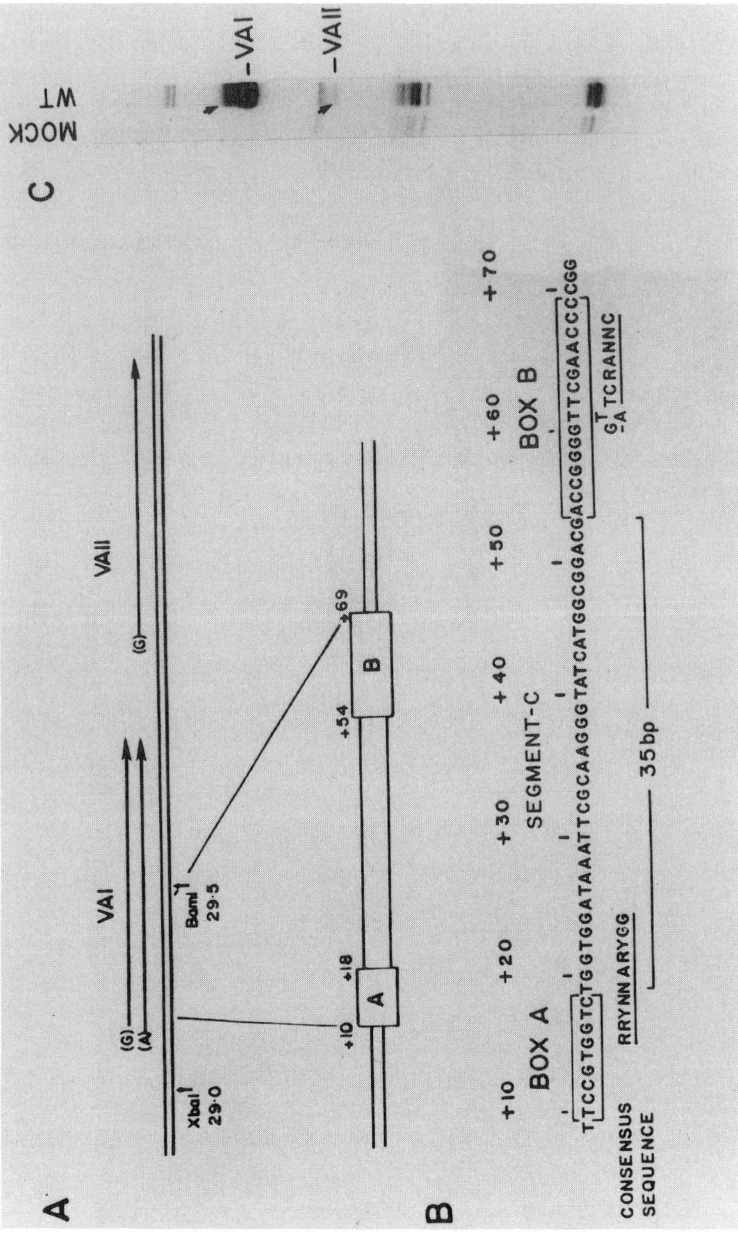
A
VAI
VAII
(G)
(A)
XbaI 29.0
BamI 29.5
+10
+18
+54
+69
B
+10 +20 +30 +40 +50 +60 +70
BOX A
SEGMENT-C
BOX B
TTCCGTGGTCTGGTGGATAAATTCGCAAGGGTATCATGGCGGACGACCGGGGTTCGAACCCCGG
CONSENSUS SEQUENCE
RRYNNARYGG
GTCRANNC
35bp
C
MOCK WT
-VAI
-VAII

virus when the B block sequences of the VAI gene are mutated.[38] The 289R protein of the E1A gene also contributes to the high-level transcription of the VAI gene at late times, for it has been shown that this transactivator can efficiently transactivate the VAI gene *in vivo* and *in vitro*.[39–41] Thus, it seems that the transcription machinery of the VAI gene is geared to produce extremely high quantities of VAI RNA at late times after infection. This phenomenon may have an important biological significance. In virus-infected cells the VAI RNA is found complexed with 45-kDa La antigen,[42,43] ribosomes,[44] and the interferon-induced p68 kinase (see section 3.3.3.).

3.1. Structure of the VA RNAs

There is only a scattered nucleotide sequence homology between VAI and VAII RNAs. Both RNAs are extremely GC-rich and can fold to form very stable secondary structures.[45] The secondary structure of VAI RNA has been experimentally determined by us[46] and others.[47,48] This structure, which is considerably different from that which was predicted by computer programs,[45] consists of two extended base-paired regions, stems I and III, which are connected by a short

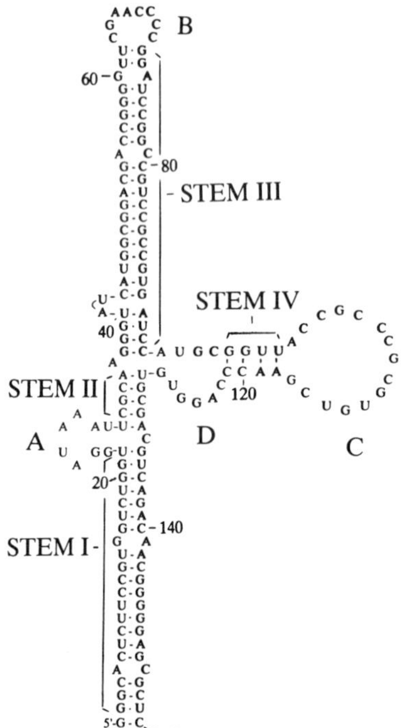

Figure 3. Secondary structure model of the Ad2 VAI RNA. The structure is based on RNase cleavage patterns. The model has features common to that proposed by others.[47,48] A and D are looped regions.

base-paired region, stem II, at the center (Fig. 3). Stems I and II are joined by a small loop, A, and stem III contains a hairpin loop, B. At the center of the molecule and at the 3′ side, stems II and III are connected by a short stem–loop (stem IV and hairpin loop C). A fourth minor loop, D, exists between stems II and IV. Stems I and III are not perfectly base-paired duplex regions; they contain several mismatches. The secondary structure of the VAII RNA has not been determined experimentally, although the RNA sequences can potentially fold to form structures resembling that of the VAI RNA.[45] The genes that encode the VA RNAs of human Ad7 and 12[49] and also an avian Ad have been sequenced;[50,51] the RNAs transcribed from these genes can potentially fold to form stable secondary structures.

3.2. Role of VA RNAs in Adenovirus Growth Cycle

The first clue that the VAI RNA has a function in the productive Ad infection came from a study of constructed Ad mutants. To probe the function of VA RNAs, two Ad mutants were constructed.[52] One of these contained a 17-bp deletion in the intragenic promoter of VAII RNA gene (*dl*328). The second mutant, *dl*331, contained a 29-bp deletion encompassing the B block of the VAI promoter. As a result, the VA genes in each of these mutants were not transcribed. Mutant *dl*328, which lacked the minor VAII species, showed no growth defects in cultured human cells, indicating that at least in rapidly growing cultured human cells the minor species did not play a role in virus replication. Mutant *dl*331, which lacked the major VAI species, on the other hand, grew more slowly in 293 cells (a transformed human embryonic kidney cell line[53]) with a 20-fold reduced growth titer. A comparison of viral DNA synthesis, mRNA splicing, polyadenylation, capping, and transport of viral mRNAs from the nucleus to the cytoplasm between *dl*331 and the wild-type infections showed that each of these steps occurred normally in *dl*331 infections. The only major defect in *dl*331-infected cells was a dramatic reduction in the rate of viral and host polypeptide synthesis. It was further demonstrated that the translational defect in *dl*331-infected cells was due to a defective initiation step.[44] In summary, these studies established that the VAI RNA was required for the efficient initiation of translation of viral and host mRNAs at late times in Ad infections.

The question of whether VAII RNA has any role in the growth cycle of the virus has not been addressed rigorously. Because of its structural similarities, it seems likely that this RNA may also be functionally similar to VAI RNA. Indeed, a mutant that lacks both VAI and VAII RNAs is much more defective than VAI$^-$/VAII$^+$ mutants with respect to polypeptide synthesis, growth yield, and eIF-2 phosphorylation.[46] The VAI$^-$/VAII$^+$ mutants synthesize about tenfold more VAII RNA compared to wild type. These mutants can grow about fivefold better than the double mutant (VAI$^-$/VAII$^-$), indicating that VAII RNA can partially complement for the VAI RNA function in virus infections.[38]

3.3. Involvement of Interferon-Induced p68 Kinase in VAI RNA-Mediated Translational Control

3.3.1. Regulation of Translation by Phosphorylation of eIF-2

Several lines of evidence suggest that the cessation of protein synthesis in *dl*331- infected cells is due to the phosphorylation of the translation initiation factor 2 (eIF-2). In eukaryotic cells one major mechanism by which translation is controlled is by phosphorylation of eIF-2. The eIF-2, the first of the initiation factors, binds to GTP and the initiator tRNA, methionyl-$tRNA_F$, and forms a ternary complex (reviewed in Safer[54]). The ternary complex then binds to the 40S ribosomal subunit and associates with an mRNA to form a preinitiation complex. In the subsequent step, when the 60S ribosomal subunit is attached to the preinitiation complex, the GTP moiety is hydrolyzed to GDP and inorganic phosphate and the eIF-2 with GDP moiety bound to it is released. For the eIF-2 to function in a new round of protein synthesis, the GDP has to be replaced by GTP. This step is accomplished by a large enzymatic complex called guanosine nucleotide exchange factor (GEF, also known as eIF-2B). The eIF-2B is present in the cell in limiting quantities.[54] The eIF-2 consists of three subunits: alpha, beta, and gamma. Two enzymes have been known to phosphorylate the serine-51 of the alpha subunit of eIF-2. One of these, the heme-controlled repressor (HCR), is present in reticulocytes and is activated when cells are deprived of heme (reviewed in Ochoa[55]). The second enzyme, the p68 kinase (the terms p68 and p68 kinase are used interchangeably in this chapter; this protein is also referred to as eIF-2alpha kinase, DAI, and DsI) is present in most eukaryotic cells and is induced by interferon (reviewed in Hovanessian[56]). When the eIF-2 is phosphorylated at serine-51 by either of these enzymes it cannot function catalytically, because the small amounts of eIF-2B present in the cell bind irreversibly to the phosphorylated form of eIF-2, thus trapping the limiting amounts of eIF-2B in an inactive complex incapable of sustaining continued initiation of protein synthesis, which eventually comes to a halt. Figure 4 shows a schematic representation of the steps involved in protein synthesis initiation in eukaryotic cells, the consequence of phosphorylation of eIF-2 by the p68 kinase, and the role of VAI RNA in protecting the translation apparatus of the host cell in Ad infections.

3.3.2. Properties of the p68 Kinase

The p68 kinase has an apparent molecular weight of 68–70 kDa.[57–61] The predicted molecular weight of this protein, based on the amino acid sequence derived from the cDNA clone is 62 kDa.[62] The enzyme is largely bound to the ribosome in its latent form and its synthesis is induced by alpha-interferon[63,64] (also reviewed in Lengyel[65]). It requires small amounts of double-stranded RNA (dsRNA) for activation, but high concentrations of dsRNA inhibit its activ-

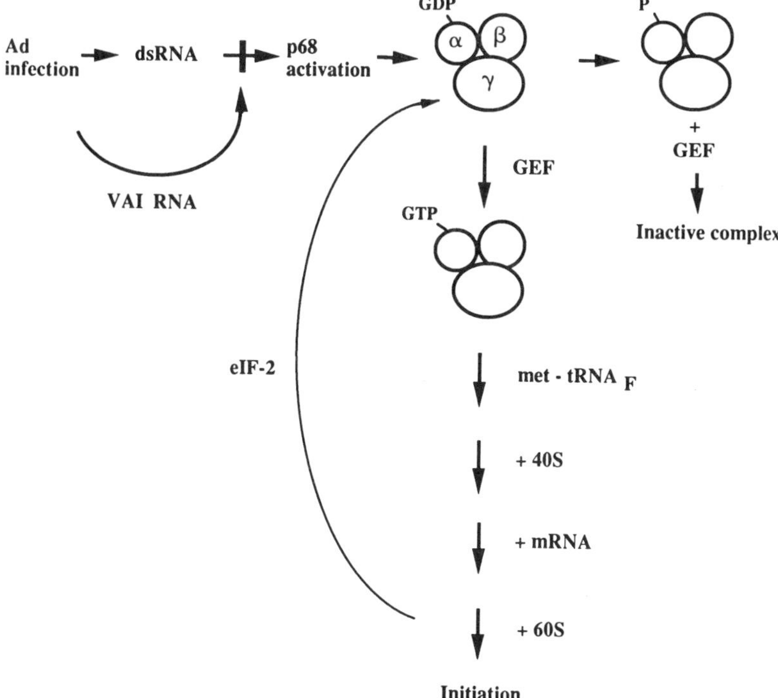

Figure 4. Schematic representation of protein synthesis initiation pathway in eukaryotic cells, effect of p68 activation on the initiation of protein synthesis, and the mechanism of VAI RNA function in Ad infection.

ity.[57,66–69] This activation step requires perfectly duplexed RNAs with a minimal length of 50bp; a 50-bp RNA molecule with a single mismatch fails to activate the enzyme.[69] It is not clear how this autophosphorylation takes place. One study showed that autophosphorylation is an intermolecular event in that two p68 molecules come in contact with each other in the presence of dsRNA with one molecule phosphorylating the other in an ATP-dependent reaction.[61] Another study, on the other hand, suggested that the autophosphorylation is an intramolecular event, that is, a molecule of p68 phosphorylates itself in the presence of dsRNA and ATP.[70] Regardless of the mechanism, the latent p68 must be activated by an autophosphorylation event before it can phosphorylate the alpha subunit of eIF-2. Thus, the p68 kinase contains two distinct kinase activities. In addition to eIF-2, p68 kinase also can phosphorylate several other proteins, including the P1 subunit of the ribosome, the histones, and protamines.[71] The significance of phosphorylation of these substrates is not clear. Thus, one of the mechanisms that cells have developed as a defense against virus infection is activation of the p68, which phosphorylates the eIF-2 and inactivates the host translation apparatus. Viruses therefore must overcome this barrier to translate their messages and have developed various strategies to do so. Adenoviruses use the VAI RNA in combating this antiviral response.

3.3.3. Relationship between VAI RNA-Mediated Translational Control and the p68 Kinase

Several lines of evidence suggest that the inhibition of translation in cells infected with the VAI-negative mutant *dl*331 is due to the increased phosphorylation of eIF-2 as a consequence of the activation of the p68 kinase and that the VAI RNA prevents the activation of this enzyme. This evidence can be summarized as follows: (1) Cell extracts prepared from *dl*331-infected cells accumulate an inhibitor of translation, and translation can be restored in these extracts by the addition of eIF-2 or GEF.[72] (2) Mutant *dl*331-infected cells show a large increase in the levels of eIF-2 phosphorylation.[73,74] These cells also show a large increase in the levels of phosphorylated p68 kinase indicative of the activation of the kinase.[71,75] (3) The p68 kinase can bind to the VAI RNA in virus-infected cells.[61,71] (4) The VAI RNA can block the autophosphorylation of the p68 kinase *in vitro*, whereas other small RNAs, such as tRNA, 5S RNA, and functionally defective VAI RNAs, fail to do so[75,76] (also G. D. Ghadge and B. Thimmapaya, manuscript in preparation). (5) Finally, cell lines were constructed by neomycin selection in which two mutant forms of eIF-2 were expressed.[77] These are phosphorylation-resistant mutants in which serine at position 48 or 51 is changed to alanine and can function efficiently *in vivo*. An Ad mutant in which both VAI and VAII genes are deleted (*dl-sub*720) was capable of synthesizing its polypeptides to wild-type levels in these cells and its growth curve was comparable to that of the wild-type Ad.[77] Thus, a translational defect due to a lack of VA RNAs was corrected by mutant alleles of eIF-2, providing genetic evidence for the involvement of eIF-2 in the VA-negative phenotype.

It should be mentioned, though, that the relationship between activation of the p68 kinase and the concomitant phosphorylation of eIF-2 and the role of VAI RNA in downregulation of p68 activity may be more complicated than it seems. For example, in HeLa and KB cells, the VAI$^-$ mutants do not show the translational defect phenotype even when the p68 is induced by interferon treatment, indicating that there may be other factors that may compensate for the VAI RNA function in certain cases.[78]

As stated above, activation of p68 requires small amounts of dsRNA. The source of dsRNA in the virus-infected cell is unclear. Some of this probably arises due to symmetrical transcription of the viral genome, as both *r* and *l* strands of the viral genome are transcribed in virus infections.[79] It has been shown that VAI RNA itself may activate p68 kinase at low concentrations.[70] This has not been confirmed. It is unlikely that the VAI RNA could serve to activate the p68 kinase, as in addition to having duplexes that fall short of the prescribed size, there are a number of mismatches in the duplex regions.[46,47]

VAI RNA-mediated translational stimulation of viral and cellular genes has been demonstrated in transient assays using cloned genes.[80,81] Further, this stimulation has been shown to be due to inhibition of p68 kinase activity by the VAI RNA.[82] p68 presumably is activated in these assays by the dsRNA generated by symmetrical transcription of the plasmid sequences from fortuitous promoters in the plasmids.

3.4. Structural Requirements of the VAI RNA for Function

A clear understanding of how the VAI RNA functions *in vivo* demands an understanding of the structural features and the sequence elements of the molecule that are important for its function. It was believed that the extended duplex regions of the VAI RNA are essential for function. The discovery of the p68 kinase as the target for the VAI RNA laid further functional emphasis on the duplex regions of the molecule. Surprisingly, however, mutational analysis showed that the long duplex regions were dispensable for function. Function correlated with the integrity of the elements present in the central region of the molecule. In one study, mutations were introduced into the stem-and-loop regions of the VAI gene in plasmid constructs and assayed for their ability to rescue the translational defect of *dl*331.[47] In a study performed in our laboratory, the structure–function relationship of the VAI RNA was probed at the virus level by construction and analysis of a series of Ad mutants with deletion and linker-scan substitution mutations which span the entire length of the VAI coding sequences.[36,46] These mutants also contained an inactive VAII gene, permitting unambiguous evaluation of the mutational effects on the VAI RNA function. In cells infected with these mutants the VAI RNA accumulated to normal levels; thus the RNA was not limiting for function. Interestingly, in denaturing polyacrylamide gels, the mobility of the VAI RNA depended on the extent of perturbation of the structure of stem III; when this stem was disrupted, the RNAs migrated at rates comparable to molecules containing large deletions. The Ad mutants which contained mutations in the VAI gene were analyzed for polypeptide synthesis, growth yield, and phosphorylation of eIF-2. Based on the capacity of the mutants to synthesis viral polypeptides (hexon) and their growth yield data, the VAI mutants were grouped into two classes: defective and nondefective. The location of these mutations in the secondary structure of the VAI RNA is shown in Fig. 5. Several mutants with mutations in stems I or III showed no defects. All the defective mutants contained mutations in the central short stem–loop (loops C and D, stem IV), and the adjacent base-paired regions (stem II and the proximal part of stem III) except in *sub*742 (see below).

To determine the effect of mutations on the secondary structure of the RNAs, the secondary structures of a number of mutant RNAs were determined by

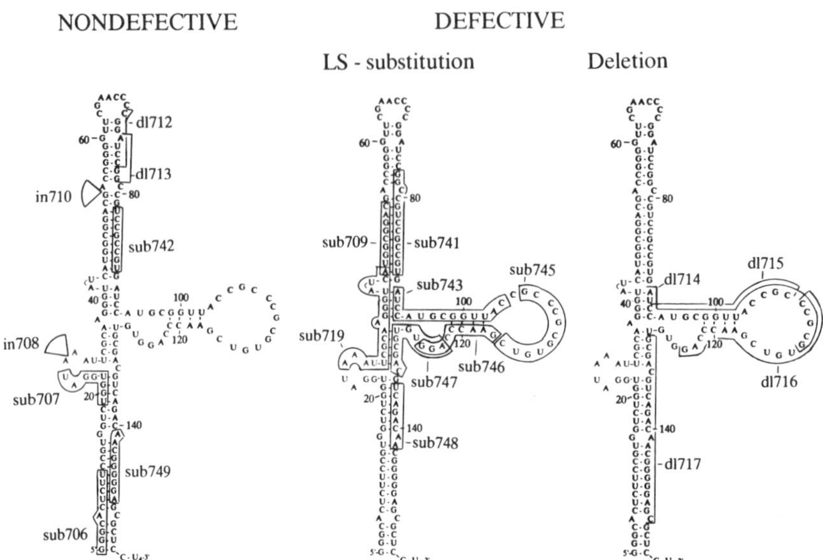

Figure 5. Location of the VAI mutation on the secondary structure of the VAI RNA. Linker-scan mutations are boxed; deletion mutations are bracketed; insertion mutants are shown by triangles. The mutants are grouped as defective or nondefective based on their phenotype. Reproduced from Furtado *et al.*[46]

subjecting the end-labeled RNAs to single-strand-specific ribonucleases. The RNA sequences of the mutants were folded in such a way that the bases that were sensitive to ribonucleases were not allowed to base pair. It was found that all the functionally competent mutant VAI RNAs had their central short stem–loop structure intact, although a part of the duplex regions of the stems I and III were perturbed. Similarly, in all of the defective molecules, sequences in the short stem–loop were rearranged. These mutant molecules had either stem I or III intact and yet they were unable to function *in vivo* (Fig. 6) (G. D. Ghadge and B. Thimmapaya, manuscript in preparation). Thus, it seems that function correlates strictly with the integrity of the central part of the molecule. Based on these results, it appears that this part of the molecule is the functional domain as far as its translation regulation function is concerned. Even though the mutations in *sub*742 overlap with those of *sub*741, *sub*742 shows near normal phenotype, whereas *sub*741 is defective. On the basis of sensitivity to single-strand-specific RNases, we can derive a novel secondary structure for this mutant RNA in which a portion of the sequences may fold to form a structure that resembles the central part of the wild-type molecule.[46] It is conceivable that the enzyme p68 kinase recognizes this part of the *sub*742 VAI RNA, which may substitute for the functional domain found in the wild-type molecule.

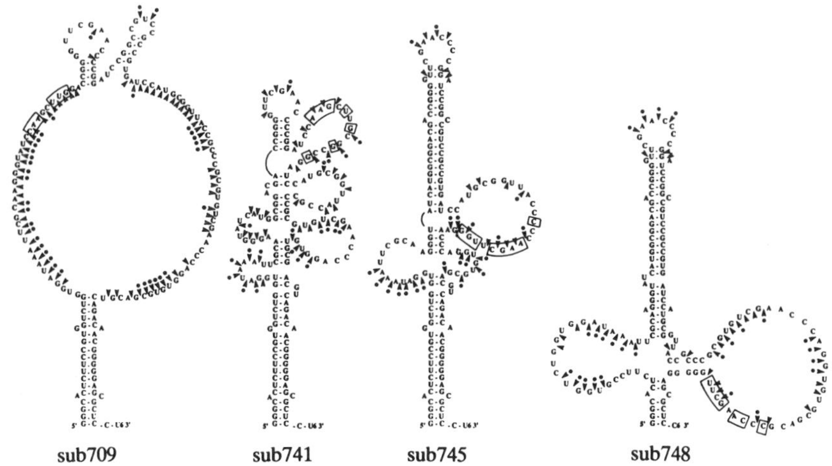

Figure 6. Experimentally derived secondary structures of the VAI RNAs of *sub*709, *sub*741, *sub*745, and *sub*748 based on single-strand-specific RNase cleavage patterns.[83] The arrowheads denote the RNase cleavages. Reproduced from Ghadge *et al.*[83]

3.5. p68 Binding Domain in the VAI RNA

A direct role for the VAI RNA in the downregulation of the function of the p68 kinase was further demonstrated by the isolation of the VAI RNA–p68 kinase complexes from Ad-infected human cells. In the experiments performed by Katze *et al.*,[71] type 293 cells were infected with Ad, labeled with ^{32}P inorganic phosphate, and lysed 20 hr postinfection with NP40. The lysates were reacted with an anti-p68 monoclonal antibody and the p68 was immuoprecipitated. It was found that large amounts of VAI RNA coimmunoprecipitated with the p68, indicating that in virus-infected cells the VAI RNA exist as a complex with p68. Purified ^{32}P-labeled VAI was also shown to bind to the p68 kinase which had been immobilized on antibody Sepharose columns.

To understand how this small RNA can inhibit the autophosphorylation of p68, we must identify the domain in the RNA with which the p68 interacts and the structural features of the RNA essential for p68 binding, and determine whether binding of the VAI RNA to the p68 correlates with its function. To address these questions, we examined the interactions between p68 and various defective and nondefective mutant VAI RNAs in living cells and *in vitro*.[83] These mutants VAI RNAs were encoded by various Ad mutants characterized previously in our laboratory (described above), and proved ideal for this purpose. Human cells were infected with Ad mutants that express functionally defective or nondefective VAI RNAs and labeled with ^{32}P, and p68–VAI RNA complexes were immunoprecipitated by an anti-p68 monoclonal antibody. The labeled VAI RNAs were recovered from the immunoprecipitated complexes and analyzed on DNA sequencing gels. The VAI RNAs synthesized in cells infected with nondefective (*sub*706, *sub*707, *in*708, *dl*713, and *sub*749) and defective Ad mutants (substitution mutants 709, 719, 741, 743, 745, 746, 747, and 748) and the labeled RNAs recovered after immunoprecipitations of the same lysates are shown in Figs. 7A and 7B, respectively. These results show that the mutant RNAs that functioned efficiently in virus-infected cells bound p68 efficiently, whereas functionally impaired mutants failed to bind to the p68. These results were reproducible in *in vitro* assays. p68 from interferon-treated 293 cells was purified with monoclonal antibody (mAb)–Sepharose and incubated with equal quantities of gel-purified, uniformly ^{32}P-labeled VAI RNAs obtained by *in vitro* transcription.[83] In parallel, labeled RNAs were also incubated with mAb–Sepharose that did not contain p68. After 20 min of incubation at 30°C, the p68–VAI complexes were washed, and bound RNAs were recovered and analyzed on denaturing polyacrylamide gels. Again, the nondefective RNAs (wild type and *sub*749) bound to p68 efficiently, whereas

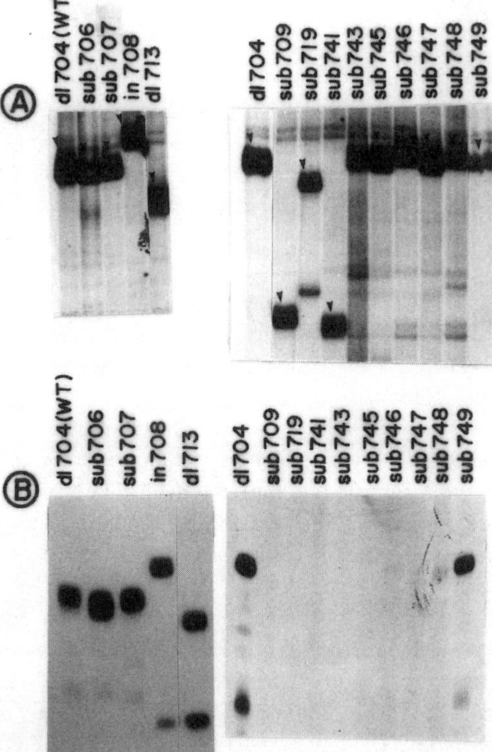

Figure 7. Binding of p68 to wild-type (WT), defective, and nondefective VAI RNAs in Ad-infected human cells. p68–VAI complexes were immunoprecipitated from virus-infected cells by an anti-p68 monoclonal antibody[71] and the RNAs were recovered from the complexes and were analyzed in denaturing 8% polyacrylamide gels.[46] **(A)** Electrophoretic analysis of VAI RNAs isolated from infected cell lysates used in immunoprecipitation analysis. Arrowheads show the positions of VAI RNAs. **(B)** VAI RNAs recovered from the p68–VAI complexes formed by Ad5 mutants in virus-infected cells. Only relevant portions of the autoradiograms are shown. Reproduced from Ghadge *et al.*[83]

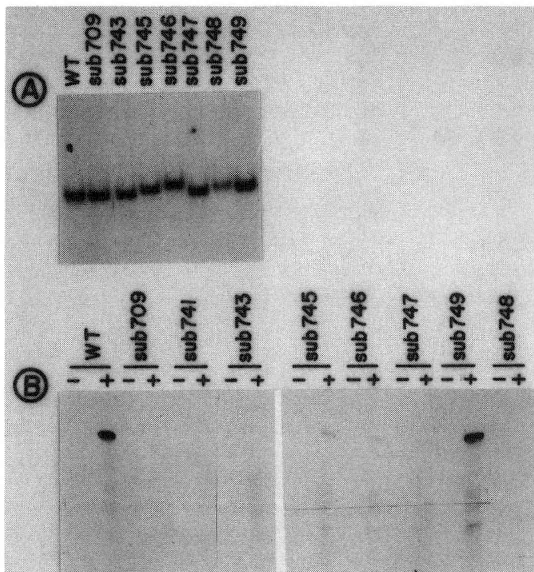

Figure 8. Binding of the p68 to wild-type (WT) and mutant VAI RNAs *in vitro*. **(A)** Electrophoretic analysis of the WT and mutant VAI RNAs transcribed *in vitro*. **(B)** Electrophoretic analysis of RNAs bound to p68 *in vitro*. VAI RNAs were recovered from p68–VAI complexes and analyzed in a denaturing polyacrylamide gel. Only relevant portions of the gel are shown. Lanes correspond to experiments in which RNA was added to mAb–Sepharose that was not previously reacted with interferon-treated cell extracts (−) or to experiments with interferon-treated cell extracts (+). Reproduced from Ghadge *et al.*[83]

the defective mutant VAI RNAs bound p68 to negligible levels. Figures 8A and 8B show the electrophoretic analysis of the *in vitro* transcribed wild-type and mutant forms of the VAI RNAs before and after binding to p68, respectively.[83] Secondary structure analysis of several mutant RNAs showed that the binding of VAI RNA to p68 was dependent on the integrity of the central part of the molecule, a domain that was shown to be critical for function. In several of the mutant RNAs analyzed, stem I (substitution mutants 709, 741, 745, and 748) or stem III (substitution mutants 745 and 748) remained intact and yet these RNAs failed to bind the kinase indicating that the p68 did not bind to the long duplex regions or the apical stem–loop.[83]

In another study, binding of the mutant VAI RNAs to p68 was examined by *in vitro* approaches such as filter binding, UV cross-linking, and immunoprecipitation of VAI RNA-p68 complexes *in vitro* with biochemically purified p68 kinase.[84] Several mutant RNAs that failed to function *in vivo* bound the kinase *in vitro* with significant efficiency, whereas some mutants which functioned efficiently *in vivo* failed to bind to p68 *in vitro*. Further, a fragment of VAI RNA that could form an apical stem–loop with a stem with as small as six perfect base pairs bound to the biochemically purified p68. It was concluded that p68 binding to VAI RNA did not correlate with function. These authors proposed that the apical stem–loop (stem III) may bind to the dsRNA binding site of the p68 and facilitate the interaction of the central short stem–loop structure (identified as the functional domain in both virus studies and transient assays) with the active site of the enzyme and inhibit its activation. Clearly, the two studies employed two different approaches to study the VAI RNA interaction with p68. Isolation of p68–VAI RNA complexes from virus-infected cells was done in a physiologically relevant environment.[83] It is therefore likely that these results reflect the *in vivo* situation. That is, binding of the VAI RNA would inactivate the enzyme and such a binding would require the functional domain. These results raise an important question with regard to the site in the p68 with which the VAI RNA would interact. Since several of the mutant VAI RNAs which retained the intact stem III and the associated loop structure did not bind to the p68 *in vivo* or *in vitro*, it is conceivable that the VAI RNA would inhibit p68 activation by binding to the kinase at sites other than the dsRNA binding site. It is possible that VAI RNA binding may lead to a conformational change in the molecule such that the activator (dsRNA) cannot bind to it. Exactly what features of the functional domain are recognized by p68 will require further mutational analysis of the RNA as well as knowledge of the three-dimensional structure of the RNA and its target, p68. A cDNA clone for p68 has been reported recently.[62] It should now be possible to construct a structure–function map of this protein which may provide explanations to some of these questions. Also not clear is the significance of the extremely large amounts of VAI RNA present in infected cells at late times and whether VAI RNA interacts with other host proteins as well in virus-infected cells. A recent report suggests that the VAI RNA can bind to a protein present in rabbit reticulocyte lysates.[85] Such a protein is not found, however, in human cells, casting doubt on its role in VAI RNA function.

3.6. Does the VAI RNA Have Other Functions in Virus Infection?

A recent study identified a defect in splicing of one of the mRNAs encoded by the major late transcription unit in *dl*331-infected cells at late times.[86] This is most likely due to one or more cellular or viral proteins present in limiting amounts in *dl*331-infected cells and the VAI RNA may not be directly involved in splicing. The reason that such a defect was not observed in the earlier study[52] may be due to the fact that the authors in the latter study[86] used fewer infectious virus particles per cell, and this difference in the multiplicity of infection could explain the discrepancy. The VAI RNA has also been shown to stabilize the mRNA coding for the chloramphenicol acetyl transferase enzyme in transient assays.[87] In an unrelated study, it was observed that in CV-1 cells infected with a recombinant SV40 (SV-VA) that expresses large amounts of VAI RNA, large-T mRNA accumulated to three- to fourfold higher levels than that of wild-type SV40 infections.[88] The transcription rate of the large-T gene was unaffected in SV-VA virus (R. A. Bhat and B. Thimmapaya, unpublished results). Although not shown, the VAI RNA most likely affected the stability of large-T mRNA in cells infected with SV-VA. Initial studies have not provided evidence for the role of VAI RNA in stabilization of viral mRNAs in *dl*331 infections.[52] Nonetheless, a role for VAI RNA in stabilization of one or more viral mRNAs in Ad infections cannot be entirely ruled out.

4. OTHER VIRAL STRATEGIES TO COMBAT THE INTERFERON-INDUCED HOST ANTIVIRAL RESPONSE

Alpha-interferon inhibits the growth of VAI-negative Ad mutants, but not that of wild-type Ad.[76] Growth yield of *dl*331 is decreased even further in cells that are pretreated with interferon, whereas yield of the wild-type virus under these conditions is unaffected, indicating that the VAI RNA protects Ad against interferon effects (Fig. 9).[76] Production of VAI RNA allows Ad to combat the interferon-induced antiviral response and permits the virus to maintain long-term infections within host animals despite their ability to produce interferon. Interferon-induced host antiviral defense is a problem that viruses must overcome in establishing a successful infection in animals. To combat this antiviral response,

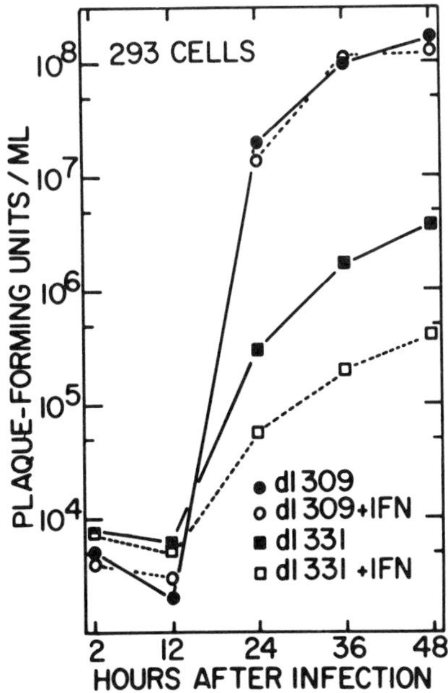

Figure 9. Effect of interferon on the growth of wild-type (WT) and a VAI-negative (*dl*331) mutant. The type 293 cells were treated with alpha A subspecies of leukocyte interferon (IFN) and then infected with *dl*309 (phenotypically WT variant) and *dl*331 at 3 PFU/cell. Control experiments were carried out without interferon treatment. Virus yield was assayed by plaque assay on 293 cells. Reproduced from Kitajewski *et al.*[76]

viruses have, as one might expect, evolved various strategies. Such strategies have been identified in vaccinia[89–94] and influenza[95–97] viruses and there is indication that other viruses, such as mengovirus,[98] reovirus,[99] Epstein–Barr virus,[100–104] and simian virus 40,[105] also may have mechanisms to counteract this antiviral response. Influenza virus is able to translate its messages in human cells infected with *dl*331, although translation of host and Ad messages is blocked as a result of p68 activation.[95] Apparently, influenza virus has developed a mechanism to block the activation of p68 and recent results suggest that the virus induces a cellular protein that directly or indirectly downregulates the p68 activity.[97] The mechanism by which this protein can prevent the activation of p68 will be illuminating. Vaccinia virus grows normally in interferon-treated cells and it can also confer resistance on other interferon-sensitive viruses such as vesicular stomatitis virus and picorna virus.[89] Growth of these viruses is normally inhibited by interferon; coinfection with vaccinia virus protects them from interferon effects. Consistent with these results, it has been shown that vaccinia virus-infected cells accumulate a protein which can inhibit p68 kinase activity.[91–94] The mechanism by which this protein is able to inhibit the activity of p68 is not clear, since this protein has not been isolated or characterized. One possibility is that this protein may be an RNA binding protein and may bind dsRNA present in vaccinia virus-infected cells that is required for the activation of the p68.[93,94] Epstein–Barr virus (EBV) has been shown to encode two small RNAs (EBER) of about 170 nucleotides.[100] Both RNAs are found in abundant quantities in EBV-transformed B cells.[100] The first indication that these RNAs may be functionally analogous to adenovirus VA RNAs came from the construction and analysis of adenovirus substitution mutants in which the two VA genes were deleted and replaced with multiple copies of the EBER genes.[101,102] Such a mutant was found to synthesize Ad viral polypeptides to significantly higher levels than the Ad mutant lacking the VA genes and the virus grew tenfold better than an Ad mutant that lacks both VAI and VAII genes (*dl-sub*720). Cells infected with this mutant were also found to contain significantly reduced levels of phosphorylated eIF-2 compared to those infected with an Ad mutant that lacks both VA genes (M. R. Furtado and B. Thimmapaya, unpublished results). Recent studies show that one of the EBERs, EBER-1, can rescue the dsRNA-dependent inhibition of protein synthesis in reticulocyte lysates[103] and can bind the p68 *in vitro*.[104] The role of these RNAs in the EBV life cycle is an enigma because a variant of the EBV which lacks these genes can grow efficiently in the presence of interferon and transform B cells *in vitro* with the same efficiency as the wild type.[106] Nonetheless, a role for the EBERs in protection against interferon-induced p68 effects in natural EBV infections cannot be ruled out at present. SV40 may also have developed mechanisms to combat the p68 activation. When CV-1p cells are infected with *dl-sub*720, translation of viral mRNAs is severely inhibited, a phenotype that is similar to that observed in human 293 cells for this mutant.[105] If CV-1p cells are preinfected with wild-type SV40, the translation of Ad mRNAs is restored to normal levels, indicating that SV40 encodes a function that can parallel the adenovirus VA RNA function. This function very likely is associated with large-T, because the complementation effect can be observed in an SV40 cell line, COS-1, that constitutively synthesizes large-T. Further, translation defect of the double mutant can be rescued in CV-1p cells by the transient expression of the cloned large-T gene (P. Rajan and B. Thimmapaya, unpublished results). Further studies of this complementation effect will be informative. Finally, mengovirus RNA can inhibit the activation of the p68, raising the possibility that the mengovirus may have evolved a mechanism to block the p68 activation.[98] Thus, it seems that adenovirus is not alone in the development of strategies to downregulate the function of the interferon-induced p68. In many of these viruses, however, the mechanism with which this takes place appears to be different and probably depends on the host in which the virus multiplies.

5. OTHER TRANSLATIONAL CONTROL MECHANISMS IN ADENOVIRUS

The VAI RNA-mediated translational control is not the only translational regulation mechanism in Ad-infected cells. The virus employs at least two other regulatory mechanisms for the efficient synthesis of the late proteins during the course of viral infection. All late protein genes are embedded in one transcription unit, the major late transcription unit. The mRNAs that encode all these proteins contain the tripartite leader segment which enhances translation of these messages at late times. It has been known for some time that at late times after infection, translation of host mRNAs is greatly reduced, whereas translation of the late viral mRNAs is unabated (also termed host shutoff). Recent results show that late in infection the cap-binding complex (eIF-4F) is inactivated by underphosphorylation of the cap-binding protein.[107] Translation of host-cell mRNAs requires active eIF-4F and thus host mRNAs are unable to translate at late times. Translation of the late mRNAs occurs efficiently in this environment since the tripartite leader-containing mRNAs are not dependent on the eIF-4F for translation.[21] The 100-kDa protein encoded in the L4 region also facilitates translation of viral mRNAs at late times after infection.[18] In cells infected with Ad variants with mutations in the 100-kDa protein gene, initiation of protein synthesis from late mRNAs is impaired. It is not clear if these three mechanisms, namely the VA RNA-mediated downregulation of p68 kinase, inactivation of eIF-4F, and the eIF-4F-independent translation of Ad messages with tripartite leaders and the facilitation of translation of late viral mRNAs by the 100-kDa protein, are interrelated. It is hoped that future work will clarify the relationship if any between these mechanisms.

6. SUMMARY AND CONCLUSIONS

The VAI and VAII RNAs are Ad-encoded, 160-nucleotide-long, RNA polymerase III transcripts that are found in abundant quantities at late times after infection. Both RNAs are highly structured with long duplex regions. The function of the major VAI RNA appears to be to downregulate the cellular interferon-induced, double-stranded RNA-dependent p68 kinase. The function of the minor VAII RNA is unclear. It is likely that this RNA also provides a function similar to that of the VAI RNA, albeit less efficiently. The p68 kinase is a ribosome-bound enzyme, which, upon activation, phosphorylates the protein synthesis initiation factor eIF-2 and shuts off protein synthesis. This is one of the ways cells defend against virus infection. Both *in vitro* and *in vivo* evidence indicates that the VAI RNA binds to p68 kinase and blocks its activation. Mutational analysis of the VAI RNA shows that the central region of the RNA, which consists of a complex short stem–loop, and the adjacent base-paired regions of the molecule are critical for function. The extended duplex regions of the molecule appear to be not critical for downregulation of the p68 activity. Binding studies indicate that the structural features of the VAI RNA critical for interaction with and the inhibition of the p68 kinase are essentially indistinguishable. Precise delineation of the structural features of the VAI RNA molecule that are recognized by the p68 kinase will have to await the elucidation of the three-dimensional structures of both the interacting molecules. Also not clear is whether the VAI RNA and dsRNA binding sites on the p68 molecule are identical. A detailed structure–function analysis of the p68 kinase should help to clarify this point. It also remains to be seen if the VAI RNA through its inactivation of the p68 kinase is responsible for differential regulation which leads to the preferential translation of viral messages over host messages in Ad-infected cells. In addition to VA RNA, translation in Ad-infected cells is modulated by the 100-kDa protein and by the tripartite leader of the late messages. The interrelationship among these control mechanisms, however, is not understood. The interferon-induced host antiviral defense mediated by p68 kinase is a problem that many viruses must overcome in establishing a successful infection. Viruses have evolved different strategies to combat this antiviral defense. Elucidation of the various strategies employed by viruses to combat this host antiviral response will be informative.

ACKNOWLEDGMENTS. The authors' work described in this review was supported by a grant from the National Institutes of Health (AI18029) and the American Cancer Society (MV 418). The fellowship support of P.R. by the Chicago Baseball Cancer Charities is gratefully acknowledged.

REFERENCES

1. Beladi, I., 1972, In: *Strains of Human Viruses* (M. Majer and S. A. Plotkin, eds.), p. 1, Karger, Basel.
2. Norrby, E., Bartha, A., Boulanger, P., Dreizin, R. S., Ginsberg, H. S., Kalter, S. S., Kawamura, H., Row, W. P., Russel, W. C., Schleinger, R. W., and Wigand, R., 1976, Adenoviridae, *Intervirology* **7**:117–125.

3. Wigand, R., Bartha, A., Dreizin, R. S., Esche, H., Ginsberg, H. S., Green, M., Hierholzer, J. C., Kalter, S. S., McFerran, J. B., Pettersson, U., Russel, W. C., and Waddel, G., 1982, Adenoviridae: Second report, *Intervirology* **18:**169–176.
4. Green, M., 1986, In: *Fundamental Virology* (B. N. Fields and D. M. Knipe, eds.), pp. 183–234, Raven Press, New York.
5. Roberts, R. J., Akusjarvi, G., Alestrom, P., Gelinas, R. E., Gingeras, T. R., Sciaky, D., and Pettersson, U., 1986, In: *Adenovirus DNA* (W. Dofler, ed.), pp. 1–51, Martinus Nijhoff, Boston.
6. Roberts, R. J., 1987, *Nucleic Acids Res. Suppl.* **15:**r189–r217.
7. van Ormondt, H., and Galibert, F., 1984, *Curr. Top. Microbiol. Immunol.* **110:**73–142.
8. Levine, A. J., 1984, *Curr. Top. Microbiol. Immunol.* **110:**143–167.
9. Ziff, E. B., 1980, *Nature* **297:**365–371.
10. Montell, C., Courtois, G., Eng, C., and Berk, A. J., 1984, *Cell* **36:**951–961.
11. Flint, J., and Shenk, T., 1989, *Annu. Rev. Genet.* **23:**141–161.
12. Stillman, B. W., Lewis, J. B., Chow, L. T., Mathews, M. B., and Smart, J. E., 1981, *Cell* **23:** 497–508.
13. Wold, W. S. M., and Gooding, L. R., 1991, *Virology* **184:**1–8.
14. Halbert, D. N., Cutt, J. R., and Shenk, T., 1985, *J. Virol.* **56:**250–257.
15. Hardy, S., Engel, D. A., and Shenk, T., 1989, *Genes Dev.* **3:**1062–1074.
16. Hasson, T. B., Soloway, P. D., Ornelles, D. A., Doerfler, W., and Shenk, T., 1989, *J. Virol.* **63:** 3612–3621.
17. Cepko, C. L., and Sharp, P. A., 1982, *Cell* **31:**407–415.
18. Hayes, B. W., Telling, G. C., Myat, M. M., Williams, J. F., and Flint, S. J., 1990, *J. Virol.* **64:** 2732–2742.
19. Akusjarvi, G., and Pettersson, U., 1979, *J. Mol. Biol.* **134:**143–158.
20. Logan, J., and Shenk, T., 1984, *Proc. Natl. Acad. Sci. USA* **81:**3655–3659.
21. Dolph, P. J., Racaniello, V., Villamarin, A., Palladino, F., and Schneider, R. J., 1988, *J. Virol.* **62:** 2059–2066.
22. Reich, P. R., Baum, S. G., Rose, J. A., Row, W. P., and Weissman, S. M., 1966, *Proc. Natl. Acad. Sci. USA* **55:**336–341.
23. Ohe, K., and Weissman, S. M., 1971, *J. Biol. Chem.* **246:**6691–7009.
24. Price, R., and Penman, S., 1972, *J. Mol. Biol.* **70:**435–450.
25. Weinmann, R., Raskas, H. J., and Roeder, R. G., 1974, *Proc. Natl. Acad. Sci. USA* **71:**3426–3430.
26. Soderlund, H., Pettersson, U., Vennstrom, B., Philipson, L., and Mathews, M. B., 1976, *Cell* **7:** 585–593.
27. Mathews, M. B., 1975, *Cell* **6:**223–229.
28. Pettersson, U., and Philipson, L., 1975, *Cell* **6:**1–4.
29. Mathews, M. B., and Pettersson, U., 1978, *J. Mol. Biol.* **119:**293–328.
30. Thimmapaya, B., Jones, N., and Shenk, T., 1979, *Cell* **18:**947–954.
31. Celma, M. L., Pan, J., and Weissman, S. M., 1977, *J. Biol. Chem.* **252:**9032–9042.
32. Celma, M. L., Pan, J., and Weissman, S. M., 1977, *J. Biol. Chem.* **252:**9043–9046.
33. Pan, J., Celma, M. L., and Weissman, S. M., 1977, *J. Biol. Chem.* **252:**9047–9054.
34. Fowlkes, D. M., and Shenk, T., 1980, *Cell* **22:**405–413.
35. Guilfoyle, R., and Weinmann, R., 1981, *Proc. Natl. Acad. Sci. USA* **78:**3378–3382.
36. Bhat, R. A., Metz, B., and Thimmapaya, B., 1983, *Mol. Cell. Biol.* **3:**1996–2005.
37. Railey, J. F., and Wu, G.-J., 1988, *Mol. Cell. Biol.* **8:**1147–1159.
38. Bhat, R. A., and Thimmapaya, B., 1984, *Nucleic Acids Res.* **12:**7377–7388.
39. Gaynor, R. B., Feldman, L. T., and Berk, A. J., 1985, *Science* **230:**447–450.
40. Berger, S. L., and Folk, W. R., 1985, *Nucleic Acids Res.* **13:**1413–1428.
41. Hoeffler, W. K., and Roeder, R. G., 1985, *Cell* **41:**955–963.
42. Lerner, M. R., Boyle, J. A., Hardin, J. A., and Steitz, J. A., 1981, *Science* **211:**400–402.

43. Mathews, M. B., and Francoeur, A. M., 1984, *Mol. Cell. Biol.* **4:**1134–1140.
44. Schneider, R. J., Weinberger, C., and Shenk, T., 1984, *Cell* **37:**291–298.
45. Akusjarvi, G., Mathews, M. B., Andersson, P., Vennstrom, B., and Pettersson, U., 1980, *Proc. Natl. Acad. Sci. USA* **77:**2424–2428.
46. Furtado, M. R., Subramanian, S., Bhat, R. A., Fowlkes, D. M., Safer, B., and Thimmapaya, B., 1989, *J. Virol.* **63:**3423–3434.
47. Mellits, K. H., and Mathews, M. B., 1988, *EMBO J.* **7:**2849–2859.
48. Monstein, H. J., and Philipson, L., 1981, *Nucleic Acids Res.* **9:**4239–4250.
49. Engler, J. A., Hoppe, M. S., and van Bree, M. P., 1983, *Gene* **21:**145–149.
50. Larsson, S., Svensson, C., and Akusjarvi, G., 1986, *J. Virol.* **60:**635–644.
51. Larsson, S., Bellett, A., and Akusjarvi, G., 1986, *J. Virol.* **58:**600–609.
52. Thimmapaya, B., Weinberger, C., Schneider, R. J., and Shenk, T., 1982, *Cell* **31:**543–551.
53. Graham, F. L., Smiley, J., Russell, U. C., and Nairu, R., 1977, *J. Gen. Virol.* **36:**59–72.
54. Safer, B., 1983, *Cell* **33:**7–8.
55. Ochoa, S., 1983, *Arch. Biochem. Biophys.* **223:**325–349.
56. Hovanessian, A. G., 1989, *J. Interferon Res.* **6:**641–647.
57. Farrell, P. J., Balkow, K., Hunt, T., Jackson, R. J., and Trachsel, H., 1977, *Cell* **11:**187–200.
58. Lebleu, B., Sen, G. C., Shaila, S., Cabrer, B., and Lengyel, P., 1976, *Proc. Natl. Acad. Sci. USA* **73:**3107–3111.
59. Roberts, W. K., Hovanessian, A., Brown, R. E., Clemens, M. J., and Kerr, I. M., 1976, *Nature* **264:**477–480.
60. Zilberstein, A., Kimchi, A., Schmidt, A., and Revel, M., 1978, *Proc. Natl. Acad. Sci. USA* **75:** 4734–4738.
61. Kostura, M., and Mathews, M. B., 1989, *Mol. Cell. Biol.* **9:**1576–1586.
62. Meurs, E., Chong, K., Galabru, J., Thomas, N. S., Kerr, I. M., Williams, B. R. G., and Hovanessian, A. G., 1990, *Cell* **62:**379–390.
63. Baglioni, C., 1979, *Cell* **17:**255–264.
64. Johnston, M. I., and Torrence, P. F., 1984, In: *Interferon 3, Mechanisms of Production and Action* (R. M. Friedman, ed.), pp. 189–298, Elsevier/North-Holland, Amsterdam.
65. Lengyel, P., 1982, *Annu. Rev. Biochem.* **51:**251–282.
66. Berry, M. J., Knutson, G. S., Lasky, S. R., Munemitsu, S. M., and Samuel, C. E., 1985, *J. Biol. Chem.* **260:**11,240–11,247.
67. Hovanessian, A. G., and Kerr, I. M., 1979, *Eur. J. Biochem.* **93:**515–526.
68. Hunter, T., Hunt, T., Jackson, R. J., and Robertson, H. D., 1975, *J. Biol. Chem.* **250:**409–417.
69. Minks, M. A., West, D. K., Benvin, S., and Baglioni, C., 1979, *J. Biol. Chem.* **254:**10,180–10,183.
70. Galabru, J., Katze, M. G., Robert, N., and Hovanessian, A. G., 1989, *Eur. J. Biochem.* **178:** 581–589.
71. Katze, M. G., DeCorato, D., Safer, B., Galabru, J., and Hovanessian, A. G., 1987, *EMBO J.* **6:** 689–697.
72. Reichel, P. A., Merrick, W. C., Siekierka, J., and Mathews, M. B., 1985, *Nature* **313:**196–200.
73. Siekierka, J., Mariano, T. M., Reichel, P. A., and Mathews, M. B., 1985, *Proc. Natl. Acad. Sci. USA* **82:**1959–1963.
74. Schneider, R. J., Safer, B., Munemitsu, S. M., Samuel, C. E., and Shenk, T., 1985, *Proc. Natl. Acad. Sci. USA* **82:**4321–4325.
75. O'Malley, R. P., Mariano, T. M., Siekierka, J., and Mathews, M. B., 1986, *Cell* **44:**391–400.
76. Kitajewski, J., Schneider, R. J., Safer, B., Munemitsu, S. M., Samuel, C. E., Thimmapaya, B., and Shenk, T., 1986, *Cell* **45:**195–200.
77. Davies, M. V., Furtado, M., Hershey, J. W. B., Thimmapaya, B., and Kaufman, R. J., 1989, *Proc. Natl. Acad. Sci. USA* **86:**9163–9167.
78. Kitajewski, J., Schneider, R. J., Safer, B., and Shenk, T., 1986, *Mol. Cell. Biol.* **6:**4493–4498.

79. Maran, A., and Mathews, M. B., 1988, *Virology* **164:**106–113.
80. Svensson, C., and Akusjarvi, G., 1984, *Mol. Cell. Biol.* **4:**736–742.
81. Svensson, C., and Akusjarvi, G., 1985, *EMBO J.* **4:**957–964.
82. Akusjarvi, G., Svensson, C., and Nygard, O., 1987, *Mol. Cell. Biol.* **7:**549–551.
83. Ghadge, G. D., Swaminathan, S., Katze, M. G., and Thimmapaya, B., 1991, *Proc. Natl. Acad. Sci. USA* **88:**7140–7144.
84. Mellits, K. H., Kostura, M., and Mathews, M. B., 1990, *Cell* **61:**843–852.
85. Rice, A. P., Kostura, M., and Mathews, M. B., 1989, *J. Biol. Chem.* **264:**20,632–20,637.
86. Svensson, C., and Akusjarvi, G., 1986, *Proc. Natl. Acad. Sci. USA* **83:**4690–4694.
87. Strijker, R., Fritz, D. T., and Levinson, A. D., 1989, *EMBO J.* **8:**2669–2675.
88. Bhat, R. A., Furtado, M. R., and Thimmapaya, B., 1989, *Nucleic Acids Res.* **17:**1159–1176.
89. Youngner, J. S., Thacore, H. R., and Kelly, M. E., 1972, *J. Virol.* **10:**171–181.
90. Whitaker-Dowling, P., and Youngner, J. S., 1983, *Virology* **131:**128–136.
91. Rice, A. P., and Kerr, I. M., 1984, *J. Virol.* **50:**229–236.
92. Paez, E., and Esteban, M., 1984, *Virology* **134:**12–48.
93. Whitaker-Dowling, P., and Youngner, J. S., 1984, *Virology* **137:**171–181.
94. Watson, J. C., Chang, H.-W., and Jacobs, B. L., 1991, *Virology* **185:**206–216.
95. Katze, M. G., Chen, Y.-T., and Krug, R. M., 1984, *Cell* **37:**483–490.
96. Katze, M. G., Detjen, B., Safer, B., and Hovanessian, A. G., 1988, *J. Virol.* **62:**3710–3717.
97. Lee, T. G., Tomita, J., Hovanessian, A. G., and Katze, M. G., 1990, *Proc. Natl. Acad. Sci. USA* **87:**6208–6212.
98. Rosen, H., Knoller, S., and Kaempfer, R., 1981, *Biochemistry* **20:**3011–3020.
99. Imani, F., and Jacobs, B. L., 1988, *Proc. Natl. Acad. Sci. USA* **85:**7887–7891.
100. Rosa, M. D., Gottlieb, E., Lerner, M. R., and Steitz, J. A., 1981, *Mol. Cell. Biol.* **1:**785–796.
101. Bhat, R. A., and Thimmapaya, B., 1983, *Proc. Natl. Acad. Sci. USA* **80:**4789–4793.
102. Bhat, R. A., and Thimmapaya, B., 1985, *J. Virol.* **56:**750–756.
103. Clarke, P. A., Sharp, N. A., and Clemens, M. J., 1990, *Eur. J. Biochem.* **193:**635–641.
104. Clarke, P. A., Sharp, N. A., and Clemens, M. J., 1991, *Nucleic Acids Res.* **19:**243–248.
105. Subramanian, S., Bhat, R. A., Rundell, M. K., and Thimmapaya, B., 1986, *J. Virol.* **60:** 363–368.
106. Swaminathan, S., Tomkinson, S. B., and Kieff, E., 1991, *Proc. Natl. Acad. Sci. USA* **88:**1546–1550.
107. Huang, J. T., and Schneider, R. J., 1991, *Cell* **65:**271–280.

Chapter 11

Translational Regulation in Adenovirus-Infected Cells

Robert J. Schneider and Yan Zhang

1. ADENOVIRUS INFECTIOUS CYCLE

Adenoviruses comprise a large group of DNA viruses that infect humans, a variety of animals, and birds.[1] The large number of different adenovirus serotypes display different tissue tropisms.[2] Respiratory tract infections are associated with types 1–7 and 14, conjunctival infections with type 8, and enteric infections with types 12, 18, 21, 40, and 41. In addition, types 1, 2, 5, and 6 are associated with latent respiratory tract infections that may persist for months and occasionally years.[3] Adenoviruses were originally studied because of their ability to induce a profound cytopathic effect and to alter basic cellular metabolism during infection. The investigation for the past 40 years of the complex cellular metabolic changes that occur during infection by adenovirus has contributed important fundamental information to the understanding of gene regulation in animal cells.

ROBERT J. SCHNEIDER AND YAN ZHANG • Department of Biochemistry and Kaplan Cancer Center, New York University Medical Center, New York, New York 10016.

Translational Regulation of Gene Expression 2, edited by Joseph Ilan. Plenum Press, New York, 1993.

1.1. Temporal and Physical Organization of the Adenovirus Genome

The adenovirus genome is temporally organized into early and late transcription units that are activated before or with the onset of viral DNA replication, respectively. The six early transcription units synthesize a variety of messenger RNAs (mRNAs) which encode polypeptides involved in functions related to establishing both productive viral replication and transformation of the infected cell. For instance, regions E1A and E1B encode proteins involved in cellular transformation and transactivation of the other viral transcription units.[4] Regions E2A and E2B encode proteins involved in adenoviral DNA replication, including the viral polymerase and DNA binding protein. Regions E3 and E4 encode polypeptides involved in a variety of early viral functions, including suppression of histocompatability antigen expression (reviewed in Wold and Gooding[5]), transcriptional transactivation,[6] and regulation of nuclear to cytoplasmic transport of cellular and viral mRNAs.[7,8] The products of the early transcription units comprise only a very minor proportion of cellular mRNA and protein synthesis. Accordingly, the early phase of the viral infectious cycle is not generally associated with dramatic alterations in cellular metabolism.

The late phase of adenovirus infection is marked by the onset of viral DNA replication, which typically begins from 10 to 16 hr after infection (reviewed in Ginsberg[9]). Unlike early viral transcription, which is initiated at promoters located at six different locations, there is a single major late promoter (MLP) located at 16.4 map units on the viral genome that is strongly activated after DNA replication commences. The MLP generates five families of late transcripts (L1–L5) by differential splicing and polyadenylation of a large primary transcript that terminates within the right end of the genome at 99 map units (reviewed in Ziff[10]). Every MLP transcribed mRNA contains an identical 5′ noncoding region of 200 nucleotides in length called the tripartite leader,[11,12] because it is derived from the splicing of three small exons located upstream of the five families of late transcripts. Most of the late adenoviral mRNAs encode structural polypeptides that are involved in packaging viral genomic DNAs and comprise the virion particle. It is therefore not surprising that synthesis of these proteins occurs in large amounts and is accompanied by the suppression of cellular protein synthesis.

During the late phase of adenovirus infection there is also the synthesis of large amounts of the viral encoded RNA polymerase III products called virion-associated (VA) RNAs I and II (reviewed in Chapter 10). The VA RNAs have been shown to be absolutely required for translation of mRNAs at late times during infection because they counter the cellular antiviral response mediated by the interferon-stimulated p68 kinase.

1.2. Changes in Cellular Metabolism during Infection

Productive infection of cultured cells (HeLa, KB, and 293 cells) typically begins shortly after adsorption of virus, lasts from 30 to 48 hr, and is accompanied in the late phase with dramatic alterations to host-cell metabolism. The early phase of the infectious cycle occurs within approximately 8–12 hr, followed by the onset of viral DNA synthesis and the increasing synthesis of late viral mRNAs and polypeptides (for a detailed review see Ginsberg[9]). As the late phase progresses, extremely large quantities of late viral polypeptides are produced. Several complex metabolic changes in host-cell metabolism also occur during this period of infection. First, infected cells no longer divide after the viral productive cycle enters the late phase. Second, there is an almost exclusive synthesis of late viral polypeptides.[13] Third, the rate of transport of cellular mRNAs from the nucleus to the cytoplasm is extensively reduced,[14] although transcription of most cellular genes is not inhibited.[15] Thus, the cellular synthesis of DNA, RNA, and protein is largely redirected toward the production and assembly of viral particles. Eventually, virions are released when cell lysis occurs, which is thought to result from extended viral inhibition of host-cell metabolism.

The inhibition of cellular division is probably indirectly related to many of the effects of adenovirus on host-cell metabolism. There is no evidence that a single adenoviral gene product prevents cellular division (reviewed in Flint[16]). For example, cessation of cell division could result from prevention of cellular DNA synthesis, which occurs as the viral productive cycle enters the late phase[13] and viral DNA synthesis begins.[17] It could also result, however, from inhibition of cellular protein synthesis, or from competition between viral and cellular DNAs for a limiting component of the replication apparatus.

The inhibition of cellular protein synthesis is observed concomitant with the late phase of adenovirus replication.[13] Late viral mRNAs generally constitute the majority (~90–95%) of those found in polyribosomes, although they represent only a fraction of the cytoplasmic pool of messages[18–20] (reviewed in Schneider and Shenk[21]). There is, therefore, preferential translation of the late, tripartite leader containing viral mRNAs and suppression of cellular mRNAs. In addition, there is also a selective transport of late viral mRNAs from the nucleus to the cytoplasm, which involves the E1B-55k/E4-34k protein complex.[7,8,22]

2. EVIDENCE FOR TRANSLATIONAL REGULATION IN LATE ADENOVIRUS-INFECTED CELLS

As the late phase of the adenovirus growth cycle begins, typically between 14 and 18 hr after infection, cells exclusively synthesize viral polypeptides.[13,23] There are several translational mechanisms which could result in the appearance of predominantly late viral polypeptides in adenovirus-infected cells at the expense of cellular proteins. In addition to translational regulation, other mechanisms include: (1) virus-mediated degradation of host-cell mRNAs, as suggested in poxvirus[24] and herpes-simplex virus-infected cells,[25] (2) transcription of vast amounts of viral mRNAs that simply dilute and outcompete cellular species, as found in vesicular stomatitis virus-infected cells,[26] (3) suppression of cellular transcription and/or host mRNA transport to the cytoplasm, which occurs to some extent in human immunodeficiency virus (HIV)-infected cells,[27] and (4) viral modifications to cellular mRNAs, such as the removal of cap structures, as occurs to some extent in influenza virus-infected cells.[28]

2.1. Cellular mRNAs Are Not Degraded in Late Adenovirus-Infected Cells

Early evidence suggested that cellular mRNAs are not degraded by adenovirus and, in fact, are translationally competent.[29] If one extracts mRNAs from late adenovirus-infected cells and uses them to program *in vitro* translation systems, the mRNAs direct the synthesis of large amounts of cellular and late viral polypeptides.[30,31] It was also shown that cell mRNAs remain capped and polyadenylated.[31] Thus, cellular mRNAs are present at near normal levels in the cytoplasm of late adenovirus-infected cells and are not processed abnormally or physically modified in a manner that would prevent their translation. Examination of individual mRNAs by cDNA cloning also showed the cytoplasmic abundance of the more stable cellular mRNAs to be about the same in adenovirus-infected and uninfected cells.[20] Therefore, cellular mRNAs are well represented in the cytoplasm at late times after infection.

2.2. Adenoviral Late mRNAs Do Not Significantly Dilute Cellular Messages

After adenovirus infection enters the late phase, the majority (90–95%) of cytoplasmic mRNAs found in polysomes are viral transcripts. These mRNAs correspond predominantly to late species transcribed from the viral major late promoter, and from two independent transcription units encoding proteins IX and IVa2.[19,32] These results, however, cannot be explained simply on the basis of large amounts of viral mRNAs diluting the concentration of cellular species. Early studies which investigated the proportion of mRNAs corresponding to viral and cellular origin determined that at late times after infection adenoviral mRNAs constitute only about 20% of the total cytoplasmic pool.[19,33,34] These results are also in agreement with later studies (described above) demonstrating the efficient *in vitro* translation of cellular mRNAs extracted from late adenovirus-infected cells. It is clear, therefore, that adenovirus promotes the synthesis of its own late mRNAs while suppressing those of the host cell.

2.3. Reduced Transport of Cellular mRNAs Does Not Account for Adenovirus Inhibition of Host Translation

As described above, with the entry of adenovirus into its late phase of infection the majority of newly synthesized RNAs which accumulate in the cytoplasm correspond to viral encoded sequences.[19,32] Flint and co-workers established that approximately 90–95% of the mRNA transported to the cytoplasm during late infection is adenoviral in origin[14,15] (reviewed in Flint[16]). Analysis of nuclear transcription rates for a variety of cellular genes demonstrated levels of expression that were generally similar in uninfected and late adenovirus-infected cells.[14,20,35,36] These results therefore excluded the possibility of accelerated rates of nuclear degradation of host mRNAs. Thus, adenovirus infection does not prevent transcription of cellular genes, but blocks the transport of cellular mRNAs from the nucleus to the cytoplasm. The reduced transport of cellular mRNAs was shown to involve the adenovirus early E1B-55, and E4-34k protein complex.[7,8,22] Adenovirus mutants which fail to synthesize either of these two proteins are unable to selectively transport viral mRNAs during late infection.

It would seem reasonable to propose that adenovirus inhibition of cellular translation is related to the block in transport of host mRNAs. Nevertheless, the results from a number of studies do not support this conclusion. Since adenovirus does not induce the degradation of cellular mRNAs, the cytoplasmic abundance of host messages at late times after infection is controlled by the intrinsic half-life of each transcript. The steady-state level of a number of short-lived mRNAs was found to be reduced dramatically during the late phase of infection[35,37] and this could account for the failure to synthesize corresponding polypeptides. A large proportion of host mRNAs are long-lived, however, and their steady-state cytoplasmic levels are only slightly reduced in late adenovirus-infected cells,[20,36,38] although translation of these mRNAs has been inhibited. It is also important to note that the inhibition of cellular protein synthesis progresses rapidly as adenovirus enters it late phase (at 13–15 hr). Inhibition of cellular translation therefore occurs before the block in transport significantly reduces the cytoplasmic pool of host mRNAs. This is underscored by the recent observation that some cellular mRNAs, such as β-tubulin, escape the transport block in late adenovirus-infected cells and accumulate to near normal levels in the cytoplasm, but are still not translated.[39] Finally, it was recently shown that adenovirus inhibition of cellular protein synthesis can be prevented by the drug 2-aminopurine without relieving the normal block in transport of host mRNAs.[40] These results all imply that translation of cellular mRNAs is specifically prevented during late adenovirus infection.

3. ADENOVIRUS INHIBITION OF CELLULAR PROTEIN SYNTHESIS MAY INVOLVE SEVERAL DISTINCT MECHANISMS OF TRANSLATIONAL CONTROL

In the past several years three independent lines of investigation have provided insights into potential mechanisms for the regulation of translation in late adenovirus-infected cells. Each line of research apparently reveals a different aspect of adenovirus translational regulation. Accordingly, each proposed mechanism involves the activity of different adenoviral genes and proposes a different model for translational control. It is clear that each of the adenovirus genes studied is vital for the successful translation of viral mRNAs at late times after infection. It is unclear, however, whether they act in an integrated manner to achieve virus-specific translation, or whether there are several distinct mechanisms that ensure viral domination of the host translational machinery. In this section we will review the different lines of investigation and explore the potential mechanisms that may account for adenovirus translational control.

3.1. Translational Regulation by Adenovirus 100k Protein

The 100k protein is a late adenovirus polypeptide encoded by the L4 transcription unit. The 100k protein is one of the first late polypeptides to be expressed with the onset of the late phase of viral infection,[41] and represents one of the most abundant viral polypeptides.[42] This protein is found associated with the viral capsid hexon protein, participates in morphogenesis of the hexon protein,[42,43] and is essential for production of viral particles.[43,44]

3.1.1. 100k Protein Binds Strongly to Cytoplasmic mRNAs

A number of studies showed that the 100k protein is associated with mRNAs, in addition to its association with hexon protein.[45–47] The association is similar to that described for classic messenger ribonucleoproteins (mRNPs), in that it is resistant to dissociation by high concentrations of salt.[45] It was therefore suggested that a second potential role for 100k protein could include the regulation of translation during the late phase of adenovirus infection.[45] In particular, it was suggested that the binding of 100k protein to cytoplasmic mRNAs might reduce their translational efficiencies. It was suggested that late viral mRNAs could partially overcome the disadvantage if they also possessed an intrinsically superior ability to initiate translation. Thus, translation of cellular mRNAs would be extinguished, but late viral mRNAs would be reduced, effectively inhibiting host protein synthesis with some impairment of viral translation as well.

3.1.2. 100k Protein Is Required for Translation of Late Adenovirus mRNAs

Evidence directly linking the 100k protein to a role in late viral translation has been obtained recently by Flint and co-workers.[48] In the course of screening temperature-sensitive (ts) mutants for defects in adenovirus gene expression, Hayes *et al.*[48] discovered that an adenovirus type 5 variant, H5ts-1, possessed a severe defect in late viral protein synthesis at the restrictive temperature. The ts lesion was mapped to L4 100k protein and shown to be a single-base-pair mutation that converted Ser-466 to Pro.

A full characterization of the H5ts-1 phenotype showed that the reduced translation of late viral mRNA occurs during initiation and could be attributed solely to the effects of the ts mutation. First, all second-site revertants of the ts-1 phenotype mapped in the L4 100k gene, converting the Pro-466 to Thr, Leu, or His. Second, the synthesis and steady-state level of late viral mRNAs are unchanged at the nonpermissive temperature. Examination of polysome profiles indicated that at the restrictive temperature late viral mRNAs bound significantly fewer ribosomes, a result consistent with reduced efficiency of initiation. Third, translation of most early adenoviral mRNAs at late times after infection was not reduced, in contrast to that of late messages. Taken together, these results clearly established a role for 100k protein in efficient translation initiation of late viral mRNAs. Most surprisingly, however, was the observation that 100k protein is not required for inhibition of cellular protein synthesis during late infection.[48] Translation of cellular mRNAs was as efficiently blocked in H5ts-1-infected cells at the restrictive temperature as in wild-type-infected cells at the normal temperature. This study suggested that the shutoff of cellular protein synthesis and the preferential translation of late adenoviral mRNAs may involve separate mechanisms. It is not known why the ts-1 mutation generally prevents protein synthesis, and it may be unrelated to the normal mechanism by which the virus inhibits host translation.

3.1.3. Model for Selective Translation of Late Adenoviral mRNAs by L4 100k Protein

The 100k protein contains a sequence motif identified as an RNA binding moiety in many RNA binding proteins[48] and corresponds quite well to a consensus sequence determined experimentally.[49] The ts-1 mutation, insertion of a Pro residue at position 466, lies near an important RNA binding element.[48] Whether the mutation is likely to disrupt the RNA binding ability of the 100k protein at the destabilizing (elevated) temperature has not been investigated. It was suggested, however, that the RNA binding activity of 100k protein is a vital part of its ability to facilitate translation of late viral mRNAs. Several models can be envisioned in which the RNA binding activity of 100k protein contributes to selective translation of late mRNAs. In one model, 100k protein would be expected to bind strongly to a common sequence or structure found within all adenoviral transcripts that are translated at late times, but much less well to cellular mRNAs. The 100k protein might then enhance mRNA translation by reducing the requirement for a limiting translation factor. In an alternate model, it was proposed that 100k protein might direct late viral mRNAs to a translation "compartment" within the cell established by the virus. Whether either of these models accounts for the translation function of 100k protein will first depend upon the demonstration that the protein preferentially associates with late viral transcripts. Previous studies, described above, have not yet established differential binding of 100k protein to viral rather than cellular mRNAs.

3.2. Translational Regulation by Adenovirus VAI RNA

An extensive number of studies have demonstrated a role for VAI RNA in late adenovirus translation regulation. Several excellent reviews have been published recently detailing the role of the VA RNAs in translational control[50] (and this volume, Chapter 10). Thus, we will only briefly review the evidence establishing a role for VA RNA in translational regulation, and instead concentrate on potential mechanisms by which VAI RNA may be involved in the shutoff of cellular protein synthesis and the preferential translation of late adenovirus mRNAs.

3.2.1. VAI RNA Prevents Phosphorylation (Inactivation) of eIF-2

Adenovirus-infected cells contain large amounts of two distinct 160-nucleotide (nt) RNA species, referred to as VAI and VAII RNAs.[51,52] The VA RNAs possess a very stable secondary structure that has been verified experimentally, and in the case of VAI RNA, shown to be critical for its translation activity.[53,54] VAI RNA is the major species and its synthesis increases rapidly during late infection.[55] Initial evidence that VAI RNA is involved in adenovirus late translation was revealed by adenovirus mutant *dl*331, which was engineered to be incapable of synthesizing VAI RNA.[31] Cells infected with this virus showed a global inhibition in translation of *both* viral and cellular mRNAs at late times after infection, which was not attributable to defects in viral DNA replication or mRNA accumulation. Subsequent investigations indicated that the translation defect occurred during an early step in initiation,[56,57] which was localized to the activity and phosphorylation state of initiation factor eIF-2.[57,58]

The phosphorylation of eIF-2 is a central controlling event in protein synthesis, and one of the best-studied steps in initiation. This is largely because a number of systems have been shown to employ the phosphorylation and activity of eIF-2 in controlling translation, including exposure of cells to interferon, heat shock, serum, or nutrient deprivation, and viral infection (reviewed in Jagus *et al.*[59]). Although eIF-2 consists of three subunits (α, β, γ), only phosphorylation of the α subunit alters its activity. eIF-2 complexes with GTP and the initiating methionyl-tRNA to form a ternary complex, which then directs the anticodon recognition of the 40S ribosomal subunit with the initiation codon.

The creation of the 80S initiating ribosome upon joining of the 60S subunit is accompanied by the hydrolysis of GTP and the release of inactive eIF-2–GDP. The catalytic "recycling" of GDP to GTP on eIF-2 is carried out by the guanine nucleotide exchange factor (GEF), also called translation factor eIF-2B (reviewed in Safer[60]). Thus, the recycling of eIF-2–GDP by GEF regulates the rate of translation initiation by controlling the entry of eIF-2–GTP into the free pool. Phosphorylation of the α subunit of eIF-2–GDP greatly enhances its affinity for the limited amounts of GEF, resulting in sequestration of the exchange factor into an inactive complex and inhibition of initiation.

Phosphorylation of eIF-2 is mediated in animals cells by a protein kinase which is induced by interferon, cell stress responses, or virus infection, and is activated by low concentrations of double-stranded (ds) RNA. Historically, this 68-kDa kinase is known as the dsRNA-activated inhibitor (DAI) because synthesis of an inactive form is induced by interferon, but activated by dsRNA (reviewed in Hovenessian[61]). The role of dsRNA is to bind to and drive the pairwise interaction of kinase molecules, which then activate each other by phosphorylation (reviewed by Mathews and Shenk[50]). VAI RNA prevents the activation of DAI kinase by directly binding to the enzyme and blocking its activation by dsRNA.[62,63] Mutational and functional studies have confirmed that an interaction between DAI kinase and a region of VAI RNA is essential for the blocking activity.[53,64,65]

A considerable amount of evidence indicates that the function of VAI RNA is to block the antiviral response mediated by DAI kinase during late adenovirus infection. First, protein synthesis is inhibited in most cells at late times after infection with *dl*331 (VAI^-) virus, but can be restored in cell extracts supplemented with either eIF-2 or GEF.[57,58,66] Second, adenovirus *dl*331 grows normally in cell lines deficient in DAI kinase[58,67] and significantly more poorly in cells treated with α-interferon.[58,67–69] Third, symmetrical transcription of the adenovirus genome normally produces dsRNA during the late phase of infection, which seems to serve as the activating signal for DAI kinase in adenovirus-infected cells.[70]

3.2.2. Model for Selective Translation of Late Adenoviral mRNAs by VAI RNA

Adenovirus VAI RNA and the DAI kinase could potentially participate as discriminatory factors that preferentially translate late viral mRNAs and shut off host-cell protein synthesis. A potential role of VAI RNA and DAI kinase in selective translation is based on a number of intriguing but circumstantial observations, as listed below.

1. *Adenovirus cannot inhibit host translation in cells deficient in DAI kinase activity*.[67,71] Examination of polypeptide profiles and viral growth curves indicates that adenovirus synthesizes its late polypeptides and produces infectious particles (albeit at lower levels) in DAI-deficient cells, but does not shut off cellular protein synthesis.

2. *Partial activation of DAI kinase and phosphorylation of eIF-2 occurs in at least some cell lines during late infection by adenovirus*[40,67] (Fig. 1). Typically, up to 20–25% of eIF-2α may be phosphorylated in wild-type-adenovirus-infected cells if DAI kinase is activated. Nevertheless, under these conditions late viral mRNAs are translated despite seemingly inhibitory levels of DAI kinase activity and eIF-2α phosphorylation.

3. *DAI kinase is generally found associated with polysomes in the cell*,[72]

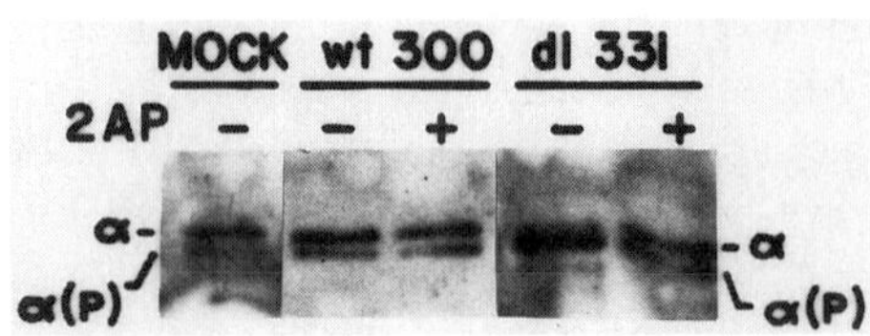

Figure 1. *In vivo* phosphorylation of eIF-2α in cells infected with *wt*300 or *dl*331 adenovirus without (−) or with (+) 2-aminopurine treatment. Cell lysates were prepared from type 293 cells infected at high multiplicities of infection (>100 plaque-forming units) by lysis directly into ampholine–urea buffer, subjected to slab isoelectric focusing, transferred to nitrocellulose, and probed with labeled antibody directed against eIF-2α. The lower form represents the acidic (phosphorylated) α subunit.

as are many translation factors[73,74] *and possibly some VAI RNA.*[56] Thus, the physical location of these components is at least consistent with the notion of mRNA-specific translational regulation.

4. *It has been proposed that the level of eIF-2α phosphorylation may contribute to the specificity of mRNA translation.*[75] For example, a number of systems display inhibitory levels of active DAI kinase and phosphorylated eIF-2α without the complete extinction of protein synthesis. These include translation of influenza virus mRNAs in cells coinfected with adenovirus *dl*331,[76] translation of late adenovirus mRNAs in some infected cells,[40,67] translation of poliovirus mRNA in infected cells,[77,78] translation of some mRNAs in cells transfected by plasmids,[75] and translation of certain mRNAs in cell-free systems.[79] Thus, there may be a "window" or range in eIF-2α phosphorylation levels which promotes specificity in mRNA translation, thereby restricting mRNAs which require a larger pool of active eIF-2 for their translation.

A model for late adenovirus translation has been proposed incorporating these observations, in which specific inhibition of cellular mRNA translation is mediated by activated DAI kinase and elevated levels of phosphorylated eIF-2α.[67] In its simplest form, this model suggests that late viral mRNAs, by virtue of a potential ability to bind VAI RNA,[80] might block local activation of ribosome-associated DAI kinase. The problem with this simple model, however, is that local activation of DAI kinase should lead to the global inhibition of translation through the release of eIF-2α(P) into the free pool of factor, and subsequent sequestration of GEF.

To explain the paradoxical translation of late viral mRNAs in the presence of inhibiting levels of phosphorylated eIF-2α, it was further proposed that during late adenovirus infection the translational machinery becomes compartmentalized in the cell[67] (reviewed in Mathews and Shenk[50]). According to this scheme, a functionally active compartment would contain, of necessity, VAI RNA, late viral mRNA, the active (nonphosphorylated) pool of eIF-2, and probably GEF (Fig. 2).

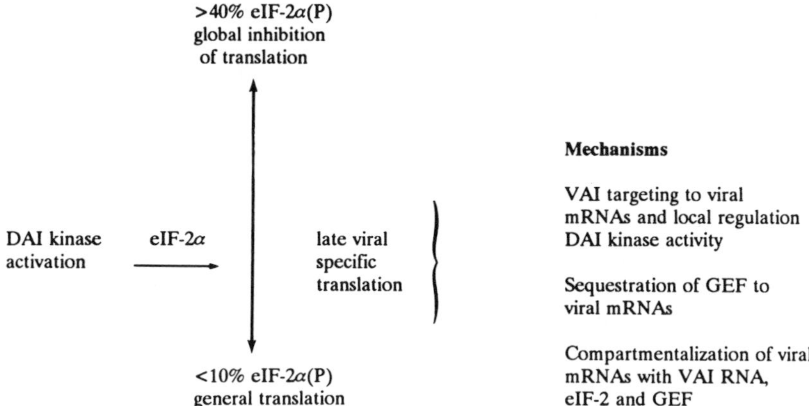

Figure 2. Model for the selective translation of late adenovirus mRNAs involving VAI RNA, DAI kinase, eIF-2, and GEF. The selective translation of late viral mRNAs is depicted, mediated by increased levels of phosphorylated eIF-2α. The role of VAI RNA in this model is to regulate the activity of DAI kinase rather than fully suppress it. Several mechanisms are proposed to explain ongoing translation of viral mRNAs in the presence of inhibiting levels of eIF-2α(P).

The functionally inactive compartment would therefore be depleted of active eIF-2 and GEF. It may be particularly relevant, then, that several studies have found GEF and phosphorylated eIF-2α to be differently distributed in cells, potentially through an interaction with 60S ribosomal subunits.[81–83]

An essential element of this model is the elevated and measurable phosphorylation of eIF-2α in late adenovirus-infected cells. Inhibiting levels of 20–30% in eIF-2α phosphorylation have been occasionally reported.[40,67] Most studies, however, have not observed any increase in eIF-2 phosphorylation during late virus infection, despite inhibition of cellular protein synthesis and preferential translation of late viral mRNAs.[48,58,66,69] In addition, the drug 2-aminopurine was shown to prevent the shutoff of cellular translation during late adenovirus infection without altering DAI kinase activity or the level or eIF-2α phosphorylation.[40]

There is clearly an important body of experimental evidence linking VAI RNA, DAI kinase, and eIF-2 to events in selective translation of late viral mRNAs. The poor correlation with phosphorylation of eIF-2α in a number of studies, however, suggests that, irrespective of the proposed model, these factors cannot fully account for translational discrimination during late infection. Rather, some other modification of the components involved in cellular protein synthesis must occur during the late phase of adenovirus infection to suppress host and promote viral mRNA translation.

3.3. Translational Regulation by the Tripartite Leader 5′ Noncoding Region

3.3.1. The Tripartite Leader Is Required for Translation of mRNAs during Late Adenovirus Infection

As described earlier, at late times after infection the majority of viral mRNAs are transcribed from the major promoter (MLP), giving rise to five families of 3′ coterminal mRNAs. All of these mRNAs contain an identical 5′ noncoding region, 200 nucleotides in length, called the tripartite leader.[11] A mutational study was conducted in which large segments of a tripartite leader cDNA were deleted, then reconstructed into the left end of viral genomes as a duplicate and nonessential copy fused to a reporter mRNA. These studies demonstrated that the intact leader was required for translation of mRNAs at late, but not early times after adenovirus infection.[84,85] In addition, the tripartite leader was shown to enhance translation of mRNAs in transfected cells,[86] which also required all three leader segments. It was therefore suggested that the tripartite leader is involved in preferential translation during late virus infection.[84,85]

3.3.2. The Tripartite Leader Reduces or Eliminates a Requirement for Cap-Binding Protein Complex (eIF-4F)

Several studies suggested that the translation properties of late adenovirus mRNAs may be unusual. In particular, it was found that translation of late viral mRNAs is resistant to inhibition by superinfecting poliovirus.[41,87] In poliovirus-infected cells, inhibition of cap-dependent cellular protein synthesis was correlated with the proteolytic degradation of a 220-kDa polypeptide (p220).[88,89] The p220 polypeptide is a component of initiation factor eIF-4F, a cap-dependent RNA helicase which stimulates protein synthesis by unwinding the 5′ end of mRNAs (reviewed in Sonenberg[90]), thereby promoting the binding of 40S ribosomes.[91–96] This factor was shown to contain three proteins: (1) eIF-4E, a 24-kDa protein which specifically binds cap structures, (2) eIF-4A, a 45-kDa ATP-dependent RNA helicase, and (3) p220, a 220-kDa protein of unknown function (reviewed in Rhoads[97] and Thach[98]). The degradation of p220 during poliovirus infection is thought to prevent or alter the normal cap-dependent RNA helicase activity associated with eIF-4F.[90] Factor eIF-4F is not essential for translation of picornaviral mRNAs because they internally bind ribosomes,[99,100] although eIF-4F has also been found to stimulate internal initiation.[101]

Considerable experimental evidence indicates that stable 5′ proximal secondary structure acts as an energy barrier to translation initiation. In particular, stable structure in the 5′ noncoding region proximal to the cap decreases ribosome binding, whereas reduced secondary structure promotes it.[91,102–106] Stable secondary structure in the 5′ noncoding region correlates with an increased requirement and decreased binding of eIF-4F,[91,93,96] while relaxed 5′ structure enhances the binding and decreases the requirement for eIF-4F.[91,107,108] Translation in the absence of eIF-4F activity, or in the presence of minimal amounts of this factor, is therefore unique to only a small number of mRNAs. This translation property has been established only for picornaviral and several cellular mRNAs that internally initiate translation[99,100,109,110] and for a few mRNAs with minimal 5′ secondary structure, such as alflafa mosaic virus 4 (AMV 4) mRNA[108] and heat-shock hsp70 and hsp83 mRNAs.[111,112]

Studies conducted by Dolph *et al.*[113] demonstrated that the adenovirus tripartite leader confers translation independent of eIF-4F activity in poliovirus-infected cells. It was found that in late adenovirus-infected cells superinfected with poliovirus, the low level of translation normally observed for cellular and early viral mRNAs was abolished, while mRNAs containing the tripartite leader were translated efficiently (Fig. 3). Factor eIF-4F was inactivated (or modified) by poliovirus superinfection because p220 was proteolyzed. This study also showed that eIF-4F-independent translation by the tripartite leader does not require any adenovirus gene products. The tripartite leader directed efficient eIF-4F-independent translation when it was attached to mRNAs expressed from transfected plasmids in the presence of infecting poliovirus. Thus, the tripartite leader was shown to be sufficient in confering translation independent of eIF-4F activity.

3.3.3. Elements Which Direct eIF-4F-Independent Translation by the Tripartite Leader

The mechanism by which the tripartite leader reduces or eliminates the requirement for eIF-4F was investigated in detail by Dolph *et al.*[114] Three potential mechanisms were proposed for tripartite leader activity: (1) Promotion of internal initiation through an internal ribosome entry site (IRES), as described for picornaviral mRNAs (reviewed in Jackson[115]). (2) Translation initiating at an unstructured 5′ end, as shown for hsp70, hsp83, and AMV 4 mRNAs (reviewed in Thach[98]). (3) Translation utilizing a unique nucleotide sequence or secondary structure that directs a prokaryotic-type interaction between 18S ribosomal RNA (rRNA) and the mRNA. In this regard, the nucleotide sequence of the tripartite leader was found to possess a surprising feature. Each leader segment contains a significant complementarity to a conserved hairpin structure found in the 3′ end of 18S rRNA, suggesting the potential for a prokaryotic-type Shine–Dalgarno interaction. Although such an interaction in eukaryotic cells is unprecendented, it has been suggested as a means to facilitate the binding of ribosomes to certain conventional and IRES-containing mRNAs.[116–120]

The ability of the tripartite leader to direct internal translation was examined by determining whether it could promote translation of a second (internal) cistron when constructed into a dicistronic mRNA.[114] Dicistronic mRNA constructs were used previously to demonstrate that picornaviral mRNAs contain an IRES element.[99,100,121] Translation of dicistronic mRNAs was studied *in vivo* by transfection

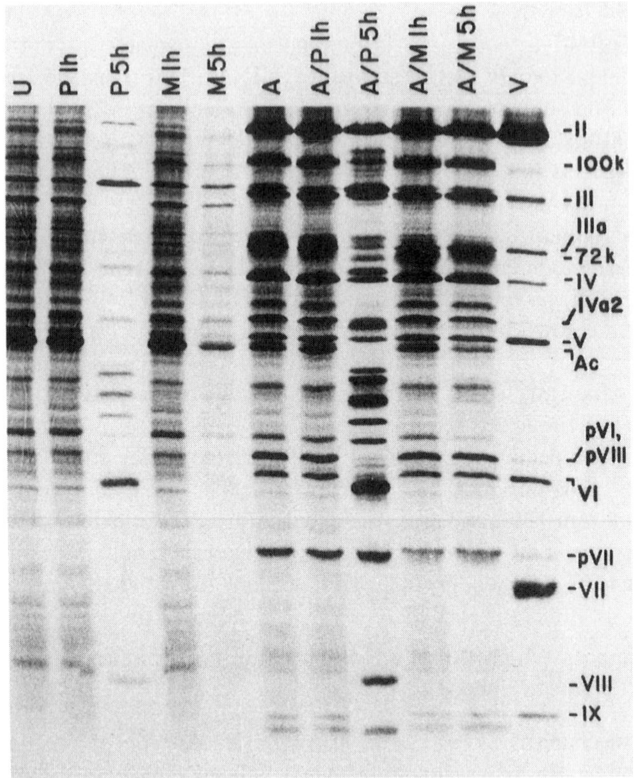

Figure 3. Electrophoretic analysis of polypeptides synthesized in cells infected by adenovirus and poliovirus. Cells were labelled with ^{35}S-methionine at early or late times after infection with viruses (see Dolph *et al.*[113] for details). Lanes: U, uninfected; P, poliovirus; A, adenovirus; M, poliovirus mutant 2A-2, defective for cleavage of p220; V, adenovirus marker proteins. Cells were labeled at 1 or 5 hr after infection or superinfection with poliovirus.

of plasmids into cells. Whereas the picornaviral IRES element was found to efficiently direct translation at an internal cistron, the tripartite leader did not.[114] It was concluded that the tripartite leader does not likely promote internal ribosome binding.

Although the tripartite leader possesses a striking complementarity to 18S rRNA, and to a lesser degree to VAI RNA, no evidence was found to suggest that direct RNA interactions play a role in tripartite leader translational activity. Specific mutations in the tripartite leader were constructed in an attempt to identify elements responsible for eIF-4F independence and translation efficiency. Translation was studied by measuring the level of protein expressed from variant mRNAs in the absence or presence of infecting poliovirus.[114] It was found that leader sequences complementary to 18S rRNA could be deleted without significantly altering the translational efficiency of the tripartite leader or its independence from eIF-4F activity. Tripartite leader sequences complementary to VAI RNA could also be deleted without affecting translation in the presence or absence of cotransfected VAI RNA. Thus, despite the intriguing complementarity to 18S rRNA and VAI RNA, there is little reason to suspect that translation of the tripartite leader involves specific RNA interactions. It was not determined whether complementary sequences in the tripartite leader play a role in its translation during adenovirus infection, as opposed to translation of mRNAs expressed from transfected plasmids. Consequently, a potential role for tripartite leader interactions with 18S rRNA and VAI RNA during adenovirus infection cannot be excluded at the current time.

Some evidence was presented that the 5′ end of the tripartite leader may possess an unstructured conformation. Duplication of the 5′ end of the tripartite leader (nt 1–33) in the antisense orientation was shown to create a stable hairpin structure and to reduce its translation. Creation of 5′ proximal secondary structure decreased the efficiency of tripartite leader translation by fivefold, and most importantly, made the mRNA fully dependent on eIF-4F activity. These results suggested that the primary element controlling tripartite leader independence from eIF-4F activity is located in the 5′ proximal section of the leader, which may provide an unstructured conformation.

Secondary structure analysis of the tripartite leader was carried out by Zhang *et al.*[122] and supported the notion that the 5′ end may be unstructured. The structure of the tripartite leader was determined alone and when attached to the body of an mRNA by probing the susceptibility to single-strand-specific nucleases. It was concluded that the tripartite leader probably adopts a conformation which is not greatly influenced by the mRNA to which it is attached, because similar structures were found in both cases. The primary observation of this study, however, was that the first 22 nt of the tripartite leader are fully susceptible to cleavage by single-strand-specific nucleases, indicative of an unstructured conformation.

3.3.4. Phosphorylation of eIF-4E (CBP) and Activity of eIF-4F Are Significantly Reduced during Late Adenovirus Infection

The preferential translation of tripartite leader containing mRNAs at the expense of most other species during late adenovirus infection suggested that the discrimination between cellular and late viral transcripts could involve modification or inactivation of translation factors. Two potential players, adenovirus VAI RNA and DAI kinase, were already implicated in the inhibition of cellular protein synthesis and promotion of late viral mRNA translation, at least in some infected cell lines. Nevertheless, experimental evidence indicated that factors other than VAI RNA and DAI kinase are probably involved in selective viral translation, since host protein synthesis is inhibited in many cell lines without an increase in eIF-2α phosphorylation, and the drug 2-aminopurine prevented the shutoff of cellular protein synthesis without affecting the activity of DAI kinase.[40]

Given the ability of the tripartite leader to confer translation apparently independent of eIF-4F activity, it would seem reasonable to expect that initiation factor eIF-4F might be inactivated by adenovirus during the late phase of infection. It was shown, however, that eIF-4F is not inactivated by proteolysis of the p220 component in late adenovirus-infected cells.[113] Nevertheless, translation mediated by eIF-4F is also regulated by the phosphorylation state of the eIF-4E (CBP) component. Reduced phosphorylation of eIF-4E correlates with an inability to associate with 40S ribosomes,[123] and with inhibition of translation during heat shock[112,124–126] and mitosis.[127] Increased phosphorylation of eIF-4E correlates with enhanced translation after activation of cells by mitogens or serum[128–130] and in cells transformed with the *Src* oncogene.[131] Overexpression of eIF-4E in cells also results in aberrant and enhanced cell growth,[132] including transformation.[133,134] Again, phosphorylation of eIF-4E is required for this effect. The phosphorylation of eIF-4E does not alter its ability to bind specifically to cap structures.[124,127,135,136] Rather, there is some evidence that phosphorylation of eIF-4E may be required for eIF-4F to interact with 40S ribosomes or to enhance ribosomes association with mRNA mediated by eIF-4F.[123]

The phosphorylation of eIF-4E was investigated at late times after infection by Huang and Schneider,[71] by quantitating the level of eIF-4E which could be labeled *in vivo* with ^{32}P-orthophosphate.[71] Compared to uninfected cells, the level of ^{32}P-labeled eIF-4E was found to be reduced 10- to 20-fold at late times after infection (Fig. 4). The reduction was also shown to result from decreased phosphorylation of eIF-4E rather than a decrease in the stability or steady-state level of the protein. The underphosphorylation of eIF-4E was shown to occur with the onset of the late phase of infection, and therefore correlated with suppression of cellular protein synthesis.

It had been previously demonstrated that 2-aminopurine prevented the shut-off of cellular translation during late adenovirus infection.[40] This study also demonstrated that 2-aminopurine largely prevented the dephosphorylation of eIF-4E without altering the steady-state level of the protein.[71] The distribution of total eIF-4E protein between phosphorylated and nonphosphorylated forms was investigated by two-dimensional isoelectrophoresis and immunoblot analysis.[71] The amount of phosphorylated eIF-4E measured by this analysis was found to support the results from *in vivo* labeling of cells with ^{32}P-orthophosphate (Fig. 5). The proportion of phosphorylated eIF-4E in exponentially growing cells was approximately 60% that of the total eIF-4E protein, and within the range of 30–60% reported by others.[124,135,136] In late adenovirus-infected cells the fraction of phosphorylated eIF-4E was found to be reduced to about 5% that of the total protein, which was partially blocked by treatment of cells with 2-aminopurine (Fig. 5). In addition, studies have demonstrated the inability of adenovirus to shut off cellular protein synthesis in cells lacking detectable DAI kinase activity.[67]

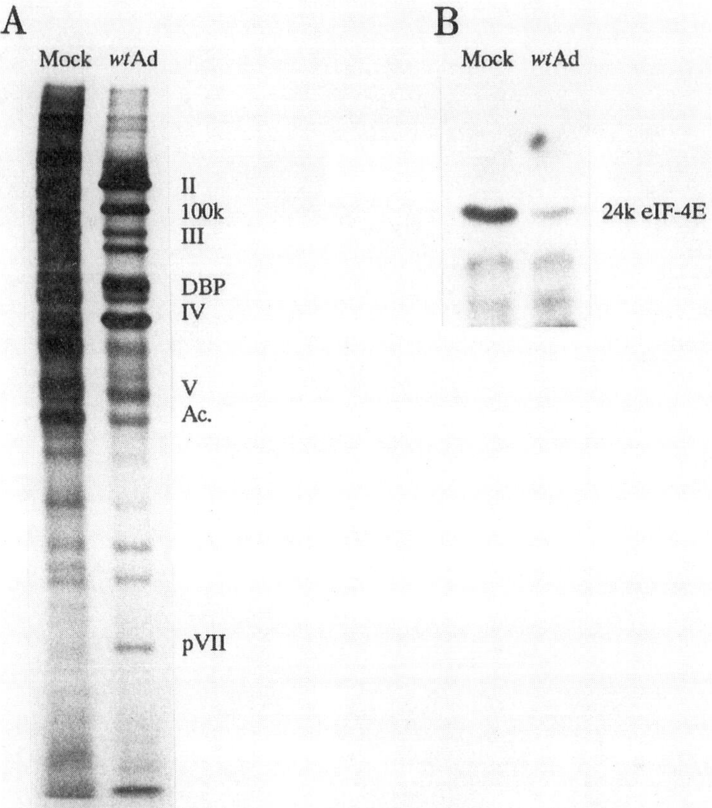

Figure 4. *In vivo* phosphorylation of eIF-4E in 293 cells at late times after infection with adenovirus. Cells were labeled at 18 hr after infection and extracts prepared.[71] **(A)** Sodium dodecyl sulfate–polyacrylamide gel analysis of ^{35}S-methionine-labeled polypeptides, demonstrating selective translation of late adenovirus mRNAs. **(B)** Cells were labeled with ^{32}P-orthophosphate, lysates prepared, and subjected to m^{7}GTP–Sepharose affinity chromatography. Eluted proteins were then resolved by electrophoresis and autoradiography.

Consistent with these observations, eIF-4E was found to remain phosphorylated during the late phase of adenovirus infection in these cells.[71] The results from this study suggested that DAI kinase is either involved in the control of eIF-4F activity or that the activities of both eIF-4F and DAI kinase may be coordinately regulated by unknown factors.

3.3.5. Model for Selective Translation of Late Adenoviral mRNAs Mediated by the Tripartite Leader

A model was proposed by Huang and Schneider[71] involving the tripartite leader and inactivation of eIF-4F to account for the selective translation of late

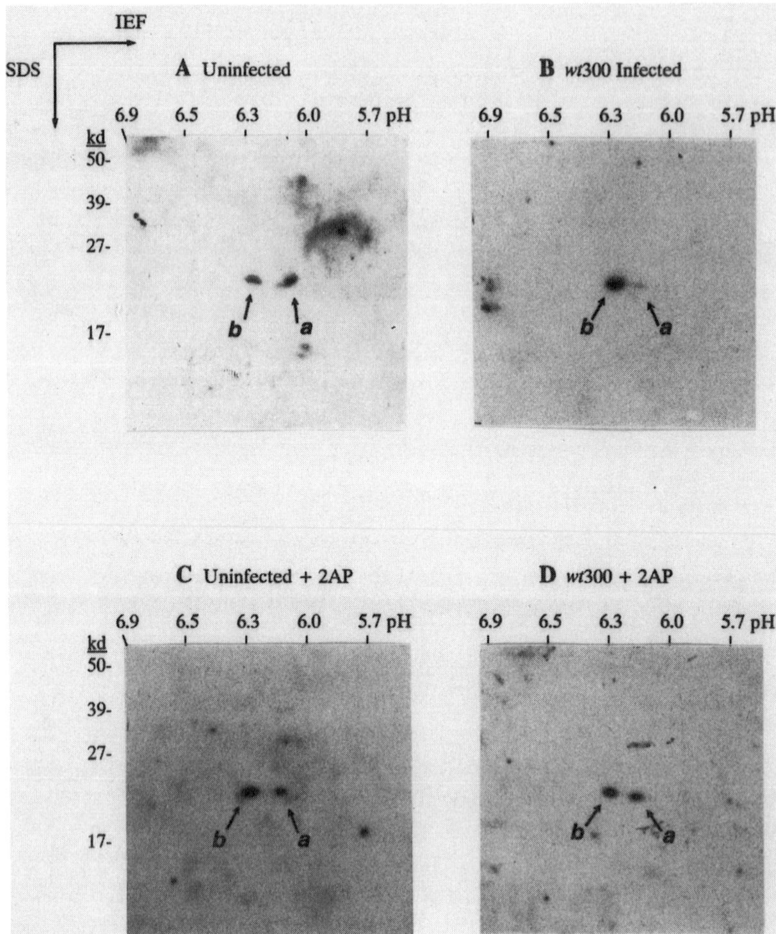

Figure 5. Two-dimensional isoelectric focusing (IEF) gel electrophoretic analysis of eIF-4E phosphorylation in adenovirus-infected cells. Cell extracts were prepared and subjected to m^{7}GTP–Sepharose affinity chromatography.[71] Lysates were subjected to two-dimensional IEF gel electrophoresis and Western immunoblot analysis using antisera directed against eIF-4E. *a*, the acidic (phosphorylated) form of eIF-4E; *b*, the basic (nonphosphorylated) eIF-4E; 2AP, 2-aminopurine.

adenovirus mRNAs (Fig. 6). It was proposed that during late viral infection the reduced phosphorylation of eIF-4E would greatly diminish or eliminate the availability of active eIF-4F. The elimination or severe reduction in cap-dependent RNA helicase activity would provide a means for discrimination between late Ad mRNAs that contain the tripartite leader and those that do not.

Several predictions of this model have been tested experimentally and appear to be consistent with the premise that inactivation or reduction in eIF-4F activity is

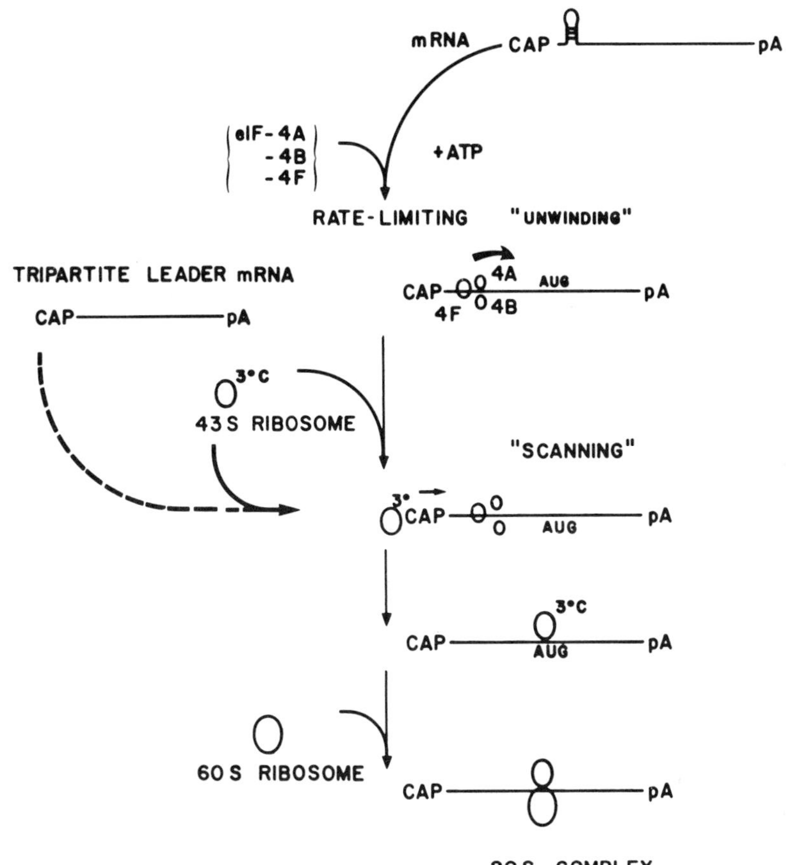

Figure 6. A model for the suppression of cellular protein synthesis and selective translation of late adenovirus mRNAs. Cap recognition by eIF-4F (eIF-4E + eIF-4A + p220) is generally considered a rate-limiting step in translation initiation. Dephosphorylation of eIF-4E during late adenovirus infection severely reduces or eliminates the pool of active eIF-4F. Tripartite leader mRNAs either recruit the small amounts of active eIF-4F very efficiently or translate independently of the factor, effectively bypassing a limiting step in translation of host-cell mRNAs.

involved in the selective translation of late viral mRNAs and the shutoff of host-cell protein synthesis. First, infection of a number of cell lines demonstrated a clear and consistent correlation between the shutoff of host translation and decreased phosphorylation of eIF-4E.[71] Second, the inhibition of translation during metaphase in mitotic cells results from impaired eIF-4F activity due to dephosphorylation of eIF-4E,[127] but is otherwise translationally competent. Adenovirus infection of metaphase arrested cells demonstrated the efficient translation of tripartite leader mRNAs despite a significant inhibition of cellular protein synthesis. Therefore, the inhibition or severe reduction of eIF-4F activity through underphosphorylation of eIF-4E restricts translation of cellular, but not tripartite leader mRNAs during mitosis and in late adenovirus infection. Third, inactivation of eIF-4F activity during late adenovirus infection would be expected to restrict the translation of all viral mRNAs which lack the tripartite leader. Some experimental evidence has been obtained which is consistent with this prediction. If the shutoff of host protein synthesis is prevented with 2-aminopurine, an enhanced translation of cellular and several nontripartite leader viral mRNAs is observed.[40] In particular, the translation of adenovirus DNA binding protein (DBP) and pIX mRNAs was elevated by 10- to 20-fold in the presence of 2-aminopurine.

4. CONCLUDING REMARKS

It seems clear that reduction or inactivation of eIF-4F activity through underphosphorylation of eIF-4E contributes to the inhibition of cellular protein synthesis and selected translation of late viral mRNAs. It also seems likely that dephosphorylation of eIF-4E in late adenovirus-infected cells serves to restrict greatly the availability of eIF-4F, rather than to prevent fully its activity. For instance, during late adenovirus infection some 5% of eIF-4E always remains phosphorylated.[71] Translation of tripartite leader mRNAs may be able to recruit more efficiently the limited amounts of eIF-4F due to its high affinity for the unstructured 5′ end of capped mRNAs, as shown for AMV 4 mRNA.[108]

Although the reduced phosphorylation of eIF-4E is a primary event in the regulation of translation in late adenovirus-infected cells, elucidation of the role of eIF-4F does not complete the story. It is not yet known whether the activity of other translation factors is affected by adenovirus infection. In addition, there is currently no mechanistic understanding of the role played by L4 100k protein in late adenovirus translation.

Finally, it is not understood why adenovirus fails to shut off cellular protein synthesis in cells which lack DAI kinase activity. Perhaps activation of DAI kinase, phosphorylation of eIF-2α, and dephosphorylation of eIF-4E are all controlled through similar pathways, but promote late adenovirus translation through different mechanisms. It would seem plausible that adenovirus might possess several different mechanisms to achieve translational dominance, given the wide variety of cell types normally infected by this virus. Research must be directed toward understanding the contribution and interactions of all three pathways in establishing selective translation of late adenoviral mRNAs.

ACKNOWLEDGMENTS. The authors' work described in this review was supported by a grant from the National Institutes of Health (CA-42357) to R.J.S.

REFERENCES

1. Pereira, H. G., Huebner, R. J., Ginsberg, H. S., and van der Vern, J., 1963, *Virology* **20:** 613–620.
2. Beladi, I., 1972, In: *Strains of Human Viruses* (M. Majer and S. A. Plotkin, eds.), Karger, Basel.
3. Stewart, W. D., 1979, In: *The Interferon System*, Springer-Verlag, New York.
4. Flint, J., and Shenk, T., 1989, *Annu. Rev. Genet.* **23:**141–161.
5. Wold, W. S. M., and Gooding, L. R., 1991, *Virology* **184:**1–8.
6. Hardy, S., Engel, D. A., and Shenk, T., 1989, *Genes Dev.* **3:**1062–1074.
7. Babiss, L. E., and Ginsberg, H. S., 1984, *J. Virol.* **50:**202–212.
8. Halbert, D. N., Cutt, J. R., and Shenk, T., 1984, *J. Virol.* **56:**250–257.
9. Ginsberg, H. S., 1984, In: *The Adenoviruses*, Plenum Press, New York.
10. Ziff, E. B., 1980, *Nature* **287:**491–499.
11. Berget, S. M., Moore, C., and Sharp, P., 1977, *Proc. Natl. Acad. Sci. USA* **74:**3171–3175.
12. Broker, T. R., Chow, L. T., Dunn, A. R., Gelinas, R. E., Hassell, J. A., Klessig, D. G., Lewis, J. B., Roberts, R. J., and Zain, B. S., 1977, *Cold Spring Harbor Symp. Quant. Biol.* **42:**531–553.
13. Ginsberg, H. S., Bello, L. S., and Levine, A. J., 1967, In: *The Molecular Biology of Viruses* (J. S. Cotter, and W. Paranchych, eds.), Academic Press, New York.
14. Beltz, G. A., and Flint, S. J., 1979, *J. Mol. Biol.* **131:**353–373.
15. Castiglia, C. L., and Flint, S. J., 1983, *Mol. Cell. Biol.* **3:**662–671.
16. Flint, S. J., 1984, Adenovirus cytopathology, *Comp. Virol.* **19:**297–358.
17. Hodge, L. D., and Scharff, M. D., 1969, *Virology* **37:**554–564.
18. Lucas, J. L., and Ginsberg, H. S., 1971, *J. Virol.* **8:**203–213.
19. Lingberg, U., and Sunquist, B., 1974, *J. Mol. Biol.* **86:**451–468.
20. Babich, A., Feldman, C. T., Nevins, J. R., Darnall, J. E., and Weinberger, C., 1983, *Mol. Cell. Biol.* **3:**1212–1221.
21. Schneider, R. J., and Shenk, T., 1987, *Annu. Rev. Biochem.* **56:**317–332.
22. Pilder, S., Logan, J., and Shenk, T., 1986, *Mol. Cell. Biol.* **6:**470–476.
23. Bello, L. J., and Ginsberg, H. S., 1967, *J. Virol.* **1:**843–850.
24. Rice, A. P., and Roberts, B. E., 1983, *J. Virol.* **47:**529–539.
25. Nishioka, Y., and Silverstein, S., 1977, *Proc. Natl. Acad. Sci. USA* **74:**2370–2374.
26. Lodish, H. F., and Porter, M., 1980, *J. Virol.* **36:**719–733.
27. McCune, J. M., 1991, *Cell* **64:**351–363.
28. Krug, R. M., Broni, B. R., and Bouloy, M., 1979, *Cell* **18:**329–334.
29. Petterson, V., and Philipson, L., 1975, *Cell* **6:**1–14.
30. Anderson, C. W., Lewis, J. B., Atkins, J. F., and Gesteland, R. F., 1979, *Proc. Natl. Acad. Sci. USA* **71:**2756–2760.
31. Thimmappaya, B., Weinberger, C., Schneider, R. J., and Shenk, T., 1982, *Cell* **31:**543–551.
32. Bhaduri, S., Raskas, H. J., and Green, M., 1972, *J. Virol.* **10:**1126–1129.
33. Price, R., and Penman, S., 1972, *J. Virol.* **9:**621–626.
34. Tal, J. T., Craig, E. A., and Raskas, H. J., 1975, *J. Virol.* **15:**137–144.
35. Flint, S. S., Beltz, G. A., and Linzer, D. I. H., 1983, *J. Mol. Biol.* **167:**335–359.
36. Yoder, S. S., Robberson, B. L., Leys, E. J., Hook, A. G., Al-Ubaidi, M., Yeung, C. Y., Kellems, R. E., and Berget, S. M., 1983, *Mol. Cell. Biol.* **3:**819–828.
37. Flint, S. J., Plumb, M. A., Yang, U. C., Stein, S., and Stein, J. L., 1984, *Mol. Cell. Biol.* **4:**1363–1371.
38. Singer, R. H., and Penman, S., 1973, *J. Mol. Biol.* **78:**321–334.
39. Moore, M., Schaack, J., Baim, S. B., Morimoto, R. I., and Shenk, T., 1987, *Mol. Cell. Biol.* **7:** 4505–4512.
40. Huang, J., and Schneider, R. J., 1990, *Proc. Natl. Acad. Sci. USA* **87:**7115–7119.
41. Bablanian, R., and Russell, W. C., 1974, *J. Gen. Virol.* **24:**261–279.

42. Oosterom-Dragon, E. A., and Ginsberg, H. S., 1980, *J. Virol.* **33:**1203–1207.
43. Cepko, C. L., and Sharp, P. A., 1983, *Virology* **129:**137–154.
44. Oosterdom-Dragon, E. A., and Ginsberg, H. S., 1981, *J. Virol.* **40:**491–500.
45. Adam, S. A., and Dreyfuss, G., 1987, *J. Virol.* **61:**3276–3283.
46. Lindberg, V., and Sundquist, B., 1974, *J. Mol. Biol.* **86:**451–468.
47. Sundquist, B., Persson, T., and Lindberg, V., 1977, *Nucleic Acids Res.* **4:**899–915.
48. Hayes, B. W., Telling, G. C., Myat, M. M., Williams, J. F., and Flint, S. J., 1990, *J. Virol.* **64:** 2732–2742.
49. Query, C. C., Bentley, R. C., and Kenne, J. D., 1989, *Cell* **57:**89–101.
50. Mathews, M. B., and Shenk, T., 1991, *J. Virol.* **65:**5657–5662.
51. Mathews, M., 1975, *Cell* **6:**223–229.
52. Mathews, M. B., and Peterson, U., 1978, *J. Mol. Biol.* **119:**293–328.
53. Furtado, M. R., Subramanian, S., Bhat, R. A., Wowlkes, D. M., Safer, B., and Thimmappaya, B., 1989, *J. Virol.* **63:**3423–3434.
54. Mellits, K. H., and Mathews, M. B., 1988, *EMBO J.* **7:**2849–2859.
55. Söderlund, H., Petterson, U., Vennstrom, B., Philipson, L., and Mathews, M. B., 1976, *Cell* **7:** 585–593.
56. Schneider, R. J., Weinberger, C., and Shenk, T., 1984, *Cell* **37:**291–298.
57. Reichel, P. A., Merrick, W. C., Siekierka, J., and Mathews, M. B., 1985, *Nature* **313:**196–200.
58. Schneider, R. J., Safer, B., Munemitsu, S., Samuel, C. E., and Shenk, T., 1985, *Proc. Natl. Acad. Sci. USA* **82:**4321–4325.
59. Jagus, R., Anderson, W. F., and Safer, B., 1981, *Prog. Nucleic Acid Res. Mol. Biol.* **25:**127–185.
60. Safer, B., 1983, *Cell* **33:**7–8.
61. Hovenessian, A. G., 1989, *J. Interferon Res.* **6:**641–647.
62. Katze, M. G., DeCorato, D., Safer, B., Galabru, J., and Hovanessian, A. G., 1987, *EMBO J.* **6:** 689–697.
63. Kostura, M., and Mathews, M. B., 1989, *Mol. Cell. Biol.* **9:**1576–1586.
64. Mellits, K. H., Kostura, M., and Mathews, M. B., 1990, *Cell* **61:**843–852.
65. Ghadge, G. D., Swaminathan, S., Katze, M. G., and Thimmappaya, B., 1992, *Proc. Natl. Acad. Sci. USA* **88:**7140–7144.
66. Siekierka, J., Mariano, T. M., Reichel, P. A., and Mathews, M. B., 1985, *Proc. Natl. Acad. Sci. USA* **82:**1959–1963.
67. O'Malley, R. P., Duncan, R. F., Hershey, R. W. B., and Mathews, M. B., 1989, *Virology* **168:** 112–118.
68. Kitajewski, J., Schneider, R. J., Safer, B, and Shenk, T., 1986, *Mol. Cell. Bio.* **6:**4493–4498.
69. Kitajewski, J., Schneider, R. J., Safer, B., Munemitsu, S. M., Samuel, C. E., and Shenk, T., 1986, *Cell* **45:**195–200.
70. Maron, A., and Mathews, M. B., 1988, *Virology* **164:**106–113.
71. Huang, J., and Schneider, R. J., 1991, *Cell* **65:**271–280.
72. Zilberstein, A., Federmann, P., Schulman, L., and Revel, M., 1976, *FEBS Lett.* **68:**119–124.
73. Duncan, R., and Hershey, J. W. B., 1983, *J. Biol. Chem.* **258:**7228–7235.
74. Howe, J. G., and Hershey, J. W. B., 1984, *Cell* **37:**85–93.
75. Kaufman, R. J., Davies, M. V., Pathak, V. K., and Hershey, J. W. B., 1989, *Mol. Cell. Biol.* **9:** 946–958.
76. Katze, M. G., Detjen, B. M., Safer, B., and Krug, R. M., 1986, *Mol. Cell. Biol.* **6:**1741–1750.
77. O'Neall, R. E., and Racaniello, V. R., 1989, *J. Virol.* **63:**5069–5075.
78. Black, T. L., Safer, B., Hovanessian, A., and Katze, H. G., 1989, *J. Virol.* **63:**2244–2251.
79. Jacobsen, H., Epstein, D. A., Friedman, R. M., Safer, B., and Torrenie, P. F., 1983, *Proc. Natl. Acad. Sci. USA* **80:**41–45.
80. Mathews, M. B., 1980, *Nature* **285:**575–577.
81. Gross, M., Redman, R., and Kaplansky, D. A., 1985, *J. Biol. Chem.* **260:**9491–9500.

82. Thomas, N. S. B., Matts, R. L., Levin, D. H., and London, I. M., 1985, *J. Biol. Chem.* **260:** 9860–9866.
83. DeBenedetti, A., and Baglioni, C., 1985, *J. Biol. Chem.* **260:**3135–3139.
84. Logan, J., and Shenk, T., 1984, *Proc. Natl. Acad. Sci. USA* **81:**3655–3659.
85. Berkner, K. E., and Sharp, P. A., 1985, *Nucleic Acids Res.* **13:**841–857.
86. Kaufman, R. J., 1985, *Proc. Natl. Acad. Sci. USA* **82:**689–693.
87. Castrillo, J. L., and Carrasco, L., 1987, *J. Biol. Chem.* **262:**7328–7334.
88. Etchison, D., Milburn, S. C., Edery, I., Sonenberg, N., and Hershey, J. W. B., 1982, *J. Biol. Chem.* **257:**14806–14810.
89. Grifo, J. A., Tahara, S. M., Morgan, M. A., Shatkin, A. J., and Merrick, W. C., 1983, *J. Biol. Chem.* **258:**5804–5810.
90. Sonenberg, N., 1987, *Adv. Virus Res.* **33:**175–204.
91. Lawson, T. G., Ray, B. K., Dodds, J. T., Grifo, J. A., Abramson, R. D., Merrick, W. C., Betsch, D. F., Weith, H. L., and Thach, R. E., 1986, *J. Biol. Chem.* **261:**13979–13989.
92. Ray, B. K., Brendler, T. G., Adya, S., McQueen, S. D., Miller, J. K., Hershey, J. W. B., Grifo, J. A., Merrick, W. C., and Thach, R. E., 1983, *Proc. Natl. Acad. Sci. USA* **80:**663–667.
93. Ray, B. K., Lawson, T. G., Kramer, J. C., Cladaras, M. H., Grifo, J. A., Abramson, R. D., Merrick, W. C., and Thach, R. E., 1985, *J. Biol. Chem.* **260:**7651–7658.
94. Abramson, R. D., Dever, T. E., Lawson, T. G., Ray, B. K., Thach, R. E., and Merrick, W. C., 1987, *J. Biol. Chem.* **262:**3826–3832.
95. Lee, K. A. W., and Sonenberg, N., 1982, *Proc. Natl. Acad. Sci. USA* **79:**3447–3451.
96. Sonenberg, N., Guertin, D., and Lee, K. A. W., 1982, *Mol. Cell. Biol.* **2:**1633–1638.
97. Rhoads, R. E., 1988, *Trends Biochem. Sci.* **13:**52–56.
98. Thach, R. E., 1992, *Cell* **68:**177–180.
99. Pelletier, J., and Sonenberg, N., 1988, *Nature* **334:**320–325.
100. Jang, S. K., Krausslich, H. G., Nicklin, M. J. H., Duke, G. M., Palmenberg, A. C., and Wimmer, E., 1988, *J. Virol.* **62:**2636–2643.
101. Anthony, D. D., and Merrick, W. C., 1991, *J. Biol. Chem.* **266:**10218–10226.
102. Gehrke, L., Auron, P. E., Quigly, G. J., Rich, A., and Sonenberg, N., 1983, *Biochemistry* **22:** 5157–5164.
103. Kozak, M., 1980, *Cell* **19:**79–90.
104. Kozak, M., 1986, *Proc. Natl. Acad. Sci. USA* **83:**2850–2854.
105. Pelletier, J., and Sonenberg, N., 1985, *Mol. Cell. Biol.* **5:**3222–3230.
106. Pelletier, J., and Sonenberg, N., 1985, *Cell* **40:**515–526.
107. Browning, K. S., Fletcher, L., and Ravel, J. M., 1988, *J. Biol. Chem.* **263:**8380–8383.
108. Fletcher, L., Corbin, S. D., Browning, K. G., and Ravel, J. M., 1990, *J. Biol. Chem.* **265:**19582–19587.
109. Sarnow, P., 1989, *Proc. Natl. Acad. Sci. USA* **86:**5795–5799.
110. Macejak, D. G., and Sarnow, P., 1991, *Nature* **353:**90–94.
111. Lindquist, S., and Peterson, R., 1991, *Enzyme* **44:**147–166.
112. Zapata, J. M., Maroto, F. G., and Sierra, J. M., 1991, *J. Biol. Chem.* **266:**16007–16014.
113. Dolph, P. J., Racaniello, V., Villamarin, A., Palladino, F., and Schneider, R. J., 1988, *J. Virol.* **62:** 2059–2066.
114. Dolph, P. J., Huang, J., and Schneider, R. J., 1990, *J. Virol.* **64:**2669–2677.
115. Jackson, R. J., 1991, *Nature* **353:**14–15.
116. Azad, A. A., and Deacon, N. J., 1980, *Nucleic Acids Res.* **8:**4365–4375.
117. Gallie, D. R., and Kado, C. I., 1989, *Proc. Natl. Acad. Sci. USA* **86:**129–132.
118. Hagenbuchle, O., Santer, M., Steitz, J. A., and Mans, R. J., 1978, *Cell* **13:**551–563.
119. Nakashima, K., Darzynkiewicz, E., and Shatkin, A. J., 1980, *Nature* **286:**226–230.
120. Pilipenko, E. V., Guyl, A. P., Maslova, S. V., Suitkin, Y. V., Sinyakov, A. N., and Agol, V. I., 1992, *Cell* **68:**119–131.

121. Bienkowska-Szewczyk, K., and Ehrenfeld, R., 1988, *J. Virol.* **62:**3068–3072.
122. Zhang, Y., Dolph, P. J., and Schneider, R. J., 1989, *J. Biol. Chem.* **264:**10679–10684.
123. Joshi-Barve, S., Rychlik, W, and Rhoads, R. E., 1990, *J. Biol. Chem.* **265:**2979–2983.
124. Duncan, R., Milburn, S. C., and Hershey, J. W. B., 1987, *J. Biol. Chem.* **262:**380–388.
125. Panniers, R., Stewart, E. B., Merrick, W. C., and Henshaw, E. C., 1985, *J. Biol. Chem.* **260:** 9648–9653.
126. Lamphear, B. J., and Panniers, R., 1990, *J. Biol. Chem.* **265:**5333–5336.
127. Bonneau, A. M., and Sonenberg, N., 1987, *J. Virol.* **61:**986–991.
128. Kaspar, R., Rychlik, W., White, M. W., Rhoads, R. E., and Morris, D. F., 1990, *J. Biol. Chem.* **265:**3619–3622.
129. Marino, M. W., Pfeffer, L. M., Guidon, P. T., and Donner, D. B., 1989, *Proc. Natl. Acad. Sci. USA* **86:**8417–8421.
130. Morely, S. M., and Trough, J. A., 1989, *J. Biol. Chem.* **264:**2401–2404.
131. Frederickson, R. M., Montine, K. S., and Sonenberg, N., 1991, *Mol. Cell. Biol.* **11:**2896–2900.
132. Smith, M. R., Saramllo, M., Liv, L.-L., Dever, T. E., Merrick, W. C., Kung, H. F., and Sonenberg, N., 1990, *New Biol.* **2:**648–654.
133. Lazaris-Karatzas, A., Montine, K. S., and Sonenberg, N., 1990, *Nature* **345:**544–547.
134. DeBenedetti, A., and Rhoads, R. E., 1990, *Proc. Natl. Acad. Sci. USA* **87:**8212–8216.
135. Hiremath, L. S., Hiremath, S. T., Rychlik, W., Joshi, S., Domier, L. L., and Rhoads, R. E., 1989, *J. Biol. Chem.* **264:**1132–1138.
136. Rychlik, W., Gardner, P. R., Vanaman, T. C., and Rhoads, R. E., 1986, *J. Biol. Chem.* **261:** 71–75.

Chapter 12

Concerted Gene Expressions in Elicited Fibroin Synthesis

G. C. Candelas, G. Arroyo, C. Carrasco, E. Carrasquillo, A. Plazaola, and M. Irizarry

Our studies show that the making of fibroin by mechanically stimulated spider glands seems to require a series of well-orchestrated gene expressions. Monitoring of the process through time sequence has revealed four transient waves of molecular syntheses. The last and most dramatic of these events is the synthesis of the full-size fibroin product, which is preceded by a wave which generates template RNA by a 60-min interval. The other two events generate small RNAs. Analyses of the first of the small RNA-generating bouts, consistently of higher magnitude than the subsequent one, displays upgrading of 5S RNA, to a higher extent of Ul snRNA, and a dramatic boost in alanine transfer RNA (tRNA) accumulation. This tRNA resolves into two isoforms, one of which is gland-specific and quantitatively correlated to its fibroin-synthesizing activity. The second of these waves serves to optimize the gland's translational milieu through the differential expression of the tRNAs cognate to the most preponderant amino acids of the gland's fibroin product in a similar proportion to that in which these appear in the fibroin. Worthy of note is the disproportionate accumulation of

---

G. C. Candelas, G. Arroyo, C. Carrasco, E. Carrasquillo, A. Plazaola, and M. Irizarry • Department of Biology, University of Puerto Rico, Rio Piedras, Puerto Rico 00931.

Translational Regulation of Gene Expression 2, edited by Joseph Ilan. Plenum Press, New York, 1993.

alanine-tRNA which is produced primarily within the first wave of small RNA syntheses and which selectively enriches the system with a tissue-specific isoacceptor species in a proportion of 4:1 to its constitutive counterpart. The nucleotide sequence of this isoform endows it with structural features which foster its possible performance in other than elongation functions during the synthesis of fibroin.

Fibroins stand out among nature's largest protein products. In addition to their size, they are unusual in their amino acid contents. As natural fibers, they display great strength and elasticity.[1] The fibroin of our particular interest is the luminar product of the large ampullate glands of the spider *Nephila clavipes*. The fate of this protein has been the subject of several investigators, quoted later in this paper.

These large, tissue-specific protein products perform a series of biological services, the best known of which is the cocoon of the silkworm, primarily because of its commercial uses. Orb-web spiders, on the other hand, do not restrict the use of fibroin to a particular developmental stage, nor to one sole service. They rely heavily on fibroins and contain several sets of silk-producing glands. Within the natural history of these organisms, silks are used for anchorage, transportation, defense, feeding by swathing of prey, reproduction, and encasing of the eggs, among other uses. They play critical roles during the entire adult life of the female spider. Some of the orb-web's fibroin products are under not only constant but rather high demands. Such is the case of the product of the large ampullates, which is under critical demand during the entire life span of the adult female. The fate of this gland's product, which we extract from its lumen by removing the epithelium, has been reported to contribute, not equate, to the organism's dragline.[2–4] In fact, Peakall[4] claims more than one fate for the large ampullate's product, while Lucas[3] claims more than one source for the dragline. The production of complex tissue products such as these fibroins, which fulfill a wide gamut of biological functions and in some cases are in constant and high demands, requires accurate timing and should be expected to entail complex mechanisms and well-concerted molecular strategies.

Our studies, conducted on the stimulus-elicited fibroin synthesis in the large ampullate glands of the orb-web spider, *Nephila clavipes*, have served not only to confirm the complexity of the process, but also to reveal some of the orchestrated sequence of events involved in the production of the large tissue-specific product. The purpose of this chapter is to summarize and update our investigations on some of the tissue-specific expressions which seem to contribute toward the appropriate setting for the process.

1. GENE EXPRESSION IN FIBROIN SYNTHESIS

By way of review, we offer some comments on the system and its proper handling, since this information has been amply described, published, and previously reviewed.[5–8] The fibroin synthetic activity expressed by the glands can be significantly altered through simple manipulations. The large ampullates from organisms deprived of food and space for web-building display virtually no fibroin-synthesizing activity, while organisms whose contents of silks have been mechanically removed display high levels of activity.[5,6] The two contrasted metabolic states, previously monitored biochemically, may now be visualized in the ultrastructural features of the organelles known to be involved in protein-synthetic processes. Figure 1 shows cross sections from the secretory epithelium of the tails of stimulated and unstimulated glands, highlighting in this particular case the endoplasmic reticulum. The contrast between the well-distended cisternae in the stimulated gland versus the flattened structures in the unstimulated gland not only confirms our biochemical data, but serves to evidence visually the success of our manipulations and thus our previous claims. Similar contrasts between the two states of the glands have been shown in the Golgi apparatus.[9] A gradual shift in the structure of both of these organelles progressing through stimulation has been shown in sections from glands subjected to time sequence studies.[10]

Figure 2 contains a summary of the tissue-specific events provoked by stimulation, previously reported.[5–8] The four discrete transient waves of activity are reproducible with fidelity both in their timing and relative amplitude. As we

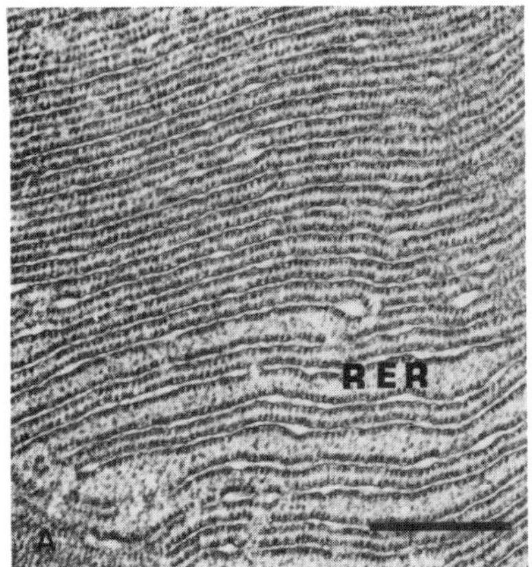

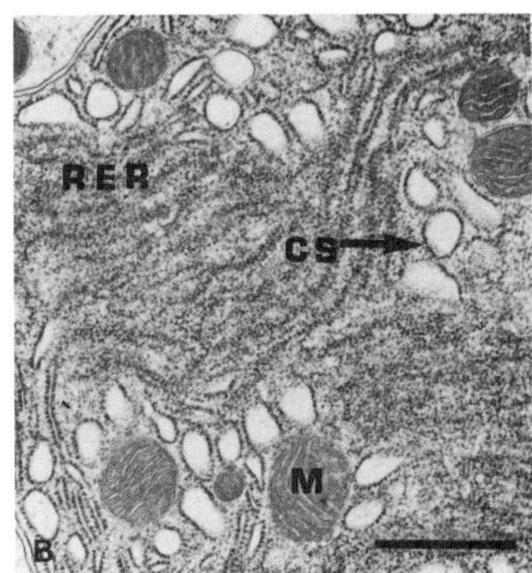

Figure 1. Ultrastructural features of the rough endoplasmic reticulum of fibroin-secreting cells of the large ampullate gland. Cross sections through the tail region highlighting the rough endoplasmic reticulum of fibroin-secreting cells. The flattened appearance of this organelle in the glands from unstimulated spiders **(A)** contrasts with the well-distended cisternae (indicated by arrows) in those from stimulated spiders **(B)**. RER, rough endoplasmic reticulum; Cs, cisternae; M, mitochondrion. Magnification bar = 5 μm.

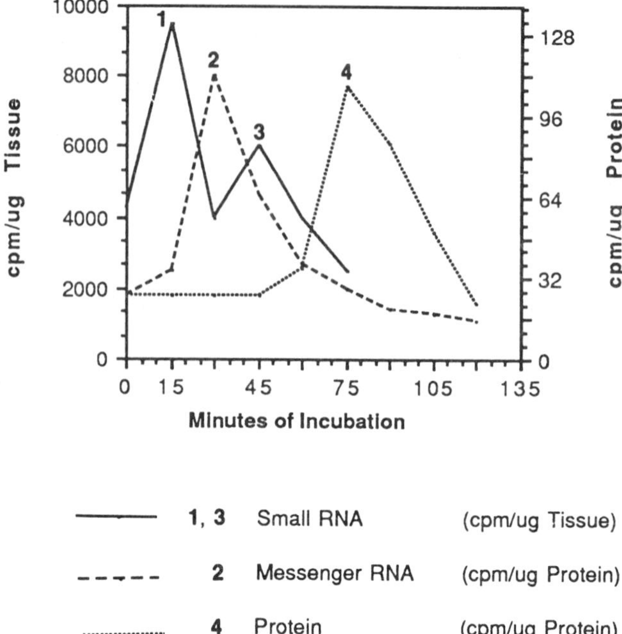

Figure 2. Expression of stimulus-elicited tissue-specific events. Time course studies on instantaneous rates of synthesis of small RNAs, messenger RNA, and protein monitored in the large ampullate glands of stimulated organisms through pulses with the appropriate labeled precursors display four peaks of transient activities, as labeled. Details of methods and nonglandular tissue controls can be found in the original publications.[5,6,8]

have previously shown, the dramatic production of the full-size fibroin product peaks after 90 min of incubation (peak 4) of the stimulated glands, and is preceded by the synthesis of the template (peak 2) by a 60-min interval.[6] Our main focus is on the other two waves in this figure (peaks 1 and 3), both of which have been shown to generate small RNAs.[8]

The tRNAs generated during the wave which peaks after 45 min incubation (peak 3), of lesser magnitude than the first small RNA-generating wave (peak 1), have been titrated through aminoacylation with labeled amino acids[8] and the results are summarized in Table I. Shown is the production of the tRNAs cognate to the fibroin product's most abundant amino acids.[11,12] During this wave (peak 3), there seems to be no particular bias and the tRNAs studied are upgraded in equal magnitudes. This exerts a shift in the tissue's tRNA population, providing a pool enriched in these tRNAs, and thus optimizing the gland's translational milieu for the decodification of a template rich in selected codons. The timing of this tissue-specific selective upgrading of tRNAs[8] can be considered adaptive within the scheme of events of the eukaryotic protein-synthesizing machinery. This process

Table I. Concentration of tRNAs Cognate to the Abundant Amino Acids in Spider Fibroin

	tRNA Concentration (picomole/0.5 OD at 260 nm)				
		Stimulated glands			
	Unstimulated	15-min incubation		45-min incubation	
Amino acid	glands[a]	Total	Net increase[b]	Total	Net increase
Ala	9	22	13	32	10
Gly	4	5	1	14	9
Pro	2	2	—	12	10

[a]The unstimulated glands were incubated for 45 min.
[b]The net increase in tRNA concentration was determined by subtracting the total values of each particular time interval from the total values of the preceding one.

occurs at the time of silk production in the silkworm and has been exhaustively studied by the Lyon investigators.[13–17] This activity, the only one detected thus far for the products of this wave, imposes a translational level of control on the final expression of the fibroin product. Control of gene expression at the level of translation is a recurrent and widespread biological resource. One of the earliest cases mentioned in the literature is the control of the gene expression mediated by the juvenile hormone in *Tenebrio molitor*.[18]

Two features of the earliest of the responses consisting of a wave of small RNA synthesis deem it worthy of further enquiries: its magnitude and the diversity of the small RNAs it seems to contain. Figure 3 shows the uppermost domain of a gel separation of small RNAs. Lane 1 contains RNA extracted from stimulated glands and lane 2 shows that from nonstimulated glandular tissues. Indicated with an arrow is the band corresponding to 5S RNA; all the tRNAs identified by us migrate beyond this band. What can be concluded from these types of experiments is that the stimulus elicits the expression of diverse small RNAs and that this merits further investigation. We have thus initiated this venture through Northern blots, starting with the 5S RNA.

In view of the established role of 5S RNA in the assembly of the ribosomes and what we have thus far observed in the stimulated state of the large ampullate glands, it is reasonable to expect upgrading of these molecules as a response to stimulation into protein synthesis. The 5S RNA upgrading accompanies the early phase of vitellogenesis[19] and during midblastula transition during the embryogenesis of *Xenopus*.[20] These cases as well as the stimulation of the large ampullates pose a demand for ribosomes. Figure 4 shows a Northern hybridization of small RNAs of *N. clavipes* and *Lythechinus variegatus* as a positive control. Lane 1 shows the 5S accumulation in the glands from stimulated organisms at 15 min after incubation; lane 2, the same conditions, but from unstimulated organ-

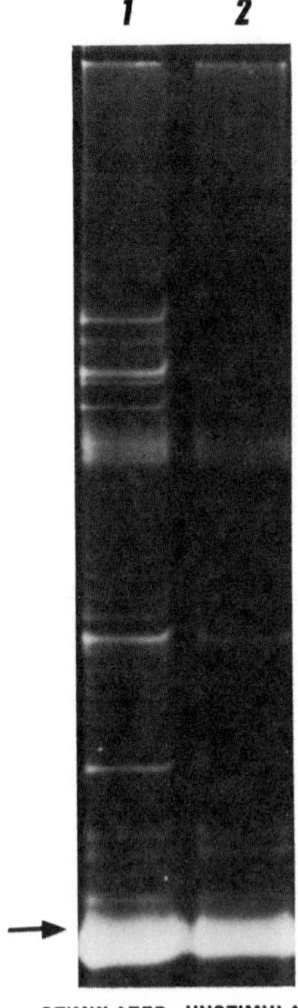

Figure 3. Electrophoretic analyses of small RNAs. This figure shows the denaturing gel separation of total RNA extracted through the ethanol perchlorate (EPR) method and fractionated by exclusion chromatography on BioGel A 1.5 m.[8] Of interest are the series of bands whose mobility occupies the domain above that of 5S RNA, characteristic of the U subset of snRNAs seen only in the stimulated glands. Lane 1, RNA from stimulated glands; lane 2, from unstimulated glands; the arrow identifies the 5S RNA.

isms; and lane 3, from midblastula sea urchin. Densitometric analyses indicate a 29% increase in accumulation of 5S RNA as a function of stimulation, which agrees with the demand of a system in preparation for protein synthesis. Prompted by the presence of variants in *Xenopus*[21] and other systems[22] and applying the same methods which allowed us to detect the *Nephila* alanine-tRNA variant, we sought but did not find 5S RNAs variants in our system.[8] The timing of the 5S RNA upgrading, an early response to the stimulation, could be deemed functional, since it anticipates and provides for the need for an enriched ribosome pool.

Another response registered during the first wave of small RNA synthesis is the upgrading of U1 snRNA. Figure 5 shows the labeled Northern blot, which displays a twofold upgrade of this molecule in the stimulated glands as revealed by densitometric analysis. Again, in view of the established function of U1

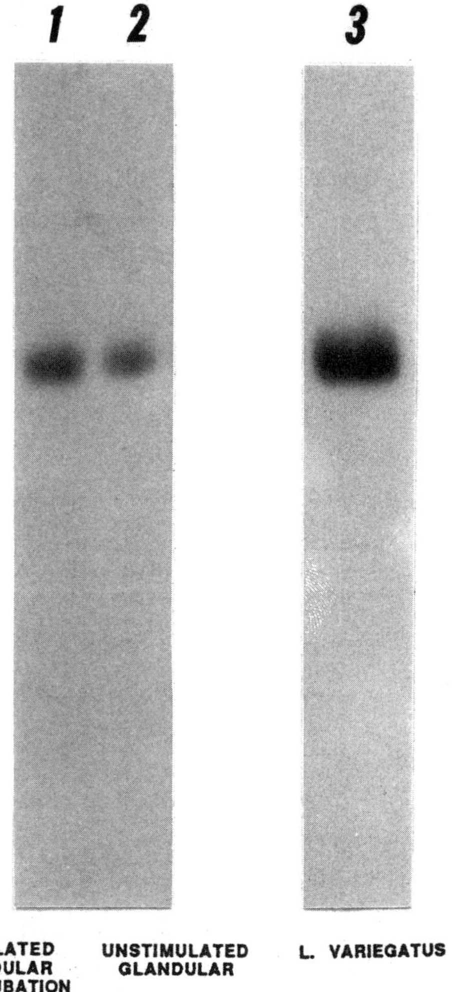

Figure 4. Northern blot assays for 5S RNA from stimulated and unstimulated glands. Lithium chloride-precipitated RNAs[45] from glands of unstimulated and stimulated organisms, incubated for 15 min, were electrophoresed as described in Fig. 3. Following transference to a nylon membrane, they were probed with nick-translated pLu103, a pACYC184 derivative containing a 1.3-kilobase fragment which carries a 5S gene from the sea urchin, *Lythechinus variegatus*. The band in lane 1, containing the stimulated gland RNA, shows upgrading over that of lane 2 from unstimulated glands. Lane 3, showing *L. variegatus* blastula 5S RNA, serves as positive control.

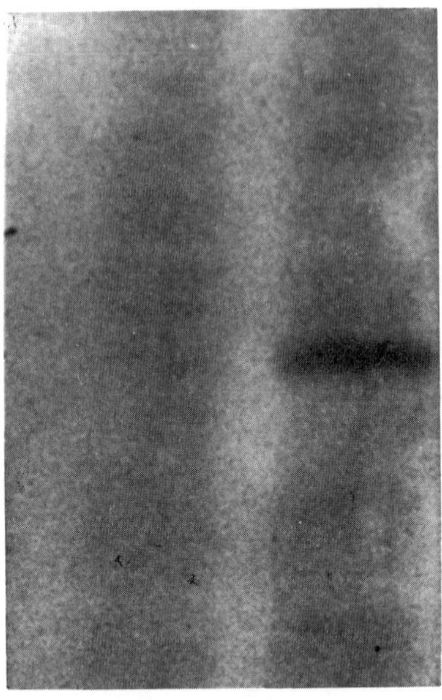

Figure 5. Northern blot analyses of U1 snRNA from stimulated and unstimulated glands. Electrophoretically separated small RNAs[8] from glandular tissues of stimulated and unstimulated organisms, incubated for 15 min, were probed for U1 snRNA with an *in vitro* transcript of a U1 snRNA sequence from the sea urchin, *Lythechinus variegatus*. Upgrading of this snRNA as a function of stimulation can be detected by the visual inspection of lane 1 containing unstimulated glands versus lane 2 from stimulated glands.

snRNA,[23,24] this upgrading may well be of functional value to the organism. An enriched and readily available pool of this small RNA can become an asset in the processing of the mRNA generated subsequently. The pool can certainly play a significant role in providing the fibroin-synthetic machinery with the augmented levels of mature messenger RNA for translation. Assessed by comparative electrophoretic mobility, the U1 snRNA of *N. clavipes* seems to be smaller than in other species.[25,26] Search for tissue-specific or developmentally expressed isoforms, as have been found in other differentiating systems,[27,28] has proven fruitless. Northern blot searches for other small RNA members of the spliceosome complex[29,30] should prove fruitful and are being undertaken.

Most outstanding within this early response is the generation of alanine-tRNA. As seen in Table I, a large bulk of the alanine-tRNA generated in the stimulated glands is produced during the first bout of activity (peak 1). In fact, with respect to tRNA, this seems to be its exclusive function. Of the total alanine-tRNA accounted for in our experiments, 56% is produced during the earliest of the responses (see Table I). Electrophoretic separation of the tRNAs revealed two alanine-tRNA species, identified through aminoacylation, of which the one displaying the highest mobility proved to be tissue-specific.[8] At the peak of the third wave of small RNA synthesis (peak 3), the ratio of the two isoforms in the activated glands is 4:1, favoring the isoform specific to the glandular tissues as measured through Cerenkov radiation.[31]

Northern blot analyses of these tRNAs, shown in Fig. 6, confirm the presence of the two isoacceptors and also their differential accumulation by the tissues. The nonglandular tissues, lanes 1 and 2, lack the fastest-moving species. Lane 3, showing tissues from unstimulated organisms, displays preferential accumulation of the constitutive isoform. After 15 min of incubation, they have activated the expression of the tissue-specific form, such that the relative intensities of the two species are almost on par with each other (lane 4). Progress through the stimulated process seems to be accompanied by a further increase in the expression of the tissue-specific isoacceptor, as may be seen in lane 5.

We have determined the nucleotide sequence of the two mature isoforms (Fig. 7) and they seem to vary from each other by only one nucleotide.[32] This is another parallelism with another silk-producing system, the glands of *Bombyx mori*. During their silk-producing stage, these glands generate two alanine-tRNA isoforms whose mature sequences display variation by only one nucleotide.[33]

What is interesting in the system is the disproportionate enrichment of the glands with alanine-tRNA and the fact that it is programmed early during the sequence of events preceding the synthesis of the fibroin. The consideration of an alternate role(s) for these tRNAs, particularly the gland-specific form which displays preferential accumulation in the active glands, becomes plausible. A search through the literature unearths a wide variety of nontranslational roles of tRNA and tRNA-like molecules.[34–36] The references included are not exhaustive on the subject. A search for an alternate, nontranslational role for *Nephila*'s alanine-tRNA, particularly the gland-specific isoform, is worthy of consideration.

Both of the isoacceptors exhibit structural features in common with those typical of initiator tRNAs, such as (1) the position of certain determined nucleotides,[37] (2) the A bases in loop IV that participate in tertiary interaction, which has been considered in the literature as unique to initiator tRNAs,[37–39] and (3) nucleotide pairing which occurs in the aminoacyl stem of initiator methionine-tRNA and which is recognized by the transformylases.[32,37] In view of this, both of the spider tRNAs qualify potentially as initiator tRNAs and perhaps act as such during fibroin production.

Unique to the tissue-specific form, however, is its ability to form a Levitt base pair between positions 15 and 48 (G_{15} and U_{48}), a characteristic also present in initiator methionine-tRNA between these positions (G_{15} and C_{48}).[37] The conformational change conferred by the presence of this pairing may enhance the rate of translation through a tighter fit, closer tRNA–tRNA interaction, and a faster mobility of the molecule in and out of the A, P, or E site in the ribosome.[40]

According to the scanning model proposed by Kozak,[41] an initiation codon's

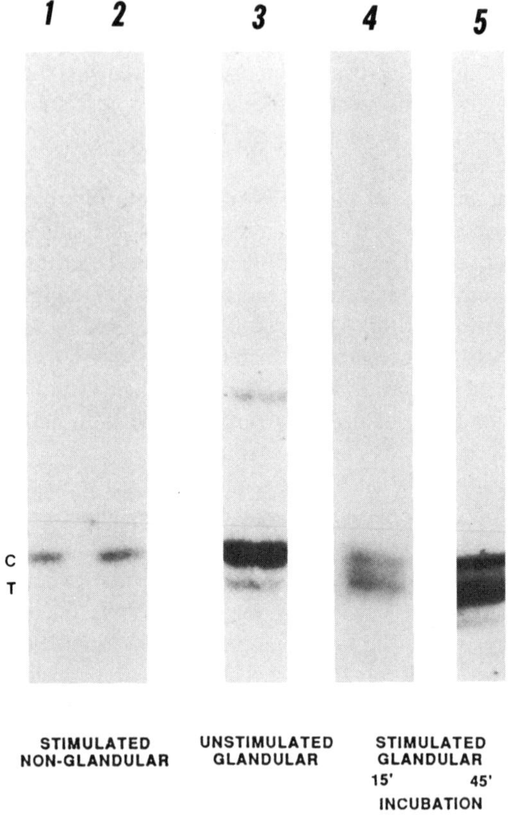

Figure 6. Northern blot analyses of alanine-tRNAs. Electrophoretically fractionated small RNA from glandular tissues of stimulated and unstimulated organisms, and from nonglandular tissues of stimulated organisms, were probed with a nick-translated plasmid containing a constitutive alanine-tRNA gene from *Bombyx mori*.[8] Lanes 1 and 2 contain stimulated nonglandular tissues; lane 3, unstimulated gland; lane 4, stimulated glands, 15 min of incubation; lane 5, stimulated glands, 45 min of incubation. Two isoacceptors have been identified in this system,[8] constitutive (C) and tissue-specific (T), not detected in the nonglandular tissues (lanes 1 and 2). While the constitutive isoform predominates in the unstimulated gland tissues (lane 3), the other isoform, evident after 15 min of incubation of the stimulated glands (lane 4), exhibits preferential accumulation, as seen after 45 min of incubation of the stimulated glands (lane 5).

flanking sequences improve its chances in its selection as initiation site. Analyses of these flanking sequences in many protein templates have revealed a conserved motif, GCC, an alanine codon. Thus, the tissue-specific alanine-tRNA may facilitate the recognition of the correct initiation site by binding to this conserved motif. Furthermore, this isoacceptor, because of its conformation, may enhance the retention of the correct reading frame by interacting with a methionine-tRNA bound to an AUG codon, flanked by the alanine codons.[41,43]

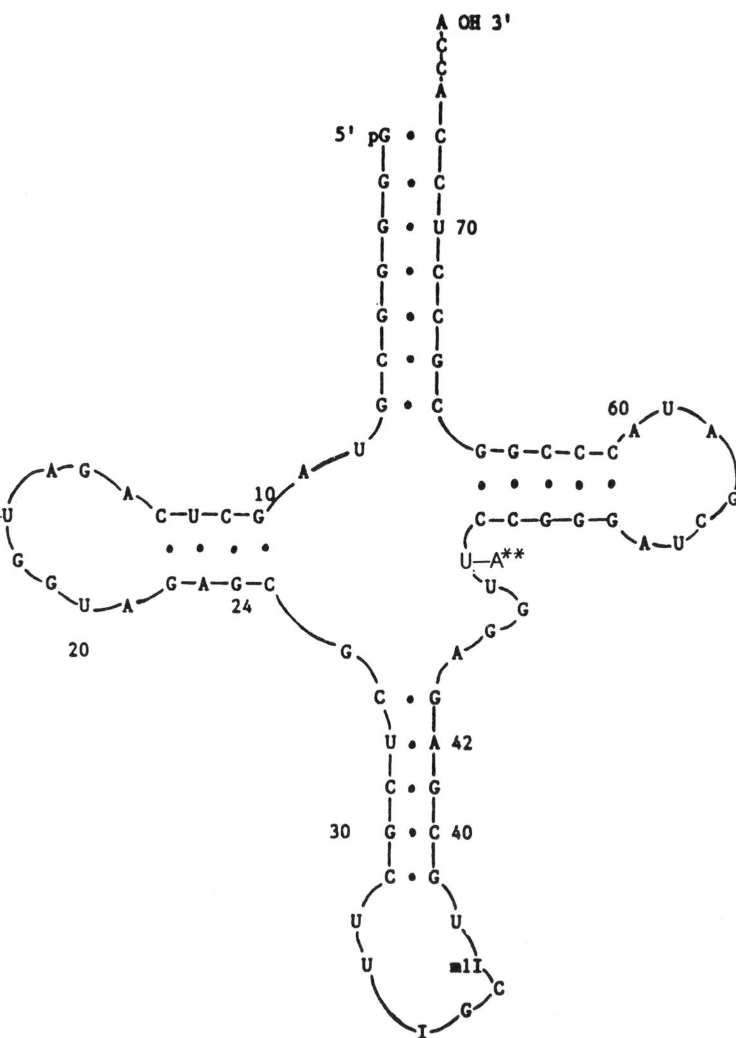

Figure 7. The alanine-tRNA isoacceptors from *Nephila clavipes* in the cloverleaf secondary structure. The sequence of the two alanine-tRNA isoacceptors was deduced by enzymatic digestion and a novel application of the methods of Sanger *et al.*[46] and Friedman and Rosbach.[47] The two isoacceptor sequences from *N. clavipes* differ in position 48, where the alanine-tRNA tissue-specific form has a uracyl and the constitutive form has an adenine.

What the results from these investigations imply is that fibroin synthesis is a complex process. In nature, the large ampullate glands are subjected to a demanding situation. The adult *N. clavipes* utilizes their product for the dragline, scaffolding of the web, and baseline of the catching spiral and also for particular web structures.[4] In another orb-web spider, *Araneus diadematus*, they have been estimated to produce protein equivalent to 10% of the gland's weight every web-building cycle.[44] Thus, in differentiating for compliance to their fated role, glands must be fully prepared for the massive synthesis of this protein product with the unusual characteristics discussed above. As our work progresses, we are adding to the list of monitored tissue-specific expressions and mechanisms observed in our manipulated model system. The list, already impressive, contains some unusual mechanisms[7] and we are currently conducting experiments to search for supplemental activities which might enrich the existing list. In spite of this, the possibility that our methods may not be discriminating enough to detect some events has to be considered. Also, it should be borne in mind that since the glands are fully differentiated when subjected to inactivation, regression to basal levels of all processes related to fibroin synthesis may not necessarily be attained. Thus, low levels of upgrading, which may play critical roles in the synthesis of the fibroin, may go undetected.

ACKNOWLEDGMENTS. We are deeply indebted to Isabel Cintron, Edgar Luciano, and Edwin Vazquez for their critical reading of the manuscript and their invaluable input in its preparation.

This work has been supported by NIH Grants RR088102 and RR0364102, NSF Grants M2586B90 and M32893021, and institutional funds. C.M. and M.I. held MARC Faculty Fellowships.

REFERENCES

1. Xu, M., and Lewis, R. V., 1990, Structure of a protein superfiber: Spider dragline silk, *Proc. Natl. Acad. Sci. USA* **87:**7120–7124.
2. Peters, V. H. M., 1955, Uber den Spinnapparat von *Nephila madagascariensis* (Radnetzspinner, Fam. Argiopidae), *Z. Naturforsch.* **10b:**395–404.
3. Lucas, F., 1964, Spiders and their silks, *Discovery* **25:**20–26.
4. Peakall, D. B., 1969, Synthesis of silk, mechanism and location, *Am. Zool.* **9:**71–79.
5. Candelas, G. C., and Cintron, J., 1981, A spider fibroin and its synthesis, *J. Exp. Zool.* **216:**1–6.
6. Candelas, G. C., and Lopez, F., 1983, Synthesis of fibroin in the cultured glands of *Nephila clavipes*, *Comp. Biochem. Physiol.* **74B:**637–641.
7. Candelas, G. C., Carrasco, C., Dompenciel, R., Arroyo, G., and Candelas, T., 1987, Strategies of fibroin production, in: *Translational Regulation of Gene Expression* (J. Ilan, ed.), pp. 209–228, Plenum Press, New York.
8. Candelas, G. C., Arroyo, G., Carrasco, C., and Dompenciel, R., 1990, Spider silkglands contain a tissue-specific alanine tRNA that accumulates *in vitro* in response to the stimulus for silk protein synthesis, *Dev. Biol.* **140:**215–220.

9. Plazaola, A., and Candelas, G. C., 1990, Stimulation of fibroin synthesis elicits ultrastructural shifts in the Golgi complex of the secretory cells, *Puerto Rico Health Sci. J.* **9:**227–229.
10. Plazaola, A., and Candelas, G. C., 1991, Stimulation of fibroin synthesis elicits ultrastructural modifications in spider silk secretory cells, *Tissue Cell* **23:**277–284.
11. Andersen, S. O., 1970, Amino acid composition of spider silks, *Comp. Biochem. Physiol.* **35:** 705–711.
12. Lucas, F., Shaw, J. T. B., and Smith, S. G., 1960, Comparative studies of fibroins. I. The amino acid composition of various fibroins and its significance in relation of their crystal structure and taxonomy, *J. Mol. Biol.* **2:**339–349.
13. Chavancy, G., Daillie, J., and Garel, J. P., 1971, Adaptation fonctionnelle des tRNA à la biosynthèse protéique dans un système cellulaire hautement differéncié IV Evolution des tRNA dans la glande serigene de *Bombyx mori* L au cours du dernier age larvaire, *Biochemie* **53:**1187–1194.
14. Chavancy, G., Chevallier, A., Fournier, A., and Garel, J. P., 1979, Adaptation of iso-tRNA concentration to mRNA codon adaptation in the eukaryote cell, *Biochemistry* **61:**71–78.
15. Chavancy, G., and Garel, J. P., 1981, Does quantitative tRNA adaptation to codon content in mRNA optimize the ribosomal translation efficiency? Proposal for a translation system model, *Biochemie* **63:**187–195.
16. Garel, J. P., Mandel, P., Chavancy, G., and Daillie, J., 1971, Adaptation fonctionelle des tRNA à la biosynthèse protéique dans un système cellulaire hautement differéncié V, *Biochemie* **53:**1195–1200.
17. Garel, J. P., 1974, Functional adaptation of tRNA population, *J. Theor. Biol.* **43:**211–225.
18. Ilan, J., and Ilan, J., 1970, Mechanism of gene expression in *Tenebrio molitor* Juvenile hormone determination of translational control through transfer ribonucleic acid determination, *J. Biol. Chem.* **25:**1275–1281.
19. Mairy, M., and Denis, H., 1972, Recherches bioquimiques sur l'oogenèse. Assemblage des ribosomes pendant le grand accroissement des oocytes de *Xenopus laevis*, *Eur. J. Biochem.* **25:** 535–543.
20. Wormington, W. M., and Brown, D. D., 1983, Onset of 5S RNA gene regulation during *Xenopus* embryogenesis, *Dev. Biol.* **99:**248–257.
21. Ford, P. J., and Brown, D. D., 1976, Sequences of 5S ribosomal RNA from *Xenopus mulleri* and the evolution of 5S gene-coding sequences, *Cell* **8:**485–493.
22. Dennis, H., and Wegnez, M., 1977, Biochemical research on oogenesis. Oocytes and liver cells of the teleost fish *Tinca tinca* contain different kinds of 5S RNA, *Dev. Biol.* **59:**228–36.
23. Reddy, R., and Busch, H., 1988, In: *Structure and Function of Major and Minor Small Nuclear Ribonucleoprotein Particles* (M. L. Birnstiel, ed.), pp. 1–35, Springer-Verlag, New York.
24. Maniatis, T., and Reed, R., 1987, The role of small nuclear ribonucleoprotein particles in pre-mRNA slicing, *Nature* **325:**673–678.
25. Reddy, R., 1988, Compilation of small RNA sequences, *Nucleic Acids Res.* **16:**71–85.
26. Reddy, R., and Gupta, S., 1990, Compilation of small RNA sequences, *Nucleic Acids Res.* **18:**supplement.
27. Santiago, C., and Marzluff, W. F., 1990, Expression of the U1 RNA gene repeat during early sea urchin development, *Proc. Natl. Acad. Sci. USA* **86:**2572–2576.
28. Lo, P. C. H., and Mount, S., 1990, *Drosophila melanogaster* genes for U1snRNA variants and their expression during development, *Nucleic Acids Res.* **18:**6971.
29. Guthrie, C., and Patterson, B., 1988, Spliceosomal snRNAs, *Annu. Rev. Genet.* **22:**387–419.
30. Parry, H. D., Scherly, D., and Mattaj, W., 1989, 'Snurpogenesis': The transcription and assembly of U snRNP components, *Trends Biochem. Sci.* **14:**15–19.
31. Dompenciel, R., 1988, tRNA adaptation during fibroin synthesis in the cultured glands of *Nephila clavipes*, M.A. Thesis, University of Puerto Rico, Rio Piedras, Puerto Rico.
32. Carrasco, C. E., 1989, The two alanine transfer RNA in *Nephila clavipes*, Ph.D. Thesis, University of Puerto Rico, Rio Piedras, Puerto Rico.

33. Sprague, K. U., Hagenbuchle, O., and Zuniga, M. C., 1977, The nucleotide sequence of two silk gland alanine tRNAs: Implications for fibroin synthesis and for initiator tRNA structure, *Cell* **11:** 561–570.
34. Lawlor, E. J., Baylis, H. A., and Chater, K. F., 1987, Pleiotropic morphological and antibiotic deficiencies result from mutations in a gene encoding a tRNA-like product in *Streptomyces coelicolor*, *Genes Dev.* **1:**1305–10.
35. Okada, N., 1990, Transfer RNA-like structure of the human *Alu* family: Implications of its generation mechanisms and possible functions, *J. Mol. Evol.* **31:**500–510.
36. Jahn, D., Verkamp, E., and Soll, D., 1992, Glutamyl-transfer RNA: A precursor of heme and chlorophyll biosynthesis, *Trends Biochem. Sci.* **17:**215–218.
37. Kozak, M., 1983, Comparison of initiation of protein synthesis in prokaryotes, eukaryotes, and organelles, *Microbiol. Rev.* **47:**1–45.
38. Noller, H. F., 1984, Structure of ribosomal RNA, *Annu. Rev. Biochem.* **53:**119–162.
39. Sprinzl, M., Hartmann, T., Meissner, F., Moll, J., and Vorderwulbecke, T., 1987, Compilation of tRNA sequences and sequences of tRNA genes, *Nucleic Acids Res.* **15:**supplement.
40. Smith, D., and Yarus, M., 1987, tRNA–tRNA interaction alter the rate of translation, unpublished.
41. Kozak, M., 1980, Evaluation of the "scanning model" for the initiation of protein synthesis in eukaryotes, *Cell* **22:**7–8.
42. Kozak, M., 1981, Mechanism of mRNA recognition by eukaryotic ribosomes during initiation of protein synthesis, *Curr. Top. Microbiol. Immunol.* **93:**81–123.
43. Cigan, A. M., Feng, L., and Donahue, T. F., 1988, $tRNA^{met}$ functions in directing the scanning ribosome to the start site of translation, *Science* **242:**93–96.
44. Peakall, D., 1966, Regulation of protein production in the silk glands of spiders, *Comp. Biochem. Physiol.* **19:**253–58.
45. Wallace, D. M., 1987, Precipitation of nucleic acids, in: *Guide to Molecular Cloning Techniques* (G. R. Berger and A. R. Kimmel, eds.), pp. 41–48.
46. Sanger, F., Nicklen, S., and Coulson, A. R., 1977, DNA sequencing with chain terminating inhibitors, *Proc. Natl. Acad. Sci. USA* **74:**5463–5467.
47. Friedman, E. Y., and Rosbach, M., 1977, The synthesis of high yields of full-length reverse transcripts of globin mRNA, *Nucleic Acids Res.* **4:**3455–3471.

Chapter 13

Translational Control in the Circadian Regulation of Luminescence in the Unicellular Dinoflagellate *Gonyaulax polyedra*

Steven Sczekan, Dong-Hee Lee, Dieter Techel, Maria Mittag, and J. Woodland Hastings

1. OVERVIEW

Circadian (*circa*, about; *dies*, day) control is a temporal regulatory mechanism. It is ubiquitous phylogenetically and controls many different biological processes.[1] Rats and cockroaches are active at night, whereas humans and birds are at rest. Body temperature exhibits a daily rhythm in birds and mammals, and the serum level of many hormones fluctuates on a 24-hr cycle. Flowers may exude nectar only at a specific time of day, and bean leaves exhibit regular diurnal movements. In certain cyanobacteria, the enzyme nitrogenase is present at night but not during the day, and many algae are more positively phototactic at a certain time of day.

STEVEN SCZEKAN, DONG-HEE LEE, DIETER TECHEL, MARIA MITTAG, AND J. WOODLAND HASTINGS • Department of Cellular and Developmental Biology, Harvard University, Cambridge, Massachusetts 02138.

Translational Regulation of Gene Expression 2, edited by Joseph Ilan. Plenum Press, New York, 1993.

The circadian mechanism has clocklike properties; it involves a time-of-day regulation and, although it derives its phase from the light–dark cycle, it should be emphasized that it is *not* dependent on light–dark changes for its continuation. Thus, it operates under constant conditions, yet can readily be "reset" to a new phase by a single brief exposure to light. Moreover, its period is not greatly altered under different but still constant environmental conditions, notably temperature.

Regulation of gene expression, both at the transcriptional and translational levels, now appears to be of key importance in this process.[2,3] But the clock mechanism, both at the cellular and molecular levels, remains unknown.[4–6]

In many animals, both vertebrate and invertebrate, the nervous system has a prominent role in circadian rhythmicity;[7] in mammals, for example, destruction of the suprachiasmatic nuclei (SCN) leads to the loss of overt rhythmicity in the animal, which may otherwise appear normal.[1,8,9] In *Drosophila*, expression of the *per* gene is required for rhythmicity, and mutations in *per* may affect period length.[10] Moreover, expression of *per* in the brain is required for these effects on the circadian system.

Circadian regulation occurs also in unicellular organisms and plants, where there are no apparent specializations or localizations of the circadian mechanism;[11] these systems thus constitute excellent models for the study of the molecular mechanism of circadian rhythmicity. The applicability of model system studies to the more complicated systems is postulated but not yet definitively established.

2. THE CIRCADIAN SYSTEM

A commonly espoused paradigm for the circadian system is that it comprises both a clock and its hands (Fig. 1): the mechanism and the clock-controlled

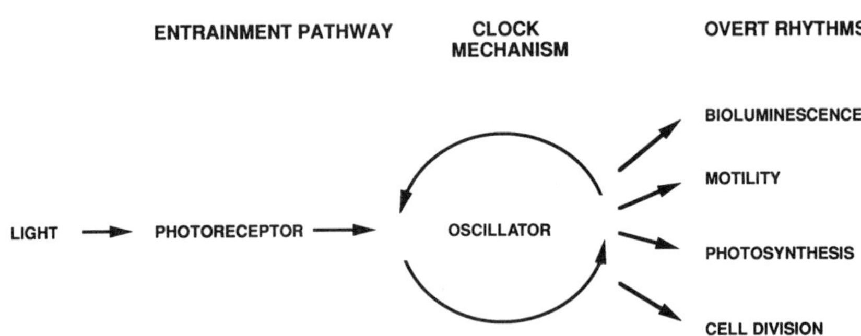

Figure 1. A paradigm for the circadian clock, represented as including three components. The oscillator refers to the clock mechanism itself, which is entrained by light and is responsible for controlling diverse cellular processes, several of which are indicated. The cellular/biochemical nature of the putative oscillator remains unknown.

processes.[4] In these terms, gene expression may be involved in both the feedback mechanism of the clock that is responsible for the generation of the ca. 24-hr period and, though possibly in a different way, in the mechanism whereby the clock turns on and off processes under its control, be it motility, bioluminescence, or enzyme activity.

2.1. Molecular Mechanism of the Clock

With regard to the clock mechanism itself, transcriptional processes may or may not be required, but translation—or the daily synthesis of one or more proteins—does appear to be essential. These conclusions are derived from studies in several systems. The giant alga *Acetabularia*, which exhibits circadian rhythmicity in photosynthesis and other processes,[12] is well known for its ability to survive for several weeks after the removal of its nucleus. Its circadian clock also continues to function under these conditions, indicating that clock function is not absolutely dependent upon daily nuclear transcription. But its clock *is* affected by inhibitors of eukaryotic protein synthesis, such as puromycin and cycloheximide,[13] but not by inhibitors of prokaryotic protein synthesis, such as chloramphenicol. Indeed, this is generally true for most or all circadian systems that have been studied in this regard, and it represents one of the features that appears to be common at all phylogenetic levels.[14,15] In *Gonyaulax polyedra*, brief (1 min to 1 hr) exposures of a drug that inhibits protein synthesis will cause a shift in the phase of the bioluminescence rhythm by as much as several hours (Fig. 2).[16,17] The possibility that transcription may be involved in the clock mechanism itself is indicated, however, by recent studies with the mollusk *Aplysia*, in which DRB (5,6-dichloro-1-β-ribobenzimidazole), an inhibitor of RNA synthesis, was shown to have a phase-shifting effect.[18]

2.2. Regulated Processes at the Molecular Level

With regard to the regulated processes—the hands of the clock—there are now a number of examples of clock-controlled changes at the cellular and molecular levels. Circadian regulation of transcription, in which the levels of specific messenger RNAs (mRNAs) change over the course of a subjective day, has been shown in several systems.[19–22] This does not constitute evidence that the synthesis of RNA is a part of the central clock mechanism; it could simply be under the control of the clock. Moreover, and indeed in general, we should emphasize that these changes are not themselves light-activated, though there are such cases, namely, the light-activated or so-called photogenes.[23] But circadian changes are defined as those that continue to cycle under constant conditions, and thus are not dependent for their continuation upon light–dark cycles, but upon the still-elusive clock mechanism. The normal daily light cycle does, however, serve to entrain the rhythm; single light pulses will phase shift the clock (Fig. 3).

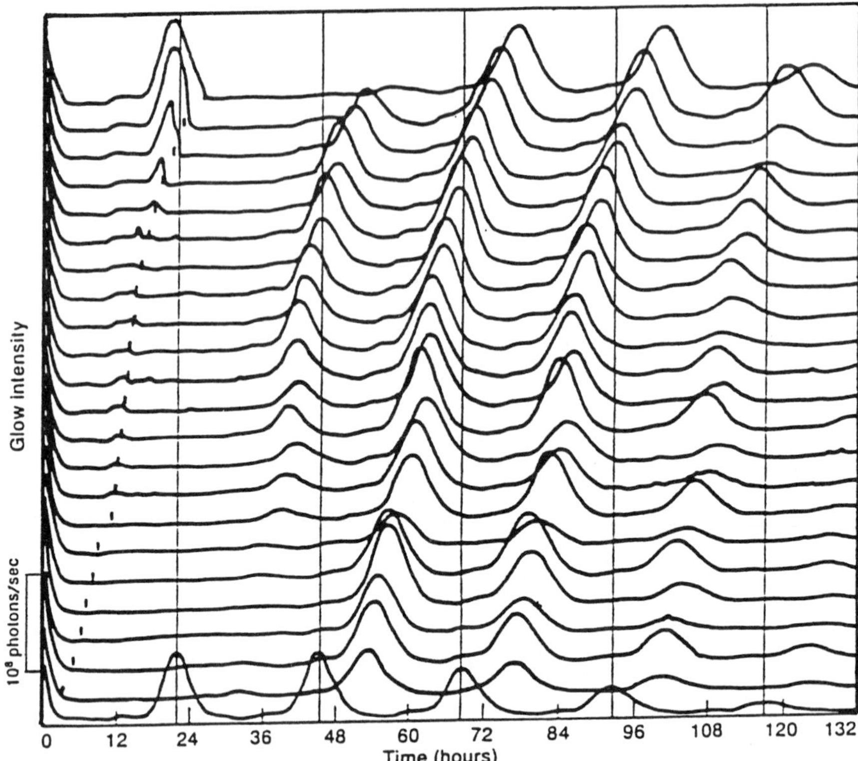

Figure 2. Phase shifting of the bioluminescence rhythm in *Gonyaulax* induced by brief exposures to anisomycin, an inhibitor of protein synthesis. All cultures were initially in phase with the control culture (bottom trace), and all remained in constant light conditions throughout the experiment. Exposures to the drug were for 1 hr, at the times indicated by the vertical black bars. The new phases generally parallel the times at which the drug was applied, indicating that the clock is reset by the drug.[16]

Regulation by the circadian mechanism also occurs at the translational level. This was first found and has been studied in some detail in the unicellular dinoflagellate *G. polyedra*.[24] Other examples include vasopression in mammals,[25] the *per* gene product in *Drosophila*,[26] and one of the *cab* genes in *Arabidopsis*.[27]

3. CIRCADIAN RHYTHMICITY IN *GONYAULAX*

3.1. Introduction

Gonyaulax exhibits rhythmicity in many of its properties, including bioluminescence, photosynthesis, motility, pattern formation, and the activities of a

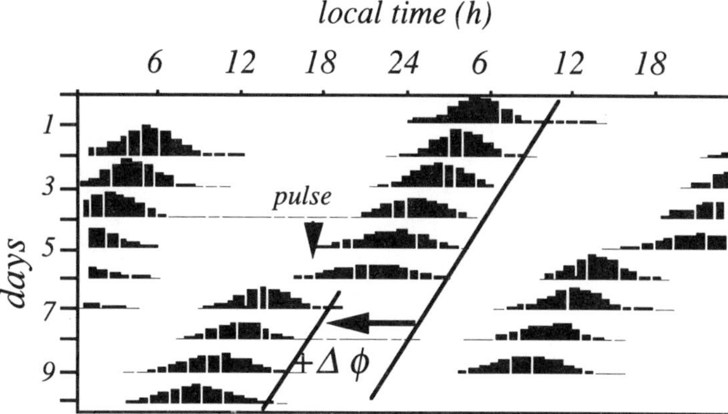

Figure 3. Phase shifting of the bioluminescence rhythm in *Gonyaulax* induced by a brief (3-hr) exposure to light. The culture was maintained in constant dim light for the first few days, exposed to bright light at the time indicated by the arrow, and then returned to constant dim light. The phase was advanced by about 7 hr, while the free-running period both before and after the pulse was about 22.9 hr. T. Roenneberg, unpublished results.

number of different enzymes. Bioluminescence and the molecular mechanism of its control have been studied most extensively,[28] but it is considered a clock-controlled process and not part of the clock mechanism itself.

3.2. Translational Regulation of Proteins in Bioluminescence

The rhythm of *Gonyaulax* bioluminescence, which is illustrated in Figs. 2 and 3, is a particularly favorable system for the study of circadian control. Its biochemical components are unique and unambiguous. These include the enzyme (*Gonyaulax* luciferase), the substrate (dinoflagellate luciferin, an open-chain tetrapyrrole), and a luciferin-binding protein (LBP) that sequesters the substrate at neutral and alkaline pH values. The reaction mechanism (Fig. 4) is also unique: upon a decrease of the pH to an acid value the luciferin is released and then enzymatically oxidized to give a product in an electronically excited state, with subsequent light emission.[29]

Both luciferase (MW 140 kDa) and LBP (MW 70 kDa) are present in cell extracts made during the night phase, but are absent in day-phase extracts.[24,30,31] This might involve regulation of protein synthesis or destruction, or both (Fig. 5). For LBP it has been shown by pulse labeling that synthesis is indeed regulated, such that it occurs only during a time period of about 4 hr spanning the late day and early night phases. But mRNA levels for LBP remain constant with time, indicating that there is translational control (Fig. 6) The clock can thus regulate a particular cellular process by directly controlling the synthesis of a particular

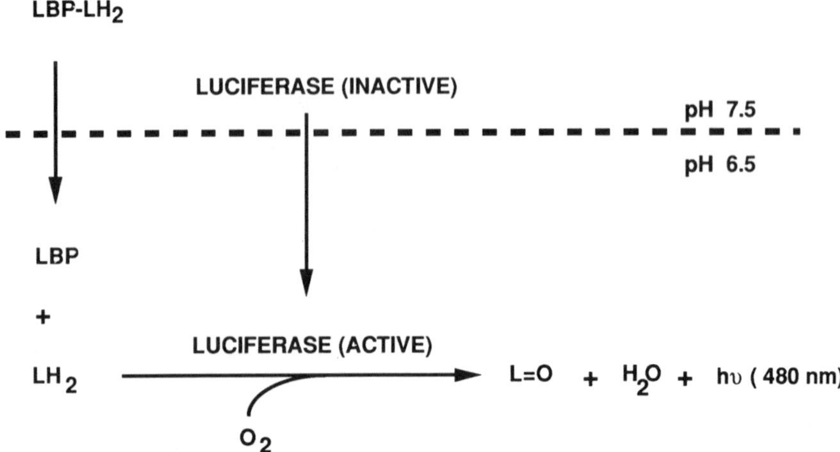

Figure 4. The biochemical components and the pH control of the bioluminescence reaction in *Gonyaulax*. The reaction occurs at acidic pH, where luciferin (LH_2) is released from its binding protein (LBP) and then oxidized by an active form of the luciferase.

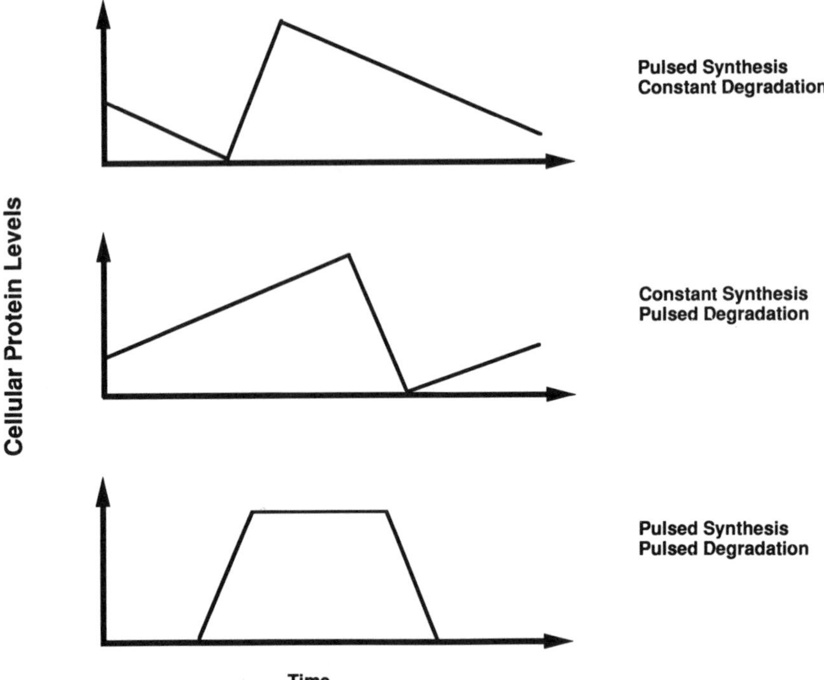

Figure 5. Schematic representation of the ways whereby *Gonyaulax* might regulate the expression of the luciferin-binding protein (LBP) in a time-dependent fashion. Since the protein is present only during the night phase, either synthesis, degradation, or both, must be regulated.

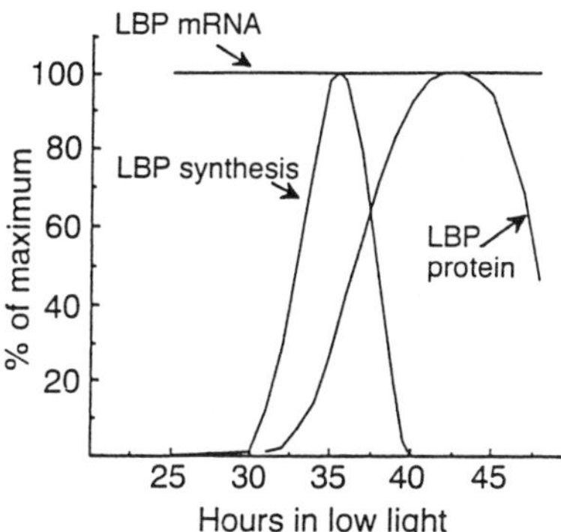

Figure 6. The circadian rhythm in the amount of extractable LBP protein; it parallels the *in vivo* bioluminescence. LBP synthesis occurs earlier, as a pulse during late subjective day and early night. The mRNA levels for LBP, however, as determined from Northern blots and *in vitro* translations of *Gonyaulax* mRNA, remain constant.

protein. In *Gonyaulax*, evidence suggests that many different proteins are controlled in this way.[32]

What possibilities for control might account for this temporal regulation of the translation of a constitutively expressed mRNA? When poly(A)$^+$ RNA isolated at different times in the circadian cycle was translated in an *in vitro* cell-free lysate (rabbit reticulocyte), the LBP protein was readily detected at comparable levels with poly(A)$^+$ RNA isolated from all time points.[24] This leads to two important conclusions. First, the mRNA is not rendered inactive in some way by covalent modification during the day, when it is translationally inactive, since it is still perfectly capable of supporting *in vitro* translation, and second, that at least for the deproteinized mRNA, its secondary structure as such does not appear to prevent protein synthesis in any direct way. The latter point is especially significant, as high levels of secondary structure have been shown in other systems to have a deleterious effect on the initiation of translation, in both prokaryotes and eukaryotes.[33–35] This could also result in competition among different RNAs for initiation factors.[36,37]

Analysis of the 5′ and 3′ untranslated regions (UTR) of the LBP mRNA indicates that there may be an unusually high level of secondary structure (Fig. 7). While the existence of such structures has not been verified by mapping, the stability of the patterns indicates that some significant structures do occur. Yet the cell-free translation system has no difficulty at all in translating this message. It must be kept in mind, however, that the native message has proteins bound to it, and that the translational machinery of *Gonyaulax* may be affected by secondary structure in a different way than is that of the rabbit reticulocyte system.

It is possible to rule out specific compartmentalization of the LBP mRNA as a regulatory mechanism. When the message was localized in the cell by means of

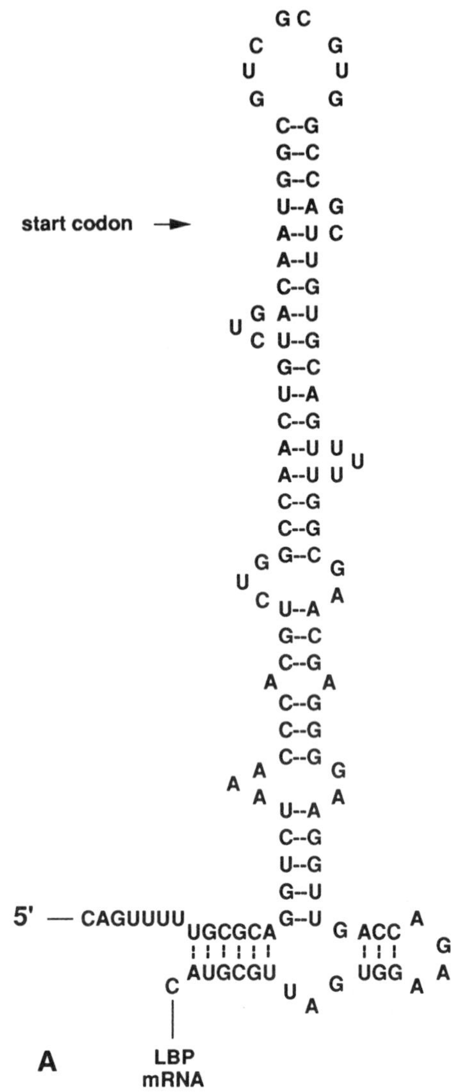

Figure 7. Theoretical secondary structure predictions of (**A**) the 5′ and (**B**) the 3′ untranslated regions of the LBP mRNA. Folding constructs were obtained by computer, using the program of Zuker.[50] For **A**, $\Delta G = -45.3$ kcal/mole; for **B**, $\Delta G = -69.6$ kcal/mole.

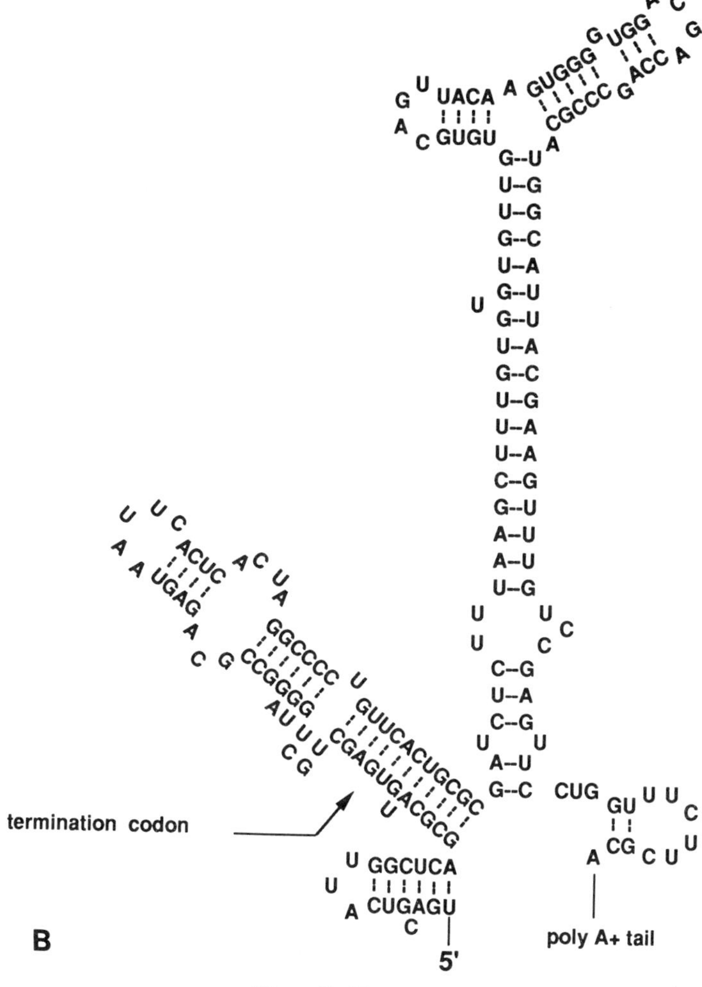

Figure 7. (*Continued*)

hybridization labeling, it was found that it is distributed evenly throughout the cell, both in night- and day-phase cells.[38] Thus, the mRNA would appear to be accessible to the translational machinery during the entire 24-hr cycle.

We can also consider the level at which the block in translation of this message is achieved. It could occur anywhere along the translational pathway, including initiation,[39,40] elongation,[41,42] or termination.[43] For example, a block at some point in the interior of the mRNA (possible related to structure) could result in the loading of a large section of the message with ribosomes, which would then remain arrested until the block was removed (e.g., at the onset of night phase). Alternatively, a block at initiation would prevent the loading of any ribosomes onto the message. To examine this, the ribosome sedimentation profiles for day and night cells were analyzed for the presence of the LBP mRNA. It was found that essentially all the message is located in the fraction lighter than the 80S ribosome monomer during the day phase, indicating that the message is unable to initiate on the translational machinery. During the night phase, by contrast, the message is distributed broadly in the polysome fractions. The block thus appears to occur at the very onset of protein synthesis, implicating the initiation process, possibly the leader sequence. It must be noted, however, that regulatory elements which affect initiation can be located virtually anywhere within the mRNA molecule.[44,45] In this context, the high level of secondary structure in the untranslated regions of the LBP mRNA becomes more relevant for repression.

4. CURRENT AND FUTURE STUDIES

What do these results imply for the regulation of the LBP mRNA? As the mRNA itself is not altered, it appears likely that there exists some cellular component, perhaps a protein, which is responsible for the specific regulation. Examples of such translational repressors are widespread, including viral RNAs,[46] ribosomal proteins,[47] and the iron-regulated proteins, ferritin and transferrin receptor.[48,49] In such cases, there is frequently found a repressor protein which binds to a regulatory portion of the mRNA and prevents translation, usually by interfering with initiation. A recurring motif in many of these proteins–RNA interactions is that the repressor is often, though not always, found to bind to a hairpin structure in the RNA.[47]

The search for putative regulatory factors (proteins) which might bind to the LBP mRNA has involved two approaches. In the first, radiolabeled transcripts of the 3′ untranslated region (synthesized *in vitro*) were incubated with extracts of *Gonyaulax* made at different times of day and then examined in gel-shift assays. In early-day extracts the autoradiograms exhibited two strong retardation bands, evidence of specific LBP mRNA–protein interactions. In early-night extracts these two bands were much weaker, and a third weak band was also detectable. Similar studies with the 5′ untranslated region are in progress.

A second approach involves the isolation of LBP mRNA in association with its native protein factors, the messenger ribonucleoprotein particles (mRNPs). Using biotinylated LBP antisense mRNA as a probe, these LBP mRNPs were specifically isolated and their protein patterns analyzed. Preliminary results indicate that LBP mRNPs associated with polysomes lack a protein found on cytoplasmic LBP mRNPs.

Although it is not possible to state at this time what the precise mechanism of repression is, it is possible that it is pervasive in *Gonyaulax*, as numerous proteins appear to be translationally controlled on a circadian basis.[32] It would be much more efficient to use a general repressor which could coregulate or inhibit synthesis of several different proteins at a specific phase of the clock, which would permit the transition from day phase to night phase to be made much more simply. Indeed, the transition could also be made more rapidly; with only a relatively few cellular signals, a large number of gene products could be regulated. Such cascade systems are widespread in biochemical regulation.

ACKNOWLEDGMENTS. The research was supported in part by NIH grant GM-19536, and ONR grant N00C14-88-K-0130 to J.W.H., U.S. National Science Foundation grant NSF OCE-891-5090 to S.S. and a DFG grant Mi373/1-1 to M.M.

REFERENCES

1. Hastings, J. W., Boulos, Z., and Rusak, B., 1991, Circadian rhythms, in: *Neural and Integrative Animal Physiology* (C. L. Prosser, ed.), pp. 435–546, Wiley-Interscience, New York.
2. Morse, D., Fritz, L., and Hastings, J. W., 1990, What is the clock? Translational regulation of circadian bioluminescence, *Trends Biochem. Sci.* **15:**262–265.
3. Dunlap, J. C., 1990, Closely watched clocks, *Trends Genet.* **6**(5)**:**135–168.
4. Johnson, C. H., and Hastings, J. W., 1986, The elusive mechanism of the circadian clock, *Am. Sci.* **74:**29–36.
5. Strumwasser, F., 1992, Biological timing: Circadian oscillations, cell division, and pulsatile secretion, in: *Induced Rhythms in the Brain* (E. Basar and T. H. Bullock, eds.), pp. 295–306, Birkhauser, Boston.
6. Lakin-Thomas, P. L., Coté, G. G., and Brody, S., 1990, Circadian rhythms in *Neurospora crassa*: Biochemistry and genetics, *Crit. Rev. Microbiol.* **17**(5)**:**365–416.
7. Takahashi, J. S., and Menaker, M., 1984, Circadian rhythmicity. Regulation in the time domain, in: *Biological Regulation and Development*, Volume 3B (R. F. Goldberger and Keith R. Yamamoto, eds.), pp. 285–303, Plenum Press, New York.
8. Rusak, B., and Zucker, I., 1979, Neural regulation of circadian rhythms, *Physiol. Rev.* **59:** 449–526.
9. Meijer, J. H., and Rietveld, N. J., 1989, Neurophysiology of SCN in rodents, *Physiol. Rev.* **69:** 671–707.
10. Hall, J. C., and Rosbash, M., 1988, Mutations and molecules influencing biological rhythms, *Annu. Rev. Neurosci.* **11:**373–393.
11. Hastings, J. W., 1959, Unicellular clocks, *Annu. Rev. Microbiol.* **13:**697–706.
12. Schweiger, H. G., and Schweiger, M., 1977, Circadian rhythms in unicellular organisms, *Int. Rev. Cytol.* **51:**315–342.
13. Karakashian, M. W., and Schweiger, H. G., 1976, Evidence for a cycloheximide-sensitive component in the biological clock of *Acetabularia*, *Exp. Cell Res.* **98:**303–312.
14. Karakashian, M. W., and Hastings, J. W., 1963, The effects of inhibitors of macromolecular biosynthesis upon the persistent rhythm of luminescence in *Gonyaulax*, *J. Gen. Physiol.* **47:**1–12.
15. Olesiak, W., Ungar, A., Johnson, C. H., and Hastings, J. W., 1987, Is protein synthesis inhibition and phase-shifting of the circadian clock in *Gonyaulax* correlated? *J. Biol. Rhythms* **2:** 121–138.
16. Taylor, W. R., and Hastings, J. W., 1982, Minute-long pulses of anisomycin phase-shift the biological clock in *Gonyaulax* by hours, *Naturwissenschaften* **69:**94–96.

17. Taylor, W. R., Dunlap, J. C., and Hastings, J. W., 1982, Inhibitors of protein synthesis on 80s ribosomes phase shift the *Gonyaulax* clock, *J. Exp. Biol.* **97:**121–136.
18. Raju, U., Koumenis, C., Nunez-Regueiro, M., and Eskin, A., 1991, Alteration of the phase and period of a circadian oscillator by a reversible transcription inhibitor, *Science* **253:**673–675.
19. Kloppstech, K., 1985, Diurnal and circadian rhythmicity in the expression of light-induced nuclear messenger RNAs, *Planta* **165:**502–506.
20. Loros, J. J., Denome, S. A., and Dunlap, J. C., 1989, Molecular cloning of genes under control of the circadian clock in *Neurospora*, *Science* **243:**385–388.
21. Loros, J. J., and Dunlap, J. C., 1991, *Neurospora crassa* clock-controlled genes are regulated at the level of transcription, *Mol. Cell. Biol.* **11:**558–563.
22. Nagy, F., Kay, S. A., and Chua, N.-H., 1988, A circadian clock regulates transcription of the wheat *Cab-1* gene, *Genes Dev.* **2:**376–382.
23. Thompson, W. F., and White, M. J., 1991, Physiological and molecular study of light regulation of nuclear genes in higher plants, *Annu. Rev. Plant Physiol. Plant Mol. Biol.* **42:**423–466.
24. Morse, D., Milos, P. M., Roux, E., and Hastings, J. W., 1989, Circadian regulation of the synthesis of substrate binding protein in the *Gonyaulax* bioluminescent system involves translational control, *Proc. Natl. Acad. Sci. USA* **86:**172–176.
25. Carter, D. A., and Murphy, D., 1989, Diurnal rhythm of vasopressin mRNA species in the rat suprachiasmatic nucleus: Independence of neuroendocrine modulation and maintenance in explant culture, *Mol. Brain Res.* **6:**233–239.
26. Zwiebel, L. J., Hardin, P. E., Liu, X., Hall, J. C., and Rosbash, M., 1991, A post-transcriptional mechanism contributes to circadian cycling on a *per* β-galactosidase fusion protein, *Proc. Natl. Acad. Sci. USA* **88:**3882–3886.
27. Millar, A. J., and Kay, S. A., 1991, Circadian control of cab gene transcription and messenger RNA accumulation in *Arabidopsis*, *Plant Cell* **3**(5)**:**541–550.
28. Fritz, L., Morse, D., and Hastings, J. W., 1990, The circadian bioluminescence rhythm of *Gonyaulax* is related to daily variations in the number of light emitting organelles, *J. Cell Sci.* **95:** 321–328.
29. Morse, D., Pappenheimer, A. M., and Hastings, J. W., 1989, Role of a luciferin binding protein in the circadian bioluminescent reaction of *G. polyedra*, *J. Biol. Chem.* **264:**11822–11826.
30. Dunlap, J., and Hastings, J. W., 1981, The biological clock in *Gonyaulax* controls luciferase activity by regulating turnover, *J. Biol. Chem.* **256:**10509–10518.
31. Johnson, C. H., Roeber, J., and Hastings, J. W., 1984, Circadian changes in enzyme concentration account for rhythm of enzyme activity in *Gonyaulax*, *Science* **223:**1428–1430.
32. Milos, P., Morse, D., and Hastings, J. W., 1990, Circadian control over synthesis of many *Gonyaulax* proteins is at the translational level, *Naturwissenschaften* **77:**87–89.
33. Van Duijn, L. P., Holsappel, S., Kasperaitis, M., Bunschoten, H., Konings, D., and Voorma, H. O., 1988, Secondary structure and expression *in vivo* and *in vitro* of messenger RNAs into which upstream AUG codons have been inserted, *Eur. J. Biochem.* **172:**59–66.
34. De Smit, M. H., and Van Duin, J., 1990, Secondary structure of the ribosome binding site determines translational efficiency: A quantitative analysis, *Proc. Natl. Acad. Sci. USA* **87:**7668–7672.
35. Kozak, M., 1986, Influences of mRNA secondary structure on initiation by eukaryotic ribosomes, *Proc. Natl. Acad. Sci. USA* **83:**2850–2854.
36. Pain, V. M., 1986, Initiation of protein synthesis in mammalian cells, *J. Biochem.* **235:**625–637.
37. Walden, W. E., and Thach, R. E., 1986, Translational control of gene expression in a normal fibroblast. Characterization of a subclass of mRNAs with unusual kinetic properties, *Biochemistry* **25:**2033–2041.
38. Fritz, L., Milos, P., Morse, D., and Hastings, J. W, 1991, *In situ* hybridization of luciferin-binding protein anti-sense RNA to thin sections of the bioluminescent dinoflagellate *Gonyaulax polyedra*, *J. Phycol.* **27:**436–441.

39. Larson, D. E., and Sells, B. H., 1987, The function of proteins that interact with mRNA, *Mol. Cell. Biochem.* **74:**5–15.
40. Kozak, M., 1989, The scanning model for translation: An update, *J. Cell Biol.* **108:**229–241.
41. Wolin, S. L., and Walter, P., 1988, Ribosome pausing and stacking during translation of a eukaryotic mRNA, *EMBO J.* **7**(11)**:**3559–3669.
42. Webster, C., Kim, C.-Y., and Roberts, J. K. M., 1991, Elongation and termination reactions of protein synthesis on maize root tip polyribosomes studied in a homologous cell-free system, *Plant Physiol.* **96:**418–425.
43. Valle, R. P. C., and Morch, M.-D., 1988, Stop making sense or, Regulation at the level of termination in eukaryotic protein synthesis, *Fed. Eur. Biochem. Soc.* **235**(1,2,)**:**1–15.
44. Kwon, Y. K., and Hecht, N. B., 1991, Cytoplasmic protein binding to highly conserved sequences in the 3′ untranslated region of mouse protamine 2 mRNA, a translationally regulated transcript of male germ cells, *Proc. Natl. Acad. Sci. USA* **88:**3584–3588.
45. Kruys, V. I., Wathelet, M. G., and Huez, G. A., 1988, Identification of a translation inhibitory element (TIE) in the 3′ untranslated region of the human interferon-β mRNA, *Gene* **72:**191–200.
46. Heaphy, S., Dingwall, C., Ernberg, I., Gait, J. J., Green, S. M., Karn, J., Lowe, A. D., Singh, M., and Skinner, M. A., 1990, HIV-1 regulator of viron expression (Rev) protein binds to an RNA stem-loop structure located within the REV response element region, *Cell* **60:**685–693.
47. Kozak, M., 1988, A profusion of controls, *J. Cell Biol.* **107:**1–7.
48. Koeller, D. M., Case, J. L, Hentze, M. W., Gerhardt, E. M., Chan, L.-N. L., Klausner, R. D., and Harford, J. B., 1989, A cytosolic protein binds to structural elements within the iron regulatory region of the transferrin receptor mRNA, *Proc. Natl. Acad. Sci. USA* **86:**3574–3578.
49. Leibold, E. A., and Munro, H. N., 1988, Cytoplasmic protein binds *in vitro* to a highly conserved sequence in the 5′ untranslated of ferritin heavy- and light-subunit mRNAs, *Proc. Natl. Acad. Sci. USA* **85:**2171–2175.
50. Zuker, M., 1989, On finding all suboptimal foldings of an RNA molecule, *Science* **244:**48–52.

Chapter 14

Autoregulation of the Heat-Shock Response

Susan Lindquist

1. INTRODUCTION

The heat-shock proteins (hsps) were initially defined as a small set of proteins that are rapidly and dramatically induced when cells or whole organisms are exposed to high temperatures[1–5] (see Fig. 1). The same proteins are induced by a wide variety of other types of stress—ethanol, heavy metal ions, and anoxia being among the most common inducing agents.

Over the past 20 years a large number of experiments, in many different organisms, has demonstrated that the induction of these proteins by moderate stress conditions is generally accompanied by the induction of tolerance to more severe stresses.[1,4–9] The literature contains a few compelling contradictions, cases in which tolerance is induced in the absence of hsp synthesis or hsps are induced without inducing tolerance.[10–15] In the majority of circumstances, however, it is now clear that hsps play a central role in protecting cells from stress.

The magnitude of the effect that even one of these proteins can have on survival is remarkable. When cultures of the yeast *Saccharomyces cerevisiae* are incubated at 37°C to induce thermotolerance and are then exposed to 50°C for 10 min, wild-type cells survive 1000 times as well as isogenic cells carrying

Susan Lindquist • Howard Hughes Medical Institute and Department of Molecular Genetics and Cell Biology, University of Chicago, Chicago, Illinois 60637.

Translational Regulation of Gene Expression 2, edited by Joseph Ilan. Plenum Press, New York, 1993.

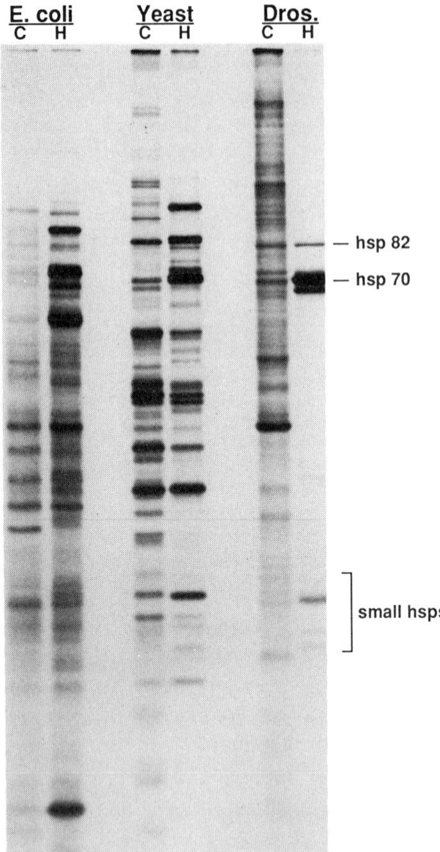

Figure 1. The heat shock responses of *Escherichia coli*, *Saccharomyces cerevisiae*, and *Drosophila melanogaster*. Cells were labeled with ^{3}H-leucine at their normal growing temperature (C) or after incubation at a heat-shock temperature (H). The strains and temperatures used were: *E. coli* strain BSJ72, 37°C and 50°C; *S. cerevisiae* strain W303, 25°C and 39°C; *D. melanogaster* S2 cells, 25°C and 36.5°C. Electrophoretically separated proteins were visualized by fluorography.

mutations in the *HSP104* gene.[16] When rat fibroblasts are transformed with a gene designed to produce hsp70 constitutively and are shifted directly from normal growing temperatures to 45°C, survival in the transformants can be as much as 1000 times higher than in the original cell line.[17]

Given the magnitude of these effects, it is not surprising that cells have evolved mechanisms to ensure that hsps are produced as rapidly as possible on exposure to stress. Indeed, the speed and intensity of the response are truly extraordinary. In *Drosophila*, the organism in which the response was first discovered, a complete change in the pattern of protein synthesis is obtained within 15 min of exposure to high temperature. Hsp70 is virtually undetectable at normal temperatures, but after 1 hr at 37°C it is one of the most abundant proteins in the cell. Expression of this protein is induced at least 10,000-fold.[18,19]

Only the concerted action of several different regulatory mechanisms can drive such a dramatic response. Mechanisms acting at the level of transcription, RNA processing, RNA turnover, and translation all come into play. These have been the subject of recent reviews.[20,21] A feature of regulation that has been less stressed in the literature is the effect that hsps themselves, particularly hsp70, have on their own synthesis.[22–25] In this chapter, I will first present a brief description of the major hsp families and review general aspects of heat-shock regulation. Appropriate to the subject of this volume, the greatest emphasis will be placed upon posttranscriptional mechanisms. I will then discuss the evidence that the heat-shock response is self-regulated and suggest a model that explains how hsps can affect their own regulation at several different levels.

2. THE HEAT-SHOCK PROTEINS

2.1. General Remarks

Hsps are among the most highly conserved proteins known. Certain hsps are induced by heat in virtually all organisms, and homologous genes have been cloned from both eukaryotes and prokaryotes. Others have been found only in certain types of organisms, but such distinctions are disappearing; some proteins that are abundantly expressed in response to heat in one organism have been found to have counterparts in other organisms that are simply less abundant or less strongly induced by heat.[26–28]

Structurally, the hsps are diverse in size and oligomeric composition. Nevertheless, for those proteins whose activities have been characterized, a unifying functional theme has emerged. Hsps modulate the folding and unfolding of other proteins, facilitate the assembly of proteins into multisubunit complexes, and promote the degradation of improperly folded or denatured proteins.[1,2,4,5,29] High temperatures, ethanol, heavy metal ions, and other inducing agents promote protein denaturation, suddenly and drastically increasing the requirement for these activities.[30] Because a key element in the function of some hsps is to prevent improper associations and activities, they can be considered members of the general class of proteins known as molecular chaperones.[31]

Within this general framework, however, hsp functions are distinct. Deletion mutations in different hsp genes produce different phenotypes and biochemical analysis reveals that the proteins bind to different substrates and interact with them in different ways. Rather than present a complete catalog, I will briefly describe only the most abundant and best understood hsps, to provide a guide to understanding the functions of hsps in regulation.

2.2. Major Hsp Families

2.2.1. Hsp70

Hsp70 is one of the most prominent hsps in most organisms. The *E. coli* genome encodes only one hsp70 protein, DnaK.[32] Eukaryotic cells, however, encode several proteins in this family, each of which shares at least 50% amino acid identity with the *E. coli* protein.[33–36] This multiplicity ensures that the proteins will be present in every major compartment of the cell, under all conditions.[2] Distinct hsp70 proteins are localized to the endoplasmic reticulum, to mitochondria, and, in plants, to chloroplasts.[37,38] Others are found in the nucleus and cytoplasm, but these vary in expression and intracellular distribution. Some are constitutive, while others are induced by heat or cold; some concentrate in the cytoplasm, while others shuttle between the nucleolus, the nucleus, and the cytoplasm.[39–41] In multicellular organisms, tissue-specific versions of the protein also exist.[42,43] Several hsp70 proteins are essential in yeast, while others are required only for growth at high or low temperatures.[35,44,45]

All members of the hsp70 family bind ATP and have a weak ATPase activity that is stimulated by substrate binding.[46–49] Hsp 70 proteins bind to small peptides, to nascent chains on polysomes, to proteins that have been targeted to the wrong cellular compartment, to certain mutant proteins, to some protein subunits that are expressed in the absence of their partners, and to certain oligomeric proteins in the process of assembly or disassembly.[2,29,47,50] In binding to these substrates, hsp70 proteins participate in a variety of protein folding, unfolding, assembly, and disassembly processes that employ the energy of ATP. For example, one hsp70 protein catalyzes the disassembly of clathrin cages,[51,52] another facilitates the transport of proteins across cell membranes,[53,54] and yet another facilitates the assembly of immunoglobulins in the endoplasmic reticulum.[55,56] The favored model to explain the role of hsp70 in these processes is that it binds to hydrophobic surfaces, preventing adventitious associations and stabilizing target proteins in a fully or partially unfolded state. At high temperatures, as proteins unfold, hsp70 binds to sequences that would normally be buried in the hydrophobic core of the protein, preventing aggregation and helping to restore native structures when temperature return to normal.[29,47,50] In only one case, that of RNA polymerase in *E. coli*, has a specific heat-related function been studied in detail. DnaK can both protect the polymerase from denaturation at high temperatures and promote the reactivation of previously heat-denatured polymerase.[57] For the former activity, ATP is not required; for the latter, it is.

2.2.2. Hsp60/GroEL (Chaperonin-60)

In bacteria, hsp60 (also known as GroEL or Chaperonin-60) is an essential protein that is one of the most abundant proteins in the cell even at normal temperatures.[58,59] In eukaryotes, Hsp60/GroEL proteins are found in mitochondria and chloroplasts.[26] Proteins in the GroEL family have not yet been clearly identified in other eukaryotic compartments. Proteins with very limited homology, but remarkably similar structures (i.e., oligomeric double rings) are found in archaebacteria and in the cytoplasm of eukaryotes.[28] The cytoplasmic proteins of eukaryotes are not strongly heat-inducible (A. Horwich, personal communication), but in the thermophilic archaebacteria, this protein is by far the most prominent hsp.[28]

Hsp60/GroEL does not have a high affinity for most hsp70 substrates. It does have a high affinity for completely denatured proteins, such as urea-denatured polypeptides, and for the individual subunits of certain proteins destined for oligomeric assembly[60–62] The functions of GroEL are dependent upon another oligomeric hsp, GroES (reviewed in refs. 29 and 31). *In vitro*, GroEL binds denatured proteins and prevents their aggregation. When GroES joins the GroEL complex, ATP hydrolysis promotes the folding of unfolded proteins on the surface of GroEL and their release from the complex.[63] *In vivo*, proteins in the hsp60/GroEL class are required for the assembly of oligomeric protein structures such as ribulose bisphosphate carboxylase (Rubisco), F_1-ATPase, and bacteriophage λ heads. In some cases, the requirement for GroEL increases with temperature. When urea-denatured Rubisco is diluted into buffer at 10°C, it spontaneously assembles into functional protein. At 25°C, it does not assemble properly and forms insoluble aggregates unless GroEL is present.[62,64]

2.2.3. Hsp100

Members of the hsp100 family have been cloned only recently.[65,66] Proteins in this group have been found in evolutionarily diverse organisms, including bacteria, fungi, trypanosomes, plants, and mammals. (*Drosophila* cells are unusual in that no abundant protein in this size class is expressed in response to high temperatures.) The yeast and *E. coli* proteins are dispensable for growth at normal temperatures, but play an important role in protecting cells from extreme temperatures and other forms of stress.[67,68,68a] Hsp100 proteins contain two regions of very high homology centered around two nucleotide-binding domains. Both of these are required for the protective functions of the protein in yeast.[66] Biochemical investigations of hsp100 functions are just beginning, but genetic investigations are highly suggestive. Overexpression of hsp70 (SSA1) can compensate in thermotolerance for the deletion of *HSP104* (Y. Sanchez, E. Craig, D. Parsell, J. Taulien, J. Vogel, E. Craig, and S. Lindquist, manuscript submitted), indicating a close functional relationship between the proteins in these two families.

2.2.4. Hsp90

Members of the hsp90 family are also found in both prokaryotic and eukaryotic organisms.[2] In prokaryotes this protein (C62.5) is dispensable for growth at normal temperatures and has only a minor effect on growth at high temperatures.[32] In yeast cells, and presumably in most eukaryotes, hsp90 is essential, and the quantity of protein that is required for growth increases with the temperature[69] (B. Kursheed, J. Taulien, and S. Lindquist, unpublished). In higher eukaryotes the family has diversified, with one member being targeted to the endoplasmic reticulum.[70,71] In most eukaryotic cells, hsp90 proteins are abundant at all temperatures and further induced by heat.

Members of the hsp90 family interact with many other cellular proteins, including steroid hormone receptors,[72] oncogenic tyrosine kinases,[73–75] the heme-regulated eIF-2alpha kinase,[76] calmodulin,[77] actin,[77] tubulin,[78] and casein kinase II.[79] The exact nature and function of these interactions are largely unknown. For the steroid hormone receptors, however, hsp90 appears to have both activating and repressing activities. Interaction with hsp90 is required for the steroid receptors to assume an activation-competent state.[80] As long as hsp90 remains associated with the receptor, however, the receptor remains inactive.[72] A difference between hsp90 and other hsps is that complexes between hsp90 and certain of its target proteins are stable for long periods *in vivo*.

2.2.5. Ubiquitin and the Ubiquitin-Conjugating Enzymes

Ubiquitin is a small, highly conserved, abundant protein that has been found in all eukaryotic cells examined, but not in any prokaryotic cells. Ubiquitin is implicated in a wide variety of cellular processes. One of its best-characterized functions is in the regulation of proteolysis. Conjugation of multiple ubiquitin residues to a protein targets it for degradation by the major ATP-dependent proteolytic complex of the cell.[79a] Both ubiquitin itself and certain of the enzymes that conjugate ubiquitin to other proteins are induced by heat. These inductions are required for cells to survive long exposures at temperatures just beyond their normal growth range.[81,82] It is envisioned that as proteins begin to unfold at high temperatures they have two possible fates. If they can be repaired by the protective, disaggregating functions of the other hsps, they will be. If not, they will be targeted by ubiquitination for degradation.

2.2.6. Cooperation between hsps

Certain activities of hsp70 are potentiated by other hsps. The clearest examples are from *in vitro* analysis of the *E. coli* DnaK protein and involve the hsps GrpE and DnaJ. For the initiation of bacteriophage λ-encoded replication, the P protein binds *E. coli* DNA helicase and recruits it to the origin. P protein must then be stripped from the complex for replication to proceed. This requires both DnaJ and DnaK, and is stimulated by GrpE.[83–85] In plasmid P1 replication, the activation of the initiator protein RepA proceeds in two steps. First, DnaJ binds inactive RepA dimers. Then, DnaK stimulates release of RepA monomers, the active form of the protein.[86] In this process, DnaJ targets the substrate protein for DnaK recognition. DnaJ and GrpE may act as allosteric effectors of DnaK, since they stimulate the ATP hydrolysis activity of DnaK by severalfold *in vitro*.[87] Hsp70 proteins may not need such a targeting mechanism to recognize heat-denatured substrates, which they seem to recognize through exposed hydrophobic surfaces.[50,57]

Some hsps may work together to provide a pathway for protein folding, assembly, and disassembly. Both hsp70 and hsp60 are required in mitochondria for the translocation of proteins into the organelle and for their assembly into functional complexes. Apparently, hsp70 keeps the proteins unfolded as they enter the organelle, maintaining them in a state that is competent for transfer to hsp60/GroEL.[29,88] Hsp60/GroEL, in turn, promotes their assembly into oligomeric structures. It is also notable that cytoplasmic hsp70 is associated, in substoichiometric quantities, with hsp90::steroid hormone receptor complexes. Antibodies against hsp70 block the association of hsp90 with the receptor *in vitro*,[89] suggesting that hsp70 plays a role in the assembly of complexes of hsp90 and its target proteins.

3. OVERVIEW OF REGULATION

3.1. Transcriptional Regulation

Transcriptional regulation is not the subject of this volume; however, it is so central to the heat-shock response and is such an integral part of autoregulation that it demands description. The reader is referred to recent reviews of the subject for more detailed treatment.[3,5]

In all organisms and most cell types examined to date, heat-shock genes are regulated at the level of transcription. In eukaryotic cells there appear to be at least three mechanisms at work. First, heat-shock genes are predisposed in an open chromatin configuration with hypersensitive sites at their 5′ ends.[90,91] In *Drosophila*, and presumably in many other organisms, RNA polymerase is transcriptionally engaged on hsp genes at normal temperatures, but elongation is inhibited by a negative regulatory mechanism that is not yet understood.[92] A shift to high temperatures releases this inhibition. At the same time, a preexisting transcriptional activator called heat-shock factor (HSF) is activated, promoting extremely rapid rates of transcription, perhaps by stimulating the release of polymerase.[93,94] Induction is extremely rapid, with new transcripts appearing within minutes of the inducing stimulus.

In yeast cells HSF is bound to chromatin prior to the shift to high temperatures. In higher eukaryotes, binding is generally observed only after heat shock.[93] In all cases, the sequence that binds HSF is a multimer of the basic heat-shock element, XGAAX. Efficient heat-shock transcription requires at least two, preferably three, repeats of the basic element in alternating orientation, e.g., XGAAXXTTCX-XGAAX. HSF binds cooperatively to these sites as a trimer and multiples of trimers.[95–98] During recovery from heat shock, HSF is inactivated and the transcription of heat-shock genes is repressed.

How is HSF activated? Heat shock stimulates the phosphorylation of HSF, but the role of this modification is unknown. Several findings suggest that HSF has an intrinsic transcription-activation potential that is negatively regulated by other factors in the cell: (1) When cloned *Drosophila* HSF is expressed in *Xenopus*, heat shock is required to activate it. When the same gene is expressed in *E. coli* at 18°C, the HSF protein exhibits maximal binding activity without further treatment. (2) HSF from *Drosophila* cells can be activated for DNA binding *in vitro* simply by reaction with a monoclonal antibody specific for HSF.[97] (3) When the carboxy termini of yeast, *Drosophila* and human HSF are deleted, the proteins are constitutively active (ref. 99 and C. Wu, personal communication). (4) Prior to heat shock, *Drosophila* HSF migrates on native gels as a 220-kDa complex, suggesting that it is a dimer or a heterooligomer (C. Wu, personal communication) (HSF monomers are 110 kDa). A particularly intriguing possibility (and one that fits the general model of regulation presented in Section 4 below) is that HSF is folded into a repressed conformation at normal temperatures or is associated with another protein that keeps it repressed. Heat disrupts this repressed structure, thereby allowing HSF to assemble into trimers (and higher multiples of trimers). During recovery, the structure is reestablished and heat-shock transcription is repressed.

In *E. coli*, heat-shock genes are regulated by special sigma factors that bind to core RNA polymerase and direct it to heat-shock promoters.[100,101] The concentration of the major heat-shock sigma factor, σ^{32}, determines the level of hsp gene transcription at normal temperatures and at temperatures as high as 50°C.[102] The gene encoding σ^{32} (*rpo*H) is constitutively transcribed by the standard, σ^{70}-containing polymerase. It is posttranscriptional regulation that determines the level of σ^{32} and thus determines the level of heat-shock transcription. High temperatures (1) increase the translational efficiency of the σ^{32} message, (2) reduce the turnover of the σ^{32} message, and (3) increase the stability of the σ^{32} protein.[102] Yet another mechanism reduces hsp transcription when cells are shifted from normal temperatures to low temperatures. In this case, the concentration of σ^{32} does not change, but its activity is reduced.[103] At very high temperatures (>50°C) another hsp regulon is induced, controlled by yet another sigma factor.[104]

3.2. RNA Processing

The effects of heat shock on RNA processing were discovered during investigations of hsp82 expression in *Drosophila*. This protein is strongly, induced when cells are shifted from 25 to 36°C. At higher temperatures, however, hsp82 expression is impaired relative to other hsps (see Fig. 2). Transcripts of the *Hsp82*

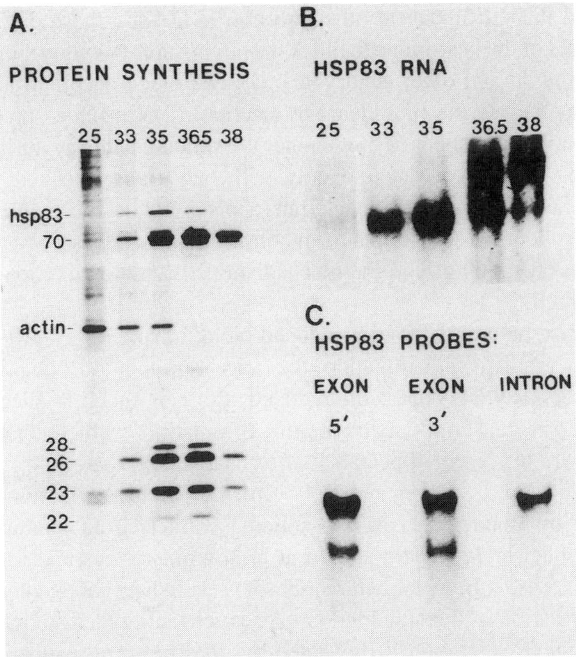

Figure 2. Hsp83 expression is blocked at high temperatures by a block in RNA splicing. **(A)** Protein synthesis at various temperatures. *Drosophila* cells grown at 25°C were shifted to the indicated temperatures. After 15 min, ^{3}H-leucine was added and incubation was continued for 45 min. **(B, C)** Northern analysis of RNAs. At the same time that proteins were harvested for panel **A**, RNAs were isolated from duplicate samples of cells. Electrophoretically separated RNAs were transferred to nitrocellulose and hybridized with ^{32}P-labeled DNAs from the entire genomic *Hsp82* clone **(B)**, or from the 5′ exon, the 3′ exon, or the intron **(C)**. See ref. 105 for further details.

gene are produced at high temperatures, but the single intervening sequence they contain is not removed.[105] The other heat-shock genes in *Drosophila* do not contain intervening sequences and their expression is not impaired.

This block in splicing is not specific to *Hsp82* transcripts. When heterologous genes are placed under the control of heat-shock promoters, so that their transcripts can be expressed at high temperatures, the splicing of these transcripts is also blocked at high temperatures. Furthermore, this general phenomenon has been demonstrated in other *Drosophila* species,[106] chickens,[107] mammals,[108] slime molds,[109] yeast,[110] and trypanosomes.[111]

Since trypanosomes are in a very ancient eukaryotic lineage, a comparison with other eukaryotes is particularly informative. Unlike the great majority of eukaryotes, trypanosomes produce their messenger RNAs by *trans*-splicing. Protein coding sequences and mRNA leader sequences are encoded at separate genetic loci and are joined posttranscriptionally. Although some splicing factors are conserved in both lineages, other are not.[112] Despite these differences, the general features of the heat-induced block in splicing are strikingly similar to those seen in *Drosophila* and other eukaryotes: (1) The block occurs at an early step in the pathway, before the first cleavage reaction. (2) Unspliced precursors that accumulate during heat shock do not reenter the splicing pathway during recovery. When splicing recovers it does so only with new transcripts that are spliced directly. (3) The splicing of heat-shock transcripts is significantly more resistant to heat than the splicing of transcripts from non-heat-shock genes.[111,113,114] Thus, the effects of heat on splicing, and the mechanism cells employ to cope with it, are highly conserved.

The precise nature of the heat-induced block in splicing is still unclear. As determined by gradient centrifugation, the ratio of protein to RNA in ribonucleoprotein particles (RNPs) is markedly reduced after heat shock.[115] Electron micrographs of nascent transcripts in *Drosophila* chromatin,[116] which contain splicing components, are altered in appearance after heat shock, suggesting a change in RNP packaging (A. L. Beyer and S.A. Amero, personal communication). As demonstrated by Bond with cell-free splicing extracts from mammalian cells, certain small nuclear RNPs are altered at high temperatures and form aberrant splicing complexes.[117] In *Drosophila* embryos, severe heat shock alters the structure or conformation of the splicing complexes, changing their antigenic properties. This antigenic alteration is prevented by mild heat pretreatments,[118] which also protect splicing.[105]

The antiquity of the phenomenon explains a long-puzzling curiosity of heat-shock-gene evolution. Most heat-inducible genes in most organisms do not contain intervening sequences. Closely related genes that are expressed at normal temperatures do have introns.[33,119] Clearly, a strong evolutionary pressure has been operating to remove introns from heat-shock genes. This makes perfect sense in light of the heat-induced block in splicing. Introns would prevent the synthesis of hsps when they are most needed, during a high-temperature heat shock. Furthermore, precursor RNAs that accumulate during heat shock enter the cytoplasm and can be translated into aberrant polypeptides.[120] Since (as discussed in section 4.1.2.) truncated hsps can have very deleterious effects on thermotolerance, intervening sequences in heat-shock genes not only would reduce hsp expression at high temperatures, but also might result in the production of abnormal, deleterious polypeptides. They have, therefore, been eliminated in most hsp genes. Why, then, has the hsp82 gene retained an intron? First, hsp82 is abundant even at normal temperatures. This might mitigate the effects of decreased production at high temperatures. Second, the single intron of hsp82 directly precedes the coding region and contains multiple stop codons. Thus the unspliced transcript would not produce an aberrant polypeptide.

Finally, the block in splicing at high temperatures might help to explain why the transcription of most other genes is so quickly repressed at high temperatures. It has recently been reported that HSF binds to previously active chromosomal loci immediately after heat shock, suggesting that HSF plays a role in repressing transcription as well as in activating it.[121] If so, the transcriptional repression of normal genes during heat shock would be an active, directed process, rather than a simple toxic effect of heat on the transcription machinery. The translational competence of unspliced RNAs and their potential for producing aberrant proteins might have provided pressure to develop such a mechanism.

3.3. Translational Regulation

3.3.1. General Remarks

Although it is not as universal as transcriptional regulation, translational regulation is an important factor in the heat-shock responses of organisms as diverse as *Tetrahymena*, mammals, and soybeans. In mammals, high temperatures cause an immediate cessation of all protein synthesis, which recovers when cells are returned to normal temperatures. With sustained heat shocks at intermediate temperatures, synthesis is only transiently repressed. Under these conditions, hsp mRNAs have a translation advantage over other messages and are found on larger polysomes.[122] The general repression of translation is apparently due to the phosphorylation of eIF-2α by a heminlike eIF-2 kinase.[123–125] When translation begins to recover, the inactivation of eIF-4F (cap-binding complex)[126] may play an important role in promoting the translation of heat-shock messages over normal cellular messages. Heat-shock messages seem to be less dependent on eIF-4F for translation (see Section 3.3.4. below).

Global translational regulation of this type is not significant in the responses of *E. coli* and yeast (see Fig. 1). In these organisms, mRNAs normally have very short half-lives. By changing the specificity of transcription, and by letting preexisting messages decay at their own rates, a rapid change in protein synthesis can be achieved without global translational suppression of preexisting messages.[127,128] Even in these organisms, however, posttranscriptional mechanisms are very important in regulating a few particular gene products. (See Section 3.1 above for a discussion of σ^{32} regulation in *E. coli* and Section 3.4. below for a discussion of hsp82 regulation in yeast.)

In few organisms is translational control as strongly exerted as in *Drosophila*. Upon a shift from 25 to 37°C, the synthesis of normal cellular proteins is completely repressed and the synthesis of hsps is vigorously and rapidly induced (see Fig. 1). This, and the fact that the response was first discovered in *Drosophila*, account for the concentration of translational studies on this organisms.

When preexisting messages are repressed in *Drosophila* cells, they are not degraded. They are retained in the cell, in a state that is fully competent for translation *in vitro* and allows them to be translated *in vivo* during recovery at normal temperatures.[127,129] Thus, physical inactivation of the messages does not account for their repression.

The speed with which repression is accomplished is remarkable, being completed within 10 min of the shift in temperature. Heat-induced gene products are not required. Repression of preexisting mRNAs occurs before hsp mRNAs appear in the cytoplasm in significant numbers.[127] Furthermore, it occurs even when the appearance of hsp mRNAs is blocked by actinomycin D[19] or when the appearance of functional proteins is blocked with amino acid analogs.[22,130]

Since high temperatures are sufficient to repress normal translation, it was suggested that a structure within normal mRNAs required for their translation might simply melt at high temperatures, preventing translation. This is clearly not the case. When preexisting mRNAs and hsp mRNAs are mixed and translated in heterologous cell-free lysates over the range of 25 to 37°C, the overall efficiency of translation varies, but at all temperatures the two classes of message behave identically.[127] Furthermore, translational specificity is retained in cell-free *Drosophila* lysates. Lysates prepared from cells growing at normal temperatures translate both preexisting mRNAs and hsp mRNAs at 30°C, while lysates prepared from heat-shocked cells translate only hsp mRNAs.[129,131,132] Finally, when heat-shocked cells are returned to normal temperatures, preexisting messages do not immediately return to normal translation. Instead, hsps continue to be the exclusive products of protein synthesis for up to several hours.[130] Thus, the translational specificity of the cell depends upon its physiological state.

3.3.3. Translation Signals on *Drosophila* hsp mRNAs

Since hsp mRNAs are translated while coexisting 25°C mRNAs are not, they must be recognizably different to the translational machinery of heat-shocked cells. A series of transformation experiments with chimeric genes demonstrates that sequences encoded in the hsp genes themselves promote their translation at high temperatures.

The critical sequences are located in the message leader or 5′ untranslated region (UTR). When the hsp70 5′ UTR is fused to heterologous coding sequences, these sequences are translated efficiently at high temperatures.[133–136] Conversely, when sequences in the 5′ leader region of either the *hsp70* gene or the *hsp22* gene[137,138] are deleted, the mRNAs behave like normal cellular messages. They are repressed at high temperatures and are translated only when cells begin to recover at normal temperatures. An example is presented in Fig. 3. In this experiment, the wild-type *hsp70* gene was first marked by a deletion in its protein-coding sequence, so that its gene product could readily be distinguished from that of the endogenous gene ("hsp"44). The protein-coding deletion did not affect translation at high temperatures; the leader deletion did.

Taken together, these experiments demonstrate that the 5′ UTRs of hsp mRNAs are both necessary and sufficient to provide translation at high temperatures. To determine if any other region of the message plays a role in translation, deletions were produced in other portions of the coding sequence and in the 3′

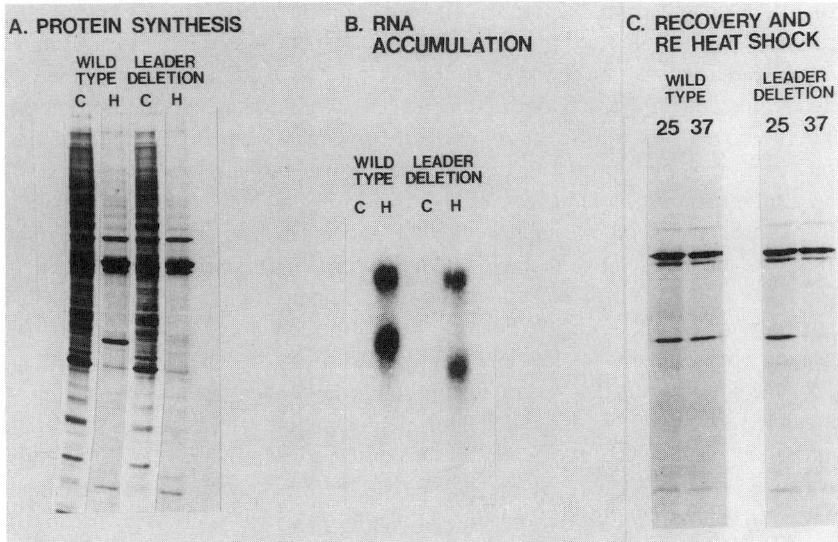

Figure 3. The deletion of sequences in the hsp70 5′ UTR prevents translation at high temperatures but not at low temperatures. *Drosophila* cells were transformed with *hsp70* genes carrying an internal deletion in the protein-coding sequence, producing a 44-kDa protein. In samples shown on the right in each panel, the transforming gene also carried a deletion of nucleotides 2–205 of the 241-nucleotide leader (5′ UTR). **(A)** Cells were incubated at 25°C (C) or at 37.5°C for 1 hr (H) and labeled with ^{3}H-leucine during the last 30 min of incubation. Deletion of the 5′ UTR, but not the coding sequence, eliminates translation at high temperatures. **(B)** RNAs extracted from duplicate aliquots of cells were electrophoretically separated, transferred to nitrocellulose, and hybridized with ^{32}P-labeled hsp70 DNA. RNAs from both constructs accumulated to the same level as the endogenous hsp70 RNA. **(C)** Cells were heat shocked at 37.5°C for 20 min and allowed to recover at 25°C for 1 hr. One aliquot of cells was then labeled with ^{3}H-leucine, the other was shifted back to 37.5°C for 10 min prior to labeling. Translation of the leader deletion is restored during recovery at normal temperatures, but is again inhibited on return to high temperatures. See ref. 137 for further details.

UTR. All of these messages were translated exactly like the endogenous hsp70 messages.[20,139] (As discussed in Section 4.1.2, the altered hsp70 proteins produced by some of these mutations did affect regulation during recovery at 25°C, but these were *trans* effects; endogenous messages were affected in the same manner.) Thus, it does not appear that any sequences other than the 5′ UTR are required for translation at high temperatures.

Unfortunately, the precise features required in the 5′ UTR are still unknown. All *Drosophila* hsp mRNA 5′ UTRs have certain properties in common: they are long (151–250 nucleotides), unusually rich in adenine residues (45–49% A), and contain two short blocks of sequence homology (one at the extreme 5′ end, the other in the middle). Length alone is not the determining feature: (1) Two mutations that delete most of the leader (from +37 to +205 of the 241-nucleotide hsp70 leader and from +27 to +242 of the 250-nucleotide hsp22 leader) are translated well at high temperatures.[138,140] (2) Hsp70 messages carrying a full-length leader with an inversion of nucleotides +2 to +205 are not translated at high temperatures.[140] (3) Two hsp70 messages carrying sequence additions at the extreme 5′ end are translated at very different rates—a message that initiates transcription 39 nucleotides upstream of the normal start site is not translated,[137] but a message with a duplication of the first 37 nucleotides of the leader is.[140]

These results might suggest that the sequence and position of the first conserved element relative to the 5′ end are the critical features. Deleting the first 23 nucleotides of the hsp70 messages, however, impairs high-temperature translation only a fewfold. Further, deleting only the second conserved element or deleting both elements is also only mildly detrimental.[137,140–142]

One possibility is that several features of the leader contribute to translation at high temperatures. Since large deletions are compatible with heat-shock translation as long as they do not extend into the conserved sequence at the 5′ end, this sequence is likely to play an important role.[138,140] Another important element may be structural. Unlike most normal cellular mRNAs, hsp mRNAs are predicted to have virtually no secondary structure. Arguing in favor of the importance of this feature, hsp70 mRNAs are efficiently translated at K^+ concentrations that inhibit the translation of most normal cellular RNAs.[143] (The inhibitory effects of K^+ in such experiments are generally ascribed to the stabilization of secondary structures in the message.) When a series of random A-rich sequences, also predicted to have little secondary structure, was used to replace the natural hsp70 leader, however, none restored high-efficiency heat-shock translation.[144] Again these results suggest that it is a combination of structural and sequence features that determines translatability.

3.3.4. Heat-Induced Changes in the *Drosophila* Translational Machinery

As drastic as the repression of normal cellular mRNAs is in heat-shocked *Drosophila* cells, it does not betoken any general damage to the translation machinery. At 37°C, ribosomes initiate on hsp70 mRNAs at a rate of 9–14 initiations per minute per message, comparable to the highest known rates of initiation in eukaryotes.[18] Nevertheless, the mechanism that inhibits normal translation must have very broad specificity. Normal cellular messages are very heterogeneous, yet all are repressed at the same rate and to nearly the same extent.[19,145] Moreover, they all return to translation during recovery with nearly the same kinetics (Fig. 4).

The primary mechanism for their repression must operate at the level of initiation. Within the first few minutes of heat shock, polysomes virtually disappear and monosomes increase in proportion.[145] The disappearance of polysomes cannot be due to premature termination, because the proteins that are synthesized during this period are full length (S. Lindquist, unpublished). A block in initia-

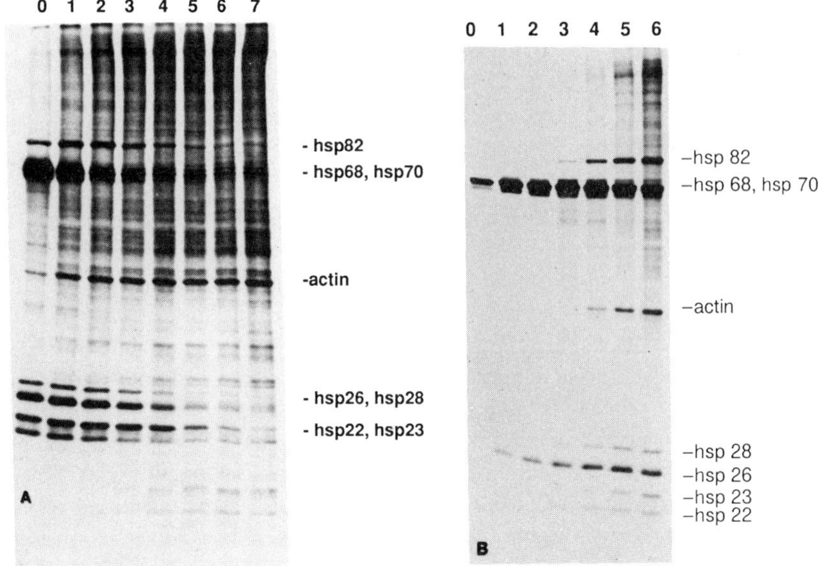

Figure 4. The rate at which normal protein synthesis is restored during recovery depends upon the severity of the preceding heat shock. *Drosophila* cells grown at 25°C were shifted to **(A)** 37°C or to **(B)** 39°C for 30 min and then returned to 25°C. Aliquots were pulse-labeled with ^{3}H-leucine at 30-min intervals beginning immediately after return to 25°C. See ref. 22 for similar experiments.

tion is also consistent with the observation that the critical sequences for heat-shock translation are upstream of the initiation codon.

It appears, however, that additional translational mechanisms are operating. Certain normal cellular mRNAs are associated with polysomes of roughly the same size at 25 and 37°C,[146] suggesting that the translation of some messages is blocked in elongation. In one *Drosophila* cell line (Kc cells), maximal hsp synthesis occurs under conditions in which normal protein synthesis is only partially impaired.[147] Here, the ability of hsp mRNAs to compete for generally limiting translation components seems to be a key factor in their translation. Considering the enormity of the change in protein synthesis that occurs in *Drosophila* with heat shock, it is not surprising that additional translational mechanisms reinforce the primary mode of regulation at the level of initiation.

How is the block in initiation on normal messages accomplished? *Drosophila* cell-free translation lysates have provided major insights. As mentioned above, lysates from heat-shocked cells discriminate between hsp mRNAs and normal mRNAs. When these lysates are mixed with lysates from control cells, translational activities are simply additive.[148] Thus, the repression of normal translation is not due to a soluble, dominant, inhibitor. Rather, as supplementation experiments demonstrate, heat shock inactivates a factor that is required for the translation of normal messages. One laboratory reported that this factor, which restores normal translation to heat-shock lysates, is found in the crude ribosomal pellets of control cells.[148] Another reported the activity in supernatant fractions. Since a higher salt concentration was employed in the latter experiments, the discrepancy can be reconciled by assuming that the fractionation properties of the factor are affected by salt.[132]

Several recent experiments strongly suggest that this critical factor is cap-binding complex, an oligomeric ATP-dependent RNA unwinding activity, of which cap-binding protein is a subunit. First, when cap analogs are added to lysates of heat-shocked *Drosophila* embryos, the translation of hsp mRNAs continues, while the residual translation of normal mRNAs is further repressed.[149] Second, antibodies raised against the 35-kDa cap-binding protein of *Drosophila* block translation of the bulk of normal mRNAs, but not of hsp70 mRNA, in lysates of embryos recovering from heat shock.[150,143] Third, purified cap-binding complex partially rescues the translation of normal mRNAs in heat-shock lysates. Finally, although heat-shocked and control embryos contain equal quantities of cap-binding protein, heat-shocked cells contain much less of the complex. Apparently, heat shock disrupts the association between cap-binding protein and other proteins in the complex.[143]

These experiments complement the results of the mutational analyses described above, suggesting a model for regulation: Preexisting, normal cellular mRNAs require cap-binding complex for efficient initiation, in part to unwind secondary structure in the leader. Heat shock inactivates this factor, causing a precipitous decline in the translation of normal messages. Additional translation mechanisms reinforce this repression. Hsp mRNAs, with an absence of secondary structure in their leaders and other features to attract ribosomes, do not require cap-binding complex.

Support for this notion has come from an unexpected quarter. In mammalian cells the endoplasmic reticulum version of hsp70, GRP78 (also known as BiP or immunoglobulin heavy-chain binding protein), is resistant to the repression of normal translation in poliovirus-infected cells. Since poliovirus mRNAs are known to initiate by a cap-independent mechanism, this suggested that GRP78 might be translated by the same mechanism. A test for such elements is to clone them into the spacer region of a dicistronic message, where they dramatically increase translation of the second cistron. The GRP78 leader sequence works very effectively in such assays.[151] Thus, a gene that is induced by heat in at least some cell types is now known to be translated by the mechanism proposed for *Drosophila* hsp mRNA cells. It is curious, however, that other mammalian hsp mRNAs may not utilize the same mechanism, or at least do not utilize it as effectively. Most mammalian hsp mRNAs are only partially resistant to the general inhibition of translation by poliovirus.[152]

3.4. RNA Turnover

Changes in the stability of *Drosophila* hsp70 mRNA are as important in regulating expression as changes in transcription and translation. Studies examining the process of recovery after heat shock provided the first clue to the importance of mRNA stability. When heat-shocked cells are returned to normal temperatures, normal protein synthesis is restored and hsp synthesis is repressed (Fig. 4). The rate of repression varies for each hsp and depends upon the severity of the preceding heat shock. In all cases, hsp70 is the first protein to be repressed. Furthermore, in all cases examined, under many different conditions, the repression of hsp70 synthesis is closely paralleled by the degradation of its message.[22,130,136] The behavior of the hsp70 message during recovery contrasts with its behavior during maintenance at high temperatures. Here, the translation of hsp70 continues at a high rate and the message is completely stable for at least 12 hr.

To determine if the hsp70 message would be unstable if expressed at normal temperatures, hsp70 sequences, including the 5′ UTR, most of the coding region, and the 3′ UTR, were expressed from a heterologous promoter. At 25°C the half-life of this message was less than 15 min.[136] When the cells were shifted to 36.5°C, however, the half-life of this messages was greater than 6 hr. Thus, a mechanism for degrading hsp70 mRNA preexists in *Drosophila* cells at normal temperatures and is disrupted by heat shock.

Sequences which target hsp70 mRNA for degradation during growth at normal temperatures and during recovery from heat shock are found in the 3′ UTR. When the hsp70 3′ UTR is used to replace the 3′ UTR of heterologous genes that normally produce very stable messages, their pattern of expression changes. The chimeric messages are unstable at normal temperatures and are stabilized by heat shock.[153] The mechanism of stabilization appears to be quite general. During extended heat shocks the full spectrum of normal cellular RNAs is retained for at least 12 hr (S. Lindquist, unpublished).

These findings explain the often contradictory results obtained in expressing foreign genes from heat-shock promoters. In some cases high levels of constitutive expression are observed, in others not. The explanation is simple. The heat-shock promoter is rather leaky. If the chimeric transcripts carry their own degradation signal, as does the hsp70 transcript, little protein will be produced until a shift to high temperatures stabilizes the message. If the transcript does not carry such a signal, and if the protein is stable, constitutive expression can be nearly as high as that obtained with heat shock.[153] To circumvent such problems, recent expression vectors have been designed to carry not only the hsp70 promoter, but also the message leader and 3′ UTF.[20,153]

The precise nature of the 3′ UTR destabilizing sequence has not yet been determined, but its sequence alone is highly suggestive. It is very rich in adenine and uracil, and contains sequence elements resembling those that target unstable mammalian messages (such as the c-*myc*, c-*fos*, and various lymphokine mRNAs) for degradation.[154] It is notable that both the c-*myc* and c-*fos* messages are stabilized by heat shock in mammalian cells[155] and that c-*myc* messages are stabilized by heat shock in *Drosophila* cells (R. Petersen and S. Lindquist, unpublished).

It appears, then, that the mechanism that degrades hsp70 mRNAs in *Drosophila* is the same highly conserved mechanism that degrades unstable messages in other systems. This mechanism keeps hsp70 expression at an extremely low level during growth at normal temperatures. (See Fig. 5; in order to detect the very small quantity of hsp70 mRNA in the low-temperature samples, the high-temperature lanes had to be grossly overexposed.) It is the inactivation of this mechanism, combined with the burst of new transcription, that allows hsp70 mRNAs to accumulate so rapidly and to such high levels at high temperatures. During recovery, hsp70 synthesis is repressed both by shutting off transcription and by reactivating this preexisting mRNA degradation mechanism.

Is this type of regulation of general importance in the heat-shock response? In *Drosophila*, selective RNA turnover plays a role in regulating the expression of the small hsps and of hsp82, although its effects on these messages are not as great as on the hsp70 message. In mammalian cells, the hsp70 message,[156] and perhaps other hsp messages, are unstable at normal temperatures and stabilized by heat shock. Finally, a similar selective degradation mechanism appears to operate

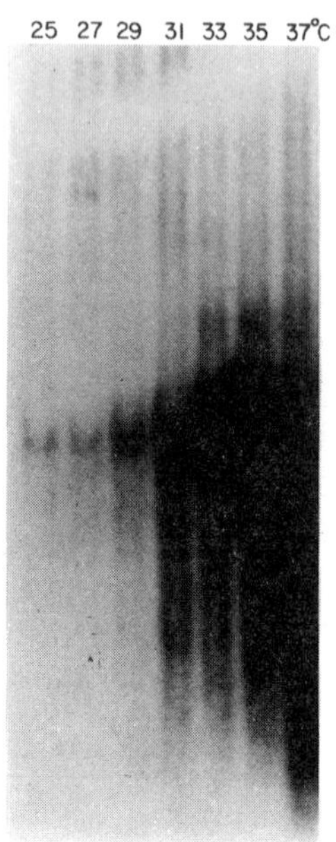

Figure 5. The induction of hsp70 message in *Drosophila* during heat shock. *Drosophila* cells grown at 23°C were shifted to the indicated temperatures for 1 hr and RNAs were extracted. Ethidium bromide staining of electrophoretically separated RNAs demonstrated that each well contained an equal quantity of total cellular RNA. The RNAs were transferred to nitrocellulose and hybridized with ^{32}P-labeled hsp70 DNA, and hybrids were visualized by autoradiography.

on the hsp82 message in yeast. Hsp82 is constitutively expressed at a much lower level than hsc82, When the *HSP82* promoter and the *HSC82* promoter are employed to regulate a β-galactosidase gene, however, both constructs are expressed at similar levels (D. Gross, personal communication). When the 3′ UTR of *HSC82* is substituted for the 3′ UTR of *HSP82*, constitutive expression of hsp82 increases severalfold (J. Vogel and S. Lindquist, unpublished), suggesting the 3′ UTR of hsp82 targets the RNA for rapid degradation at normal temperatures. Thus the rapid turnover of hsp mRNAs at low temperatures and the stabilization of these messages at high temperatures may be a very general feature of the heat-shock response in eukaryotes.

3.5. Why hsp70 Expression Is So Tightly Regulated in *Drosophila*

The number of regulatory mechanisms that are employed to regulate hsp70 expression in *Drosophila* is truly extraordinary. The net effect of these mechanisms is both to maximize the expression of hsp70 at high temperatures and to minimize its expression at normal temperatures. This suggests that hsp70 may be toxic at normal temperatures. To test this possibility, hsp70 coding sequences were placed under the regulation of heterologous promoters and transformed into tissue culture cells and into whole flies.[156a] Expression of hsp70 did not reduce viability, but it did reduce the rate of growth. Thus, while expressing hsp70 at normal temperatures may be beneficial for thermotolerance, it is detrimental to growth. Curiously, with continued maintenance of such cultures, growth resumed, despite continued expression of hsp70. Immunological staining with hsp70 antibodies revealed that during the period when growth was inhibited, the protein was diffusely distributed throughout the cell. When growth resumed, however, hsp70 had coalesced into a small number of discrete points of very high concentration. Similar "granule-like" structures were observed in both wild-type flies and in culture cells after long periods of recovery from heat shock. The formation of such structures is therefore a normal part of hsp70 metabolism. The change in the distribution of constitutively expressed hsp70 from diffuse to punctate as cells resume growth may represent yet another mechanism for controlling hsp70 activity, in this case by sequestration.

4. AUTOREGULATION OF THE RESPONSE

4.1. *Drosophila*

4.1.1. Hsps Are Required to Restore Normal Regulation

All of the changes in gene expression that are elicited by heat shock in *Drosophila* cells are fully reversible. When cells are returned to normal temperatures, normal patterns of transcription, RNA processing, message turnover, and translation are restored. The kinetics of the recovery process vary enormously, depending upon the severity of the preceding heat treatment. For any one type of treatment, however, it is remarkably reproducible.[22] Recovery is clearly divided into two phases. Immediately after return to normal temperatures, cells continue to produce hsps almost exclusively. The length of this phase is proportional to the severity of the heat treatment. For example, when cells are heat shocked at 37°C for 30 min, the repression of hsp synthesis and the recovery of normal synthesis begin within 30 min. After a 39°C heat shock for 30 min, recovery is initiated only after 5 hr at 25°C (Fig. 4). Another striking finding arose from densitometric quantification of such experiments: the half-way point for hsp70 repression always coincides with the half-way point for the restoration of normal protein synthesis.[22,145] This suggests that the two processes are mechanistically linked.

Several lines of evidence strongly suggest that it is the hsps themselves that coordinate the recovery process, by restoring regulatory mechanisms to their normal mode. The first evidence of hsp involvement came from experiments in which the accumulation of hsps was selectively reduced with inhibitors. Such experiments are possible in *Drosophila* cells because the regulatory mechanisms are so tight that virtually no proteins other than hsps are being produced at high temperatures. In different experiments, different inhibitors were employed to analyze transcription, message accumulation, and translation.

For example, in the experiment of Fig. 6, cells were heat shocked for 60 min at 36.5°C and returned to 25°C for recovery. In the absence of inhibitors, transcription of hsp70 was repressed within 1 hr. At the same time, the degradation of previously accumulated hsp70 mRNAs was initiated. When cycloheximide was added immediately after the 60-min heat treatment, it had only a small effect on this process. When the drug was added just before the heat shock, however, blocking the initial induction of hsps, hsp70 transcription continued for at least 6 hr after cells were returned to 25°C and hsp70 mRNAs continued to accumulate. Note that blocking the appearance of hsps also prevented the resumption of normal transcription. Since the heat-shock pattern of transcription continues at a vigorous rate for so many hours, the failure to return to normal transcription is not due to a general, toxic effect of heat on the transcription machinery. Rather, it appears that regulatory mechanisms recognize the absence of hsps and sustain the attempt to produce them.[22,157]

In a similar series of experiments, cells were treated with actinomycin D at different times during a 60-min heat shock and were then returned to 25°C for recovery (Fig. 7). All cells experienced the same heat shock, but the quantity of hsp mRNA they produced and, consequently, the rate at which hsps were synthesized varied. Cells that produced hsps at a slower rate produced them for a longer period before hsp repression was initiated and normal protein synthesis was reactivated. Inspection of the raw data is suggestive (Fig. 7a), but quantitative analysis is far more compelling (Fig. 7b). Since hsp70 is stable under these

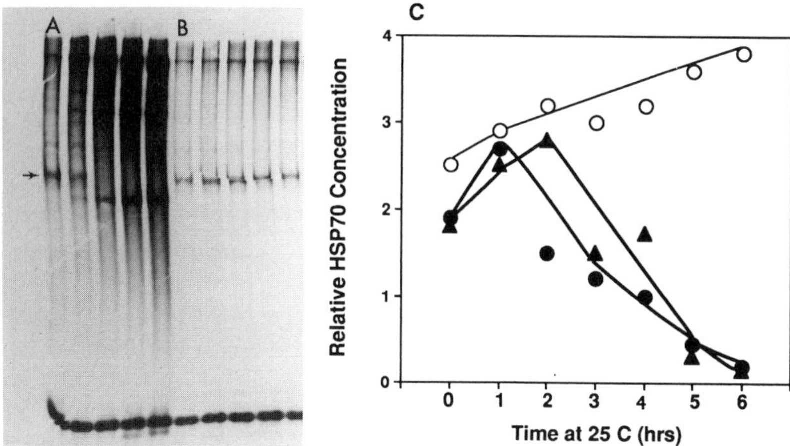

Figure 6. Blocking the synthesis of hsps extends the synthesis of hsp70 mRNA and maintains the stability of the hsp70 message. *Drosophila* cells grown at 25°C were shifted to 36.5°C for 60 min and were returned to 25°C for recovery. **(A, B)** Aliquots of the culture were pulse-labeled with ^{3}H-uridine for 30 min during the last 30 min to heat shock and at hourly intervals thereafter. **(A)** No cycloheximide; **(B)** cycloheximide added immediately prior to the shift to 36.5°C. RNAs were extracted, electrophoretically separated, and visualized by fluorography. The arrow indicates the position of hsp70 mRNA. **(C)** RNAs were extracted from unlabeled cells at the same time as for panels **A** and **B**. Electrophoretically separated RNAs were transferred to nitrocellulose and reacted with ^{32}P-labeled hsp70 DNA. Autoradiograms were quantified on an LKB laser beam scanning densitometer. (●) No additions, (○) cycloheximide added immediately prior to the shift to 36.5°C, (△) cycloheximide added immediately after the return to 25°C. See ref. 22 for similar experiments.

conditions, its rate of synthesis multiplied by the period of synthesis (i.e., the area under the curve) is an accurate reflection of hsp70 accumulation. Focusing on the point at which repression of hsp70 is initiated, it is striking that this always coincides with the accumulation of a specific quantity of hsp70. Similar experiments were performed with heat shocks of different severity. In every case, the quantity of hsp70 produced prior to the initiation of repression was precisely regulated and proportional to the severity of the preceding heat shock.[22,145]

Another set of experiments examined the effect of hsps on RNA splicing. When *Drosophila* cells are given a short, mild heat shock, which allows them to accumulate hsps, and are then exposed to a more severe temperature, splicing is protected from disruption. The degree of protection, measured as the percentage of new transcripts that are properly spliced, varies in a dose-dependent fashion with the duration of the pretreatment and with the accumulation of hsps.[105] Furthermore, the protective effect of the pretreatment is eliminated when cycloheximide is added before hsps are produced. The drug has no effect when added after hsps are produced.

These experiments, and others of the same ilk, strongly suggest that hsps are

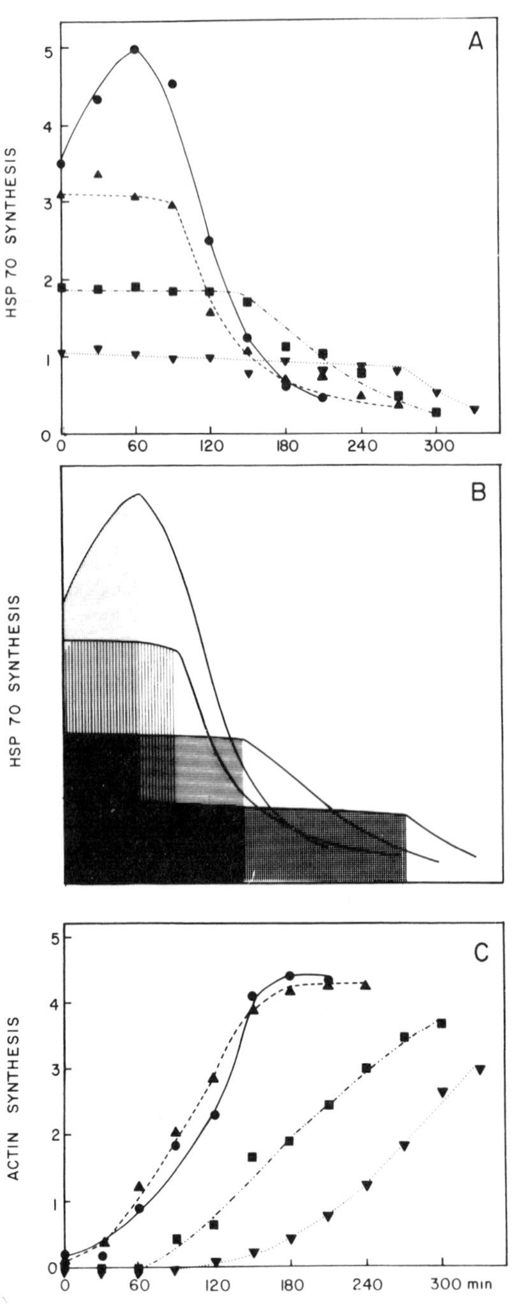

Figure 7. Limiting the concentration of hsp mRNAs extends the synthesis of HSPs and delays the recovery of normal protein synthesis. Cells grown at 25°C were shifted to 36.5°C with no additions (●), or with the addition of actinomycin D, to block further heat-shock transcription, at 8 (▼), 15(■), or 58 (▲) min. After 1 hr all cells were returned to 25°C for recovery. Protein synthesis was monitored by the incorporation of ^{3}H-leucine. Fluorographs of electrophoretically separated proteins were quantified by densitometry. **(A, B)** The synthesis of hsp70. Since the hsp70 protein is stable under these conditions, the shaded area under the curve represents the quantity of hsp70 accumulated at the point when the repression of hsp70 synthesis is initiated. **(C)** Synthesis of actin, an abundant normal cellular protein. See ref. 22 for details.

required to repress their own synthesis and to restore normal patterns of gene expression. They also indicate that hsps exert their effects on regulation at all levels, transcription, RNA processing, translation, and selective message turnover. Hsp70 seems to be particularly important in this process. It, for example, is the only hsp whose accumulation always bears a simple quantitative relationship to the initiation of recovery. (As discussed in Section 4.2 below, genetic evidence in yeast and *E. coli* strongly supports the importance of this particular hsp in autoregulation.)

The intracellular distribution of hsp70 is consistent with this hypothesis and explains what had initially been a very puzzling observation in experiments of the type just described. Once hsps have accumulated in sufficient quantities, further transcription of hsp genes is repressed, whether cells are maintained at high temperatures or returned to 25°C. The recovery of normal translation and the repression of hsp synthesis, however, requires not only the accumulation of a specific quantity of hsps, but also a return to 25°C.[22,145] When antibodies specific for hsp70 were developed it was determined that hsp70 primarily concentrates in nuclei during heat shock. It retains this localization as long as cells are maintained at high temperatures. When cells are returned to normal temperatures hsp70 gradually relocalizes to the cytoplasm, with kinetics that roughly correlate with the restoration of normal protein synthesis.[39] Thus, hsp70 is not in a position to affect cytoplasmic processes until it relocalizes to that compartment on return to 25°C.

4.1.2. The Role of hsp70

To test the role of hsp70 in the regulation of the heat-shock response more rigorously and to examine its role in providing tolerance against more extreme temperatures, tissue culture cells were transformed with a variety of constructs designed to alter hsp70 expression. The unusually tight controls on hsp70 expression in *Drosophila* provide an advantage in such experiments since aberrant polypeptides, which might have pleiotropic effects on metabolism, would not be expressed during normal growth. Experiments examining regulation were performed under much milder conditions than those examining thermotolerance, conditions that have no detectable toxic effects on any of the cell lines. The results[139] can be summarized as follows.

1. Cells that underexpress hsp70 were created by transforming *Drosophila* cells with constructs designed to produce antisense RNA. These transformants produce hsp70 at about one-third the rate of wild-type cells or cells transformed with the vector alone. Correspondingly, when cells are returned to normal temperatures after a short, mild heat shock, hsp70 is produced three times longer than in wild-type cells, other hsps are overexpressed, and the recovery of normal protein synthesis is delayed.[139,158] In thermotolerance experiments, which were performed at much higher temperatures, a slight, but reproducible reduction in survival is observed in the antisense line.

2. Cells that overexpress hsp70 were created by transforming cells with extra copies of the wild-type gene. Overexpression has a minimal effect on the regulation of the response, but a strong positive effect on survival at extreme temperatures.

3. Cells that express hsp70 in the absence of heat shock were produced by transforming cells with a chimeric gene in which *hsp70* coding sequences were placed under the control of the *metallothionein* promoter. Strong, selective induction of hsp70 is achieved in these cells at copper concentrations that have no effect on the induction of other hsps. As with the overexpressing line, preinduction of hsp70 has little effect on the regulation of the response. It does, however, have a strong positive effect on survival at extreme temperatures.

4. In an attempt to produce dominant negative mutations of hsp70, cells were transformed with deletion and frameshift mutations (Fig. 8 and 9). One mutation, *402*, deletes nearly the entire amino terminus of hsp70. It has little effect on either regulation (Fig. 8) or thermotolerance (Fig. 9). A second mutation, *110*, that deletes a smaller portion of the amino terminus has a strong negative effect on regulation, greatly extending the period of time required to restore normal gene expression (Fig. 8). This mutation, however, has little effect on the ability of cells to survive extreme temperatures (Fig. 9). A third mutation, *#300fs*, shifted the reading frame of the protein at codon 337, causing translation to terminate 53 amino acids downstream and eliminating the carboxy-terminal domain. This mutation has only a small effect on regulation, slightly delaying the recovery of normal protein synthesis.[139] It has a profound effect, however, on survival at extreme temperatures and virtually eliminated induced thermotolerance (Fig. 9).

As discussed above, hsp70 is believed to interact with unfolded proteins, preventing their aggregation and promoting renaturation. One explanation for the different effects of these mutations is that the 402 protein does not interfere with hsp70 functions, because it does not interact with other proteins, while the 401 and 300fs mutations do interfere with hsp70 functions, but do so in mechanistically different ways.

To investigate these possibilities, the behavior of the mutant proteins was analyzed by gel filtration chromatography of crude lysates.[139] In such analyses the wild-type protein elutes in two broad peaks, the first corresponding to complexes of greater than 450 kDa, the second to complexes in the range of 60–150 kDa. The 402 protein chromatographs as a monomer, indicating an absence of stable associations with any other protein. The 110 protein chromatographs exactly like the endogenous hsp70 protein, suggesting that it either binds to the same targets as hsp70 or binds to hsp70 itself. The chromatographic behavior of the 300fs protein could not be analyzed, because it pellets when lysates are subjected to clearing spins. When cells containing this protein are disrupted under very gentle conditions the protein fractionates with nuclei, perhaps because, like hsp70, it concentrates in nuclei[39] or perhaps because it simply forms very large aggregates. Unlike the endogenous hsp70 protein, however, the mutant protein cannot be released from these pellets with ATP.

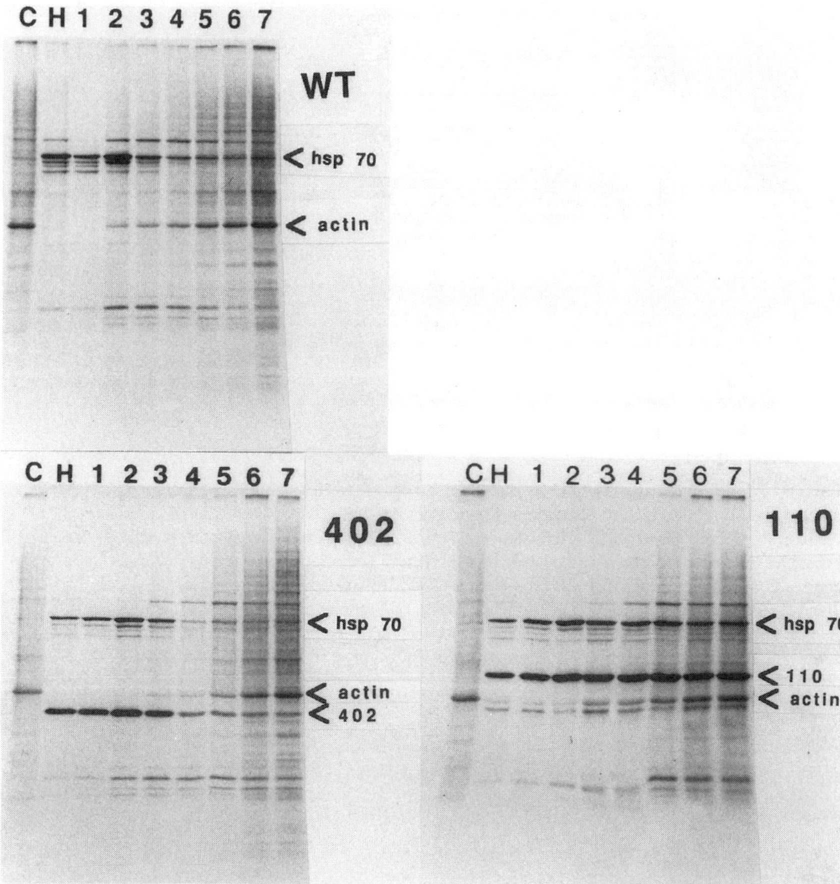

Figure 8. A mutation that deletes codons 6–337 of the hsp70 gene has a dominant negative effect on the regulation of the response. S2 cells transformed with mutant forms of *Hsp70* (*402* or *110*) or with the drug-resistance plasmid alone (WT) were maintained at 25°C (C) or heat-shocked at 36.5°C for 45 min (H) and then returned to 25°C for recovery. Cells were labeled with ^{3}H-leucine for 20 min during the last 20 min of heat shock and at hourly intervals during recovery. Total cellular proteins were prepared, electrophoretically separated on 10% sodium dodecyl sulfate (SDS)–polyacrylamide gels, and visualized by fluorography. Cells transformed with the *300fs* mutation, together with the control line, were analyzed in the same manner in a separate experiment. Similar results were obtained in four independent experiments. See ref.139 for details.

The results of these experiments support the hypothesis that hsp70 plays a major role regulating the heat-shock response as well as in protecting cells from the toxic effects of extreme heat. Surprisingly, they also indicate that these two functions can be separated. This alleviates a potential problem in interpretation. Regulation experiments were performed under very mild conditions that appeared to have no detectable toxic effects on the cells. Nevertheless, if an exact correspon-

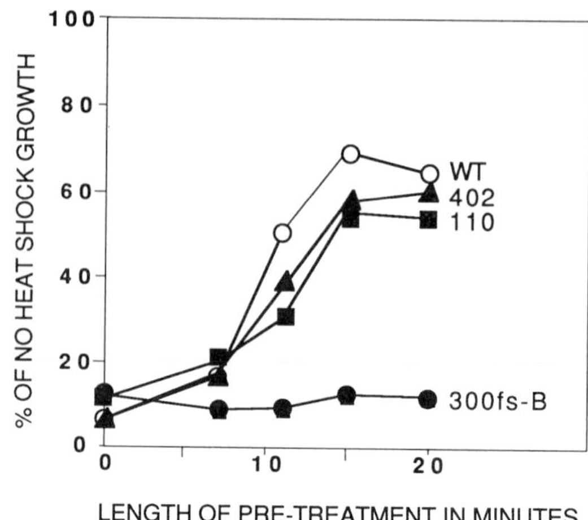

Figure 9. Induced thermotolerance in cells expressing mutant forms of Hsp70. The ability of cells to resume growth after a severe heat shock provides a qualitative measure of thermotolerance. S2 cells were transformed with the drug-resistance plasmid alone (WT) or cotransformed with the drug-resistance plasmid and one of the *Hsp70* mutations *402* (deletion of codons 6–337), *110* (deletion of codons 114–337), or *300fs* (frameshift at codon 337). Cells were diluted to a low density in microtiter dishes, pretreated at 35°C for 0, 7, 11, 15, or 20 min, and then exposed to a severe heat shock at 41.5°C for 45 min. Five days later, the relative number of cells in each well of the microtiter dish was determined by neutral red uptake. Each point is the average of six wells, normalized to the values obtained for cells not exposed to heat shock. See ref. 139 for details.

dence had existed between those hsp70 perturbations that affect regulation and those that affect survival, the regulatory affects might simply have been dismissed as resulting from reduced protection against heat.

The simplest hypothesis to explain the dual role of hsp70 in thermotolerance and regulation is that these two processes are united through the common theme of thermal protein denaturation: Changes in gene expression are induced by the thermal disruption of protein structures that maintain normal regulation (see Fig. 10). Killing, at yet higher temperatures is due to the denaturation of other, more vital structures. Thus, if the role of hsp70 is to renature denatured proteins,[47,50] it would naturally affect both thermotolerance and regulation. It would, of course, make good sense for the proteins that govern regulation to have a greater sensitivity to heat than those that are critical for viability. Thus, under normal biological conditions, as the temperature rises, regulatory denaturations would occur first, facilitating the rapid induction of hsps. By the time cells reached more extreme temperatures, hsps would be available to protect vital structures and provide protection.

That pertubations in hsp70 synthesis can affect thermotolerance and regulation so differently may seem to contradict this hypothesis. Recent analyses of hsp functions in other systems (see Section 2), however, suggest ways in which these differences can be reconciled within the framework of the above hypothesis. Two factors to be considered are (1) possible mechanistic differences in the manner in which hsp70 interacts with target proteins in regulation and in thermotolerance and (2) the roles of other hsps.

In considering mechanistic differences in what might be required of hsp70 in thermotolerance and regulation, an interesting precedent is provided by studies in *E. coli* examining the effects of purified DnaK on the heat inactivation of RNA polymerase *in vitro*.[57] When DnaK is added after the increase in temperature, it can disaggregate and reactivate the heat-denatured polymerase by a mechanism that requires ATP. The mutant protein encoded by the DnaK756 allele is not functional in this assay. When DnaK is added prior to the heat treatment, however, it prevents inactivation and aggregation of polymerase by a mechanism that does not require ATP. The DnaK756 protein is fully capable of protection in this assay.

If the thermal disruption and denaturation of regulatory complexes establishes heat-shock regulation, reestablishing normal regulation would require the renaturation and reassembly of these complexes. Thermotolerance, on the other hand, might be accomplished by preventing vital structures from becoming fully denatured or aggregated in the first place. The 110 protein interferes with regulation but not thermotolerance. Perhaps, by analogy with the effects of the DnaK756 mutation on RNA polymerase reactivation, it might function in providing protection (thermotolerance), but not in reactivating previously disrupted regulatory factors. Indeed, since the 110 protein is missing most of the ATP binding domain, it would not be expected to function in reactivation. If, as the chromatographic data suggest, 110 binds to the wild-type protein or to the same substrates as the wild-type protein, it would have a dominant negative effect on regulation because it would interfere with the reactivation of regulatory factors.

To explain the *300fs* mutation, one may speculate that hsp70 has a higher affinity for vital targets than for regulatory factors. Indeed, this would conveniently ensure that sufficient protein had accumulated to take care of vital functions before further synthesis was repressed. If the *300fs* protein binds to these targets but does not release them, it would have a specific dominant-negative effect on thermotolerance.

To provide an explanation for the results of underexpressing and overexpressing hsp70, we must consider the relationship between hsp70 and other hsps. Certain functions of the hsp70 protein of *E. coli*, DnaK, require other hsps, GrpE and DnaJ, to tag the target protein for interaction with DnaK and to stimulate ATP hydrolysis. These include the replication of bacteriophage lambda and plasmid P1 DNA[84,159–161] (see Section 2.2.6.), the reactivation of heat-denatured bacteriophage lambda repressor[162] and the regulation of σ^{32} accumulation and activity[163] (see Section 4.2). Other DnaK functions, such as protecting RNA polymerase from denaturation, do not require GrpE and DnaJ.[57] Another set of hsps, the GroEL and GroES proteins, may form an independent pathway for protein

25°C

HSF Inactivated

HSF Active

Transcription

Splicing

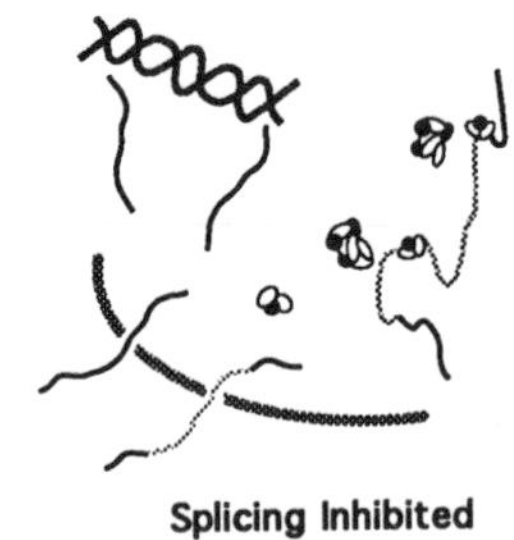

Splicing Inhibited

Intron Splicing

Normal mRNAs Active

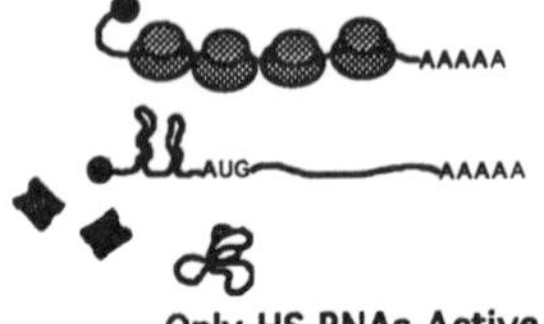

Only HS RNAs Active

Translation

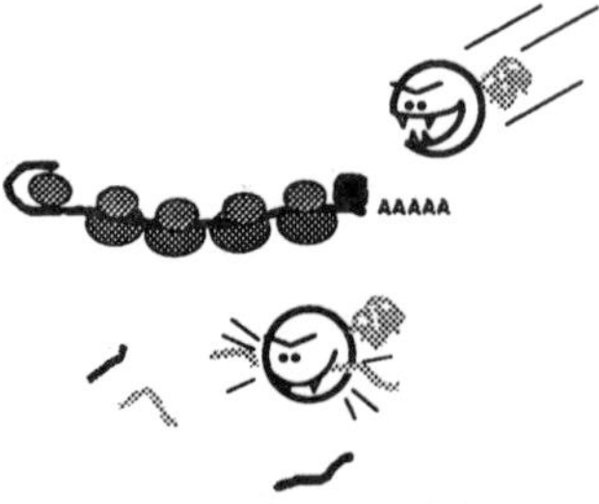

hsp70 message unstable

hsp70 message stable

mRNA Degradation

assembly with a limited capacity for renaturation. The GroEL–GroES complex can bind to denatured proteins in solution, preventing their aggregation and promoting their renaturation, but it cannot reactivate previously denatured and aggregated proteins.[62]

Cells transformed with *hsp70* antisense genes overexpress other hsps. In these cells, regulation is strongly affected, but thermotolerance is not. Perhaps other proteins can compensate for the underexpression hsp70 in protecting vital targets from irreversible denaturation and aggregation (thermotolerance). But they either do not function in the renaturation and reassembly of previously disrupted regulatory factors or they require the participation of hsp70, in stoichiometric quantities, for this function. On the other hand, the overexpression of hsp70 increases thermotolerance but does not hasten the recovery of normal regulation. This is readily explained if hsp70 cannot renature regulatory factors on its own, but requires the accumulation of another limiting factor (a DnaJ or GrpE analog).

To recap the general hypothesis: heat shock disrupts a variety of regulatory factors that are required for normal gene regulation. The heat-shock genes have evolved a variety of features that allow them to circumvent the requirement for these factors and to exploit the lack of competition to maximize their own expression (Fig. 11). As hsps accumulate and temperatures continue to rise, hsps bind to vital proteins as they are being denatured, in a form that prevents them from aggregating and promotes rapid renaturation when cells are returned to normal temperatures. Since the release of hsp70 from its target proteins requires ATP, a decline in ATP at high temperatures may stabilize these associations. As cells begin to recover from heat shock, hsp70 is released from vital targets and begins to promote the reassembly of previously disrupted regulatory factors.

4.2. Autoregulation in Other Organisms

Given the conservation of hsp functions, autoregulation by hsp70 is likely to be a universal feature of the response. In mammalian cells genetic evidence is

←

Figure 10. The effects of heat on gene expression. The cartoons depict the effects of heat on various factors involved in regulating gene expression in *Drosophila*. At 25°C, HSF is either complexed with another protein that represses it or is folded into a self-repressing conformation. At high temperatures the repressed HSF structure is disrupted, allowing HSF to trimerize and promote the transcription of heat-shock genes. At 25°C, transcripts containing intervening sequences are spliced by spliceosomes. High temperatures disrupt one or more of the spliceosome components, blocking splicing. Since most hsp mRNAs do not require splicing, their expression is not inhibited. Transcripts that are unspliced are able to enter the cytoplasm and can be translated into protein. At 25°C, normal cellular mRNAs require cap-binding complex for translation. This complex is disrupted by high temperatures, blocking normal translation. The newly induced hsp mRNAs, by virtue of special features in their leader sequences, do not require cap-binding complex. At 25°C, a nuclease degrades messages that carry special signal sequences in their 3′ UTRs. In the figure it is presumed that the nuclease interacts with a protein that is bound to the special signal sequence. Either the nuclease itself, or the protein to which it binds, is disrupted by high temperatures, blocking the degradation of unstable messages.

Figure 11. Summary: Features of HSP70 gene regulation. □, HSE, heat-shock element, binding site for HSF (heat-shock transcription factor), consensus sequence _GAA__TTC__GGA_. △, Start site for transcription initiation. ▬, Message leader (5′UTR), necessary and sufficient for translation at high temperatures—241 nucleotides long, 49% adenine, contains two sequences elements shared with the leaders of other hsp mRNAs (one at the extreme 5′ end, the other in the middle) which may influence translation, polymerase pausing at normal temperatures, and RNA turnover. ▭, Coding sequences do not appear to affect the behavior of the message in *cis*. The protein, however, plays a major role in regulating the heat-shock response at both the transcriptional and post-transcriptional levels. Thus, mutations here can affect regulation in *trans*. ◇◇◇, Message tail (3′UTR), targets messages for rapid degradation during growth at normal temperatures and during recovery from heat shock. Rich in adenine and uracil (75%) and contains consensus elements known to target other messages for rapid degradation, AUUUA. ●, Transcription termination site. The hsp70 gene is more resistant than other genes to run-on transcription at high temperatures. Note that no introns are present in the gene.

lacking, but several regulatory phenomena are sufficiently similar to *Drosophila* phenomena to suggest a common mechanism. For example, the expression of hsp70, but not of other hsps, is strictly regulated according to the severity of the heat shock. Furthermore, pre-heat treatments, which induce the synthesis of hsps, reduce the expression of hsps in a subsequent heat shock.[24]

By far the best evidence for autoregulation of the heat shock response by hsp70 in other organisms comes from experiments with *E. coli* and yeast (reviewed in ref. 25). In *E. coli*, mutations in the *DnaK* gene, the *hsp70* homologue of bacteria that is both constitutively expressed and induced by heat, have dramatic effects on regulation. They increase the constitutive expression of other hsps, prolong the expression of hsps at high temperature, and prevent the decline in hsp synthesis that normally accompanies a shift to low temperature.[23,103] Mutations in two other hsp genes, *DnaJ* and *GrpE*, have the same effect. DnaJ and GrpE physically interact with DnaK. Moreover, the three proteins function together in several other processes (see Section 2.2.6). The common point of DnaK, DnaJ, and GrpE action in regulation is the control of σ^{32} concentrations.[163–166] As discussed in Section 3.1, σ^{32} is the primary transcriptional regulator of the heat-shock response in *E. coli*. Remarkably, DnaK, DnaJ, and GrpE affect σ^{32} at three different levels. They (1) facilitate the degradation of the σ^{32} protein at low temperatures and during the extinction phase of the response at high temperatures, apparently by direct interaction with σ^{32}, (2) promote the translation of σ^{32} at high temperatures, and (3) inactivate σ^{32} when cells are shifted to low temperatures.

In fact, the primary function of DnaK in *E. coli* at normal temperatures may be to regulate the expression of other hsps. DnaK is essential for the growth of *E. coli* at both high (42°C) and low (16°C) temperatures. At 30°C, deletion mutants are viable, but grow slowly and exhibit abnormalities in cell division.[167] The 30°C defects are efficiently suppressed by σ^{32} mutations that downregulate the expression of other hsps.[168] These results suggest that the most important function of DnaK at 30°C is to maintain appropriate levels of other hsps by repressing σ^{32} expression and activity.

In yeast, mutations in members of the hsp70 family that are both constitutively expressed and heat-inducible result in overexpression of other hsps.[44] To investigate the role of hsp70 in greater detail, a heat-shock reporter plasmid was constructed from a cytochrome *c* (*CYC1*)/β-galactosidase fusion gene, replacing the upstream activating sequence of *CYC1* with a heat-shock activating sequence (HSE2). In wild-type cells this gene was strongly heat-inducible. However, if the cells also carried a constitutive hsp70 expression plasmid (a GAL 1-driven SSA1 gene), induction of the reporter plasmid was reduced at least tenfold.[169] Thus, hsp70 acts to repress transcription of hsp70 genes through their upstream activating sequences. Intriguingly, when flanking sequences were deleted from the HSE, sensitivity to excess hsp70 was lost. Apparently a sequence very close to the HSE, or overlapping the HSE, is important in mediating the sensitivity of this promoter to hsp70. Since the HSF of yeast cells binds DNA in the absence of heat shock, this suggests that a repressing cofactor may also be bound (see Section 3.1 for other evidence for HSF corepressors).

Although the mechanisms that regulate the levels of σ^{32} are themselves complex, the regulation of the heat-shock response in *E. coli* is simpler than in *Drosophila*. Over a broad range of temperatures the single control point is σ.[32] To provide a model for regulation that unifies other known functions of hsp70 with its effects on regulation, Gross and colleagues[25,163] suggest that hsps are induced at high temperatures because the denaturation of other proteins provides other targets for hsp70/DnaK binding. Thus, in *E. coli*, when DnaK is recruited to protect these other proteins, it is no longer able to repress σ.[32] On return to low temperatures, as proteins refold and DnaK is released, it is again available for σ^{32} repression. A similar scheme explains the attenuation of the response in cells maintained at high temperatures.

Evidence in support of this model comes from several sources. First, many of the other inducers of hsps have in common the property of causing protein denaturation.[50,170,171] Second, the injection of denatured protein into Xenopus oocytes[172] and the expression of a stable, unstructured protein in *E. coli*[173] is sufficient to induce hsp synthesis. Third, mutations in ubiquitin and the ubiquitin-conjugating enzymes[81,82] also lead to hsp induction. A blockade in the degradation of abnormal proteins would raise their intracellular concentration, providing additional substrates for hsp70 binding.

This model shares several features with the model for *Drosophila* regulation proposed above (see also refs. 22 and 145). There is a notable difference between the two models. Certain aspects of *Drosophila* regulation are best explained by positing that the activity of some regulatory factors is physically disrupted by heat and, furthermore, that the reactivation of these factors requires specific restoration by hsp70. These factors include certain components of normal translation, such as cap-binding complex, and nucleolytic factors involved in RNA degradation. The difference between the two models may, however, be smaller than meets the eye. As discussed in ref. 163, the simple recruitment of DnaK by other denatured proteins at high temperatures can explain most aspects of the *E. coli* response, but it cannot explain the effects that temperature has on translation of the σ^{32} message. In this instance, Gross and colleagues suggest that high temperatures may denature a translational repressor of σ^{32} mRNA. Similarly, the general *Drosophila* model does not explain transcriptional regulation very well. Here, the effect of heat on the key regulatory factor, HSF, is to activate it. As discussed in Section 3.1, it seems that another cellular factor represses the inherent transcription-activation potential of HSF. If so, there are three possible explanations for HSF activation that provide a role for hsp70 in repression: (1) HSF is normally repressed by association with hsp90. High temperatures disrupt this association and the restoration of the repressed complex requires hsp70. (Recall that hsp90 represses the DNA binding activity of steroid hormone receptors and the initial association of hsp90 with these transcription factors seems to require hsp70.[72,89]) (2) The activity of HSF is conformation dependent. At high temperatures the protein spontaneously assumes a conformation that promotes its association into trimers and its binding to DNA. At low temperatures, hsp 70 refolds HSF into its repressed state. (3) At normal temperatures, HSF forms a complex with hsp70 (or a low-temperature version of hsp70). At high temperatures this complex breaks down, either because the interactions between HSF and hsp70 are temperature sensitive or because hsp70 is recruited elsewhere. This final explanation, of course, is very similar to the model of DnaK regulation in *E. coli* proposed by Gross and collaborators.

Clearly, much work remains to be done. In bacteria, yeast, *Drosophila*, and other organisms, however, hypotheses on the regulation of heat-shock synthesis are converging. In all cases, hsp70 plays a key role. It is particularly satisfying that the roles that hsp70 proteins play in regulating the heat-shock response are completely consistent with the roles they play in normal cell biology and in helping cells to cope with the toxic effects of heat.

Note added in proof. This review was completed in January 1992. Since that time a great deal of information on the molecular functions of hsps has accumulated. The reader is referred to one recently published review and one that is in press for general updates.[174,175]

Concerning the regulation of hsps, new data are beginning to define the specific mechanisms by which hsps repress heat-shock transcription in *E. coli*. *In vitro*, DnaJ and DnaK bind σ^{32} directly and, together with GrpE, specifically suppress σ^{32}-directed transcription.[176,177] DnaJ and DnaK also enhance degradation of σ^{32} protein and repress translation of the σ^{32} message, perhaps by interacting with σ^{32} nascent chains. A recent model for hsp regulation in *E. coli* is compatible with earlier models, including that described here, but places greater emphasis on the role of DnaJ.[178] In eukaryotes, recent results provide support for the model of heat-shock transcription factor (HSF) regulation presented here and similar models have been proposed by others.[179,180] It has now been demonstrated that hsp70 associates with activated HSF as the concentration of hsp70 rises during heat shock and this association is disrupted by ATP. Furthermore, hsp70 blocks the activation of HSF *in vitro* from a non-DNA-binding state to a DNA-binding state.[181] HSF has a high affinity for hsp90 as well as hsp70.[182] These findings suggest that hsp70 and hsp90 interact with HSF and suppress transcriptional activity in a manner that is similar to their interaction with steriod hormone receptors.[183–185] However, work by Wu and colleagues suggests that HSF is autorepressed by the formation of a coiled coil between a leucine zipper repeat located in the carboxy-terminal region and another located in the amino-terminal region.[186] It is suggested that this coiled coil is disrupted by high temperatures, giving HSF an intrinsic capacity to respond to elevated temperatures by switching from a repressed conformation to an active conformation that trimerizes and binds DNA. This hypothesis is not incompatible with a role for hsp70 and hsp90 in regulating HSF after heat shock. Indeed, the dissociation of active HSF trimers into inactive monomers would seem a likely place for chaperone proteins to function.

ACKNOWLEDGMENTS. I thank J. Feder, D. Parsell, M. Singer, and M. Welte for comments on the manuscript, and S. Herder for help with the figures.

REFERENCES

1. Lindquist, S., 1986, The heat-shock response, *Ann. Rev. Biochem.* **55:**1151–1191.
2. Lindquist, S., and Craig, E. A., 1988, The heat-shock proteins, *Annu. Rev. Genet.* **22:**631–677.
3. Morimoto, R. I., Tissieres, A., and Georgopoulos, C., 1990 (eds.), *Stress Proteins in Biology and Medicine*, Cold Spring Harbor Laboratory Press, Cold Spring Harbor, New York.
4. Nover, L., 1991, *Heat Shock Response*, CRC Press, Boca Raton, Florida.
5. Maresca, B., and Lindquist, S., 1991, *Heat Shock*, Springer-Verlag, Berlin.
6. Plesset, J., Palm, C., and McLaughlin, C. S., 1982, Induction of heat shock proteins and thermotolerance by ethanol in *Saccharomyces cerevisiae*, *Biochem. Biophys. Res. Commun.* **108:** 1340–1345.
7. Hahn, G. M., and Li, G. C., 1982, Thermotolerance and heat shock proteins in mammalian cells, *Radiat. Res.* **92:**452–457.
8. Li, G. C., and Laszlo, A., 1985, Amino acid analogs while inducing heat shock proteins sensitize CHO cells to thermal damage, *J. Cell. Physiol.* **122:**91–97.
9. Hahn, G. M., and Li, G. C., 1990, Thermotolerance, thermoresistance, and thermosensitization, in: *Stress Proteins in Biology and Medicine* (R. I. Morimoto, A. Tissieres, and C. Georgopoulos, eds.), pp. 79–100, Cold Spring Harbor Laboratory Press, Cold Spring Harbor, New York.
10. Hall, B. G., 1983, Yeast thermotolerance does not require protein synthesis, *J. Bacteriol.* **156:** 1363–1365.
11. Watson, K., Dunlop, G., and Cavicchioli, R., 1984, Mitochondrial and cytoplasmic protein syntheses are not required for heat shock acquisition of ethanol and thermotolerance in yeast, *FEBS Lett.* **172:**299–302.
12. Hallberg, R. L., Kraus, K. W., and Hallberg, E. M., 1985, Induction of acquired thermotolerance in *Tetrahymena thermophila*: Effects of protein synthesis inhibitors, *Mol. Cell. Biol.* **5:** 2061–2069.

13. Widelitz, R. B., Magun, B. E., and Gerner, E. W., 1986, Effects of cycloheximide on thermotolerance expression, heat shock protein synthesis, and heat shock protein mRNA accumulation in rat fibroblasts, *Mol. Cell. Biol.* **6:**1088–1094.
14. Carper, S. W., Duffy, J. J., and Gerner, E. W., 1987, Heat shock proteins in thermotolerance and other cellular processes, *Cancer Res.* **47:**5249–5255.
15. Barnes, C. A., Johnston, G. C., and Singer, R. A., 1990, Thermotolerance is independent of induction of the full spectrum of heat shock proteins and of cell cycle blockage in the yeast *Saccharomyces cerevisiae*, *J. Bacteriol.* **172:**4352–4358.
16. Sanchez, Y., and Lindquist, S. L., 1990, HSP104 required for induced thermotolerance, *Science* **248:**1112–1115.
17. Li, G. C., Li, L. G., Liu, Y. K., Mak, J. Y., Chen, L. L., and Lee, W. M., 1991, Thermal response of rat fibroblasts stably transfected with the human 70-kDa heat shock protein-encoding gene, *Proc. Natl. Acad. Sci. USA* **88:**1681–1685.
18. Lindquist, S., 1980, Translational efficiency of heat-induced messages in *Drosophila melanogaster* cells, *J. Mol. Biol.* **137:**151–158.
19. Lindquist, S., 1980, Varying patterns of protein synthesis in *Drosophila* during heat shock: Implications for regulation, *Dev. Biol.* **77:**463–479.
20. Lindquist, S., and Petersen, R., 1990, Selective translation and degradation of heat-shock messenger RNAs in *Drosophila*, *Enzyme* **44:**147–166.
21. Yost, H. J., Petersen, R. B., and Lindquist, S., 1990, RNA metabolism: Strategies for regulation in the heat shock response, *Trends Genet.* **6:**223–227.
22. DiDomenico, B. J., Bugaisky, G. E., and Lindquist, S., 1982, The heat shock response is self-regulated at both the transcriptional and posttranscriptional levels, *Cell* **31:**593–603.
23. Tilly, K., McKittrick, N., Zylicz, M., and Georgopoulos, C., 1983, The *dnaK* protein modulates the heat-shock response of *Escherichia coli*, *Cell* **34:**641–646.
24. Mizzen, L. A., and Welch, W. J., 1988, Characterization of the thermotolerant cell. I. Effects on protein synthesis activity and the regulation of heat-shock protein 70 expression, *J. Cell Biol.* **106:** 1105–1116.
25. Gross, C., and Craig, E. A., 1991, *Trends Biochem. Sci.* **16:**135–140.
26. McMullin, T. W., and Hallberg, R. L., 1988, A highly evolutionarily conserved mitochondrial protein is structurally related to the protein encoded by the *Escherichia coli groEL* gene, *Mol. Cell. Biol.* **8:**371–380.
27. Blumberg, H., and Silver, P. A., 1991, A homologue of the bacterial heat-shock gene *DnaJ* that alters protein sorting in yeast, *Nature* **349:**627–630.
28. Trent, J. D., Nimmesgern, E., Wall, J. S., Hartl, F.-U., and Horwich, A. L., 1991, A molecular chaperone from a thermophilic archaebacterium is related to the eukaryotic protein t-complex polypeptide-1, *Nature* **354:**490–493.
29. Gething, M.-J., and Sambrook, J., 1992, Protein folding in the cell, *Nature* **355:**33–45.
30. Hightower, L. E., 1980, Cultured animal cells exposed to amino acid analogues or puromycin rapidly synthesize several polypeptides, *J. Cell. Physiol.* **102:**407–427.
31. Ellis, R. J., and van der Vies, S. M., 1991, Molecular chaperones, *Annu. Rev. Biochem.* **60:**321–347.
32. Bardwell, J. C., and Craig, E. A., 1984, Major heat shock gene of *Drosophila* and the *Escherichia coli* heat-inducible *dnaK* gene are homologous, *Proc. Natl. Acad. Sci. USA* **81:** 848–852.
33. Ingolia, T. D., and Craig, E. A., 1982, *Drosophila* gene related to the major heat shock-induced gene is transcribed at normal temperatures and not induced by heat shock, *Proc. Natl. Acad. Sci. USA* **79:**525–529.
34. Wadsworth, S. C., 1982, A family of related proteins is encoded by the major *Drosophila* heat shock gene family, *Mol. Cell. Biol.* **2:**286–292.
35. Craig, E. A., and Jacobsen, K., 1985, Mutations in cognate genes of *Saccharomyces cerevisiae* hsp70 result in reduced growth rates at low temperatures, *Mol. Cell. Biol.* **5:**3517–3524.

36. Watowich, S. S., and Morimoto, R. I., 1988, Complex regulation of heat shock- and glucose-responsive genes in human cells, *Mol. Cell. Biol.* **8:**393–405.
37. Craig, E. A., Kramer, J., Shilling, J., Werner-Washburne, M., Holmes, S., Kosic, D., Smithers, J., and Nicolet, C. M., 1989, SSC1, an essential member of the yeast HSP70 multigene family, encodes a mitochondrial protein, *Mol. Cell. Biol.* **9:**3000–3008.
38. Mizzen, L. A, Chang, C., Garrels, J. I., and Welch, W. J., 1989, Identification, characterization, and purification of two mammalian stress proteins present in mitochondria, grp 75, a member of the hsp 70 family and hsp 58, a homolog of the bacterial *groEL* protein, *J. Biol. Chem.* **264:** 20664–20675.
39. Velazquez, J. M., and Lindquist, S., 1984, hsp70: Nuclear concentration during environmental stress and cytoplasmic storage during recovery, *Cell* **36:**655–662.
40. Welch, W. J., and Feramisco, J. R., 1984, Nuclear and nucleolar localization of the 72,000-dalton heat shock protein in heat-shocked mammalian cells, *J. Biol. Chem.* **259:**4501–4513.
41. Werner, W. M., and Craig, E. A., 1989, Expression of members of the *Saccharomyces cerevisiae* hsp70 multigene family, *Genome* **31:**684–689.
42. Allen, R. L., OBrein, D. A., and Eddy, E. M., 1988, A novel hsp70-like protein (P70) is present in mouse spermatogenic cells, *Mol. Cell. Biol.* **8:**828–832.
43. Zakeri, Z. F., Wolgemuth, D. J., and Hunt, C. R., 1988, Identification and sequence analysis of a new member of the mouse HSP70 gene family and characterization of its unique cellular and developmental pattern of expression in the male germ line, *Mol. Cell. Biol.* **8:**2925–2932.
44. Craig, E. A., and Jacobsen, K., 1984, Mutations of the heat inducible 70 kilodalton genes of yeast confer temperature sensitive growth, *Cell* **38:**841–849.
45. Craig, E. A., 1989, Essential roles of 70kDa heat inducible proteins, *Bioessays* **11:**48–52.
46. Welch, W. J., and Feramisco, J. R., 1985, Rapid purification of mammalian 70,000-dalton stress proteins: Affinity of the proteins for nucleotides, *Mol. Cell. Biol.* **5:**1229–1237.
47. Rothman, J. E., 1989, Polypeptide chain binding proteins: Catalysts of protein folding and related processes in cells, *Cell* **59:**591–601.
48. Yamamoto, T., McIntyre, J., Sell, S. M., Georgopoulos, C., Skowyra, D., and Zylicz, M., 1987, Enzymology of the pre-priming steps in lambda DNA replication *in vitro*, *J. Biol. Chem.* **262:** 7996–7999.
49. Flynn, G. C., Pohl, J., Flocco, M. T., and Rothman, J. E., 1991, Peptide-binding specificity of the molecular chaperone BiP, *Nature* **353:**726–730.
50. Pelham, H. R., 1986, Speculations on the functions of the major heat shock and glucose-regulated proteins, *Cell* **46:**959–961.
51. Ungewickell, E., 1985, The 70-kd mammalian heat shock proteins are structurally and functionally related to the uncoating protein that releases clathrin triskelia from coated visicles, *EMBO J.* **4:**3385–3391.
52. Chappell, T. G., Welch, W. J., Schlossman, D. M., Palter, K. B., Schlesinger, M. J., and Rothman, J. E., 1986, Uncoating ATPase is a member of the 70 kilodalton family of stress proteins, *Cell* **45:**3–13.
53. Chirico, W. J., Waters, M. G., and Blobel, G., 1988, 70K heat shock related proteins stimulate protein translocation into microsomes, *Nature* **332:**805–810.
54. Sanchez, E. R., Toft, D. O., Schlesinger, M. J., and Pratt, W. B., 1985, Evidence that the 90-kDa phosphoprotein associated with the untransformed L-cell glucocorticoid receptor is a murine heat shock protein, *J. Biol. Chem.* **86:**1123–1127.
55. Munro, S., and Pelham, H. R., 1986, An Hsp70-like protein in the ER: Identity with the 78 kd glucose-regulated protein and immunoglobulin heavy chain binding protein, *Cell* **46:**291–300.
56. Gething, M. J., McCammon, K., and Sambrook, J., 1986, *Cell* **46:**939–950.
57. Skowyra, D., Georgopoulos, C., and Zylicz, M., 1990, The *E. coli dnaK* gene product, the hsp70 homolog, can reactivate heat-inactivated RNA polymerase in an ATP hydrolysis-dependent manner, *Cell* **62:**939–944.

58. Neidhardt, F. C., VanBogelen, R. A., and Vaughn, V., 1984, The genetics and regulation of heat-shock proteins, *Annu. Rev. Genet.* **18:**295–329.
59. Fayet, O., Ziegelhoffer, T., and Georgopoulos, C., 1989, The *groES* and *groEL* heat shock gene products of *Escherichia coli* are essential for bacterial growth at all temperatures, *J. Bacteriol.* **171:**1379–1385.
60. Hemmingsen, S. M., Woolford, C., van der Vies, S. M., Tilly, K., Dennis, D. T., Georgopoulos, C. P., Hendrix, R. W., and Ellis, R. J., 1988, Homologous plant and bacterial proteins chaperone oligomeric protein assembly, *Nature* **333:**330–334.
61. Prevelige, P., Thomas, D., and King, J., 1988, *J. Mol. Biol.* **202:**743–757.
62. Goloubinoff, P., Gatenby, A. A., and Lorimer, G. H., 1989, *GroE* heat-shock proteins promote assembly of foreign prokaryotic ribulose bisphosphate carboxylase oligomers in *Escherichia coli*, *Nature* **337:**44–47.
63. Martin, J., Langer, T., Boteva, R., Schramel, A., Horwich, A. L., and Hartl, F. U., 1991, Chaperonin-mediated protein folding at the surface of groEL through a 'molten globule'-like intermediate, *Nature* **352:**36–42.
64. Viitanen, P. V., Lubben, T. H., Reed, J., Goloubinoff, P., O'Keefe, D. P., and Lorimer, G. H., 1990, Chaperonin-facilitated refolding of ribulosebisphosphate carboxylase and ATP hydrolysis by chaperonin 60 (groEL) are K+ dependent, *Biochemistry* **29:**5665–5671.
65. Gottesman, S., 1990, Conservation of the regulatory subunit for the Clp ATP-dependent protease in prokaryotes and eukaryotes, *Proc. Natl. Acad. Sci. USA* **87:**3513–3517.
66. Parsell, D. A., Sanchez, Y., Stitzel, J. D., and Lindquist, S., 1991, Hsp104 is a highly conserved protein with two essential nucleotide-binding sites, *Nature* **353:**270–273.
67. Sanchez, Y., Taulien, J., Borkovich, K., and Lindquist, S. L., 1990, Hsp104 is required for tolerance to many forms of stress, *Science* **248:**1112–1115.
68. Squires, C. L., Pedersen, S., Ross, B. M., and Squires, C., 1991, ClpB is the *Escherichia coli* heat shock protein F84.1, *J. Bacteriol.* **173:**4254–4262.
68a. Sanchez, Y., Taulien, J., Borkovich, K. A., and Lindquist, S., 1992, Hsp104 is required for tolerance to many forms of stress, *EMBO J.* **11:**2357–2364.
69. Borkovich, K. A., Farrelly, F. W., Finkelstein, D. B., Taulien, J., and Lindquist, S., 1989, Hsp82 is an essential protein that is required in higher concentrations for growth of cells at higher temperatures, *Mol. Cell. Biol.* **9:**3919–3930.
70. Mazzarella, R. A., and Green, M., 1987, ERp99, an abundant, conserved glycoprotein of the endoplasmic reticulum, is homologous to the 90-kDa heat shock protein (hsp90) and the 94-kDa glucose regulated protein (GRP94), *J. Biol. Chem.* **262:**8875–8883.
71. Sorger, P. K., and Pelham, H. R., 1987, The glucose-regulated protein grp94 is related to heat shock protein hsp90, *J. Mol. Biol.* **194:**341–344.
72. Pratt, W. B., 1990, Interaction of hsp90 with steroid receptors: Organizing some diverse observations and presenting the newest concepts, *Mol. Cell. Endocrinol.* **74:**69–76.
73. Brugge, J. S., Erikson, E., and Erikson, R. L., 1981, The specific interaction of the Rous sarcoma virus transforming protein, pp60 src, with two cellular proteins, *Cell* **25:**363–372.
74. Lipsich, L. A., Cutt, J. R., and Brugge, J. S., 1982, Association of the transforming proteins of Rous Fujinami, and Y73 avian sarcoma viruses with the same two cellular proteins, *Mol. Cell. Biol.* **2:**2875–2880.
75. Ziemiecki, A., Catelli, M. G., Joab, I., and Moncharmont, B., 1986, Association of the heat shock protein hsp90 with steroid hormone receptors and tyrosine kinase oncogene products, *Biochem. Biophys. Res. Commun.* **138:**1298–1307.
76. Rose, D. W., Wettenhall, R. E., Kudlicki, W., Kramer, G., and Hardesty, B., 1987, The 90-kilodalton peptide of the heme-regulated eIF-2 alpha kinase has sequence similarity with the 90-kilodalton heat shock protein, *Biochemistry* **26:**6583–6587.
77. Nishida, E., Koyasu, S., Sakai, H., and Yahara, I., 1986, Calmodulin-regulated binding of the 90-kDa heat shock protein to actin filaments, *J. Biol. Chem.* **261:**16033–16036.

78. Pratt, W. B., Sanchez, E. R., Bresnick, E. H., Meshinchi, S., Scherrer, L. C., Dalman, F. C., and Welsh, M. J., 1989, Interaction of the glucocorticoid receptor with the Mr 90,000 heat shock protein: An evolving model of ligand-mediated receptor transformation and translocation, *Cancer Res.* **49:**2222s–2229s.
79. Dougherty, J. J., Rabideau, D. A., Iannotti, A. M., Sullivan, W. P., and Toft, D. O., 1987, Identification of the 90 kDa substrate of rat liver type II casein kinase with the heat shock protein which binds steroid receptors, *Biochim. Biophys. Acta* **927:**74–80.
79a. Jentsch, S., 1992, The ubiquitin-conjugation system, *Annu. Rev. Genet.* **26:**179–207.
80. Picard, D., Khursheed, B., Garabedian, M., Fortin, M. G., Lindquist, S., and Yamamoto, K. R., 1990, Reduced levels of hsp90 compromise steroid receptor action *in vivo*, *Nature* **348:**166–168.
81. Finley, D., Ozkaynak, E, and Varshavsky, A., 1987, The yeast polyubiquitin gene is essential for resistance to high temperatures, starvation, and other stresses, *Cell* **48:**1035–1046.
82. Seufert, W., and Jentsch, S., 1990, Ubiquitin-conjugating enzymes UBC4 and UBC5 mediate selective degradation of short-lived and abnormal proteins, *EMBO J.* **9:**543–550.
83. Zylicz, M., Ang, D., Liberek, K., and Georgopoulos, C., 1989, Initiation of Lambda DNA replication with purified host- and bacteriophage-encoded proteins: The role of the *dnaK*, *dnaJ* and *grpE* heat shock proteins, *EMBO J.* **8:**1601–1608.
84. Dodson, M., McMacken, R., and Echols, H., 1989, Specialized nucleoprotein structures at the origin of replication of bacteriophage Lambda. Protein association and disassociation reactions responsible for localized initiation of replication, *J. Biol. Chem.* **264:**10719–10725.
85. Alfano, C., and McMacken, R., 1989, Ordered assembly of nucleoprotein structures at the bacteriophage Lambda replication origin during the initiation of DNA replication, *J. Biol. Chem.* **264:**10699–10708.
86. Wickner, S., Hoskins, J., and McKenney, K., 1991, Function of DnaJ and DnaK as chaperones in origin-specific DNA binding by RepA, *Nature* **350:**165–167.
87. Liberek, K., Marszalek, J., Ang, D., Georgopoulos, C., and Zylicz, M., 1991, *Escherichia coli* DnaJ and GrpE heat shock proteins jointly stimulate ATPase activity of DnaK, *Proc. Natl. Acad. Sci. USA* **88:**2874–2878.
88. Horwich, A., 1990, Protein import into mitochondria and peroxisomes, *Curr. Opin. Cell. Biol.* **2:** 625–633.
89. Smith, D. F., Stensgard, B. A., Welch, W. J., and Toft, D. O., 1992, Assembly of progesterone receptor with heat shock proteins and receptor activation are ATP mediated events, *J. Biol. Chem.* **267:**1350–1356.
90. Wu, C., 1980, The 5′ ends of *Drosophila* heat shock genes in chromatin are hypersensitive to DNase I, *Nature* **286:**854–860.
91. Keene, M. A., Corces, V., Lowenhaupt, K., and Elgin, S. C., 1981, DNase I hypersensitive sites in *Drosophila* chromatin occur at the 5′ ends of regions of transcription, *Proc. Natl. Acad. Sci. USA* **78:**143–146.
92. Gilmour, D. S., and Lis, J. T., 1986, RNA polymerase II interacts with the promoter region of the noninduced hsp70 gene in *Drosophila melanogaster* cells, *Mol. Cell. Biol.* **6:**3984–3989.
93. Sorger, P. K., Lewis, M. J., and Pelham, H. R., 1987, Heat shock factor is regulated differently in yeast and HeLa cells, *Nature* **329:**81–84.
94. Wu, C., Wilson, S., Walker, B., Dawid, I., Paisley, T., Zimarino, V., and Ueda, H., 1987, Purification and properties of *Drosophila* heat shock activator protein, *Science* **238:**1247–1253.
95. Perisic, O., Xiao, H., and Lis, J. T., 1989, Stable binding of *Drosophila* heat shock factor to head-to-head and tail-to-tail repeats of a conserved 5 bp unit, *Cell* **59:**797–806.
96. Sorger, P. K., and Nelson, H. C., 1989, Trimerization of a yeast transcriptional activator via a coiled-coil motif, *Cell* **59:**807–813.
97. Clos, J., Westwood, J. T., Becker, P. B., Wilson, S., Lambert, K., and Wu, C., 1990, Molecular cloning and expression of a hexameric *Drosophila* heat shock factor subject to negative regulation, *Cell* **63:**1085–1097.

98. Xiao, H., Perisic, O., and Lis, J. T., 1991, Cooperative binding of *Drosophila* heat shock factor to arrays of a conserved 5 bp unit, *Cell* **64**:585–593.
99. Nieto, S. J., Wiederrecht, G., Okuda, A., and Parker, C. S., 1990, The yeast heat shock transcription factor contains a transcriptional activation domain whose activity is repressed under nonshock conditions, *Cell* **62**:807–817.
100. Grossman, A. D., Erickson, J. W., and Gross, C. A., 1984, The *htpR* gene product of *E. coli* is a sigma factor for heat-shock promoters, *Cell* **38**:383–390.
101. Landick, R., Vaughn, V., Lau, E. T., VanBogelen, R., Erickson, J. W., and Neidhardt, F. C., 1984, Nucleotide sequence of the heat shock regulatory gene of *E. coli* suggests its protein product may be a transcription factor, *Cell* **38**:175–182.
102. Straus, D. B., Walter, W. A., and Gross, C. A., 1987, The heat shock response of *E. coli* is regulated by changes in the concentration of sigma 32, *Nature* **329**:348–351.
103. Straus, D. B., Walter, W. A., and Gross, C. A., 1989, The activity of sigma 32 is reduced under conditions of excess heat shock protein production in *Escherichia coli*, *Genes Dev.* **3**:2003–2010.
104. Erickson, J. W., and Gross, C. A., 1989, Identification of the sigma E subunit of *Escherichia coli* RNA polymerase: A second alternate sigma factor involved in high-temperature gene expression, *Genes Dev.* **3**:1462–1471.
105. Yost, H. J., and Lindquist, S., 1986, RNA splicing is interrupted by heat shock and is rescued by heat shock protein synthesis, *Cell* **45**:185–193.
106. Blackman, R. K., and Meselson, M., 1986, Interspecific nucleotide sequence comparisons used to identify regulatory and structural features of the *Drosophila* hsp82 gene, *J. Mol. Biol.* **188**: 499–515.
107. Bond, U., and Schlesinger, M. J., 1986, The chicken ubiquitin gene contains a heat shock promoter and expresses an unstable mRNA in heat-shocked cells, *Mol. Cell. Biol.* **6**:4602–4610.
108. Kay, R. J., Russnak, R. H., Jones, D., Mathias, C., and Candido, E. P., 1987, Expression of intron-containing *C. elegans* heat shock genes in mouse cells demonstrates divergence of 3′ splice site recognition sequences between nematodes and vertebrates, and an inhibitory effect of heat shock on the mammalian splicing apparatus, *Nucleic Acids Res.* **15**:3723–3741.
109. Maniak, M., and Nellen, W., 1988, A developmentally regulated membrane protein gene in *Dictyostelium discoideum* is also induced by heat shock and cold shock, *Mol. Cell. Biol.* **8**: 153–159.
110. Yost, H. J., and Lindquist, S., 1991, Heat shock proteins affect RNA processing during the heat shock response of *Saccharomyces cerevisiae*, *Mol. Cell. Biol.* **11**:1062–1068.
111. Muhich, M. L., Hsu, M. P., and Boothroyd, J. C., 1989, Heat-shock disruption of *trans*-splicing in trypanosomes: Effect on Hsp70, Hsp85 and tubulin synthesis, *Gene* **82**:169–175.
112. Sutton, R. E., and Boothroyd, J. C., 1988, Trypanosome *trans*-splicing utilizes 2′-5′ branches and a corresponding debranching activity, *EMBO J.* **71**:1431–1437.
113. Muhich, M. L., and Boothroyd, J. C., 1989, Synthesis of trypanosome hsp70 mRNA is resistant to disruption of *trans*-splicing by heat shock, *J. Biol. Chem.* **264**:7107–7110.
114. Fini, M. E., Bendena, W. G., and Pardue, M. L., 1989, Unusual behavior of the cytoplasmic transcript of hsr omega: An abundant, stress-inducible RNA that is translated but yields no detectable protein product, *J. Cell Biol.* **108**:2045–2057.
115. Mayrand, S., and Pederson, T., 1983, Heat shock alters nuclear ribonucleoprotein assembly in *Drosophila* cells, *Mol. Cell. Biol.* **3**:161–171.
116. Beyer, A. L., and Osheim, Y. M., 1988, Splice site selection, rate of splicing, and alternative splicing on nascent transcripts, *Genes Dev.* **2**:754–765.
117. Bond, U., 1988, Heat shock but not other stress inducers leads to the disruption of a sub-set of snRNPs and inhibition of *in vitro* splicing in HeLa cells, *EMBO J.* **7**:3509–3518.
118. Wright, S. L. G., Reichlin, M., and Tobin, S. L., 1989, Alteration by heat shock and immunological characterization of *Drosophila* small nuclear ribonucleoproteins, *J. Cell Biol,* **108**:2007–2016.

119. Dworniczak, B., and Mirault, M. E., 1987, Structure and expression of a human gene coding for a 71 kd heat shock 'cognate' protein, *Nucleic Acids Res.* **15:**5181–5197.
120. Yost, H. L., and Lindquist, S., 1988, Translation of unspliced transcripts after heat shock, *Science* **242:**1544–1548.
121. Westwood, J. T., Clos, J., and Wu, C., 1991, Stress-induced oligomerization and chromosomal relocalization of heat-shock factor, *Nature* **353:**822–827.
122. Hickey, E., and Weber, L. A. (eds.), 1982, *Preferential Translation of Heat-Shock mRNAs in HeLa Cells*, Cold Spring Harbor Press, Cold Spring Harbor, New York.
123. Duncan, R., and Hershey, J. W., 1984, Heat shock-induced translational alterations in HeLa cells. Initiation factor modifications and the inhibition of translation, *J. Biol. Chem.* **259:**11882–11889.
124. Scorsone, K. A., Panniers, R., Rowlands, A. G. and Henshaw, E. C., 1987, Phosphorylation of eukaryotic initiation factor 2 during physiological stresses which affect protein synthesis, *J. Biol. Chem.* **262:**14538–14543.
125. Duncan, R. F., and Hershey, J. W., 1989, Protein synthesis and protein phosphorylation during heat stress, recovery, and adaptation, *J. Cell Biol.* **109:**1467–1481.
126. Panniers, R., Stewart, E. B., Merrick, W. C., and Henshaw, E. C., 1985, Mechanism of inhibition of polypeptide chain initiation in heat-shocked Ehrlich cells involves reduction of eukaryotic initiation factor 4F activity, *J. Biol. Chem.* **260:**9648–9653.
127. Lindquist, S., 1981, Regulation of protein synthesis during heat shock, *Nature* **293:**311–314.
128. Finkelstein, D. B., Strausberg, S., and McAlister, L., 1982, Alterations of transcription during heat shock of *Saccharomyces cerevisiae*, *J. Biol. Chem.* **257:**8405–8411.
129. Storti, R. V., Scott, M. P., Rich, A., and Pardue, M. L., 1980, Translational control of protein synthesis in response to heat shock in *D. melanogaster* cells, *Cell* **22:**825–834.
130. DiDomenico, B. J., Bugaisky, G. E., and Lindquist, S., 1982, Heat shock and recovery are mediated by different translational mechanisms, *Proc. Natl. Acad. Sci. USA* **79:**6181–6185.
131. Kruger, C., and Benecke, B. J., 1981, *In vitro* translation of *Drosophila* heat-shock and non-heat-shock mRNAs in heterologous and homologous cell-free systems, *Cell* **23:**595–603.
132. Sanders, M. M., Triemer, D. F., and Olsen, A. S., 1986, Regulation of protein synthesis in heat-shocked *Drosophila* cells. Soluble factors control translation *in vitro*, *J. Biol. Chem.* **261:**2189–2196.
133. Di Nocera, P. P., and Dawid, I. B., 1983, Transient expression of genes introduced into cultured cells of *Drosophila*, *Proc. Natl. Acad. Sci. USA* **80:**7095–7098.
134. Bonner, J. J., Parks, C., Parker, T. J., Mortin, M. A., and Pelham, H. R., 1984, The use of promoter fusions in *Drosophila* genetics: Isolation of mutations affecting the heat shock response, *Cell* **37:**979–991.
135. Klemenz, R., Hultmark, D., and Gehring, W. J., 1985, Selective translation of heat shock mRNA in *Drosophila melanogaster* depends on sequence information in the leader, *EMBO J.* **4:**2053–2060.
136. Petersen, R., and Lindquist, S., 1988, The *Drosophila* hsp70 message is rapidly degraded at normal temperatures and stabilized by heat shock, *Gene* **72:**161–168.
137. McGarry, T. J., and Lindquist, S., 1985, The preferential translation of *Drosophila* hsp70 mRNA requires sequences in the untranslated leader, *Cell* **42:**903–911.
138. Hultmark, D., Klemenz, R., and Gehring, W. J., 1986, Translational and transcriptional control elements in the untranslated leader of the heat-shock gene hsp22, *Cell* **44:**429–438.
139. Solomon, J. M., Rossi, J. M., Golic, K., McGarry, T., and Lindquist, S., 1991, Changes in Hsp70 alter thermotolerance and heat-shock regulation in *Drosophila*, *New Biol.* **3:**1106–1120.
140. McGarry, T. J., 1986, Genetic analysis of heat shock protein synthesis, Ph.D. Dissertation, University of Chicago, Chicago, Illinois.
141. Holmgren, R., Corces, V., Morimoto, R., Blackman, R., and Meselson, M., 1981, Sequence homologies in the 5′ regions of four *Drosophila* heat-shock genes. *Proc. Natl. Acad. Sci. USA* **78:** 3775–3778.

142. Lindquist, S., 1987, Translational Regulation in the Heat-Shock Response of Drosophila Cells, Translational Regulation of Gene Expression (J. Ilan, ed.), Plenum Press, New York, pp. 187–207.
143. Zapata, J. M., Maroto, F. G., and Sierra, J. M., 1991, Inactivation of mRNA cap-binding protein complex in *Drosophila melanogaster* embryos under heat shock, *J. Biol. Chem.* **266:**16007–16014.
144. Lindquist, S., and Petersen, R., 1990, Selective translation and degradation of heat shock messenger RNAs in *Drosophila*, *Enzyme* **44:**147–166.
145. Lindquist, S., and DiDomenico, B., 1985, *Coordinate and Noncoordinate Gene Expression during Heat Shock: A Model for Regulation*, Academic Press, New York.
146. Ballinger, D. G., and Pardue, M. L., 1983, The control of protein synthesis during heat shock in *Drosophila* cells involves altered polypeptide elongation rates, *Cell* **33:**103–113.
147. Jackson, R. J., 1986, The heat-shock response in *Drosophila* KC 161 cells: mRNA competition is the main explanation for reduction of normal protein synthesis, *Eur. J. Biochem.* **158:**623–634.
148. Scott, M. P., and Pardue, M. L., 1981, Translational control in lysates of *Drosophila melanogaster* cells, *Proc. Natl. Acad. Sci. USA* **78:**3353–3357.
149. Maroto, F. G., and Sierra, J. M., 1988, Translational control in heat-shocked *Drosophila* embryos. Evidence for the inactivation of initiation factor(s) involved in the recognition of mRNA cap structure, *J. Biol. Chem.* **263:**15720–15725.
150. Maroto, F. G., and Sierra, J. M., 1989, Purification and characterization of mRNA cap-binding protein from *Drosophila melanogaster* embryos, *Mol. Cell. Biol.* **9:**2181–2190.
151. Sarnow, P., 1989, Translation of glucose-regulated protein 78/immunoglobulin heavy-chain binding protein mRNA is increased in poliovirus-infected cells at a time when cap-dependent translation of cellular mRNAs is inhibited, *Proc. Natl. Acad. Sci. USA* **86:**5795–5799.
152. Munoz, A., Alonso, M. A., and Carrasco, L., 1984, Synthesis of heat-shock proteins in HeLa cells: Inhibition by virus infection, *Virology* **137:**150–159.
153. Petersen, R. B., and Lindquist, S., 1989, Regulation of Hsp70 synthesis by messenger RNA degradation, *Cell Regulat.* **1:**135–149.
154. Shaw, G., and Kamen, R., 1986, A conserved AU sequence from the 3′ untranslated region of GM-CSF mRNA mediates selective mRNA degradation, *Cell* **46:**659–667.
155. Sadis, S., Hickey, E., and Weber, L. A., 1988, Effect of heat shock on RNA metabolism in HeLa cells, *J. Cell. Physiol.* **135:**377–386.
156. Theodorakis, N. G., and Morimoto, R. I., 1987, Posttranscriptional regulation of hsp70 expression in human cells: Effects of heat shock, inhibition of protein synthesis, and adenovirus infection on translation and mRNA stability, *Mol. Cell. Biol.* **7:**4357–4368.
156a. Feder, J. H., Rossi, J. M., Solomon, J., Solomon, N., and Lindquist, S., 1992, The consequences of expressing hsp70 in *Drosophila* cells at normal temperatures, *Genes Dev.* **6:**1402–1403.
157. Bugaisky, G. E., 1981, RNA metabolism during heat shock and recovery in Drosophila, Ph.D. Dissertation, University of Chicago, Chicago, Illinois.
158. Lindquist, S., McGarry, T J., and Golic, K., 1988, *Use of Antisense RNA in Studies of the Heat-Shock Response*, Cold Spring Harbor Press, Cold Spring Harbor, New York.
159. Liberek, K., Georgopoulos, C., and Zylicz, M., 1988, Role of the *Escherichia coli* DnaK and DnaJ heat shock proteins in the initiation of bacteriophage lambda DNA replication, *Proc. Natl. Acad. Sci. USA* **85:**6632–6636.
160. Liberek, K., Osipiuk, J., Zylicz, M., Ang, D., Skorko, J., and Georgopoulos, C., 1990, Physical interactions between bacteriophage and *Escherichia coli* proteins required for initiation of lambda DNA replication, *J. Biol. Chem.* **265:**3022–3029.
161. Wickner, S. H., 1990, Three *Escherichia coli* heat shock proteins are required for P1 plasmid DNA replication: Formation of an active complex between *E. coli* DnaJ protein and the P1 initiator protein, *Proc. Natl. Acad. Sci. USA* **87:**2690–2694.

162. Gaitanaris, G. A., Papavassiliou, A. G., Rubock, P., Silverstein, S. J., and Gottesman, M. E., 1990, Renaturation of denatured λ repressor requires heat shock proteins, *Cell* **61:**1013–1020.
163. Straus, D., Walter, W., and Gross, C. A., 1990, DnaK, DnaJ, and GrpE heat shock proteins negatively regulate heat shock gene expression by controlling the synthesis and stability of sigma 32, *Genes Dev.* **4:**2202–2209.
164. Tilly, K., Spence, J., and Georgopoulos, C., 1989, Modulation of stability of the *Escherichia coli* heat shock regulatory factor sigma, *J. Bacteriol.* **171:**1585–1589.
165. Yura, T., Kawasaki, Y., Kusukawa, N., Nagai, H., Wada, C., and Yano, R., 1990, Roles and regulation of the heat shock sigma factor sigma 32 in *Escherichia coli*, *Antonie Van Leeuwenhoek* **58:**187–190.
166. Kamath, L. A. S., and Gross, C. A., 1991, Translational regulation of sigma 32 synthesis: Requirement for an internal control element, *J. Bacteriol.* **173:**3904–3906.
167. Bukau, B., and Walker, G. C., 1989, Cellular defects caused by deletion of the *Escherichia coli dnaK* gene indicate roles for heat shock protein in normal metabolism, *J. Bacteriol.* **171:**2337–2346.
168. Bukau, B., and Walker, G. C., 1990, Mutations altering heat shock specific subunit of RNA polymerase suppress major cellular defects of *E. coli* mutants lacking the DnaK chaperone, *EMBO J.* **9:**4027–4036.
169. Stone, D. E., and Craig, E. A., 1990, Self-regulation of 70-kilodalton heat shock proteins in *Saccharomyces cerevisiae*, *Mol. Cell. Biol.* **10:**1622–1632.
170. Lee, K. J., and Hahn, G. M., 1988, Abnormal proteins as the trigger for the induction of stress responses: Heat, diamide, and sodium arsenite, *J. Cell. Physiol.* **136:**411–420.
171. Hightower, L. E., 1991, Heat shock, stress proteins, chaperones, and proteotoxicity, *Cell* **66:** 191–197.
172. Ananthan, J., Goldberg, A. L., and Voellmy, R., 1986, Abnormal proteins serve as eukaryotic stress signals and trigger the activation of heat shock genes, *Science* **232:**522–524.
173. Parsell, D. A., and Sauer, R. T., 1989, Induction of a heat shock-like response by unfolded protein in *Escherichia coli*: Dependence on protein level not protein degradation, *Genes Dev.* **31:**1226–1232.
174. Harti, F. U., Martin, J., and Neupert, W., 1992, Protein folding in the cell: The role of molecular chaperones Hsp70 and Hsp60, *Annu. Rev. Biophys. Biomol. Struct.* **21:**293–322.
175. Parsell, D. A., and Lindquist, S., 1993, The function of heat-shock proteins in stess tolerance: Degradation and reactivation of damaged proteins, *Ann. Rev. Genet.* in press.
176. Gamer, J., Bujard, H., and Bukau, B., 1992, Physical interaction between heat shock proteins DnaK, DnaJ, and GrpE and the bacterial heat shock transcription factor sigma 32, *Cell* **69:**833–842.
177. Liberek, K., Galitski, T. P., Zylicz, M., and Georgopoulos, C., 1992, The DnaK chaperone modulates the heat shock response of *Escherichia coli* by binding to sigma 32 transcription factor, *Proc. Natl. Acad. Sci. USA* **89:**3516–3520.
178. Bukau, B., 1993, Regulation of the *E. coli* heat shock response, *Mol. Microbiol.* in press.
179. Morimoto, R. I., 1993, Cell in stress: Transcriptional activation of heat shock genes, *Science* **259:** 1409–1410.
180. Sorger, P. K., 1991, Heat shock factor and the heat shock response, *Cell* **65:**363–366.
181. Abravaya, K., Myers, M. P., Murphy, S. P., and Morimoto, R. I., 1992, The human heat shock protein hsp70 interacts with HSF, the transcription factor that regulates heat shock gene expression, *Genes Dev.* **6:**1153–1164.
182. Nadeau, K., Das, A., and Walsh, C. T., 1993, Hsp90 chaperonins possess ATPase activity and bind heat shock transcription factors and peptidyl prolyl isomerases, *J. Biol. Chem.* **268:**1479–1487.
183. Picard, D. *et al.*, 1990, Reduced level of hsp90 compromise steroid receptor action *in vivo*, *Nature* **348:**166–168.

184. Pratt, W. B., Scherrer, L. C., Hutchison, K. A., and Dalman, F. C., 1992, A model of glucocorticoid receptor unfolding and stabilization by a heat shock protein complex, *J. Steroid Biochem. Mol. Biol.* **41:**223–229.
185. Pratt, W. B., Hutchison, K. A., and Scherrer, L. C., 1992, Steroid receptor folding by heat-shock proteins and composition of the receptor heterocomplex, *TEM* **3:**326–333.
186. Rabindran, S. K., Haroun, R. I., Clos, J., Wisniewski, J., and Wu, C., 1993, Regulation of heat shock factor trimer formation: Role of a conserved leucine zipper, *Science* **259:**230–234.

Chapter 15

Repressor-Mediated Translational Control

The Regulation of Ferritin Synthesis by Iron

William E. Walden

1. INTRODUCTION

Initiation of translation is regulated through a variety of mechanisms. Among these are regulation of initiation factor and ribosome activity, and competition for limiting components of the translational machinery. Examples illustrating each are described elsewhere in this volume. The regulation of ferritin synthesis is the clearest example of translational regulation by a sequence-specific messenger RNA (mRNA) binding protein; a translational repressor. Ferritin is the major iron-storage protein in eukaryotes.[1–6] As part of its role in intracellular iron storage, ferritin serves to maintain iron in a soluble, nontoxic form, thus preventing the potentially damaging effects of free iron. Ferritin is a large protein composed of 24 identical, or nearly identical, subunits which are arranged in the form of a shell (apoferritin) into which as many as 4000 iron atoms can be stored. Two distinct subunits have been identified in mammals, heavy (H) and light (L). A third, middle (M), subunit has been identified in amphibians.[7,8] Apoferritin can be

William E. Walden • Department of Microbiology and Immunology, University of Illinois at Chicago, Chicago, Illinois 60612.

Translational Regulation of Gene Expression 2, edited by Joseph Ilan. Plenum Press, New York, 1993.

composed of any combination of these subunits, with the composition ultimately being determined by the tissue-specific expression pattern of the subunits.[1,2,4]

Synthesis of ferritin subunits, and hence ferritin, is regulated as a function of cellular iron status.[1,3–6,9–13] When iron is limiting, the rate of ferritin synthesis is low. When iron is in excess, the rate of ferritin synthesis is high; it is elevated by as much as 50- to 100-fold above iron limiting levels. It is now well established that regulation of ferritin synthesis in response to iron results from translational regulation.[10–13] In iron-poor cells, ferritin mRNA is found mainly in the untranslated pool (henceforth called the free mRNP).[11,13–15] Addition of iron to cells induces a shift of ferritin mRNA from the free mRNP into polysomes, demonstrating that translation initiation on ferritin mRNA increases in response to iron.[13,14] Ferritin mRNA isolated from iron-replete cells shows identical activity *in vitro* to mRNA isolated from iron-poor cells when translated in a wheat germ extract.[11] This indicates that iron does not directly affect the activity of ferritin mRNA, but acts through a *trans*-acting translational regulatory factor. Induction of ferritin synthesis by iron can be very rapid. It has been detected as early as 10 min after addition of iron to cells in culture.[9] Induction is also insensitive to inhibitors of transcription, supporting the conclusion that transcription plays little, if any, role in the regulation of ferritin synthesis by iron.

A simple model of ferritin mRNA translational regulation is shown in Fig. 1.

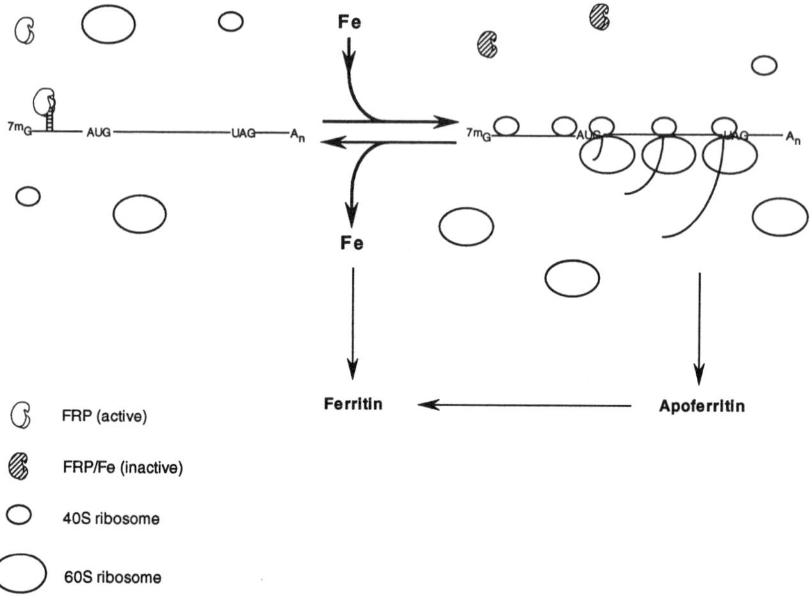

Figure 1. Scheme for translational regulation of ferritin synthesis.

In this scheme, aspects of which were originally proposed by Munro and colleagues,[13] translation of ferritin mRNA is controlled by a proteinaceous repressor which senses the iron status of the cell. In low iron conditions the repressor binds to a specific site near the 5′ end of ferritin mRNA, thereby blocking its translation. In the presence of iron the activity of the repressor is altered such that it is released from ferritin mRNA, thus allowing for its normal translation. The *cis*-acting regulatory site, which is shown as the stem–loop structure near the 5′ end of the mRNA in Fig. 1, is comprised of 28 nucleotides located within 30–50 nucleotides of the cap of ferritin mRNAs.[16–19] These 28 nucleotides have been called the iron-responsive element (IRE[17]) because they are both necessary and sufficient for iron-dependent translational regulation of ferritin mRNA.[16–18,20] The *trans*-acting factor, shown bound to the IRE in the left-hand half of Fig. 1, is an approximately 98-kDa protein. It has been variously called the IRE-binding protein,[21] ferritin repressor protein,[22] the iron regulatory factor,[23] or P90.[24] This model presents the view that free iron level is the signal which regulates ferritin mRNA translational efficiency. It is important to point out, however, that other mechanisms of induction of ferritin synthesis have been suggested.[25–28] These will be discussed later in the chapter. In either case, increased synthesis of ferritin would certainly serve to decrease free iron levels, which would presumably result in renewed repression of ferritin synthesis. In this way ferritin can be viewed as a feedback repressor of its own synthesis. The rest of this review will be devoted to discussions of what we currently know about the IRE, the regulatory protein, and the mechanisms of regulating their interaction and thus translation.

2. THE IRE: AN RNA-BASED OPERATOR

Genes and cDNAs encoding ferritin subunits of a variety of species have been cloned and sequenced.[7,8,19,29–42] Ferritin mRNAs are all approximately 1 kilobase (kb) in length and have relatively long 5′ untranslated regions (UTRs). A comparison of the 5′ UTRs of eight ferritin mRNAs is shown in Fig. 2. The IRE is easily recognized as the only highly conserved region in the 5′ UTR of all ferritin messages.[12,19] This alone initially suggested that it is involved in translational regulation of ferritin mRNA.[19] Consistent with this, deletion of the IRE renders ferritin mRNA translation unresponsive to iron.[16–18] Moreover, appending the IRE onto a heterologous mRNA renders the translation of that message responsive to iron.[16,17] Point mutations within the IRE can abolish regulation.[17,43,44] In all cases to date, these mutations also eliminate the ability of the IRE to bind to FRP.[43–47]

Computer analysis of the IRE shows that it can be folded into a moderately stable stem–loop structure (Fig. 3).[16,17,48] Evidence that this is a real structure in solution has been demonstrated by Theil and co-workers, who used chemical and enzymatic nuclease probes to analyze IRE structure.[49] The structure of the IRE consist of a six-membered loop having the sequence -CAGUG(U,C)-, a five-

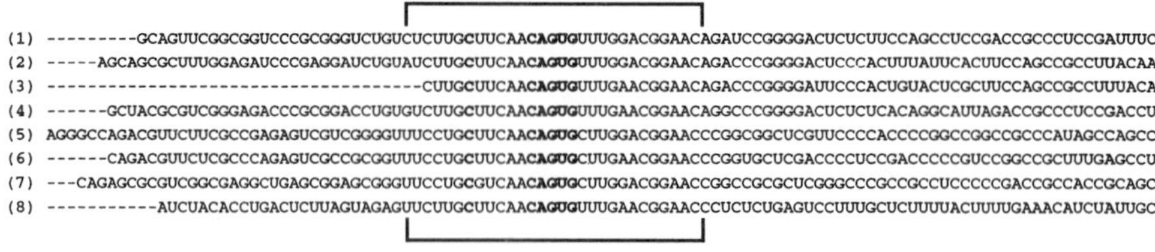

(1) CUCUCCGCUUGCAACCUCCGGGACCAUCUUCUCGGCCAUCUCCUGCUUCUGGGACCUGCCAGCACCGUUUUUGU0GGUUAGCUCCUUCUUGCCAACCAACC
(2) GUCUCUCCAGUCGCAGCCUCCGGGACCAUCUCCUUGCCGUCGGCUCCUAGGACCAGCCAGCCCGUCUUCGCGGUUAGCUCCAUACUCCGGAUCAGCC
(3) AGUCUCUCCAGUCGCAGCCUCCGGGACCAUCUCCUCGCUGCCUUCAGCUCCUAGGACCAGUCUGCACCGUCUCUUCGCGGUUAGCUCCUACUCCGGAUCAGCC
(4) CCUCUCCAGUCAGAAGCUCCGGGACCACCUCUCGGCCACCUCCUGCUCCUGGGACCAGCCCACCGCGGUUCUGUGGUCAUCUCACCGCCGCCAACC
(5) CUCCGTCACCUCUUCACCGCAGCCCUCGGGAGCUGCCCCAAGGCCCCCGCCGCCGCUCCAGCGCCGCGCAGCCACCGCCGCCGCCGCCUCUCCUUAGUCGCCGCC
(6) GAGCCCUUUGCAACUUCGUCGCUCCGCCGCUCCAGCGUCGCCUCCGCGCCUCGUCCAGCCGCCAUC
(7) AGCCGUCGUAUCCACCGCAUCUCUCUCUUUCUCUCCCGCCAGCGCC
(8) AUUACUGAUUUUUUUGCGCCACCCAAAACCACCAAACCGCUGAA

Figure 2. Ferritin mRNA 5′ UTRs. (1) Human L-chain, (2) Rat L-chain. (3) Mouse L-chain. (4) Rabbit L-chain. (5) Human H-chain. (6) Rat H-chain. (7) Chicken H-chain. (8) Bullfrog M-chain. The IRE is indicated by brackets.

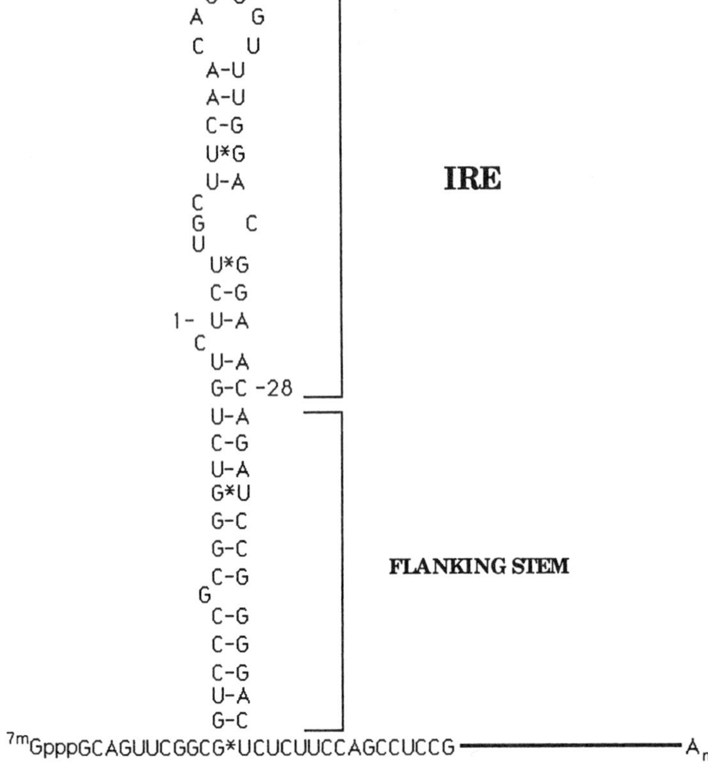

Figure 3. Structure of human L-chain ferritin IRE and flanking elements. The IRE is comprised of the 28 nucleotides, numbered from 1 to 28; forming the upper step-loop. The lower flanking stem is predicted to be present in all ferritin IRE structures.[12]

membered upper stem, and an unpaired cytosine residue (bulge C) located six residues to the 5′ side of the loop (Fig. 3).[48,49] The remaining 11 residues of the IRE form the lower stem, which in ferritin mRNAs is not a perfect duplex. In ferritin mRNA, the bulge C is actually a part of a larger bulge, as shown in Fig. 3. IREs present in other mRNAs (see next paragraph) are predicted to have only the -C- unpaired.[10] The significance of this, if any, is not clear. Evidence suggests that the elements in the upper part of the IRE are essential to its function.[45,47] Mutations which disrupt the upper stem, decrease the size of the loop, or delete the bulge C eliminate the regulatory capacity of the IRE. It is less clear what elements are important in the lower stem of the IRE. These residues are highly conserved and it has been demonstrated that FRP protects all of the 28 nucleotides of the IRE.[24] It may be that the lower stem serves only a structural role in IRE function.

It is of interest in a wider context that IRE-like sequences have been identified in other mRNAs as well.[50–54] Five IREs have been located within the 3′ UTR of the transferrin receptor (TfR) mRNA, where they are part of a larger regulatory element that serves in the control of TfR mRNA stability in response to iron.[50–52,55–57] An IRE has also been found within the 5′ UTR of the erythroid δ aminolevulinate synthase mRNA, where it regulates translation in an iron-responsive manner.[53,54] An IRE has been found in the 5′ UTR of mitrochondrial aconitase mRNA. Whether translation of aconitase mRNA is also regulated by iron is yet to be determined, but this is intriguing (see below) because the IRE-BP is itself an aconitase.[79] Thus, it is postulated that IREs, in conjunction with the protein(s) that bind to them (see below), are central to the regulation of synthesis of a variety of proteins related to iron.

3. THE FERRITIN REPRESSOR PROTEIN: A UNIQUE REGULATORY RNA BINDING PROTEIN

The translational repressor of ferritin mRNA was searched for by following two different but related activities. With the discovery of the IRE element, it was expected that the regulatory factor inferred from *in vivo* studies would exert its control through the IRE. The first evidence of an IRE-specific binding protein was provided by Leibold and Munro.[58] Using an electrophoretic RNA band shift assay, these investigators showed that cytoplasmic extracts contained an IRE binding factor. This factor was shown to consist of protein by its sensitivity to protease. UV cross-linking experiments revealed that an approximately 84- to 87-kDa protein cross-linked to the IRE when it was incubated with rat liver cytosol.[58] The rat-liver IRE-binding protein (IRE-BP) protected a 40- to 50-nucleotide fragment, which includes the IRE, from digestion by RNase T1. The affinity of the IRE-BP was used as the basis for purification of the protein.[21,23] Interestingly, the initial purification schemes employed an RNA affinity chromatography step using a TfR IRE for the affinity matrix. Nevertheless, the 90- to 100-kDa IRE-BP (also called the iron regulatory factor or IRF) binds with high affinity to both ferritin and TfR IREs.[21,23] It is interesting that two RNA–protein complexes are detected when IRE-containing transcripts are incubated with cytosol from rat cells.[58] The relationship of these complexes, designated B1 and B2 in order of distance from the top of the gel, to each other is not yet clear.

The second approach used to identify and isolate the repressor of ferritin mRNA translation was to follow translational repression *in vitro*.[22,59] Translational repression of ferritin mRNA *in vitro* was first observed in reticulocyte lysates.[59,60] When a complex mixture of mRNAs including ferritin is translated in the reticulocyte lysate, ferritin mRNA translation is specifically repressed. When the same mixture of mRNA is translated in the wheat germ lysate, ferritin mRNA is translated very efficiently.[11,59,60] The poor translation of ferritin mRNA in the reticulocyte suggested the presence of a ferritin mRNA-specific repressor. This conclusion was further supported by the finding that repression of ferritin mRNA translation could be transferred to the wheat germ lysate by adding fractions isolated from reticulocyte lysates.[59] The reticulocyte repressor was partially purified using standard protein fractionation methods and it was shown to be highly specific for ferritin mRNA; amounts of repressor which inhibit ferritin mRNA translation by greater than 90% have no effect on translation of other mRNAs. Through a series of experiments using transcripts consisting of the IRE as competitors, it was shown that such transcripts could relieve repression of ferritin mRNA by purified repressor, demonstrating that specificity derives from recognition of the IRE.[59,61] The translational repression assay was used as a basis for the purification of the ferritin repressor protein (FRP) from rabbit liver.[22] This also yielded a 90- to 100-kDa protein, though initial estimates of the molecular weight of the reticulocyte protein in crude extracts subjected to size exclusion chromatography suggested a molecular weight of approximately 180 kDa. The basis for the abnormal behavior of the reticulocyte protein in gel filtration columns is unclear, but it can be eliminated by increasing the ionic strength of the chromatography buffers.[59]

Cloning of cDNAs for FRP, IRE-BP, and IRF has demonstrated that these are all the same protein.[62–66] This is illustrated in Fig. 4, which shows a sequence comparison among the rabbit FRP[63] and the human,[62,64] mouse,[66] and rat[65] IRE-BPs. The high degree of sequence homology throughout the protein among species is striking, which indicates significant evolutionary conservation in this protein. IRE-BPs have been detected in extracts of an even wider variety of species, ranging from insects to mammals.[67] In each case the IRE-BP appears to be a 90- to 100-kDa protein based on UV cross-linking results.[67] A most interesting finding has been the homology between the TCA cycle enzyme aconitase and the IRE-BP.[68,69] The homology is most striking between *Escherichia coli* aconitase and the IRE-BP, which share nearly 60% identity (Fig. 4),[70] but there is conservation with mitochondrial aconitase as well.[68,69] Recently it has been demonstrated that the IRE-BP is a cytosolic aconitase in mammalian cells.[79] The most intriguing aspect

```
MFRP  SLSPGSGVVTYYLRESGVMPYLSQLGFDVVGYGcMTcIGNSGPLPEPVVEAITQGDLVAVGVLSGNRNFEGRVHPNTRAN YLASPPLVIAYAIAGTVRIDFEKEPLGVNAQGRQVFLKD
RTFP  SLSPGSGVVTYYLRESGVMPYLSQLGFDVVGYGcMTcIGNSGPLPEPVVEAITQGDLVAVGVLSGNRNFEGRVHPNTRAN YLASPPLVIAYAIAGTVRIDFEKEPLGVNAQGQQVFLKD
RFRP  SLSPGSGVVTYYLRESGVMPYLSQLGFDVVGYGcMTcIGNSGPLPEPVVEAITQGDLVAVGVLSGNRNFEGRVHPNTRAN YLASPPLVIAYAIAGTIRIDFEKEPLGTNAKGQQVFLRD
HFRP  SLSPGSGVVTYYLQESGVMPYLSQLGFDVVGYGcMTcIGNSGPLPEPVVEAITQGDLVAVGVLSGNRNFEGRVHPNTRAN YLASPPLVIAYAIAGTIRIDFEKEPLGVNAKGQQVFLKD
EACO  SLAPGSKVVSDYLAKAKLTPYLDELGFNLVGTGcTTcIGNSGPLPDPIETAIKKGDLTVGAVISGNRNFEGRIHPLVKTN WLASPPLVVAYALAGNMNINLASEPIGHDRKGDPVYLKD
YACO  TVTPGSEQIRATIERDGQLETFKEFGGIVLANAcGPcIGQWDR        RDIKKGDKNTI VSSYNRNFTSRNDGNPQTHAFVASPELVTAFAIAGDLRFNPLTDKL KDKDGNEFMLKP
PACO  TITPGSEQIRATIERDGYAQVLRDVGGIVLANAcGPcIGQWDR        KDIKKGEKNTI VTSYNRNFTGRNDANPETHAFVTSPEIVTALAIAGTLKFNPETDFL TGKDGKKFKLEA

MFRP  IWPTRDEIQAVERQHVIPGMFKEVYQKIETVNKSWNALAAPSEKLYAWNPKSTYIKSPPFFESLTLDLQPPKSIVDAYVLLNLGDSVTTDHISPAGNIARNSPAARYLTNRGLTPREFNS
RTFP  IWPTRDEIQEVERKYVIPGMFKEVYQKIETVNKSWNALAAPSEKLYAWNPKSTYIKSPPFFESLTLDLQPPKSIVDAYVLLNLGDSVTTDHISPAGNIARNSPAARYLTNRGLTPRDFNS
RFRP  IWPTREEIQAVERQYVIPGMFTEVYQKIETVNASWNALAAPSDKLYLWNPKSTYIKSPPFFENLTLDLQPPKSIVDAYVLLNLGDSVTTDHISPAGNIARNSPAARYLTNRGLTPREFNS
HFRP  IWPTRDEIQAVERQYVIPGMFKEVYQKIETVNESWNALATPSDKLFFWNSKSTYIKSPPFFENLTLDLQPPKSIVDAYVLLNLGDSVTTDHISPAGNIARNSPAARYLTNRGLTPREFNS
EACO  IWPSAQEIARAVEQ VSTEMFRKEYAEVFEGTAEWKGINVTRSDTYGWQEDSTYIRLSPFFDEMQATPAPVEDIHGARILAMLGDSVTTDHISPAGSIKPDSPAGRYLQGRGVERKDFNS
YACO    PHGRWFASKEVMMLVRTLTK  LHLQTVATVEVKVSPTSDRLQLLKP          FKPWD     GKDAKDMPILIKAVGKTTTDHISMAG             PWLKYRG    HLEN
PACO    PDADELPRAE FDPGQDTYQ  HPPKDSSGQRVDVSPTSQRLQLLEP          FDKWD     GKDLEDLQILIKVKGKCTTDHISAAG             PWLKFRG    HLDN

MFRP  YGSRRGNDAIMARGTFANIRLLNKFL NKQAPQTVHLPSGETLDVFDAAERYQQAGLPLIVLAGKEYGSGSSRDWAAKGPFLLGIKAVLAESYERIHRSNLVGMGVIPLEYLPGETADSL
RTFP  YGSRRGNDAIMARGTFANIRLLNKFL NKQAPQTVHLPSGETLDVFDAAERYQQAGLPLIVLAGKEYGSGSSRDWAAKGPFLLGIKAVLAESYERTHCSNLVGMGVIPLEYLPGETADSL
RFRP  YGSRRGNDAIMARGTFANIRLLNRFL NKQAPQTIHLPSGETLDVFDAAERYQQEGHPLIVLAGKEYGSGSSRDWAAKGPFLLGIKAVLAESYERIHRSNLVGMGVIPLEYLPGENADSL
HFRP  YGSRRGNDAVMARGTFANIRLLNRFL NKQAPQTIHLPSGEILDVFDAAERYQQAGLPLIVLAGKEYGAGSSRDWAAKGPFLLGIKAVLAESYERIHRSNLVGMGVIPLEYLPGENADAL
EACO  YGSRRGNHEVMMRGTFANIRIRNEMVPGVEGGMTRHLPDSDVVSIYDAAMRYKQEQTPLAVIAGKEYGSGSSRDWAAKGPRLLGIRVVIAESFERIHRSNLIGMGILPLEFPQGVTRKTL
YACO  ISNN      YMIGAINAE          NKKANCVKNVYTGEYKGVPDTARDYRDQGIKWVVIGDENFGEGSSREHAALEPRFLGGFAIITKSFARIHETNLKKQGLLPLNFKNPADYDK
PACO  ISNN      LLIGAINIE          NRKANSVRNAVTQEFGPVPDTARYYKQHGIRWVVIGDENYGEGSSREHRALEPRHLGGRAIITKSFARIHETNLKKQGLLPLTFADPADYNK

MFRP  GLTGRERYTINIPEDLKPRMTVQIKLDTGKTFQAVM    RFDTDVELTYFHNGGILNYMIRKMAQ
RTFP  GLTGRERYTIHIPEHLKPRMKVQIKLDTGKTFQAVM    RFDTDVELTYFHNGGILNYMIRKMAQ
RFRP  GLTGRERYTIIIPENLTPRMHVQVKLDTGKTFQAVI    RFDTDVELTYLHNGGILNYMIRKMAK
HFRP  GLTGQERYTIIIPENLKPQMKVQVKLDTGKTFQAVM    RFDTDVELTYFLNGGILNYMIRKMAK
EACO  GLTGEEKIDIGDLQNLQPGATVPVTLTRADGSQEVVPCRCRIDTATELTYYQNDGILHYVIRNMLK
YACO   INPDDRIDILGLAELAPGKPVTMRVHPKNGKPWDAVLTHTFN DEQIEWFKYGSALNKIKADEKK
PACO   IHPVDKLTIQGLKDFAPGKPLKCIIKHPNGTQETILLNHTFN ETQIEWFRAGSALNRMKELQQK

MFRP  MKNP FAHLAEPLDAAQPGKRFFNLNKLEDS  RYGRLPFSIRVLLEAAVRNCDEFLVKKNDIENILNWNVMQHKNIEVPFKPARVILQDFTGVPAVVDFAAMRDAVKKLGGNPEKINPV
RTFP  MKNP FAHLAEPLDPAQPGKKFFNLNKLEDS  RYGRLPFSIRVLLEAAVRNCDEFLVKKNDIENILNWSIMQHKSIEVPFKPARVILQDFTGVPAVVDFAAMRDAVKKLGGNPEKINPV
RFRP  MSNP FAYLAEPLDPAQPGKKFFNLNKLDYS  RYGRLPFSIRVLLEAAVRNCDKFLVKKEDIENILNWNVTQHMNIEVPFKPARVILQDFTGVPSVVDFAAMRDAVKKLGGDPEKINPI
HFRP  MSNP FAHLAEPLDPVQPGKKFFNLNKLEDS  RYGRLPFSIRVLLEAAVRNCDEFLVKKQDIENILNWNVTQHKNIEVPFKPARVILQDFTGVPAVVDFAAMRDAVKKLGGDPEKINPV
EACO  MSSTLREASKDTLQAKDKTYHYYSLPLAAKSLGDITRLPKSLKVLLENLLRWQDGNSVTEEDIHALAGWLKNAHADREIAYRPARVLMQDFTGVPAVVDLAAMREAVKRLGGDTAKVNPL
YACO        MLSARSAIKRPIVRGLATVSN    LTRDSKVNQNLLEDHSFINYKQNVETLDIVRKRLNRPFTYAEKILYGHLDDPH GQDIQRGVSYLKLRPDRVACQDATAQMAILQFM
PACO  MAP  YSLLVTRLQ KALGVRQYHVASV    LCQRAKVAMSHFEPHEYIRYDLLEKNIDIVRKRLNRPLTLSEKIVYGHLDDPA NQEIERGKTYLRLRPDRVAMQDATAQMAMLQFI

MFRP  CP  ADLVIDHSIQVD FNRRADSLQKNQDLEFERNKERFEFLKWGSQAFCNMRIIPPGSGIIHQVNLEYLARVVFD  QDG  CYYPDSL VGTDSHTTMIDGLGVLGWGVGGIEAEAV
RTFP  CP  ADLVIDHSIQVH FNRRADSLQKNQDLEFERNRERFEFLKWGSQAFCNMRIIPPGSGIIHQVNLEYLARVVFD  QDG  CYYPDSL VGTDSHTTMIDGLGVLGWGVGGIEAEAV
RFRP  CP  VDLVIDHSIQVD FNRRADSLQKNQDLEFERNRERFEFLKWGSKAFRNMRIIPPGSGIIHQVNLEYLARVVFD  QDG  YYYPDSL VGTDSHTTMIDGLGVLGWGVGGIEAEAV
HFRP  CP  ADLVIDHSIQVD FNRRADSLQKNQDLEFERNRERFEFLKWGSQAFHNMRIIPPGSGIIHQVNLEYLARVVFD  QDG  YYYPDSL VGTDSHTTMIDGLGILGWGVGGIEAEAV
EACO  SP  VDLVIDHSVTVDRFG DDEAFEENVRLEMERNHERYVFLKWGKQAFSRFSVVPPGTGICHQVNLEYLGKAVWSELQDGEWIAYPDTL VGTDSHTTMINGLGVLGWGVGGIEAEAA
YACO  SAGLPQVAKPVTVHCDHLIQAQVGGEKDLKRAIDLNKEVYDFLASATAKY NMGFWKPGSGIIHQIVLEN              YAFPGALIIGTDSHTPNAGGLGQLAIGVGGADAVDV
PACO  SSGLPKVAVPSTIHCDHLIEAQLGGEKDLRRAKDINQEVYNFLATAGAKY GVGFWRPGSGIIHQIILEN              YAYPGVLLIGTDSHTPNGGGLGGICIGVGGADAVDV

MFRP  MLGQPISMVLPQVIGYKLMGKPHPLVTSTDIVLTITKHLRQVGVVGKFVEFFGPGVAQLSIADRATIANMCPEYGATAAFFPVDEVSIAYLLQTGREEDKVKHIQKYLQAVGMFRDFNDT
RTFP  MLGQPISMVLPQVIGYKLMGKPHPLVTSTDIVLTITKHLRQVGVVGKFVEFFGPGVAQLSIADRATIANMCPEYGATAAFFPVDDVSIAYLVQTGREEDKVKHIKRYLQAVGMFRDFSDS
RFRP  MLGQPISMVLPQVIGYRLMGKPHPLVTSTDIVLTITKHLRQVGVVGKFVEFFGLGVAQLSIADRATIANMCPEYGATATFFPVDEVSIKYLVQTGRDESKVKQIRKYLQAVGMFRDYSDP
HFRP  MLGQPISMVLPQVIGYRLMGKPHPLVTSTDIVLTITKHLRQVGVVGKFVEFFGPGVAQLSIADRATIANMCPEYGATAAFFPVDEVSITYLVQTGRDEEKLKYIKKYLQAVGMFRDFNDP
EACO  MLGQPVSMLIPDVVGFKLTGKLREGITATDLVLTVTQMLRKHGVVGKFVEFYGDGLDSLPLADRATIANMSPEYGATCGFFPIDAVTLDYMRLSGRSEDQVELVEKYAKAQGMWR   NP
YACO  MAGRPWELKAPKILGVKLTGKMNGWTSPKDIILKLAGITTVKGGTGKIVEYFGDGVDTFSATGMGTICNMGAEIGATTSVFPFNKSMIEYLEATGRGK      IADFAKL  YHKDLLSA
PACO  MAGIPWELKCPKVIGVKLTGSLSGWTSPKDVILKVAGILTVKGGTGAIVEYHGPGVDSISCTGMATICNMGAEIGATTSVFPYNHRMKKYLSKTGRAD      IANLADE  F KDHLVP

MFRP  SQDPDFTQVVELDLKTVVPCCSGPKRPQDKVAVSEMKKDFESCLGAKQGFKGFQVAPDRHNDRKTFLYSNSEFTLAHGSVVIAAITTcTNTSNPSVMLGAGLLAKKAVEAGLSVKPYIKT
RTFP  SQDPDFTQVVELDLKTVVPCCSGPKRPQDKVAVSEIEKDFESCLGAKQGFKGFQVAPDHHNDHKTFIYNDSEFTLAHGSVVIAAITScTNTSNPSVMLGAGLLAKKAVEAGLNVKPYVKT
RFRP  SQDPDFTQVVELDLKTVVPCCSGPKRPQDKVAVSDMKKDFESCLGAKQGFKGFQVAPDHHNDHKTFIYNDSEFTLSHGSVVIAAITScTNTSNPSVMLGAGLLAKKAVDAGLNVKPYVKT
HFRP  SQDPDFTQVVELDLKTVVPCCSGPKRPQDKVAVSDMKKDFESCLGAKQGFKGFQVAPEHHNDHKTFIYDNTEFTLAHGSVVIAAITScTNTSNPSVMLGAGLLAKKAVDAGLNVMPYIKT
EACO  GDEPIFTSTLELDMNDVEASLAGPKRPQDRVALPDVPKAFAAS      NELEVNATHKDRPVDYVMNGHQYQLPDGAVVIAAITScTNTSNPSVLMAAGLLAKKAVTLGLKRQPWVKA
YACO  DKDAEYDEVVEIDLNTLEPYINGPFTPDLATPVSKMKE                       VAVANNWPL   DVRVGLIGScTNSSYEDMSRSAS IVKDAAAHGLKSKTIF
PACO  DPGCHYDQVIEINLSELKPHINGPFTPDLAHPVAEVGS                       VAEKEGWPL   DIRVGLIGScTNSSYEDMGRSAA VAKQALAHGLKCKSQF
```

Figure 4. Multiple alignment of FRP and aconitase amino acid sequences. MFRP, mouse FRP. RTFP, rat FRP. RFRP, rabbit FRP. HFRP, human FRP. EACO, *E. coli* aconitase. YACO, yeast aconitase. PACO, pig aconitase.

of this is that aconitase is an iron–sulfur protein whose activity is regulated by whether its iron–sulfur cluster is 3Fe or 4Fe.[71] Thus, this suggest the possibility that the activity of the IRE-BP is regulated through cluster state.[68,69,72,73]

An examination of the amino acid sequence of the IRE-BP reveals no obvious RNA binding motifs (Fig. 3). A current view is that RNA recognition elements evolved onto the basic aconitase structure. Thus, binding elements may have evolved in different parts of the molecule which are then brought together by the tertiary fold of the protein.

Interaction of the regulatory protein with the IRE has been investigated in several laboratories.[45,46,74] In all cases, this interaction has been found to be extremely tight. Measured dissociation constants Kd for the interaction range from 20 to 100 pM. This is nearly two orders of magnitude stronger than the binding of previously studied prokaryotic regulatory proteins. The structural basis for the tight binding of the FRP to the IRE is not understood. It is not possible to decipher information from the amino acid sequence of the protein, since functional features of RNA binding proteins of this sort have yet to be determined. It is clear, however, that the oxidation state of the protein is critical to its ability to bind to the IRE.[46,75] *In vitro*, optimal binding is observed only in the presence of sulfhydryl reducing agents, while oxidized FRP has a much lower affinity for the IRE.[46] These findings have been interpreted to suggest the participation of sulfhydryls in the binding of FRP to the IRE, analogous to that suggested for the interaction of the R17 phage coat protein with its RNA binding site.[75] Furthermore, the sensitivity of FRP to sulfhydryl reagents has led to the proposal that oxidation–reduction of sulfhydryls is part of the mechanism of iron regulation of FRP activity *in vivo*.[75]

The importance of various features of the IRE to its interaction with FRP has been investigated by introducing mutations and analyzing binding to the FRP. Natural variation in IRE sequence occurs in the IREs of TfR and ferritin mRNAs. Thus, from an analysis of these, it can be concluded that the nucleotide sequence in the stem regions of the IRE is relatively unimportant to FRP–IRE interaction since it is not conserved. On the other hand, the unpaired C residue in the stem as well as the sequence of the nucleotides in the loop are conserved in all IREs.

4. MECHANISM OF TRANSLATIONAL INHIBITION VIA IRE–FRP INTERACTIONS

The molecular mechanism by which FRP inhibits translation of ferritin mRNA is not yet clear, but it appears certain that FRP inhibits initiation of translation of ferritin message. The fact that ferritin message is in the free, untranslated mRNP fraction in iron-poor cells supports this notion.[11,13–15] In reticulocyte lysates, where translation of ferritin mRNA is repressed by the endogenous FRP,[59] ferritin mRNA does not become polysome associated during a translation reaction.[60] In contrast, ferritin message does become polysome associated during translation in the wheat germ extract. These findings are fairly strong evidence for the augument that initiation is the translation step inhibited by FRP. Translation initiation in eukaryotes is a complex process, involving a multitude of steps, including the interaction of mRNA with a number of initiation factors. Within this scheme, it is expected that FRP binding to ferritin mRNA will block one or only a few critical steps of the initiation process, presumably an early step. Results from a study of ferritin synthesis in cultured mouse fibroblasts led to the proposal that the ferritin repressor interferes with the ability of ferritin mRNA to compete for the limiting translation initiation factor, which is presumed to be eIF-4F, the cap recognition factor.[15] In such a scheme, FRP would essentially compete with eIF-4F for access to ferritin mRNA; the binding of one precludes the binding of the other. An interesting consequence of this is that by increasing available eIF-4F, ferritin synthesis would be stimulated. This has in fact been observed in the kinetic analysis of ferritin synthesis in cultured fibroblasts.[15]

Recently, Hentze and colleagues have investigated this issue using a molecular genetic approach.[43,44] The distance between the 5′ cap and the IRE was varied in a variety of artificial gene constructs. The ability to repress translation *in vitro* and the magnitude of the iron response *in vivo* were then examined. The conclusion from these studies was that the distance of the IRE from the cap is a critical factor in the regulation of translation by FRP. When the IRE is greater than about 60 nucleotides from the 5′ cap, repression *in vitro* and regulation of translation by iron *in vivo* are severly diminished. These results are consistent with a couple of different mechanisms of translational inhibition by FRP. First of all, the results are consistent with the model discussed in the preceding paragraph which proposes that FRP interferes with initial cap recognition.[15] On the other hand, FRP could interfere with 40S ribosome binding. In either case these results demonstrate that the IRE–FRP complex is a poor regulator of translation when it is placed well downstream of the 5′ end of the mRNA. Thus, even though IRE–FRP interaction is very tight, it is not able to block the scanning ribosome and/or the unwinding machinery of higher eukaryotic translation.

5. MECHANISM OF IRON INDUCTION

The exposure of eukaryotic cells to excess iron results in induction of ferritin synthesis, presumably as a result of an alteration in the activity of FRP. The FRP activity is affected in accordance with the iron status of the cell.[45,46,58,74,75] In iron-starved cells the activity of FRP is very high, as detected by RNA binding. In iron-replete cells this activity is significantly reduced. The recognition that FRP shares homology with aconitase, an iron–sulfur protein, has provided insight into one possible mechanism of regulating FRP activity directly through iron.[68,69,72,73] It has been postulated that FRP is regulated by conversion of its iron-sulfur center between a 3Fe cluster (active) and a 4Fe cluster (inactive). An alternative view is that cluster conversion is between 0Fe (apo-protein; active) and 4Fe (holo-protein; inactive). In either case FRP activity would be regulated through formation of a 4Fe cluster. Evidence consistent with this has been obtained.[72,73] Purified FRP has been loaded with iron *in vitro*. Following this, FRP has a detectable aconitase activity (suggesting formation of a 4Fe cluster)[73] and it fails to bind to the IRE.[72]

While conversion of a dynamic iron-sulfur cluster may be a mode of regulating FRP activity, it is also likely that FRP is regulated through other mechanisms; in particular, other forms of iron. For instance, hemin will inactivate FRP when treated *in vitro*.[26,74] This was initially detected by Lin *et al.*, who showed that pretreatment of FRP with hemin resulted in loss of repression of ferritin mRNA in wheat germ extracts.[26] Derepression of ferritin mRNA translation by hemin *in vitro* is specific for hemin and a close analog, Co^{+}-protoporphyrin IX. Other metallo-protoporphyrins and protoporphyrin IX have relatively little effect on FRP activity in translational repression. From these results, Lin *et al.*, proposed that hemin might be a form of iron which regulates FRP activity *in vivo*. In this regard, it is clear that the effects of hemin on FRP activity are consistent with its being a regulator of FRP activity. In addition to the study reported by Lin *et al.*, it has been reported that hemin treatment inhibits FRP binding to the IRE.[74] Inhibition of RNA binding by hemin is dose dependent. The measured affinity of FRP–IRE interaction is unchanged, however, in the presence of hemin, only the total binding activity is changed.[74] In addition, hemin causes a breakdown of a preformed FRP–IRE complex in a dose-dependent manner. The effects of hemin on FRP appear to occur in two stages. First, there is a specific binding site on FRP for hemin;[28] this interaction thus prevents FRP–IRE binding.[74] Second, hemin induces a rapid degradation of FRP in cells.[25]

It is also possible that FRP activity is modulated by the redox state of the cell, which could be affected by iron status. FRP in extracts of iron-overloaded cells is much less active than that in extracts of cells which have been incubated with an iron chelator.[46] Hentze *et al.* showed that FRP from iron-replete cells could be completely reactivated by treatment with high concentrations of reducing agents, such as 2-mercaptoethanol.[75] In contrast, FRP in extracts of iron-poor cells was unaffected by treatment with reducing agents, and it appeared that it was fully active. Based on these results, Hentze *et al.* proposed that FRP activity might be regulated through a redox mechanism, termed the "sulfhydryl switch," in which critical sulfhydryls within the repressor protein become oxidized in iron-replete cells.[75] In this scheme, oxidation of the protein results in its inactivation and this is reversed by reduction.

6. CONCLUDING REMARKS

Regulation of ferritin synthesis in response to iron status has provided the clearest example of control of translation by a *trans*-acting RNA binding protein. Such translational repressors are likely to be widespread in eukaryotic cells and undoubtedly will be shown to play critical roles in a variety of cellular processes.[15] Also, it should be noted that while the focus of this chapter has been on IRE/IRE-BP(FRP)-mediated regulation, other elements of the ferritin mRNA, particularly the flanking stem, contribute to the final magnitude of translational regulation observed in the cell.[80]

REFERENCES

1. Drysdale, J. W., 1988, Human ferritin gene expression, *Prog. Nucleic Acid Res. Mol. Biol.* **35:** 127–155.
2. Drysdale, J. W., Ambrosio, P., Adelman, T., Hazard, J. T., and Brooke, D., 1975, Isoferritins in normal and diseased states, in: *Proteins of Iron Storage and Transport in Biochemistry and Medicine* (R. R. Crichton, ed.), pp. 359–366, North-Holland, Amsterdam.
3. Munro, H. N., 1990, Iron regulation of ferritin gene expression, *J. Cell. Biochem.* **44:**107–115.
4. Munro, H. N., and Linder, M. C., 1978, Ferritin: Structure, biosynthesis, and role in iron metabolism, *Physiol. Rev.* **58:**317–396.
5. Theil, E. C., 1987, Ferritin: Structure, gene regulation, and cellular function in animals, plants, and microorganisms, *Annu. Rev. Biochem.* **56:**289–315.
6. Theil, E. C., 1990, The ferritin family of iron storage proteins, *Adv. Enzymol. Mol. Biol.* **63:**421–449.
7. Dickey, L. F., Sreedharan, S., Theil, E. C., Didsbury, J. R., Wang, Y.-H., and Kaufman, R. E., 1987, Differences in the regulation of messenger RNA for housekeeping and specialized-cell ferritin, *J. Biol. Chem.* **262**:7901–7907.
8. Didsbury, J. R., Theil, E. C., Kaufman, R. E. and Dickey, L. F., 1986, Multiple red cell ferritin mRNAs, which code for an abundant protein in the embryonic cell type, analyzed by cDNA sequence and by primer extension of the 5′-untranslated regions, *J. Biol. Chem.* **261:**949–955.
9. Chu, L. L. H., and Finberg, R. A., 1969, On the mechanism of iron-induced synthesis of apoferritin in HeLa cells, *J. Biol. Chem.* **244:**3847–3854.
10. Leibold, E. A., and Guo, B., 1992, Iron-dependent regulation of ferritin and transferrin receptor expression by the iron responsive element binding protein, *Annu. Rev. Nutr.* **12:**325–348.
11. Schull, G. E., and Theil, E. C., 1982, Translational control of ferritin synthesis by iron in embryonic reticulobytes of the bullfrog, *J. Biol. Chem.* **257:**14187–14191.
12. Theil, E. C., 1990, Regulation of ferritin and transferrin receptor mRNAs, *J. Biol. Chem.* **265:** 4771–4774.
13. Zahringer, J., Baliga, B. S., and Munro, H. N., 1976, Novel mechanism for translational control in regulation of ferritin synthesis by iron, *Proc. Natl. Acad. Sci. USA* **73:**857–861.
14. Aziz, N., and Munro, H. N., 1986, Both subunits of rat liver ferritin are regulated at a translational level by iron induction, *Nucleic Acids Res.* **14:**915–927.
15. Walden, William, E., and Thach, Robert E., 1986, Translational control of gene expression in a normal fibroblast: characterization of a sub-class of mRNAs with unusual kinetic properties, *Biochemistry* **25:**2033–3041.
16. Aziz, N., and Munro, H. N., 1987, Iron regulates ferritin mRNA translation through a segment of its 5′ untranslated region, *Proc. Natl. Acad. Sci. USA* **84:**8478–8482.
17. Hentze, M. W., Caughman, S. W., Rouault, T. A., Barriocanal, J. G., Dancis, A., Harford, J. B., and Klausner, R. D., 1987, Identification of the iron responsive element for the translational regulation of human ferritin mRNA, *Science* **238:**1570–1575.
18. Hentze, M. W., Rouault, T. A., Caughman, S. W., Dancis, A., Harford, J. B., and Klausner, R. D., 1987, A *cis* acting element is necessary and sufficient for translational regulation of human ferritin expression in response to iron, *Proc. Natl. Acad. Sci. USA* **84:**6730–6734.
19. Murray, M. T., White, K., and Munro, H., 1987, Conservation of ferritin heavy subunit gene structure: Implications for the regulation of ferritin gene expression, *Proc. Natl. Acad. Sci. USA* **84:**7438–7442.
20. Caughman, S. W., Hentze, M. W., Rouault, T. A., Harford, J. B., and Klausner, R. D., 1988, The iron responsive element is the single element responsible for iron dependent translational regulation of ferritin biosyntheis. Evidence for function as the binding site for a translational repressor, *J. Biol. Chem.* **263:**19048–19052.
21. Rouault, T. A., Hentze, M. W., Haile, J. D., Harford, J. B., and Klausner, R. D., 1989, The iron-responsive element binding protein: A method for the affinity purification of a regulatory RNA-binding protein, *Proc. Natl. Acad. Sci. USA* **86:**5768–5772.

22. Walden, W. E., Patino, M. M., and Gaffield, L., 1989, Purification of a specific repressor of ferritin mRNA translation from rabbit liver, *J. Biol. Chem.* **264:**13765–13769.
23. Neupert, B., Thompson, N. A., Meyer, C., and Kuhn, L. C., 1990, A high affinity purification method for specific RNA-binding proteins: Isolation of the iron regulatory factor from human placenta, *Nucleic Acids Res.* **18:**51–55.
24. Harrell, C. M., McKenzie, A. R., Patino, M. M., and Theil, E. C., 1991, Ferritin mRNA: Interactions of iron regulatory element with translational regulator protein P-90 and the effect on base-paired flanking regions, *Proc. Natl. Acad. Sci. USA* **88:**4166–4170.
25. Goessling, L. S., Daniels-McQueen, S., Bhattacharyya-Pakrasi, M., Lin, J.-J., and Thach, R. E., 1992, Enhanced degradation of the ferritin repressor protein during induction of ferritin messenger RNA translation, *Science* **256:**670–673.
26. Lin, J.-J., Daniels-McQueen, S., Patino, M. M., Gaffield, L., Walden, W. E., and Thach, R. E., 1990, Derepression of ferritin messenger RNA translation by hemin *in vitro*, *Science* **247:**74–77.
27. Lin, J.-J., Patino, M. M., Gaffield, L., Walden, W. E., and Thach, R. E., 1990, Specificity of induction of ferritin synthesis by hemin, *Biochem. Biophys. Acta* **1050:**146–150.
28. Lin, J.-J., Patino, M. M., Gaffield, L., Walden, W. E., and Thach, R. E., 1991, Hemin spontaneously crosslinks to a specific site on the 90 kDa ferritin repressor protein, *Proc. Natl. Acad. Sci. USA* **88:**6068–6071.
29. Boyd, D., Jain, S. K., Crampton, J., Barret, K. J., and Drysdale, J. W., 1984, Isolation and characterization of a cDNA clone for human ferritin heavy chain, *Proc. Natl. Acad. Sci. USA* **81:** 4751–4755.
30. Boyd, D., Vecoli, C., Belcher, D. M., Jain, S. K., and Drysdale, J. W., 1985, Structural and functional relationships of human ferritin H and L chains deduced from cDNAs clones, *J. Biol. Chem.* **260:**11755–11761.
31. Brown, A. J. P., Leibold, E. A., and Munro, H. N., 1983, Isolation of cDNA clones for the light subunit of rat liver ferritin: Evidence that the light subunit is encoded by a multigene family, *Proc. Natl. Acad. Sci. USA* **80:**1265–1269.
32. Chou, C. C., Gatti, R. A., Fuller, M. L., Concannon, P., Wong, A., Chada, S., Davis, R. C., and Salser, W. A., 1986, *Mol. Cell. Biol.* **6:**566–573.
33. Constanzo, F., Colombo, M., Staempfli, S., Santoro, C., Marone, M., Frank, R., Delius, H., and Cortese, R., 1986, Structure of gene and pseudogene of human apoferritin H, *Nucleic Acids Res.* **14:**721–736.
34. Constanzo, F., Santoro, C., Colantuoni, V., Bensi, G., Raugei, G., Romano, V., and Cortese, R., 1984, Cloning and sequencing of a full length cDNA coding for a human apoferritin H chain: Evidence for a multigene family, *EMBO J.* **3:**25–27.
35. Daniels-McQueen, S., Ray, A., Walden, W. E., Ray, B. K., Brown, P. H., and Thach, R. E., 1988, Nucleotide sequence of cDNA encoding rabbit ferritin L chain, *Nucleic Acids Res.* **16:**7741.
36. Dorner, M. H., Salfeld, J., Will, H., Leibold, E. A., Vass, J. K., and Munro, H. N., 1985, Structure of human ferritin light subunit messenger RNA: Comparison with heavy subunit message and functional implications, *Proc. Natl. Acad. Sci. USA* **82:**3139–3143.
37. Hentze, M. W., Keim, S., Papadopoulos, P., O'Brien, S., Mosi, W., Drysdale, J., Leonard, W. J. Harford, J. B., and Klausner, R. D., 1986, *Proc. Natl. Acad. Sci. USA* **83:**7226–7230.
38. Jain, S. K., Barret, K. J., Boyd, D., Favreau, M. F., Crampton, J., and Drysdale, J. W., 1985, Ferritin H and L Chains are derived from different multigene families, *J. Biol. Chem.* **260:**11762–11768.
39. Leibold, E. A., and Munro, H. N., 1987, Characterization and evolution of the expressed rat ferritin light subunit gene and its pseudogene family, *J. Biol. Chem.* **262:**7335–7341.
40. Santoro, C., Marone, M., Ferrone, M., Constanzo, F., Colombo, M., Minganti, C., Cortese, R., and Silengo, L., 1986, Cloning of the gene coding for human L apoferritin, *Nucleic Acids Res.* **14:** 2863–2876.
41. Stevens, P. W., Dodgson, J. B., and Engel, J. D. 1987 Structure and expression of the chicken ferritin H-subunit gene, *Mol. Cell. Biol.* **7:**1751–1758.

42. Torti, S. V., Kwak, E. L., Miller, S. C., Langdon, L. M., Ringold, G. M., Myambo, K. B., Young, A. P., and Torti, F. M., 1988, The molecular cloning and characterization of murine ferritin heavy chain, a tumor necrosis factor-inducible gene, *J. Biol. Chem.* **263:**12638–12644.
43. Gooseen, B., Caughman, S. W., Harford, J. B., Klausner, R. D., and Hentze, M. W., 1990, Translational repression by a complex between the iron responsive element of ferritin mRNA and its specific cytoplasmic binding protein is position *in vivo*, *EMBO J.* **9:**4127–4133.
44. Goossen, B., and Hentze, M. W., 1992, Position is the critical determinant for function of iron-responsive elements as translational regulators, *Mol. Cell. Biol.* **12:**1959–1966.
45. Barton, H. A., Eisenstein, R. S., Bomford, A., and Munro, H. N., 1990, Determinants of the interaction between the iron-responsive element binding protein and its binding site in rat L-ferritin mRNA, *J. Biol. Chem.* **265:**7000–7078.
46. Haile, D. J., Hentze, M. W., Rouault, T. A., Harford, J. B., and Klausner, R. D., 1989, Regulation of interaction of the iron responsive element binding protein with iron-responsive RNA elements, *Mol. Cell. Biol.* **9:**5055–5061.
47. Leibold, E. A., Laudano, A., and Yu, Y., 1990, Structural requirements of iron-responsive elements for binding of the protein involved in both transferrin receptor and ferritin mRNA prot-transcriptional regulation, *Nucleic Acids Res.* **18:**1819–1824.
48. Hentze, M. W., Caughman, S. W., Casey, J. L., Koeller, D. M., Rouault, T. A., Harford, J. B., and Klausner, R. D. 1988 A model for the structure and functions of the iron responsive elements, *Gene* **72:**201–208.
49. Wang, Y.-H., Sczekan, R. S., and Theil, E. C., 1990, Structure of the 5′ untranslated regulatory region of ferritin mNRA studied in solution, *Nucleic Acids Res.* **18:**4463–4468.
50. Casey, J. L., Di Jeso, B., Rao, K., Klausner, R. D., and Harford, J. B., 1988, Two genetic loci participate in the regulation by iron of the gene for the human transferrin receptor, *Proc. Natl. Acad. Sci. USA* **85:**1787–1791.
51. Casey, J. L., Hentze, M. W., Koeller, D. M., Caughman, S. W., Rouault, T. A., Klausner, R. D., and Harford, J. B., 1988, Iron-responsive elements: Regulatory RNA sequences that control mRNA levels and translation, *Science* **240:**924–928.
52. Casey, J. L., Koeller, D. M., Ramin, V. C., Klausner, R. D., and Harford, J. B., 1989, Iron regulation of transferrin receptor mRNA levels requires iron-responsive elements and a rapid turnover determinant in the 3′ untranslated region of the mRNA, *EMBO J.* **8:**3693–3699.
53. Cox, T. C., Bawden, M. J., Martin, A., and May, B. K., 1991, Human erytroid 5-aminolevulinate synthase: Promoter analysis and identification of an iron-responsive element in the mRNA, *EMBO J.* **10:**1891–1902.
54. Dandekar, T., Stripecke, R., Gray, N. K., Goossen, B., Constable, A., Johansson, H. E., and Hentze, M. W., 1991, Identification of a novel iron-responsive element in murine and human erytroid δ aminolevulinic acid synthase mRNA, *EMBO J.* **10:**1903–1909.
55. Mullner, E. W., and Kuhn, L. C., 1988, A stem-loop in the 3′ untranslated region mediates iron-dependent regulation of transferrin receptor mRNA stability in the cytoplasm, *Cell* **53:**815–825.
56. Mullner, E. W., Neupert, B., and Kuhn, L. C., 1989, A specific mRNA binding factor regulates the iron-dependent stability of cytoplasmic transferrin receptor mRNA, *Cell* **58:**373–382.
57. Owen, D., and Kuhn, L. C., 1987, Noncoding 3′ sequences of the transferrin receptor gene are required for mRNA regulation by iron, *EMBO J.* **6:**1287–1293.
58. Leibold, E. A., and Munro, H. N., 1988, Cytoplasmic protein binds *in vitro* to a highly conserved sequence in the 5′ untranslated region of ferritin heavy- and light-subunit mRNAs, *Proc. Natl. Acad. Sci. USA* **85:**2171–2175.
59. Walden, W. E., Daniels-McQueen, S., Brown, P. H., Russell, D. A., Bielser, D., Bailey, L. C., and Thach, R. E., 1988, Translational repression in eukarytes: Partial purification and characterization of a repressor of ferritin mRNA translation, *Proc. Natl. Acad. Sci. USA* **85:**9503–9507.
60. Dickey, L. F., Wang, Y.-H., Shull, G. E., Wortman, I. A., and Theil, E. C., 1988, The importance of the 3′ untranslated region in the control of ferritin mRNA, *J. Biol. Chem.* **263:**3071–3074.
61. Brown, P. H., Daniels-McQueen, S., Walden, W. W., Patino, M. M., Gaffield, L., Bielser, D., and

Thach, R. E., 1989, Requirements for the translational repression of ferritin transcripts on wheat germ extracts by a 90-kDa protein from rabbit liver, *J. Biol. Chem.* **23:**13883–13886.

62. Hirling, H., Emery-Goodman, A., Thompson, N., Neupert, B., Seiser, C., and Kuhn, L. C., 1992, Expression of active iron regulatory factor from a full-length human cDNA by *in vitro* transcription/translation, *Nucleic Acids Res.* **20:**33–39.
63. Patino, M. M., and Walden, W. E., 1992, Cloning of a functional cDNA for the rabbit ferritin repressor protein: Demonstration of a tissue specific pattern of expression, *J. Biol. Chem.* **267:**19011–19016.
64. Rouault, T. A., Tang, C. K., Kaptain, S., Burgess, W.H., Haile, D.J., Samaniego, F., McBride, O. W., Harford, J. B., and Klausner, R. D., 1990, Cloning of the cDNA encoding an RNA regulatory protein—the human iron-responsive element-binding protein, *Proc. Natl. Acad. Sci. USA* **87:** 7958–7962.
65. Yu, Y., Radisky, E., and Leibold, E. A., 1992, *J. Biol. Chem.* **267:**19005–19010.
66. Philpott, C. C., Rouault, T. A., and Klausner, R. D., 1991, Sequence and expression of the murine iron-responsive element binding protein, *Nucleic Acids Res.* **19:**6333.
67. Rothernberger, S., Mullner, E., and Kuhn, L. C., 1990, The mRNA-binding protein which controls ferritin and transferrin receptor expression is conserved during evolution, *Nucleic Acids Res.* **18:**1175–1179.
68. Hentze, M. W., and Argos, P., 1991, Homology between IRE-BP, a regulatory RNA-binding protein, aconitase, and isopropylmalate isomerase, *Nucleic Acids Res.* **19:**1739–1740.
69. Rouault, T. A., Stout, C. D., Kaptain, S., Harford, J. B., and Klausner, R. D., 1991, Structural relationship between an iron-regulated RNA-binding protein (IRE-BP) and aconitase: Functional implications, *Cell* **64:**881–883.
70. Prodromou, C., Artymiuk, P. J., and Guest, J. R., 1992, The aconitase of *Escherichia coli*, *Eur. J. Biochem.* **204:**599–609.
71. Robbins, A. H., and Stout, C. D., 1989, The structure of aconitase, *Proteins* **5:**289–312.
72. Constable, A., Quick, S., Gray, N. K., and Hentze, M. W., 1992, Modulation of the RNA-binding activity of a regulatory protein by iron *in vitro*: Switching between enzymatic and genetic function? *Proc. Natl. Acad. Sci. USA* **89:**4554–4558.
73. Kaptain, S., Downey, W. E., Tang, C., Philpott, C., Haile, D., Orloff, D. G., Harford, J. B., Rouault, T. A., and Klausner, R. D., 1991, A regulated RNA binding protein also possesses aconitase activity, *Proc. Natl. Acad. Sci. USA* **88:**10109–10113.
74. Swenson, G. R., Patino, M. M., Beck, M. M., Gaffield, L., and Walden, W. E., 1991, Characteristics of the interaction of the ferritin repressor protein with the iron-responsive element, *Biol. Metals* **4:**48–55.
75. Hentze, M. W., Rouault, J. B., Harford, J. B., and Klausner, R. D., 1989, Oxidation–reduction and the molecular mechanism of a regulatory RNA–protein interaction, *Science* **244:**357–358.
76. Hentze, M. W., Seuanez, H. N., O'Brien, S. J., Harford, J. B., and Klausner, R. D., 1989, Chromosomal localization of nucleic acid-binding protein by affinity mapping: Assignment of the IRE-binding protein gene to chromosome 9, *Nucleic Acids Res.* **17:**6103–6108.
77. Koeller, D. M., Casey, J. L., Hentze, M. W., Gerhardt, E. M., Chan, L. L., Klausner, R. D., and Harford, J. B., 1989, A cytosolic protein binds to structural elements within the iron regulatory region of the transferrin receptor mRNA, *Proc. Natl. Acad. Sci. USA* **86:**3574–3578.
78. Rouault, T. A., Hentze, M. W., Caughman, S. W., Harford, J. B., and Klausner, R. D., 1988, Binding of a cytosolic protein to the iron-responsive element of human ferritin messenger RNA, *Science* **241:**1207–1210.
79. Kennedy, M. C., Mende-Mueller, L., Blondin, G. A., and Beinert, H., 1992, Purification and characterization of cytosolic aconitase from beef liver and its relationship to the iron-responsive element binding protein, *Proc. Natl. Acad. Sci. USA* **89:**11730–11734.
80. Dix, D. J., Lin, P.-N., Kimata, Y., and Theil, E. C., 1992, The iron regulatory region of ferritin mRNA is also a positive control element for iron-dependent translation, *Biochemistry* **31:**2818–2822.

Chapter 16

Control of Ribosomal Protein Synthesis in Eukaryotic Cells

Roger L. Kaspar, David R. Morris, and Michael W. White

1. INTRODUCTION

Ribosomal assembly requires three or four separate ribosomal RNA (rRNA) molecules as well as ~50–80 ribosomal proteins (r-proteins); the exact numbers depend on the species (reviewed in Wool[1]). These components comprise a large portion of the total cellular RNA and protein and are synthesized in roughly equimolar amounts which are rapidly assembled into ribosomes.

Coordinating the production of ribosomes requires a synchronization of the synthesis and/or the degradation of the individual components. The best understood mechanisms of coordinate control arise from studies in *Escherichia coli*. In *E. coli*, r-protein genes are organized in operons. The resulting polycistronic messages are subject to autogenous translational regulation, such that when concentrations of free r-proteins rise, key individual r-proteins bind and directly prevent the translation of their respective multicistronic messenger RNAs (mRNAs). This feedback mechanism synchronizes the rate of synthesis of r-proteins with the demand for assembly on nascent rRNA (for a review see ref. 2). This

---

Roger L. Kaspar and David R. Morris • Department of Biochemistry, University of Washington, Seattle, Washington 98133. Michael W. White • Department of Veterinary Molecular Biology, Montana State University, Bozeman, Montana 59717.

Translational Regulation of Gene Expression 2, edited by Joseph Ilan. Plenum Press, New York, 1993.

coordinate regulatory mechanism does not involve the transcription of r-protein mRNAs and is consistent with these mRNAs being constitutively produced regardless of the level of rRNA transcripts, with only a few exceptions.[3]

No equivalent mechanism of autogenous translational control has been found in eukaryotes. Gene dosage experiments in various eukaryotes[4–7] have shown that an unbalanced rate of r-protein synthesis can occur, indicating no direct link to rRNA levels. Unassembled r-proteins, however, do not accumulate, but are rapidly degraded. This degradative mechanism is extraordinarily fast, with estimated half-lives of 2 min or less for some yeast r-proteins.[8–10] The lack of synchronization of rRNA and r-protein production is striking in *Xenopus*, where the synthesis of different ribosomal components may occur in different developmental stages.[4,11]

While the balance between rRNA and r-protein levels in eukaryotes is not tightly maintained, the control of the synthesis of the ~70–80 r-proteins is, with few exceptions, directly linked via common regulatory mechanisms. R-proteins and their corresponding mRNAs are produced in equimolar amounts in eukaryotes despite differences in gene copy number as is found in yeast[6] or the lack of synchronization with rRNA as is found in *Xenopus*. In this review, we will emphasize the coordinated regulation of r-protein synthesis. Any individual r-proteins produced in excess of the others appear to be degraded.[4,6] The mechanisms controlling r-protein synthesis involve the control of mRNA levels and/or mRNA utilization. R-protein mRNAs are relatively abundant messages of which there are over 70 different species. Selective translation of these mRNAs is a primary mode of controlling r-protein synthesis in *Xenopus* and *Drosophila* development, differentiating mouse myoblasts, and growth-regulated mammalian cells and represents one of the major translational regulatory mechanisms in these cells. In the following review we will discuss the different mechanisms of regulating the synthesis of eukaryotic r-proteins, although we will emphasize the recent studies concerning the translational control of r-proteins in vertebrate systems.

2. TRANSLATIONAL CONTROL OF RIBOSOMAL PROTEIN SYNTHESIS

The synthesis of vertebrate r-proteins has been shown to be under translational control during *Xenopus* embryogenesis,[4] differentiation of mouse myoblasts into fibers,[12] mitogen-activated T-lymphocytes[13] and fibroblasts,[14–17] and glucocorticoid-responsive lymphosarcoma cells.[18] In each case the translational regulation involves the movement of r-protein mRNAs between untranslated subribosomal particles and active polysomes. This regulation is characterized by a distinct bimodal distribution of r-protein mRNA, i.e. the mRNA is either fully loaded with ribosomes or not translated at all, indicating that once initiation occurs, the r-protein mRNA becomes efficiently and fully translated. Research in the last few years has shown that the 5′-untranslated region (UTR) is involved in mediating the translational state of r-protein mRNAs and that this region may specifically interact with distinct cytoplasmic factors and possibly translational initiation factors.

2.1. Control of Vertebrate Ribosomal Protein mRNA Translation Requires the 5′-Untranslated Region

Recently, several groups have shown that the 5′-leader sequence of several vertebrate r-protein mRNAs is sufficient to confer translational regulation on an unregulated reporter mRNA. Replacement of the promoter and 5′-UTR of the gene encoding chloramphenical acetyltransferase (CAT) with the corresponding region of *Xenopus* r-protein S19 resulted in a mRNA that was translationally regulated during *Xenopus* development in a manner similar to endogenous r-protein mRNAs.[19] The 5′-UTR of murine r-protein L30 mRNA or the first 29 nucleotides of S16 mRNA is sufficient to confer translational control on the human growth hormone mRNA in glucocorticoid-regulated murine lymphosarcoma cells,[20] and the first 31 nucleotides of S16 mRNA has also been shown to impart translational regulation to the SV-GALK mRNA during mouse myoblast differentiation.[21]

All vertebrate r-protein mRNAs analyzed to date have been shown to contain between 8 and 14 pyrimidine residues at the 5′ end with no apparent consensus sequence.[22–26] This polypyrimidine mRNA element is part of a larger sequence in the gene consisting of approximately 20 pyrimidine nucleotides which spans the transcriptional start site and has previously been suggested to be involved in precisely determining the transcriptional start site of murine S16 mRNA.[27] Deletion of the polypyrimidine region in a murine L32 construct resulted in a mRNA that was no longer translationally regulated.[13] Moreover, substituting purines for pyrimidines or shortening the pyrimidine tract resulted in a S16/growth hormone hybrid mRNA that was unregulated.[20] These studies strongly suggest that the polypyrimidine tract is necessary for translational regulation and are consistent with the studies cited above in which a segment of vertebrate r-protein 5′-UTR as small as 29 nucleotides which includes the polypyrimidine tract is sufficient to confer translational control to an unregulated reporter mRNA.[20,21] The involvement of the polypyrimidine tract in vertebrate r-protein translational control does not rule out the possibility, however, that other sequences or RNA structures in the 5′-UTR may also be needed. Secondary structure analysis of the r-protein 5′-UTRs does not predict any stable RNA stem–loop structures using conventional Watson and Crick base pairing, although as a result of the paucity of adenine residues there are many potential G–U base pairings[21] (also R. L. Kaspar, D. R. Morris, and M. W. White, unpublished data) which might be stabilized through interactions with other factors. In addition, it is not clear if the 5′-UTR can work in any context. The work of Levy and co-workers[20] and Hammond *et al.*[21] suggests that the polypyrimidine tract must be adjacent to the 5′-cap site and that the first residue must be a cytosine. This contradicts the results of Kaspar *et al.*,[13] in which a translationally regulated L32 construct mRNA contains a small Rous sarcoma virus (RSV) sequence between the 5′-cap site and the r-protein 5′-UTR.

The above results with vertebrate r-protein mRNAs strongly suggest that the polypyrimidine tract is necessary, although not necessarily sufficient, for differential translation of various r-protein mRNAs, depending on the growth or developmental state of the cells. Further experiments are needed to test whether an oligopyrimidine tract in the absence of other functional elements can transfer translational control to an unregulated mRNA. It is interesting to note that there are other vertebrate mRNAs which contain polypyrimidine sequences at the 5′ end which appear to be translationally regulated similarly to the r-protein mRNAs.[28–32] Thus, r-protein mRNAs may belong to an even larger class of mRNAs, all of which may be regulated in a similar fashion.

2.2. Cytoplasmic Factors That Interact with the 5′-Untranslated Region

In vitro binding studies suggest cytoplasmic factors may specifically interact with the r-protein 5′-UTR region.[13] RNA transcripts containing the polypyrimidine tract derived from the murine L32 5′-UTR interacted with a factor(s) present in lymphocyte and fibroblast cytoplasmic protein extracts as shown by gel retardation assays. Binding activity was apparently independent of the growth state of the cells from which the extracts were prepared, suggesting that binding of this protein to the RNA is not regulated at least in an all-or-none fashion. A similar RNA missing the polypyrimidine region showed no such interaction, suggesting that the polypyrimidine tract was necessary for binding. Excess unlabeled transcript containing the polypyrimidine tract competed away binding of the labeled transcript, while addition of the same molar excess of RNA lacking this element had no effect. Addition of the nonspecific competitors transfer RNA (tRNA) or heparin did not inhibit the gel-shift band nor did capping the RNA affect the binding pattern. Label transfer experiments, in which radiolabeled RNA is cross-linked via UV irradiation to cytoplasmic protein extracts, demonstrated that a 56-kDa factor (p56) specifically interacts with RNA derived from the L32 5′-UTR, but not to a similar RNA missing the polypyrimidine tract. Addition of an excess of unlabeled poly(U) or poly(C) inhibited both formation of the gel-shift band and cross-linking of the probe to the 56-kDa protein. Approximately 100-fold more poly(C) than poly(U) was needed to inhibit binding of labeled probe, suggesting that p56 should bind more tightly to oligo(U) than oligo(C) tracts in mRNA. These experiments strongly suggest that the polypyrimidine tract, which is involved in the *in vivo* regulation of r-protein mRNA translation, is also necessary for p56 binding and that the ratio of U to C residues in the tract may determine the strength of binding. Interestingly, although no sequence specificity has been seen in the r-protein polypyrimidine tracts, approximately equal amounts of uracil and cytosine residues are found in each of 12 vertebrate r-protein mRNAs whose cap sites have been determined.[20]

The lack of detectable differences in binding of p56 to the polypyrimidine tract, regardless of the growth state of the cells from which the extracts were made, suggests that simple binding of this protein may not regulate sequestration of the r-protein mRNAs into subpolysomal particles. While the interactions between p56 and other components of the protein synthesis machinery are not known, the signaling pathways leading to changes in the translational state of r-protein mRNAs are beginning to be understood. One might expect one or more of the components of this complex to respond to developmental and growth stimuli, modulating the translatability of the mRNA. At least one signal transduction pathway appears to involve calcium, as exhibited by the lack of recruitment of L32, L30, and S16 mRNAs into polysomes when quiescent T cells are activated in the absence of exogenous calcium or in the presence of the calmodulin-dependent kinase inhibitor W-7. These treatments have no effect on the distribution of actin mRNA, which, in each of these cases, is found associated with large polysomes. Furthermore, treatment of resting lymphocytes with the calcium ionophore ionomycin results in the shift of r-protein mRNAs into polysomes (R. L. Kaspar, M. W. White, and D. R. Morris, unpublished data). Thus, at least one of the r-protein translational regulatory factors may be regulated by intracellular calcium levels. The precise role, however, of p56 and other regulatory factor(s) and how they interact with the signal transduction pathways and the translational machinery resulting in modulation of translation remain to be elucidated.

The lack of apparent regulation of p56 binding to r-protein mRNAs contrasts with the regulatory behavior of the 87-kDa ferritin repressor protein, which is the only proven example of a eukaryotic translational repressor protein.[33] Binding of the ferritin repressor protein to the iron-responsive element, located in the 5′-UTR of ferritin mRNA, inhibits translation (reviewed in ref. 34). Unlike regulation of ferritin mRNA translation, which occurs through modulation of repressor protein binding, it seems that events other than direct modulation of binding of p56 to the polypyrimidine tract must regulate r-protein mRNA translation. One possible explanation may be that binding of p56 creates a nucleation site, around which other regulated factors might assemble in a regulated fashion. Alternatively, during times of limiting translation components, p56 may inhibit the interaction of r-protein mRNAs with certain initiation factors resulting in translational inhibition.

In recent years, a growing list of examples of polypyrimidine elements involved in RNA metabolism in eukaryotic cells has emerged. The region of the 3′-splice site of mammalian introns contains a polypyrimidine element which is required for proper splicing and has recently been demonstrated by two independent laboratories to specifically bind 62-kDa[35] and 65-kDa polypeptides[36] from nuclear extracts. Cap-independent internal translational initiation of picornaviral mRNAs requires pyrimidine-rich RNA elements[37–39] within the internal ribosome entry site. A 52-kDa protein interacts with this region of the poliovirus 5′-UTR,[40] but has not been shown to bind to the pyrimidine-rich element. Despite the similarities in the sizes of p56 and the 62-kDa splicing protein (pPTB), they seem to be distinct proteins based on purification properties, mobilities in SDS gels, and intracellular localization. For example, cross-linking of extracts containing either pPTB or p56 to an identical probe clearly results in RNA–protein complexes that have different mobilities in SDS gels. In addition, antiserum raised against pPTB does not cross-react with p56 (T. Kakegawa and R. L. Kaspar, unpublished results). These examples suggest that a new family of polyribopyrimidine binding proteins of 56–65 kDa size may be emerging whose members are involved in various aspects of RNA metabolism. The common element in each case is the formation of large multicomponent complexes leading to the assembly of active spliceosomes, translational repression structures, and possibly cap-independent translational initiation complexes.

2.3. Role of Initiation Factors in Regulating Ribosomal Protein mRNA Translation

Since the binding of p56 to the polypyrimidine tract does not correlate with the translational state of r-protein mRNA in cells from which extracts were prepared, one suspects that other regulatory factors might be involved. Consistent with this idea, Hammond and co-workers[21] have shown that initiation factor eIF-4F is limiting for the translation of S16 mRNA in reticulocyte lysates. Furthermore, in these same experiments, translation of S16 mRNA seemed to show a selective sensitivity to the cap analog m^{7}GDP. The possible regulatory role of the 25-kDa cap-binding protein (eIF-4E), which is a component of eIF-4F, has received much attention in recent years. This phosphoprotein is thought to be the first factor to interact with mRNA during the initiation process and is found in limiting concentrations with respect to other initiation factors. In addition, a positive correlation between phosphorylation of this factor and translation rates *in vivo* has been shown in a number of circumstances,[41] including the translation of r-protein L32 mRNA after serum activation of murine fibroblasts.[17] Together, these results are consistent with eIF-4E being involved in r-protein translational control. Whether this interaction is the rate-limiting step *in vivo* or whether eIF-4E interacts with the p56 polypyrimidine binding protein remain to be determined.

Cap-binding protein has been implicated in two closely related models of eukaryotic translational initiation,[42–45] which are based on the idea that under conditions of limiting initiation factor concentrations, mRNAs with low-affinity interactions would be selectively outcompeted and preferentially untranslated. The r-protein mRNAs, however, do not appear precisely to fit these models. As was previously mentioned, the mammalian r-protein mRNAs L32, S16, and L30 show a distinct bimodal distribution; they are either found in subpolysomal particles or on polysomes of approximately 4–5 ribosomes, which is consistent with an optimal loading of the mRNAs based on their coding sizes.[13,18] These data indicate that once r-protein mRNAs are recruited into the translated pool, they are translated very efficiently. Treatment of quiescent T-lymphocytes (M. W. White,

R. L. Kaspar, and D. R. Morris, unpublished results) with cycloheximide, which at low levels preferentially inhibits peptide elongation, resulting in an increase in the number of initiations per mRNA molecule,[45] can at best only partially shift the subpolysomal r-protein mRNAs into polysomes. For comparison, the inefficiently translated mRNA encoding ornithine decarboxylase is completely shifted out of subpolysome particles by the above treatment.[46] Thus, the explanation for the high degree of translational repression of the r-protein mRNAs in T-lymphocytes does not appear to be that they are poor competitors for limiting initiation factors. Several observations may suggest clues to the nature of the translational repression of these mRNAs. The subpolysomal particles containing r-protein mRNAs from mammalian cells cosediment with large particles of 35–45S in sucrose gradients[21] (also M. W. White, R. L. Kaspar, and D. R. Morris, unpublished data). This could be consistent with the association of these mRNAs with either the preinitiation complex containing the 40S ribosomal subunit or a similarly complex multicomponent particle. Moreover, Hammond *et al.*[21] have also shown that in addition to eIF-4F, the initiation factor eIF-3 also stimulates entry of r-protein S16 mRNA into polysomes in reticulocyte lysates. The addition of initiation factors eIF-1A, eIF-2, eIF-4A, and eIF-4B had no effect on the distribution of S16 mRNA in polysomes. These data are consistent with a model whereby the r-proteins are translationally blocked prior to the joining of the 60S ribosomal subunit, which may be regulated by the interaction of a number of factors, including eIF-4E and eIF-3, with the p56 polypyrimidine binding protein.

2.4. Control of Ribosomal Protein mRNA Translation in Nonvertebrate Eukaryotes

The synthesis of r-proteins has also been shown to be translationally controlled in nonvertebrate eukaryotes. R-protein synthesis in *Drosophila* is translationally regulated during embryogenesis by a change in distribution of these mRNAs between subpolysomal particles and polysomes (reviewed in ref. 47). The structure of *Drosophila* r-protein mRNAs also shows many similarities to that of vertebrates, including a conserved polypyrimidine element at the 5′ end of the mRNA.[48,49] The shift of r-protein mRNAs into subpolysomal particles after the initiation of differentiation in *Dictyostelium*[50] is another example of nonvertebrate r-protein translational control. Replacement of the 5′-UTR of a translationally unregulated *Dictyostelium* mRNA, discoidin 1α, with the 5′-UTR of r-protein 1024 results in a mRNA that is translationally regulated,[51] suggesting that the necessary sequences reside in the 5′-UTR, similar to what is seen in *Xenopus* development, myoblast differentiation, and growth-stimulated mammalian cells. Some species of r-protein mRNAs from *Dictyostelium* contain polypyrimidine RNA elements that could play a role in translational regulation analogous to the situation in vertebrates. The mRNA encoding r-protein 1024 shows multiple transcription start sites which result in different lengths of 5′-UTR,[52] some of which do not have the polypyrimidine element. By examining the polysome distribution of the various 1024 mRNA species, Steel and Jacobson[51] have shown that the different transcripts show slightly different distributions after sucrose gradient centrifugation, indicating different utilization by the translational machinery. One set of transcripts starts in a polypyrimidine tract similar to that of the vertebrate mRNAs and appears to show the highest degree of translational repression. How or whether polypyrimidine elements are involved in this regulation remains to be determined.

3. NONTRANSLATIONAL MEANS OF REGULATING RIBOSOMAL PROTEIN SYNTHESIS

While translational control of r-protein synthesis is a major mechanism for synchronizing the production of these proteins in eukaryotic cells, it is not the sole mechanism. Many eukaryotes utilize mechanisms which control the levels of r-protein mRNAs as a means to regulate r-protein synthesis. In some cell types such as yeast, the control of r-protein mRNA levels is the only controlling mechanism, while others employ both translational and transcriptional regulatory modes.

R-proteins are produced in equimolar amounts in *Saccharomyces cerevisiae*,[53] indicating that the synthesis of r-proteins is coordinately regulated. Unlike vertebrate r-protein synthesis, however, transcription of r-protein genes has proven to be a primary regulatory mechanism, with at least one notable exception (see below). No translational control of expression of the genes has been demonstrated in yeast. Furthermore, the polypyrimidine tract found at the 5′ end of vertebrate r-protein mRNAs is not present.[54] Transcription from the r-protein promoters is collectively controlled by growth rate.[55,56] The structures of yeast r-protein genes have begun to be characterized and appear to fall into two classes, based on the structures of their promoters (for a review see Mager and Planta[57]). One class of yeast r-protein promoters contains a nucleotide sequence (RPG box) which is essential for basal and activated transcription.[58,59] An abundant 92-kDa DNA-binding protein has been identified[60–62] and cloned,[63] which specifically binds the RPG box and has been designated RAP1. RAP1 contains no known DNA-binding motifs. Members of another class of yeast r-protein promoters contain in common a DNA element which is bound by the factor ABF1.[64,65] The ABF1 element is necessary for transcriptional activation induced by carbon source shifts, but does not appear to be sufficient in isolation for regulated transcription. What other factors work in concert with ABF1 have yet to be defined. The strength of binding of ABF1, a 80-kDa zinc-finger protein,[66–68] correlates with the different levels of expression of the duplicated yeast r-protein genes L2A and L2B.[69–71] The molecular details of how these two classes of r-protein promoters are regulated according to the growth rate of the cell and how the relative activities of these promoters are balanced are clearly fertile areas of future investigation.

Although production of r-proteins in mammalian cells shows a large component of translational control (see above), transcription of the r-protein genes should be coordinated as well. Similar promoter strengths have been seen for the murine r-protein genes (L32, L30, and S16) analyzed to date.[72] The mouse r-protein promoter region is characterized by the lack of a canonical TATA box and the major site of transcriptional initiation occurs at a cytosine residue embedded within a polypyrimidine tract flanked by GC-rich sequences.[73] A similar architecture is seen in the r-protein promoters from *Xenopus* (reviewed in ref. 4). Sequences of the 5′-UTR and first intron of mammalian r-protein genes have been identified which bind nuclear factors and are necessary for maximal promoter activity.[27,72,74–76] With the exception of Sp1, these factors have not yet been characterized, nor has it been demonstrated how these factors participate in coordinating transcription of r-protein genes.

The yeast r-protein L32 gene shows an interesting form of regulation, in addition to transcriptional control, that may be unique among this group of genes. Unspliced L32 transcripts accumulated when this gene was overexpressed on multicopy plasmids[77] and the accumulation of unspliced L32 RNA seemed to be regulated under certain physiological manipulations of the cells, resulting in variations in the level of mature L32 mRNA.[78] Of the group of yeast r-protein genes studied, this behavior seemed to occur only with L32. Sequences in the 5′-exon and exon/intron junction are involved in this regulation[79] and it has been suggested that this region of the pre-mRNA may form a relatively unstable secondary structure, which could interact with free L32, preventing the interaction with the U1 snRNP and thereby splicing.[77,79] Although the control of L32 mRNA levels through splicing is the only example among the yeast r-proteins studied, a similar mechanism may also be involved in the regulation of *Xenopus* r-protein L1.[80] The reasons for this apparent unique behavior for these two r-protein genes is not understood. These two exceptional genes may represent a class of "key" r-proteins that could have some particular functional or regulatory significance in the synthesis or assembly of the ribosomal components.

4. SUMMARY

Regulation of r-protein synthesis in eukaryotes involves diverse mechanisms at the levels of transcription, splicing, and mRNA translation. Often, multiple levels of regulation can occur in a particular cell type. Any excess r-proteins produced appear to be degraded, resulting in nearly equimolar levels of each individual r-protein in the cell. In contrast, prokaryotes appear to control the levels of r-proteins by feedback inhibition of translation. Thus, both mechanisms appear to prevent accumulation of free r-proteins, which, in excess, might be detrimental to the cell due to their basic nature.

Regulation at the level of translation plays a major role in specifically controlling r-protein synthesis in a number of circumstances, including differentiation and growth control. Recent advances implicate the 5′-UTR of vertebrate r-protein messages, and particularly the polypyrimidine tract at the 5′ end, in determining the translational state. At this point, no sequence specificity in this RNA element appears critical other than the requirement for pyrimidines. Future experiments should better define whether this polypyrimidine tract acts alone or in combination with other mRNA elements. The recent discovery of a 56-kDa protein which binds to this tract will provide clues as to how r-protein synthesis might be coordinately regulated. One possible model is that binding of p56 may form a nucleation site for other regulatory proteins resulting in a complex which either represses or promotes translation. Although binding of p56 to the polypyrimidine tract of the r-protein mRNAs seems not to be regulated, other interactions of the protein might be controlled, thus modulating the accessibility of these mRNAs to the protein synthetic apparatus in response to changes in cellular physiology.

ACKNOWLEDGMENTS. The recent work described herein was supported in part by National Institutes of Health (NIH) research grants GM42791 (M.W.W.) and CA39053 (D.R.M.) as well as training grant HD07183 (R.L.K.).

REFERENCES

1. Wool, I. G., 1979, The structure and function of eukaryotic ribosomes, *Annu. Rev. Biochem.* **48:** 719–754.
2. Nomura, M., Gourse, R., and Baughman, G., 1984, Regulation of the synthesis of ribosomes and ribosome components, *Annu. Rev. Biochem.* **53:**75–117.
3. Lindahl, L., Archer, R., and Zengel, J. M., 1983, Transcription of the S10 ribosomal protein operon is regulated by an attenuator in the leader, *Cell* **33:**241–248.
4. Amaldi, F., Bozzoni, I., Beccari, E., and Pierandrei-Amaldi, P., 1989, Expression of ribosomal protein genes and regulation of ribosome biosynthesis in *Xenopus* development, *Trends Biochem. Sci.* **14:**175–178.
5. Bowman, L. H., 1987, The synthesis of ribosomal proteins S16 and L32 is not autogenously regulated during mouse myoblast differentiation, *Mol. Cell. Biol.* **7:**4464–4471.
6. Warner, J., 1989, Synthesis of ribosomes in *Saccharomyces cerevisiae*, *Microbiol. Rev.* **53:**256–271.
7. Lucioli, A. Presutti, C., Ciafre, S., Caffarelli, E., Fragapane, P., and Bozzoni, I., 1988, Gene dosage alteration of L2 ribosomal protein genes in *Saccharomyces cerevisiae*: Effects on ribosome synthesis, *Mol. Cell. Biol.* **8:**4792–4798.
8. El-Baradi, T. T., van der Sande, C. A., Mager, W. H., Raue, H. A., and Planta, R. J., 1986, The cellular level of yeast ribosomal protein L25 is controlled principally by rapid degradation of excess protein, *Curr. Genet.* **10:**733–739.
9. Maicas, E., Pluthero, F. G., and Friesen, J. D., 1988, The accumulation of three yeast ribosomal proteins under conditions of excess mRNA is determined primarily by fast protein decay, *Mol. Cell. Biol.* **8:**169–175.
10. Tsay, Y., Thompson, J. R., Rotenberg, M. O., Larkin, J. C., and Woolford, J. L., 1988, Ribosomal protein synthesis is not regulated at the translational level in *Saccharomyces cerevisiae*: Balanced accumulation of ribosomal proteins L16 and rp59 is mediated by turnover of excess protein, *Genes Dev.* **2:**664–676.

11. Wormington, W. M., 1988, Expression of ribosomal protein gene during *Xenopus* development, in: *Developmental Biology: A Comprehensive Synthesis*, Volume 5 (L. E. Browder, ed.), pp. 227–240, Plenum Press, New York.
12. Agrawal, M. G., and Bowman, L. H., 1987, Transcriptional and translational regulation of ribosomal protein formation during mouse myoblast differentiation, *J. Biol. Chem.* **262:**4868–4875.
13. Kaspar, R. L., Kakegawa, T., Cranston, H., Morris, D. R., and White, M. W., 1992, A regulatory *cis* element and a specific binding factor involved in the mitogenic control of murine ribosomal protein L32 translation, *J. Biol. Chem.* **267**:508–514.
14. DePhilip, R. M., Rudert, W. A., and Lieberman, I., 1980, Preferential stimulation of ribosomal protein synthesis by insulin and in the absence of ribosomal and messenger ribonucleic acid formation, *Biochemistry* **19:**1662–1669.
15. Geyer, P. K., Meyuhas, O., Perry, R. P., and Johnson, L. F., 1982, Regulation of ribosomal protein mRNA content and translation in growth-stimulated mouse fibroblasts, *Mol. Cell. Biol.* **2:** 685–693.
16. Tushinski, R. J., and Warner, J. R., 1982, Ribosomal proteins are synthesized preferentially in cells commencing growth, *J. Cell. Physiol.* **112:**128–135.
17. Kaspar, R. L., Rychlik, W., White, M. W., Rhoads, R. E., and Morris, D. R., 1990, Simultaneous cytoplasmic redistribution of ribosomal protein L32 mRNA and phosphorylation of eukaryotic initiation factor 4E after mitogenic stimulation of Swiss 3T3 cells, *J. Biol. Chem.* **265:**3619–3622.
18. Meyuhas, O., Thompson, E. A., and Perry, R. P., 1987, Glucocorticoids selectively inhibit translation of ribosomal protein mRNAs in P1798 lymphosarcoma cells, *Mol. Cell. Biol.* **7:**2691–2699.
19. Mariottini, P., and Amaldi, F., 1990, The 5′ untranslated region of mRNA for ribosomal protein S19 is involved in its translational regulation during *Xenopus* development, *Mol. Cell. Biol.* **10:** 816–822.
20. Levy, S., Avni, D., Hariharan, N., Perry, R. P., and Meyuhas, O., 1991, Oligopyrimidine tract at the 5′ end of mammalian ribosomal protein mRNAs is required for their translational control, *Proc. Natl. Acad. Sci. USA* **88:**3319–3323.
21. Hammond, M. L., Merrick, W., and Bowman, L. H., 1991, Sequences mediating the translation of mouse S16 ribosomal protein mRNA during myoblast differentiation and *in vitro* and possible control points for the *in vitro* translation, *Genes Dev.* **5:**1723–1736.
22. Wagner, M., and Perry, R. P., 1985, Characterization of the multigene family encoding the mouse S16 ribosomal protein: Strategy for distinguishing an expressed gene from its processed pseudogene counterparts by an analysis of total genomic DNA, *Mol. Cell. Biol.* **5:**3560–3576.
23. Dudov, K. P., and Perry, R. P., 1984, The gene family encoding the mouse ribosomal protein L32 contains a uniquely expressed intron-containing gene and an unmutated processed gene, *Cell* **37:** 457–468.
24. Wiedemann, L. M., and Perry, R. P., 1984, Characterization of the expressed gene and several processed pseudogenes for the mouse ribosomal protein L32 gene family, *Mol. Cell. Biol.* **4:**2518–2528.
25. Wool, I. G., Endo, Y., Chan, Y. L., and Gluck, A., 1990, Structure, function, and evolution of mammalian ribosomes, in: *The Structure, Function and Evolution of Ribosomes* (W. E. Hill, ed.), pp 203–214, American Society for Microbiology, Washington, D.C.
26. Mariottini, P., Bagni, C., Annesi, F., and Amaldi, F., 1988, Isolation and nucleotide sequences of cDNAs for *Xenopus laevis* ribosomal protein S8: Similarities in the 5′ and 3′ untranslated regions of mRNAs for various r-proteins, *Gene* **67:**69–74.
27. Hariharan, N., and Perry, R. P., 1990, Functional dissection of mouse ribosomal protein promoter: Significance of the polypyrimidine initiator and an element in the TATA-box region, *Proc. Natl. Acad. Sci. USA* **87:**1526–1530.
28. Yenofsky, R., Cereghini, S., Krowczynska, A., and Brawerman, G., 1983, Regulation of mRNA

utilization in mouse erythroleukemia cells induced to differentiate by exposure to dimethyl sulfoxide, *Mol. Cell. Biol.* **3:**1197–1203.

29. Chitpatima, S. T., Makrides, S., Bandyopadhyay, R., and Brawerman, G., 1988, Nucleotide sequence of a major messenger RNA for a 21 kilodalton polypeptide that is under translational control in mouse tumor cells, *Nucleic Acids Res.* **16:**2350.
30. Kerfelec, B., LaForge, K. S., Vasiloudes, P., Puigserver, A., and Scheele, G. A., 1990, Isolation and sequence of the canine pancreatic phospholipase A_2 gene, *Eur. J. Biochem.* **190:**299–304.
31. Pinsky, S. D., LaForge, K. S., and Scheele, G., 1985, Differential regulation of trypsinogen mRNA translation: Full length mRNA sequences encoding two oppositely charge trypsinogen isoenzymes in the dog pancreas, *Mol. Cell. Biol.* **5:**2669–2676.
32. Makrides, S., Chitpatima, S., Bandyopadhyay, R., and Brawerman, G., 1988, Nucleotide sequence for a major messenger RNA for a 40 kilodalton polypeptide that is under translational control in mouse tumor cells, *Nucleic Acids Res.* **16:**2349.
33. Leibold, E. A., and Munro, H. N., 1988, Cytoplasmic protein binds *in vitro* to a highly conserved sequence in the 5′ untranslated region of ferritin heavy- and light-subunit mRNAs, *Proc. Natl. Acad. Sci. USA* **85:**2171–2175.
34. Klausner, R. D., and Harford, J. B., 1989, *Cis–trans* models for posttranscriptional gene regulation, *Science* **246:** 870–872.
35. Garcia-Blanco, M. A., Jamison, S. F., and Sharp, P. A., 1989, Identification and purification of a 62,000-dalton protein that binds specifically to the polypyrimidine tract of introns, *Genes Dev.* **3:** 1874–1886.
36. Zamore, P. D., and Green, M. R., 1989, Identification, purification, and biochemical characterization of U2 small nuclear ribonucleoprotein auxiliary factor, *Proc. Natl. Acad. Sci. USA* **86:**9243–9247.
37. Kuhn, R., Luz, N., and Beck, E., 1990, Functional analysis of the internal translation initiation site of foot-and-mouth disease virus, *J. Virol.* **64:**4625–4631.
38. Jang, S. K., and Wimmer, E., 1990, Cap-independent translation of encephalomyocarditis virus RNA: Structural elements of the internal ribosomal entry site and involvement of a cellular 57-kD RNA-binding protein, *Genes Dev.* **4:**1560–1572.
39. Nicholson, R., Pelletier, J., Le, S., and Sonenberg, N., 1991, Structural and functional analysis of the ribosome landing pad of poliovirus: *In vivo* translation studies, *J. Virol.* **65**:5886–5894.
40. Meerovitch, K., Pelletier, J., and Sonenberg, N., 1989, A cellular protein that binds to the 5′-noncoding region of poliovirus RNA: Implications for internal translation initiation, *Genes Dev.* **3:**1026–1034.
41. Hershey, J. W. B., 1991, Translational control in mammalian cells, *Annu. Rev. Biochem.* **60:** 717–755.
42. Lodish, H. F., 1974, Model for the regulation of mRNA translation applied to haemoglobin synthesis, *Nature* **251:**385–388.
43. Lodish, H. F., 1976, Translational control of protein synthesis, *Annu. Rev. Biochem.* **45:**39–72.
44. Godefroy-Colburn, T., and Thach, R. E., 1981, The role of mRNA competition in regulating translation, *J. Biol. Chem.* **256:**11762–11773.
45. Lodish, H. F., 1971, Alpha and beta globin messenger ribonucleic acid, *J. Biol. Chem.* **246:**7131–7138.
46. White, M. W., Kameji, T., Pegg, A. E., and Morris, D. R., 1987, Increased efficiency of translation of ornithine decarboxylase mRNA in mitogen-activated lymphocytes, *Eur. J. Biochem.* **170:**87–92.
47. Jacobs-Lorena, M., and Fried, H. M., 1987, Translational regulation of ribosomal protein gene expression in eukaryotes, in: *Translational Regulation of Gene Expression* (J. Ilan, ed.), pp. 63–85, Plenum Press, New York.
48. Itoh, N., Ohta, K., Ohta, M., Kawasaki, T., and Yamashina, I., 1989, The nucleotide sequence of a gene for a putative ribosomal protein S31 of *Drosophila*, *Nucleic Acids Res.* **17:**2121.

49. Qian, S., Zhang, J., Kay, M. A., and Jacobs-Lorena, M., 1987, Structural analysis of the *Drosophila* rpA1 gene, a member of the eucryotic 'A' type ribosomal protein family, *Nucleic Acids Res.* **15:**987–1003.
50. Steel, L. F., and Jacobson, A., 1987, Translational control of ribosomal protein synthesis during early *Dictyostelium discodeum* development, *Mol. Cell. Biol.* **7:**965–972.
51. Steel, L. F., and Jacobson, A., 1991, Sequence elements that affect mRNA translational activity in developing *Dictyostelium* cells, *Dev. Genet.* **12:**98–103.
52. Steel, L. F., Smyth, A., and Jacobson, A., 1987, Nucleotide sequence and characterization of the transcript of a *Dictyostelium* ribosomal protein gene, *Nucleic Acids Res.* **15:**10285–10298.
53. Kim, C. H., and Warner, J. R., 1983, The mRNA for ribosomal proteins in yeast, *J. Mol. Biol.* **165:** 79–89.
54. Maicas, E., and Friesen, J. D., 1990, A sequence pattern that occurs at the transcription initiation region of yeast RNA polymerase II promoters, *Nucleic Acids Res.* **18:**3387–3393.
55. Warner, J. R., and Gorenstein, C., 1977, The synthesis of eukaryotic ribosomal proteins *in vitro*, *Cell* **11:**201–212.
56. Donovan, D. M., and Pearson, N. J., 1986, Transcriptional regulation of ribosomal proteins during a nutritional upshift in *Saccharomyces cerevisiae*, *Mol. Cell. Biol.* **6:**2429–2435.
57. Mager, W. H., and Planta, R. J., 1990 Multifunctional DNA-binding proteins mediate concerted transcription activation of yeast ribosomal protein genes, *Biochim. Biophys. Acta* **1050:**351–355.
58. Rotenberg, M. O., and Woolford, J. L., 1986, Tripartite upstream promoter element essential for expression of *Saccharomyces cerevisiae* ribosomal protein genes, *Mol. Cell. Biol.* **6:**674–687.
59. Mager, W. H., 1988, Control of ribosomal protein gene expression, *Biochim. Biophys. Acta* **949:**1–15.
60. Huet, J., Cottrelle, P., Cool, M., Vignais, M. L., Thiele, D., Marck, C., Buhler, J. M., Sentenac, A., and Fromageot, P., 1985, A general upstream binding factor for genes of the yeast translational apparatus, *EMBO J.* **4:**3539–3547.
61. Vignais, M. L., Woudt, L. P., Wassenaar, G. M., Mager, W. H., Sentenac, A., and Planta, R. J., 1987, Specific binding of TUF factor to upstream activation sites of yeast ribosomal protein genes, *EMBO J.* **6:**1451–1457.
62. Shore, D., Stillman, D. J., Brand, A. H., and Nasmyth, K. A., 1987, Identification of silencer binding proteins from yeast: Possible roles in SIR control and DNA replication, *EMBO J.* **6:** 461–467.
63. Shore, D., and Nasmyth, K. A., 1987, Purification and cloning of a DNA-binding protein that binds to both silencer and activator elements, *Cell* **51:**721–732.
64. Herruer, M. H., Mager, W. H., Doorenbosch, T. M., Wessels, P. L., Wassenaar, T. M., and Planta, R. J., 1989, The extended promoter of the encoding ribosomal protein S33 in yeasts consists of multiple binding elements, *Nucleic Acids Res.* **17:**7427–7439.
65. Hamil, K. G., Nam, H. G., and Fried, H. M., 1988, Constitutive transcription of yeast ribosomal protein gene TCM1 is promoted by uncommon *cis-* and *trans-*acting elements, *Mol. Cell. Biol.* **8:** 4328–4341.
66. Rhode, P. R., Sweder, K. S., Oegema, K. F., and Campbell, J. L., 1989, The gene encoding ARS-binding factor 1 is essential for the viability of yeast, *Genes Dev.* **3:**1926–1939.
67. Halfter, H., Kavety, B., Vandekerckhove, J., Kiefer, F., and Gallwitz, D., 1989, Sequence, expression and mutational analysis of BAF1, a transcriptional activator and ARS1-binding protein of the yeast *Saccharomyces cerevisiae*, *EMBO J.* **8:**4265–4272.
68. Diffley, J. F., and Stillman, B., 1989, Similarity between the transcriptional silencer proteins ABF1 and RAP1, *Science* **246:**1034–1038.
69. Buchman, A. R., Kimmerly, W. J., Rine, J., and Kornberg, R. D., 1988, Two DNA-binding factors recognize specific sequences at silencers, upstream activating sequences, autonomously replicating sequences, and telomeres in *Saccharomyces cerevisiae*, *Mol. Cell. Biol.* **8:**210–225.
70. Diffley, J. F., and Stillman, B., 1988, Purification of a yeast protein that binds to origins of DNA replication and a transcriptional silencer, *Proc. Natl. Acad. Sci. USA* **85:**2120–2124.

71. Seta, F. D., Ciafre, S. A., Marck, C., Santoro, B., Presutti, C., Sentenac, A., and Bozzoni, I., 1990, The ABF1 factor is the transcriptional activator of the L2 ribosomal protein genes in *Saccharomyces cerevisiae*, *Mol. Cell. Biol.* **10:**2437–2441.
72. Hariharan, N., and Perry, R. P., 1989, A characterization of the elements comprising the promoter of the mouse ribosomal protein gene RPS16, *Nucleic Acids Res.* **17:**5323–5337.
73. Dudov, K. P., and Perry, R. P., 1986, Properties of a mouse ribosomal protein promoter, *Proc. Natl. Acad. Sci. USA* **83:**8545–8549.
74. Moura-Neto, R., Dudov, K. P., and Perry, R. P., 1989, An element downstream of the cap site is required for transcription of the gene encoding mouse ribosomal protein L32, *Proc. Natl. Acad. Sci. USA* **86:**3997–4001.
75. Chung, S., and Perry, R. P., 1989, Importance of introns for expression of mouse ribosomal protein gene rpL32, *Mol. Cell. Biol.* **9:**2075–2082.
76. Atchison, M. L., Meyuhas, O., and Perry, R. P., 1989, Localization of transcriptional regulatory elements and nuclear factor binding sites in mouse ribosomal protein gene rpL32, *Mol. Cell. Biol.* **9:**2067–2074.
77. Dabeva, M. D., Post-Beittenmiller, M. A., and Warner, J. R., 1986, Autogenous regulation of splicing of the transcript of a yeast ribosomal protein gene, *Proc. Natl. Acad. Sci. USA* **83:**5854–5857.
78. Wittekind, M., Kolb, J. M., Dodd, J., Yamagishi, M., Memet, S., Buhler, J., and Nomura, M., 1990, Conditional expression of RPA190, the gene encoding the largest subunit of yeast RNA polymerase I: Effects of decreased rRNA synthesis on ribosomal protein synthesis, *Mol. Cell. Biol.* **10:**2049–2059.
79. Eng, F. J., and Warner, J. R., 1991, Structural basis for the regulation of splicing of a yeast messenger RNA, *Cell* **65:**797–804.
80. Bozzoni, I., Fragapane, P., Annesi, F., Pierandrei-Amaldi, P., Amaldi, F., and Beccari, E., 1984, Expression of two *Xenopus laevis* ribosomal protein genes in injected frog oocytes. A specific splicing block interferes with the L1 RNA maturation, *J. Mol. Biol.* **180:**987–1005.

Chapter 17

Translational Regulation in Reticulocytes

The Role of Heme-Regulated eIF-2α Kinase

Jane-Jane Chen

1. INTRODUCTION

Rabbit reticulocyte lysate is the first and is still the most efficient system for cell-free protein synthesis *in vitro*. It has not only provided us with much of our understanding of translational regulation, but also provided a unique system to study the coordination of the biosynthesis of heme and globins. The finding that heme is required for the initiation of protein synthesis is the stepping stone to our current knowledge of the translational regulation of initiation by the phosphorylation of eukaryotic initiation factor 2 (eIF-2).

In heme deficiency, protein synthesis in reticulocytes is inhibited due to activation of the heme-regulated inhibitor (HRI). HRI is a cAMP-independent protein kinase which phosphorylates specifically the α subunit of eIF-2 (eIF-2α). Phosphorylation of eIF-2α results in the binding and sequestration of eIF-2B (also designated as guanine nucleotide exchange factor, GEF, or reversing factor, RF) which is required for the recycling of eIF-2. The unavailability of eIF-2B results in the cessation of protein synthesis (for reviews see refs. 1–4).

JANE-JANE CHEN • Harvard–Massachusetts Institute of Technology Division of Health Sciences and Technology, Massachusetts Institute of Technology, Cambridge, Massachusetts 02139.

Translational Regulation of Gene Expression 2, edited by Joseph Ilan. Plenum Press, New York, 1993.

Inhibition of initiation via phosphorylation of eIF-2α in reticulocyte lysates is also observed upon the addition of a low level of double-stranded RNA (dsRNA) (1–50 ng/ml)[5] or oxidized glutathione (50–500 mM).[6] Inhibition by oxidized glutathione occurs in the presence of heme and is most likely due to activation of HRI. Inhibition by dsRNA, however, is the result of the activation of a dsRNA-dependent eIF-2α kinase,[7,8] which is distinct from HRI.[9]

The HRI molecule is the focus of this chapter. The last comprehensive review devoted to HRI was published more than 10 years ago.[10] In this chapter a brief historical account on the discovery of HRI is presented followed by recent studies on the regulation of HRI activity by heme, sulfhydryl reagents, and heat-shock proteins. Finally, the structure of HRI deduced from its cDNA and the homology of HRI to other protein kinases are presented and discussed.

2. DISCOVERY OF HEME-REGULATED INHIBITOR

2.1. Requirement of Heme in the Synthesis of Globin

In the 1950s it was discovered that inorganic iron stimulates protein synthesis in immature erythroid cells.[11–13] In 1965 it was reported that hemin enhances globin synthesis in rabbit reticulocytes.[14] Further study indicated that heme is not merely serving as an iron source, but rather it exerts its effect directly, since the iron-chelating agent desferrioxamine does not block the stimulatory effect of heme.[15] In reticulocytes, iron deficiency causes a disaggregation of polyribosomes which is prevented by the addition or iron or hemin.[15–17] These observations suggested that inhibition of protein synthesis occurs in heme deficiency.

The development of a cell-free reticulocyte lysate for active protein synthesis[18,19] facilitated the study of the regulation of protein synthesis by heme. Mitochondria, which are required for heme biosynthesis, are removed in the preparation of reticulocyte lysates. Thus, reticulocyte lysates are incapable of synthesizing heme and become dependent on exogeneous hemin. In the absence of added hemin, protein synthesis continues for the first 5–10 min, followed by an abrupt decline in the rate of synthesis (shutoff). Addition of hemin permits protein synthesis to continue for 90 min.[19,20] Consistent with whole-cell studies, the inhibition of protein synthesis in reticulocyte lysates during heme deficiency is accompanied by dissociation of polyribosomes.[18] In addition, hemin added after shutoff of protein synthesis is capable of restoring protein synthesis and polyribosomes.[21]

2.2. Activation of the Heme-Regulated Inhibitor

In 1969 it was shown that the inhibition of protein synthesis in heme deficiency was due to the activation of an inhibitor.[22] The activation of this inhibitor occurred by incubating lysates at 34°C for 45 min. Reconstition experiments using fractions containing ribosome-free supernatants (S-100) and ribosomes from both normal and inhibited lysates indicated that the inhibitor is present in the S-100.[22,23] The inhibitor in the S-100 is activated on incubation at elevated temperatures (34–43°C); neither ribosomes nor protein synthesis is required for its activation.[24] Addition of hemin to the S-100, however, prevents the activation of the inhibitor.[24]

In the absence of hemin, the linear rate of protein synthesis in reticulocyte lysates declined abruptly (shutoff period). The time to reach this shutoff point decreases with increasing temperature from 29°C to 40°C.[20] This observation is consistent with an earlier finding that the rate of inhibitor formation is increased with increasing temperature.[24] At lower temperatures, a lag of inhibitor formation is observed.[25] The presence of hemin from the beginning of incubation essentially inhibits the formation of the inhibitor. Hemin added after the end of the lag period, however, only reverses the inhibitor and protein synthesis partially.[25,26] This observation suggested the presence of an intermediate during the activation of the inhibitor, and that the action of this intermediate form of inhibitor is inhibited by hemin (heme-reversible inhibitor).

Prolonged incubation of reticulocyte lysates at 30–40°C results in the formation of the heme-irreversible inhibitor.[22,24,25] The formation of the irreversible inhibitor is prevented by the presence of hemin. Once formed, however, it is not inactivated by hemin. At higher temperatures (40–45°C), the activation of heme-irreversible inhibitor is more rapid.[27] In addition to heme deficiency, activation of HRI can be achieved by other conditions, such as high hydrostatic pressure,[10] sulfhydryl-reactive agents,[28] oxidized glutathione,[6] heavy metal ions,[29] ethanol,[30] or heat shock.[31,32] The mechanism of the activation of HRI by these conditions will be discussed below.

2.3. Heme-Regulated Inhibitor Is an eIF-2α Kinase

With the shutoff of protein synthesis in heme deficiency, there is a marked decrease in the steady-state level of 40S·Met-tRNA$_f^{met}$·GTP initiation complex.[33] Since the concentrations of the 40S ribosomal subunit and Met-tRNA$_f^{met}$·GTP are unaltered in heme deficiency, it is likely that the binding of Met-tRNA$_f^{met}$ to 40S is affected. This notion is consistent with subsequent observations that the addition of purified preparations of initiation factor 2 (eIF-2) prevents as well as reverses the inhibition of protein synthesis.[34–36]

The notion that HRI is an eIF-2α kinase became apparent when it was observed that inhibition of protein synthesis during heme deficiency was potentiated by ATP[27,37] and was reversed by cAMP[37,38] and GTP.[27,37] On purification both HRI and dsI have been shown to be cAMP-independent protein kinases which specifically phosphorylate the α subunit (38,000 daltons) of eIF-2.[7,8,39–42] In addition, an increase in the phosphorylation of eIF-2α is observed in reticulocyte lysates under conditions of heme deficiency, and upon addition of dsRNA or oxidized glutathione (GSSG).[43–46]

The mechanism of inhibition of initiation by the phosphorylation of eIF-2α has been extensively studied, and the subject has been reviewed in several recent articles.[1–4] I will summarize briefly the current model for this regulation. The recycling of eIF-2 in initiation requires the exchange of bound GDP for GTP. Under physiological conditions eIF-2 has a 400-fold greater affinity for GDP than for GTP. The exchange of tightly bound GDP for GTP requires eIF-2B, which is rate limiting and is present at 15–25% of the amount of eIF-2. when eIF-2α is phosphorylated the binding of eIF-2(αP)·GDP to eIF-2B is much tighter than the binding of eIF-2·GDP to eIF-2B. Accordingly, eIF-2B is effectively sequestered and nonfunctional. In this manner, once the amount of phosphorylated eIF-2 exceeds the amount of eIF-2B, the shutoff of protein synthesis occurs. In reticulocyte lysates phosphorylation of about 30% of eIF-2α is sufficient to sequester all eIF-2B and to shutoff initiation of protein synthesis.

3. REGULATION OF THE KINASE ACTIVITIES OF HEME-REGULATED eIF-2α KINASE

3.1. Characteristics of HRI

The heme-irreversible HRI activated by heat or *N*-ethylmaleimide (NEM) treatment has been purified extensively in several laboratories.[7,39,47,48] Highly purified HRI is a cAMP-independent protein kinase which specifically phosphorylates the α subunit of eIF-2, but not the β or γ subunit. The mechanism of the action of purified HRI on protein synthesis is very similar to that of heme deficiency. Purified HRI, when added to reticulocyte lysates, produces the characteristic biphasic kinetics of shutoff of protein synthesis as well as the disaggregation of polysomes; this inhibition of protein synthesis is reversed by eIF-2.[36,48–50] Heme-reversible HRI was purified to near homogeneity in 1978.[51] When analyzed by sodium dodecyl sulfate–polyacrylamide gel electrophoresis (SDS–PAGE) (15% polyacrylamide gel) heme-reversible HRI is a 95-kDa polypeptide. Purified heme-reversible HRI undergoes autophosphorylation. This observation is consistent with the finding that phosphorylation of HRI is not affected by dilution.[10,52] In contrast to heme-irreversible HRI, heme-reversible HRI becomes inactive in inhibiting protein synthesis when it is preincubated with hemin. The autophosphorylation and phosphorylation of eIF-2α by heme-reversible HRI are also inhibited by preincubating with hemin.[51]

As described above, HRI is present in hemin-supplemented lysates in an inactive form (pro-inhibitor) and is activated in the absence of heme. Gel filtration studies indicate that both pro-inhibitor and activated inhibitor elute at the same position with the apparent molecular weight of approximately 300 kDa.[10,28]

Similarly, there is no change in the molecular size of heme-reversible HRI whether it is treated with hemin[51,53] or is phosphorylated.[53] Therefore, activation of HRI does not appear to be accompanied by a significant change in the molecular size of HRI. When the molecular weight of HRI was estimated by sucrose-gradient centrifugation, a value of 150,000 kDa was obtained.[10,48,51] A frictional ratio of 1.73 was obtained for HRI from studies of gel filtration and density gradient centrifugation, suggesting that HRI is highly elongated in shape.[10] Therefore, it is likely that HRI exists as a dimer. Indeed, chemical cross-linking of phosphorylated HRI gives a product of approximately twice the molecular weight of HRI.[10,53]

In heme-deficient lysates, inhibition of protein synthesis is accompanied by the phosphorylation of both HRI and eIF-2α.[44] This observation was confirmed recently by immunoprecipitation of ^{32}P-HRI with anti-HRI monoclonal antibody from hemin-deficient and hemin-supplemented reticulocytes.[54] Copurification of HRI activity and an 90-kDa phosphopolypeptide (10% polyacrylamide gel) have been observed.[7,51,52] We have observed a good correlation between the autophosphorylation HRI and the phosphorylation of eIF-2α by HRI throughout the purification procedure of heme-reversible HRI.[55] In addition, when partially purified HRI is incubated with ATP, there is an increase in the binding of HRI to DEAE–Sepharose.[56] These findings indicate that autophosphorylation is an intrinsic property of HRI. Purified HRI has been shown to undergo multiple autophosphorylations; approximately 5 mole of phosphate is estimated to be incorporated per mole of 90-kDa subunit.[52,57] The degree of HRI autophosphorylation correlates well with the degree of the conversion of pro-inhibitor to inhibitor[52] and with the extent of eIF-2α phosphorylation by HRI.[57] In addition, both polyclonal[9] and monoclonal[54] antibodies to HRI inhibit the phosphorylation of purified HRI and the phosphorylation of eIF-2α. These findings indicate a positive correlation between autophosphorylation of HRI and its eIF-2α kinase activity.

Studies using antibodies to HRI have demonstrated that HRI is principally responsible for the inhibition of protein synthesis during heme deficiency. A preparation of anti-HRI polyclonal antibody has been shown to slow down the inhibition of protein synthesis in heme deficiency, reverse inhibition of protein synthesis in heme deficiency, decrease eIF-2α phosphorylation, and reduce 43S preinitiation complex.[58] Note, however, that this anti-HRI antibody is more effective in reversing protein synthesis in heme deficiency at 25°C where inhibitor formation is slower and the shutoff of protein synthesis is less complete. These observations indicate that once HRI is activated it is a very potent inhibitor of protein synthesis and anti-HRI antibody does not neutralize all the inhibitory activity. It is of interest that an excess of pro-inhibitor has little effect on the ability of this anti-HRI antibody to neutralize the eIF-2α kinase activity of heat-activated heme-irreversible HRI. Therefore, it appears that the antigenic determinants recognized by this preparation of polyclonal antibody are exposed only during the conversion of pro-inhibitor to inhibitor. A monoclonal antibody (mAB) to HRI, mAB F, obtained recently from our laboratory, appears to recognize both latent and activated HRI, since mAB F immunoprecipitates equal amounts of HRI polypeptide from both hemin-supplemented and hemin-deficient lysates.[54] In addition, mAB F immunoprecipitates substantially more phosphorylated HRI from heme-deficient lysates, NEM-treated lysates, and GSSG-treated lysates than from hemin-supplemented lysates. These results are consistent with the earlier observations on the extent of the phosphorylation of HRI and the extent of HRI activation in the lysates under these conditions. The removal of HRI from lysates by mAB F results in a similar rate of protein synthesis in heme-deficient lysates and in hemin-supplemented lysates.

3.2. Effects of Sulfhydryl-Reagents and the Oxidative State on HRI Activity

Heme-irreversible HRI is formed rapidly when reticulocyte lysates are incubated with 1–5 mM NEM which alkylates free sulfhydryl (SH) groups.[28] At higher concentrations, however, NEM fails to activate HRI; rather, it inactivates HRI. The presence of ATP protects the inactivation of HRI by high concentrations of NEM.[10] These studies imply that there are two kinds of SH groups which are involved in the regulation of HRI activity; one kind is involved in activation, whereas the other kind is involved in inactivation. Studies of the activation of HRI at different pH levels show that HRI is activated by increasing pH from 7.5 to 9.0 with the apparent pK of 8.3–8.5, which corresponds to the pK of cysteine.[10] Activation of HRI by GSSG[59] and heavy metal ions[29,60] also points to the critical role of SH groups in the mechanism of activation of HRI. It has been shown recently that toxic heavy metals inhibit the capacity of hemin-supplemented lysate to reduce disulfide bonds.[60] Maintenance of a high rate of initiation requires a functional NADPH-generating system, thioredoxin reductase, and thioredoxin.[61–63] The absence of any of these components leads to oxidation of SH groups of several lysate proteins, inhibition of initiation, and increased eIF-2α phosphorylation.[62] The effects of GSSG and heavy metals on lysate HRI activity are very different from the effects of these agents on partially purified HRI. GSSG has no significant effect on activity of purified HRI;[53] heavy metal ions inhibit the activity of partially purified HRI.[60] Therefore, it is possible that the activation of HRI involves the oxidation of sulfhydryl(s) of a regulatory protein rather than of HRI itself.[60,62] This explanation is consistent with our observations that free SH groups of purified HRI are essential for HRI activity *in vitro*. The alkylation of the SH groups by NEM (J.-J. Chen, unpublished observation) and the disulfide oxidation of SH groups catalyzed by hemin inactivate HRI.[53] The presence of dithiothreitol (DTT) has been reported by Gross to prevent the activation of lysate HRI.[64] On the other hand, Hunt[10] and I (J.-J. Chen, unpublished observation) have found that DTT is necessary for maintaining activity of purified HRI. Therefore, it appears that the preferred redox state of purified HRI is quite different from that of HRI in the lysate. These results clearly demonstrate that the pro-inhibitor in reticulocyte lysates is somewhat different from purified heme-reversible HRI. Lysate HRI may be present in a different conformation from purified heme-reversible HRI since during purification HRI is always kept under reducing condition with 1–2 mM DTT which is necessary in maintaining HRI kinase activities. It is also possible that lysate HRI is associated with other protein(s) which are involved in the regulation of HRI activation and are dissociated from HRI during purification. These additional regulatory proteins may be the target of the disulfide oxidation or NEM activation discussed above. There is recent evidence that heat-shock proteins, particularly hsp90, may be involved in the regulation of HRI activity. This topic will be addressed later in this chapter.

3.3. Mechanism of the Regulation of HRI by Heme

Although it has been known for two decades that HRI is activated in reticulocyte lysates during heme deficiency, the molecular mechanism by which heme regulates HRI has not been established. The major obstacle which hinders progress on this subject has been the rather low abundance of HRI protein in reticulocyte lysates.

As described above, pure heme-reversible HRI becomes inactive in inhibiting protein synthesis when it is preincubated with hemin.[51] In addition, the autophosphorylation and eIF-2α phosphorylation are also inhibited by hemin.[51] These studies suggest strongly that hemin regulates HRI directly. Indeed, the binding of hemin to purified HRI has been demonstrated using difference spectroscopy at 418 nm.[57] The binding of hemin to proteins has been shown to cause a shift in the absorption spectrum in the Soret region, sharpening of the absorption band, and a large enhancement of intensity.[65,66] These spectral changes are unique to heme-binding proteins; no change is observed with glyceraldehyde-3-phosphodehydrogenase, pyruvate kinase, or β-galactosidase.[57] The binding of hemin to HRI is proportional to the amount of HRI added, and is increased with increasing concentrations of hemin. These findings suggest the presence of saturable heme-binding sites in HRI. It is interesting that phosphorylated HRI seems to bind less hemin.[57]

Since sulfhydryl groups are involved in the activation of HRI, we have investigated the effect of the binding of hemin on the thiol status of HRI. The inhibition of HRI autophosphorylation by hemin is very similar to that produced by thiol oxidation by diamide.[53] Treatment of phosphorylated HRI with hemin results in a significant shift in the mobility of [^{32}P]HRI from monomeric 90 kDa to approximately twice that size under nonreducing SDS–PAGE conditions. No difference in the migration of [^{32}P]HRI is observed upon hemin treatment under reducing SDS–PAGE conditions. This observation indicates that hemin promotes intersubunit disulfide bond formation in phosphorylated HRI, and thus results in the shift of migration of phosphorylated HRI (Fig. 1). Hemin also promotes intersubunit disulfide formation of latent HRI without prior phosphorylation.[135]

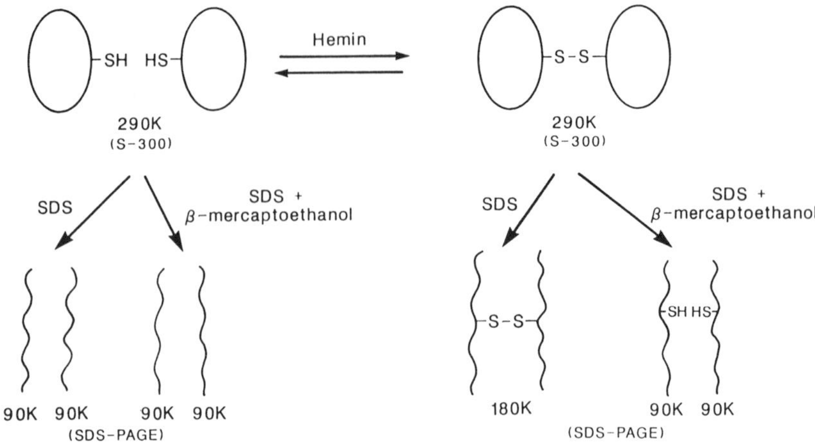

Figure 1. Schematic illustration of the regulation of HRI by covalent dimer formation.

A similar dimer of HRI is formed when HRI is treated with 1,6-bismaleimido-hexane (Bis-NEM), a double sulfhydryl cross-linking agent. In agreement with earlier observations of others,[10,28,51] we found that unphosphorylated HRI, phosphorylated HRI, hemin-treated HRI, hemin-treated prephosphorylated HRI, and Bis-NEM cross-linked HRI were all eluted identically on Sephacryl S-300 column chromatography with an apparent molecular weight of 290 kDa. Treatment of HRI with Bis-NEM results in the loss of autokinase and eIF-2α kinase activities of HRI. In addition, pretreatment of HRI with hemin inhibits the cross-linking of HRI by Bis-NEM. These findings indicate that either hemin and Bis-NEM react with the same set(s) of sulfhydryl groups of HRI or that the binding of hemin to HRI alters its conformation so that the previously reactive sulfhydryl groups are not accessible to Bis-NEM. Hemin-promoted disulfide formation in HRI occurs under quasi-physiological conditions (30°C, 5–10 μM hemin). The schema in Fig. 1 summarizes a current model of the mechanism of regulation of HRI by heme via disulfide oxidation. For simplicity, HRI is presented as a homodimer. The possibility of a heterodimer, however, cannot be ruled out. HRI is active as a dimer held by noncovalent interactions between two subunits. In the presence of hemin the intersubunit disulfide bond is formed and HRI becomes inactive (as a disulfide-linked dimer). This dimer can be demonstrated by SDS–PAGE under nonreducing conditions. Similarly, a Bis-NEM cross-linked inactive dimer of HRI is formed when it is reacted with Bis-NEM. This model is particularly attractive in interpreting the inactivation of the activated HRI and the reversal of protein synthesis during delayed addition of hemin. Phosphorylated HRI has been found to turn over very slowly in lysates.[10] Under our conditions, phosphorylated HRI when treated with hemin or Bis-NEM displays rather low eIF-2α kinase activity.[53]

Therefore, it is possible to inactivate HRI without dephosphorylating HRI *in vitro*. It is also plausible, however, that HRI in lysates may not undergo extensive multiple phosphorylation as does purified HRI. The multiple phosphorylation of HRI *in vitro* may prevent the dephosphorylation of the primary phosphorylated site *in situ*.

Purified HRI has been shown to bind ATP.[55,67] ATP binding to HRI is inhibited by preincubation of HRI with hemin (1–10 μM) prior to photo-cross-linking of [α-^{32}P]ATP to HRI.[55] This finding indicates that the inhibition of the autokinase and eIF-2α kinase activity of HRI by heme is due to inhibition of ATP binding to HRI. Since hemin promotes disulfide bond formation of HRI and inhibits ATP binding, one may infer that free sulfhydryl groups of HRI are important in ATP binding to HRI. This interpretation is consistent with an earlier finding[10] that ATP prevented the inactivation of HRI by high concentrations of NEM, an observation suggesting that free sulfhydryl groups are near the active center. Thus, hemin inhibits the ATP binding of HRI by bringing the essential free sulfhydryl groups in close proximity and promoting disulfide bond formation of these sulfhydryl groups. The metalloporphyrin IX compounds, which are able to promote disulfide bond formation, are the most potent inhibitors of HRI kinase activities and are most capable of maintaining and reversing the inhibition of protein synthesis in reticulocyte lysates. The metal-free protoporphyrin IX is less potent in inhibiting HRI kinase activities, in maintaining protein synthesis, and in reversing protein synthesis in heme-deficiency.[135]

Regulation of kinase activity through sulfhydryl oxidation is not unique to HRI. Free SH groups are essential for the catalytic activity of the cAMP-dependent protein kinase[68] and the cysteine residue is in the vicinity of the active site.[69] Activation of the kinase activities of all receptor tyrosine kinases appears to be mediated by receptor dimerization (reviewed in ref. 70). The ligand binding of colony stimulating factor-1 (CSF-1)[71] and platelet-derived growth factor (PDGF)[72] to their respective receptors stimulates both noncovalent dimerization as well as disulfide-linked dimerization of the receptors. The disulfide-linked dimerization of the CSF-1 receptor is thought to be involved in ligand-induced receptor internalization.[71] The exact role of disulfide-linked dimerization of PDGF and CSF-1 receptor tyrosine kinases remains to be investigated. It is interesting that HRI and these two receptor kinases both possess kinase insertion sequences (see below, Section 4.3). Recently a putative novel transmembrane protein tyrosine kinase, Ltk, has been shown to undergo disulfide-linked oligomerization with concomitantly increased kinase activity.[73]

Regulation of the activity of proteins through redox of free sulfhydryl and disulfide has also been demonstrated in the interactions of iron-responsive element (IRE) of ferritin mRNA and the IRE-binding protein,[74] the interactions of cellular protein with the 5′ untranslated region of the poliovirus genome,[75] the interactions of c-*fos* and c-*jun* with DNA,[76] and the reactions of many enzymes involved in the cellular metabolism (reviewed in ref. 77).

3.4. Heat Shock and Heat-Shock Proteins

Incubation of hemin-supplemented reticulocyte lysates at 42°C results in inhibition of protein synthesis, disaggregation of polysomes, and decreased 40S·Met-tRNA$_f$ initiation complex.[31,78,79] The inhibition of protein synthesis on heat shock is prevented by cAMP (10 mM), GTP (2 mM), eIF-2, and eIF-2B.[32,80] Increased phosphorylation of eIF-2α is also observed in heat-shocked reticulocyte lysates.[32] This increase in eIF-2α phosphorylation[32] is blocked by an anti-HRI polyclonal antibody.[9] These findings, therefore, indicate that HRI is activated during heat shock of reticulocyte lysates.

The major heat-shock protein, hsp90, has been shown to copurify with HRI.[81] eIF-2α kinase activity of HRI is increased with addition of phosphorylated hsp90, but not dephosphorylated hsp90.[82] The stimulation by hsp90 is about twofold with at least a 20-fold molar excess of hsp90.[82,83] Addition of purified HeLa hsp90 to hemin-supplemented reticulocyte lysates results in inhibition of protein synthesis.[83] The inhibition is not biphasic, however, as is typically observed in heme deficiency, and is only partially reversed by the addition of eIF-2. On the other hand, association of hsp90 with latent HRI in hemin-supplemented reticulocyte lysates has also been reported.[84] Thus, the mechanism of the activation of HRI during heat shock and the role of interactions of hsp90 and HRI in the regulation of HRI activity are unclear.

Hsp90 has been shown to be transiently associated with the Src family tyrosine protein kinase.[85,86] Src is hypophosphorylated and is unable to undergo tyrosine autophosphorylation when it is associated with hsp90. Activation of tyrosine kinase activity occurs comcomitantly with the dissociation of hsp90 and insertion of pp60$^{v\text{-}Src}$ into the plasma membrane.[85] Hsp90 is also reported to be associated with most of the receptors of steroid hormones (review in ref. 87); the activation of steroid receptors to a high-affinity binding for DNA is accompanied by the dissociation of hsp90.

3.5. HRI-Like eIF-2α Kinase in Nonerythroid Cells

The role of eIF-2α phosphorylation in the regulation of protein synthesis in nonerythroid cells has also been studied. In Ehrlich ascites cells starved of essential amino acids or heat shocked, the formation of 43S initiation complex is decreased[88–90] and the phosphorylation of eIF-2α is increased.[46] A decrease in the formation of 43S initiation complex is also observed in perfused rat liver deprived of essential amino acids[91] and in skeletal muscle from either starved or diabetic rats.[92,93] Serum-depleted HeLa cells exhibit a reduced rate of initiation of protein synthesis;[94] the inhibition of initiation is accompanied by the increased phosphorylation of eIF-2α, and upon supplement of fresh serum, eIF-2α phosphorylation is decreased.[95] Heat shock abolishes protein synthesis in HeLa cells with disaggregation of polysomes and an increase in the phosphorylation of eIF-2α.[96] The inhibition of protein synthesis is reversed by return to lower temperature with a concomitant decrease in the phosphorylation of eIF-2α. Activation of the eIF-2α kinase in heat-shocked HeLa cells is inhibited by an anti-HRI antibody.[97] Inhibitors of initiation of protein synthesis with properties similar to those of reticulocyte HRI have been isolated and purified to various extents from Ehrlich ascites cells,[98] Krebs ascites cells,[99] perfused rat liver,[100] and freshly isolated hepatocytes depleted of heme with allylisopropylacetamide.[101] We have examined the presence of HRI in various rabbit tissues using mAB F. Our study indicates that HRI may be erythroid-specific; HRI is detected only in reticulocytes and anemic bone marrows.[54] It remains to be determined whether eIF-2α kinase activity in nonerythroid cells is mediated by HRI or by other, as yet uncharacterized eIF-2α kinases.

4. STRUCTURAL FEATURES OF THE cDNA OF HEME-REGULATED eIF-2α KINASE

4.1. Cloning of HRI cDNA

We have obtained the amino acid sequences of three tryptic peptides of HRI.[55] HRI peptide P-52 contains the sequence Asp-Phe-Gly, which is the most highly conserved short stretch of amino acids in the catalytic domain VII of protein kinases as defined by Hanks *et al.*[102] In addition, the first eight N-terminal amino acids of P-52 share extensive homology to CDC2 and CDC28 protein kinases and kinase-related transforming proteins (ros, fms, raf, kit, erb B, Src, and abl). The last five amino acids of P-52, however, are unique to HRI. The 14 N-terminal amino acids of HRI peptide P-74 show 50–60% identity to the conserved catalytic domain IX of kinase-related transforming proteins (raf, mht, Src, and kit). Again, the last six amino acids of P-74 are unique to HRI. The amino acid sequence of HRI peptide P-56 is unique to HRI.

The homology of the amino acid sequences of HRI tryptic peptides P-52 and P-74 to the conserved domains VII and IX of protein kinases makes it possible to predict that P-52 is positioned to the N-terminal side of P-74. This information permitted us to design primers for polymerase chain reaction (PCR) amplification of a partial HRI cDNA. Using these two primers, we obtained an amplified cDNA fragment which was 234 base pairs (bp) in length.[103] Excluding the 15-bp *Eco*R1 restriction sites present on both primers, the remaining 219-bp sequence encodes an open reading frame for 73 amino acids. The newly obtained 38-amino acid sequence of HRI deduced from this cDNA sequence contains the consensus sequence (Gly-Thr/Ser-X-X-Tyr/Phe-X-Ala/Ser-Pro-Glu) of serine/threonine protein kinases located in the conserved domain VIII. This observation is consistent with the finding that HRI is a serine protein kinase which phosphorylates eIF-2α at serine-51.[104] Furthermore, the amino acid sequences of HRI among conserved domains VII, VIII, and IX are unique to HRI.

Using this 234-bp probe of HRI, we isolated the HRI cDNA of rabbit reticulocytes. A schematic illustration of HRI cDNA (2729 nucleotides) is presented in Fig. 2. There are 112 nucleotides preceding the first ATG. Starting from this first ATG (nucleotide 113), the open reading frame continues to nucleotide 1990 encoding 626 amino acids followed by multiple stop codons in the 3′ untranslated region of 739 nucleotides. The overlapping repeat of the AATAAA polyadenylation signal is found at nucleotides 2689–2698, 11 nucleotides from the poly(A) tail. The deduced amino acid sequence of the HRI cDNA contains the exact amino acid sequences of the three tryptic peptides of HRI previously obtained by microsequencing.[55] P-52 is located in domain VII, P-56 in domain I, and P-74 in domain IX.

In vitro translation of mRNA transcribed from the HRI cDNA in nuclease-treated rabbit reticulocyte lysates yields a predominant 90-kDa polypeptide as observed in SDS–PAGE. The nucleotide sequence data demonstrate that the 5′ untranslated leader sequence is extremely G–C rich (80%) with the potential to form significant secondary structure. Secondary structure at the 5′ terminus of mRNAs is known to diminish mRNA translational efficiency.[105] Indeed, we found that the HRI mRNA was not translatable in a wheat germ extract. Unlike the reticulocyte lysate, the wheat germ extract does not contain an endogenous HRI

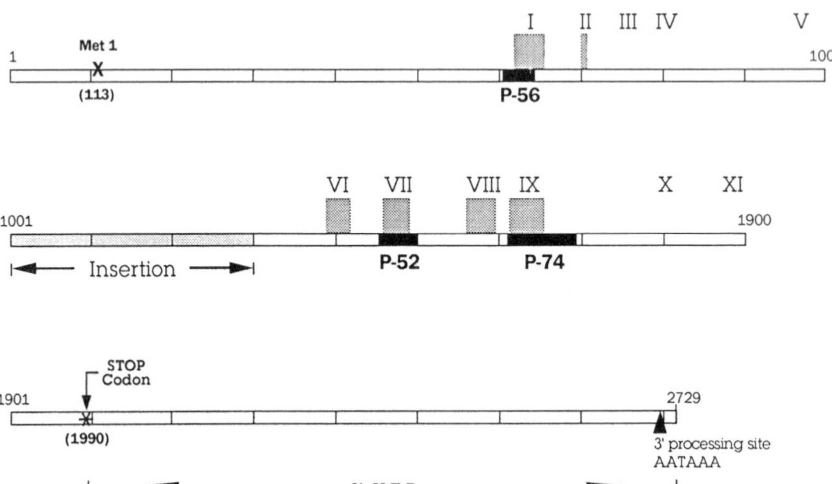

Figure 2. Schematic illustration of the structural features of HRI cDNA. The kinase conserved domains are marked by the shaded boxes. The locations of HRI tryptic peptides (P-52, P-56, and P-74) are shown by solid bars.

enzyme; therefore, expression of the HRI protein in the wheat germ system should facilitate analysis of kinase activity in the HRI translation products. The translational efficiency of mRNA transcripts can be increased by the use of untranslated leader sequences of some plant viral RNAs, such as tobacco mosaic virus (TMV).[106,107] Accordingly, we replaced the G–C-rich HRI untranslated leader sequence with that of TMV. The chimeric TMV–HRI mRNA was translated with approximately tenfold greater efficiency than HRI mRNA in the reticulocyte lysate and translation in the wheat germ extract is also clearly detectable. In all cases, the translated product of HRI mRNA migrated slightly faster than authentic purified phosphorylated HRI. This slight difference in mobility is most likely due to a lower level of phosphorylation in the translation products.

Translational products of HRI cDNA from both reticulocyte lysates and wheat germ extracts exhibit eIF-2α kinase activity. It should be noted that there is no mammalian eIF-2α kinase activity in the wheat germ extracts and our purified reticulocyte HRI phosphorylates purified wheat germ eIF-2α very inefficiently (J.-J. Chen unpublished observation). In addition, the 90-kDa polypeptide expressed from HRI cDNA is recognized by monoclonal antibodies to HRI both by Western blotting and immunoprecipitation. We conclude the HRI cDNA is expressed as a 90-kDa protein with eIF-2α kinase activity.

4.2. Conserved Catalytic Domains of Protein Kinases in HRI cDNA

Hanks *et al.*[102] have compared and aligned the protein sequences of 65 different protein kinases. They have identified 11 domains of protein kinases with invariant amino acid residues in each domain. HRI cDNA contains all 11 catalytic domains with invariant amino acid residues (Fig. 2). the consensus ATP-binding sequence Gly-X-Gly-X-X-Gly and the invariant valine residue located two positions downstream of the Gly-X-Gly-X-X-Gly are conserved in HRI. In domain II, the invariant Lys residue has been shown to be indispensable and to be involved in the phosphotransferase activity of protein kinases (reviewed in ref. 102). In HRI this invariant residue is Lys-199. HRI possesses Asp-Leu-Lys-Pro-Arg-Asn in domain VI, which is characteristic of Ser/Thr protein kinases.[102] Asp-Phe-Gly located in domain VII is the most conserved short stretch in the catalytic domains of protein kinases and is probably involved in ATP binding; it is found in HRI as Asp(−456)-Phe(−457)-Gly(−458). In domain VIII the Ala/Ser-Pro-Glu consensus sequence essential for catalytic activity of protein kinases is also found in HRI. Domain VIII of HRI contains the other consensus sequence for Ser/Thr protein kinases, Gly-Thr-Cys-Leu-Tyr. The sites of autophosphorylation of many protein kinases are located within 20 amino acids of the conserved Ala/Ser–Pro–Glu sequence in catalytic domain VIII (e.g., Thr-197 of cAMP-dependent protein kinase). The HRI equivalent of the Thr-197 of cAMP-dependent protein kinase is Thr-483. In addition, there are two serine and three more threonine residues in the vicinity of Thr-483. Since HRI can undergo multiple phosphorylation *in vitro*,[57] the availability of HRI cDNA will facilitate the further study of the sites and the role of autophosphorylation in the activation of HRI. The conserved amino acids in domain IX are also found in HRI. Thus, the homology of the deduced amino acid sequence of HRI cDNA to the conserved domains of other Ser/Thr protein kinases provides confirmatory evidence that HRI cDNA encodes a Ser/Thr protein kinase.

4.3. Kinase Insertion Sequence

As shown in Fig. 2, HRI cDNA contains an insertion of approximately 140 amino acids between catalytic domains V and VI (amino acids 276–413). Similar large inserts in this location have been reported for subclass III and IV receptor tyrosine kinases PDGF receptor, CSF-1 receptor, and c-*kit* protooncogene in which the kinase domains are divided into two halves by insertion of up to 100 mostly hydrophilic amino acid residues (reviewed in ref. 70). Yeast GCN2 protein kinase[108] and HRI are the first examples of Ser/Thr protein kinases known to contain such a large kinase insertion sequence. Since kinase insertion sequences are highly conserved among species for each specific receptor, the kinase insert may play an important role in the action of receptor kinases. Indeed the αPDGF receptor kinase insert contains an autophosphorylation site (Tyr-751), and mutation of Tyr-751 to Phe or Gly blocks association of the αPDGF receptor with phosphatidylinositol kinase and three other cellular proteins.[109] Similarly, mutation of Tyr-731 or Tyr-742 of the kinase insertion sequence of βPDGF receptor markedly impairs the association of the receptor with phosphatidylinositol kinase.[110] Kinase insert deletion mutants of mouse and human CSF-1 receptor are defective in their ability to phosphorylate and associate with phosphatidylinositol kinase.[111,112] In the case of HRI, heme binds to HRI and regulates its kinase activities.[2,57] It will be of interest to determine whether the kinase insertion sequence of HRI is involved in the binding of heme and the regulation of the autokinase and eIF-2α kinase activities. Since the interactions of HRI and hsp90 have been demonstrated both in reticulocyte lysates and in purified HRI as described above, it will also be of interest to determine whether the kinase insertion sequence of HRI is involved in the interaction with hsp90.

4.4. Extensive Homology of HRI to Yeast GCN2 Protein Kinase and Human dsRNA-Dependent eIF-2α Kinase

We have searched the Gene Bank for homology of the amino acid sequence of HRI as deduced from its cDNA to other protein sequences. Of the ten proteins with the highest scores, nine are Ser/Thr protein kinases, and of these, three are involved in regulation of the cell cycle (Nim A, Wee 1, and CDC 2).

It is especially noteworthy that GCN2 protein kinase of yeast[113,114] displays more homology to HRI than does dsI, the other known eIF-2α kinase. The scores of homology of HRI to GCN2 and dsI are significantly higher than those to other protein kinases. The cDNA of human dsI has recently been cloned.[115] The dot-matrix homology analysis reveals extensive homology of these three proteins in the protein kinase catalytic domains I–X except for domain V, where HRI has a large kinase insertion sequence. It should be emphasized that the amino acid sequences in domains III, IV, and X are very divergent among different classes of protein kinases.[102] Thus, the significant homology of both eIF-2α kinases around domains III and IV and between domains IX and X suggests that these regions may be involved in eIF-2 binding and the phosphorylation of eIF-2α. In addition, HRI synthetic peptide P-74, which resides around domain IX, inhibits the eIF-2α kinase activity of HRI.[55] Recently, the crystal structure of the catalytic subunit of cAMP-dependent protein kinase has been solved.[116,117] These studies indicate that protein kinases are composed of two lobes. The first and smaller lobe contains domains I–IV. The second and larger lobe contains domains V–X. The substrate sits between two lobes and contacts the kinase at domains IX and X. This crystal structure of cAMP-dependent protein kinase with peptide substrate fits very well with our observation that P-74 is involved in eIF-2 binding and the phosphorylation of eIF-2α.

GCN2 protein kinase of yeast displays very significant homology to HRI, especially in domains IX and X, where considerable homology is observed only in eIF-2α kinases. GCN2 protein kinase stimulates the expression of amino acid biosynthetic genes under conditions of amino acid starvation by derepressing GCN4, a transcriptional activator of these genes (reviewed in ref. 118). The derepression of GCN4 by GCN2 protein kinases occurs at the level of translation of GCN4 mRNA.[119,120] The activation of the translation of GCN4 mRNA coincides with a decrease in the rate of general polypeptide chain initiation at the level of eIF-2-dependent 43S preinitiation complex formation.[121] Furthermore, a yeast strain that overexpresses GCN2 protein kinase has been reported to have a lower rate of protein synthesis.[121] Thus, the effect of GCN2 protein kinase on protein synthesis is very similar to that of HRI. The molecular cloning of yeast eIF-2α[122] reveals 58% homology of its amino acid sequence to human eIF-2α.[123] In addition, consensus phosphorylation site Ser-51 is conserved in yeast eIF-2α, and the phosphorylation of yeast eIF-2α has been demonstrated.[122] The possibility that GCN2 protein kinase may phosphorylate eIF-2 has been raised by Cigan *et al.*[122] and Tzamarias *et al.*[121] The alignment of the amino acid sequences of HRI and GCN2 indicates 52% similarity and 28% identity in the kinase moiety of GCN2. This extensive homology of HRI and GCN2 affords further support that GCN2 is an eIF-2α kinase in yeast. Recently, it has been demonstrated that GCN2 stimulates GCN4 translation by phosphorylating eIF-2α.[136]

Recently, GCN2 has been shown to be associated with ribosomal subunits and polysomes.[108] When polysomes are dissociated to 40S and 60S subunits, GCN2 is mostly associated with 60S ribosomal subunits. GCN2 is dissociated from 60S subunits by 0.5 M KCl wash. This property of GCN2 is very similar to that of other initiation factors which are loosely associated with ribosomes and are not integral ribosomal proteins. The preferential association of GCN2 with 60S ribosomes is very interesting with regard to the site of the phosphorylation of eIF-2α. eIF-2 has been reported to be associated with 60S ribosomes after joining of the 48S initiation complex with 60S ribosomal subunits.[124,125] In addition, dsI of rabbit reticulocytes is also associated with ribosomes.[7,8] The carboxyl terminus of GCN2 is important for the association with ribosomes. In contrast, HRI is present mostly in the S-100 of reticulocytes. HRI is about half the size of GCN2 and is devoid of the carboxyl sequence of GCN2.

The similarity of GCN2 and dsI in the kinase insertion sequence has been reported recently by Ramirez *et al.*[108] Similarly, the kinase insertion sequences of HRI and GCN2 display significant homology, although not as good as that observed in kinase conserved domains IX and X or other conserved domains. Both HRI and GCN2 contain a much larger kinase insertion sequence (120–140 amino acids) than does dsI (40 amino acids). Although it is possible that part of the insertion sequence of HRI and GCN2 may be involved in the interaction with eIF-2, other portions of the kinase insertion sequence may be involved in the interactions with other proteins or regulators as discussed above.

GCN2 has an apparent molecular weight of 180 kDa and contains a 530-amino acid sequence adjacent tot he C terminus of protein kinase domains[114] which show 50% homology to HRI. This 530-amino acid sequence of GCN2 is closely related to histidyl-tRNA synthetase of *S. cerevisiae*, humans, and *E. coli*, and is required for the translational activation of GCN4.[126] Thus, the His-tRNA synthetase-related sequence has been proposed as a positive-acting domain under conditions of amino acid starvation. It is interesting that amino acid starvation in Ehrlich ascites cells also leads to increased eIF-2α phosphorylation.[46,127] In addition, increased eIF-2α phosphorylation is observed under nonpermissive temperature of Chinese hamster ovary cells which contain the temperature-sensitive aminoacyl-tRNA synthetase activities.[128] A revertant of ts-leucyl-tRNA synthetase reverses all inhibition of eIF-2 function together with the recovery of synthetase activity.[129] The mechanism by which eIF-2α phosphorylation is increased under these conditions is unknown. The homology of HRI to GCN2 may be relevant to the activation of mammalian eIF-2α kinase during amino acid starvation.

4.5. Homology to Protein Kinases Involved in Cell Division

It is intriguing that HRI has significant homology to three yeast protein kinases (Nim A. Wee 1, and CDC2) involved in cell division. This finding raises the possibility that HRI may also play a role in erythroid proliferation and differentiation.

We have found that mouse 3T3-F442A cells in culture spontaneously produce and secrete β-interferon and exhibit a pattern of dsRNA-dependent phosphorylation of dsI which is related to stages of their growth and differentiation into adipocytes. The dsRNA-dependent phosphorylation of dsI increases with cell growth until the culture becomes confluent. After confluence there is a rapid and marked decrease in phosphorylation of dsI.[130,131] It is possible that a high level of dsI activity may be related to attainment of the resting state necessary for the differentiation of fibroblasts to adipocytes. These studies lend support to the hypothesis of a physiological role for dsI, an eIF-2α kinase, in the regulation of growth and differentiation of cells.

It will be important to determine when HRI is expressed during erythroid differentiation. It has been shown that HRI is no longer present in mature erythrocytes.[132] Recently, an HRI-like eIF-2α kinase activity has been partially purified from both uninduced and induced mouse erythroleukemia (MEL) cells.[133] Upon further characterization, the HRI-like eIF-2α kinase obtained from uninduced cells is chromatographically different from HRI of induced cells, and is not capable of inhibiting protein synthesis of reticulocyte lysates. The HRI-like activity of induced MEL cells when added to reticulocyte lysates inhibits protein synthesis. In addition, protein synthesis in intact induced MEL cells is inhibited when iron is depleted by desferrioxamine.[134] Therefore, it appears that the HRI-like eIF-2α kinase from induced MEL cells behaves very similarly to HRI of reticulocytes. The eIF-2α kinase activity of uninduced MEL cells, however, is quite different; the exact identity of this eIF-2α kinase remains to be determined.

5. CONCLUDING REMARKS

Recently we have learned a great deal about the structure of both HRI and dsI through the molecular cloning of the cDNAs of these two eIF-2α kinases. The availability of these cDNAs opens new dimensions for more detailed structure–function studies of both eIF-2α kinases in the near future. HRI cDNA will also serve as a very useful tool in the understanding of the identity and the regulation of an HRI-like eIF-2α kinase in nonerythroid cells. In fact, the homology of HRI and dsI with yeast GCN2 protein kinase has provided us with linkage to the eIF-2α kinase activated during amino acid starvation of yeast. Further complementation studies of HRI cDNA to GCN2 mutants in yeast will reveal more information about the regulation of initiation of protein synthesis. Future studies of the interaction of HRI with heat-shock proteins, particularly the constitutively expressed hsp90 and hsp70, may be helpful in understanding the identity and components involved in the activation of HRI from pro-inhibitor to intermediate heme-reversible HRI and finally to heme-irreversible HRI. The homology of HRI to three proteins kinases involved in cell division suggests the possibility that HRI may have substrate(s) other than eIF-2α and that HRI may also be involved in the regulation of cell growth and differentiation of erythroid cells.

ACKNOWLEDGMENTS. The author is grateful to Drs. Irving M. London, Lee Gehrke, Robert Matts, and Daniel Levin for their review of this manuscript. Work carried out in the author's laboratory was supported by grants from the National Institutes of Health (DK-16272) and the National Science Foundation (DMB-9105907).

REFERENCES

1. Pain, V. M., 1986, Initiation of protein synthesis in mammalian cells, *Biochem. J.* **235:**625–637.
2. London, I. M., Levin, D. H., Matts, R. L., Thomas, N. S. B., Petryshyn, R., and Chen, J.-J., 1987, Regulation of protein synthesis, in: *The Enzymes*, 3d ed., Volume XVII (P. D. Boyer, and E. G. Krebs, eds.), pp. 359–380, Academic Press, New York.
3. Hershey, J. W. B., 1991, Translational control in mammalian cells, *Annu. Rev. Biochem.* **60:**717–755.
4. Jackson, R. J., 1991, Binding of Met-tRNA, in: *Translation in Eukaryotes* (H. Trachsel, ed.), pp. 193–229, CRC Press, Boca Raton, Florida.
5. Ehrenfeld, E., and Hunt, T., 1971, Double-stranded polio virus RNA inhibits initiation of protein synthesis by reticulocyte lysates, *Proc. Natl. Acad. Sci. USA* **68:**1075–1078.
6. Kosower, N. S., Vanderhoff, G. A., and Kosower, E. M., 1972, The effect of glutathione disulfide on initiation of protein synthesis, *Biochim. Biophys. Acta* **272:**623–637.
7. Farrell, P., Balkow, K., Hunt, T., Jackson, R. J., and Trachsel, H., 1977, Phosphorylation of initiation factor eIF-2 and the control of reticulocyte protein synthesis, *Cell* **11:**187–200.
8. Levin, D. H., and London, I. M., 1978, Regulation of protein synthesis: Activation by double-stranded RNA of a protein kinase that phosphorylates eukaryotic initiation factor 2, *Proc. Natl. Acad. Sci. USA* **75:**1121–1125.
9. Petryshyn, R., Trachsel, H., and London, I. M., 1979, Regulation of protein synthesis in reticulocyte lysates: Immune serum inhibits heme-regulated protein kinase activity and differentiates heme-regulated protein kinase from double-stranded RNA-induced protein kinase, *Proc. Natl. Acad. Sci. USA* **76:**1575–1579.
10. Hunt, T., 1979, The control of protein synthesis in rabbit reticulocyte lysates, in: *Miami Winter Symposium: "From Gene to Protein"*, Volume 16 (T. R. Russel, K. Brew, J. Schultz, and H. Haber, eds.) pp. 321–345, Academic Press, New York.
11. Kruh, J., and Borsook, G., 1956, Hemoglobin synthesis in rabbit reticulocytes *in vitro*, *J. Biol. Chem.* **220:**905–915.
12. Kassenaar, A., Morell, H., and London, I. M., 1957, The incorporation of glycine into globin and the synthesis of heme *in vitro* in duck erythrocytes, *J. Biol. Chem.* **229:**423–435.
13. Morell, H., Savoie, J. C., and London, I. M., 1958, The biosynthesis of heme and the incorporation of glycine into globin in rabbit bone marrow *in vitro*, *J. Biol. Chem.* **233:**923–929.
14. Bruns, G. P., and London, I. M., 1965, The effect of hemin on the synthesis of globin, *Biochem. Biophys. Res. Commun.* **18:**236–242.
15. Grayzel, A. I. P., Horchner, P., and London, I. M., 1966, The stimulation of globin synthesis by heme, *Proc. Natl. Acad. Sci. USA* **55:**650–655.
16. Waxman, H. S., and Rabinovitz, M., 1965, Iron supplementation *in vitro* and the state of aggregation and function of reticulocyte ribosomes in hemoglobin synthesis, *Biochem. Biophys. Res. Commun.* **19:**538–545.
17. Waxman, H. S., and Rabinovitz, M., 1966, Control of reticulocyte polyribosome content and hemoglobin synthesis by heme, *Biochim. Biophys. Acta* **129:**369–379.
18. Zucker, W. V., and Schulman, H. M., 1968, Stimulation of globin-chain initiation by hemin in the reticulocyte cell-free system, *Proc. Natl. Acad. Sci. USA* **59:**582–589.
19. Adamson, S. D., Herbert, E., and Godchaux, W., 1968, Factors affecting the rate of protein synthesis in lysate systems from reticulocytes, *Arch. Biochem. Biophys.* **125:**671–683.
20. Hunt, T., Vanderhoff, G. A., and London, I. M., 1972, Control of globin synthesis: The role of heme, *J. Mol. Biol.* **66:**471–481.
21. Adamson, S. D., Herbert, E., and Kemp, S. F., 1969, Effects of hemin and other prophyrins on protein synthesis in a reticulocyte lysate cell-free system, *J. Mol. Biol.* **42:**247–258.
22. Maxwell, C. R., and Rabinovitz, M., 1969, Evidence for an inhibitor in the control of globin synthesis by hemin in a reticulocyte lysate, *Biochem. Biophys. Res. Commun.* **35:**79–85.
23. Howard, G. A., Adamson, S. D., and Herbert, E., 1970, Studies on cessation of protein synthesis in reticulocyte lysate cell-free system, *Biochim. Biophys. Acta* **213:**237–240.

24. Rabinovitz, M., Freedman, M. L., Fisher, J. M., and Maxwell, C. R., 1969, Translational control in hemoglobin synthesis, *Cold Spring Harbor Symp. Quant. Biol.* **34:**567–578.
25. Maxwell, C. R., Kamper, C. S., and Robinovitz, M. J., 1971, Hemin control of globin synthesis: An assay for the inhibitor formed in the absence of hemin and some characteristics of its formation, *J. Mol. Biol.* **58:**317–327.
26. Gross, M., and Rabinovitz, M., 1972, Control of globin synthesis in cell-free preparations of reticulocytes by formation of a translational repressor that is inactivated by hemin, *Proc. Natl. Acad. Sci. USA* **69:**1565–1568.
27. Balkow, K., Hunt, T., and Jackson, R. J., 1975, Control of protein synthesis in reticulocyte lysates: The effect of nucleotide triphosphates on formation of the translational repressor, *Biochem. Biophys. Res. Commun.* **67:**366–374.
28. Gross, M., and Rabinovitz, M., 1972, Control of globin synthesis by heme: Factors influencing formation of an inhibitor of globin chain initiation in reticulocyte lysates, *Biochim. Biophys. Acta* **287:**340–352.
29. Hurst, R., Schatz, J. R., and Matts, R. L., 1987, Inhibition of rabbit reticulocyte lysate protein synthesis by heavy metal ions involves the phosphorylation of the α-subunit of the eukaryotic initiation factor 2, *J. Biol. Chem.* **262:**15939–15945.
30. Wu, J. M., 1981, Control of protein synthesis in rabbit reticulocytes: Inhibition of polypeptide synthesis by ethanol, *J. Biol. Chem.* **256:**4164–4167.
31. Bonanou-Tzedaki, S., Smith, K. E., Sheeran, B. A., and Arnstein, H. R. V., 1978, Reduced formation of initiation complexes between Met-$tRNA_f$ and 40 S ribosomal subunits in rabbit reticulocyte lysates incubated at elevated temperatures, *Eur. J. Biochem.* **84:**601–610.
32. Ernst, V., Baum, E. Z., and Reddy, R., 1982, Heat shock, protein phosphorylation and the control of translation in rabbit reticulocytes, reticulocyte lysates and HeLa cells, in: *Heat Shock: From Bacteria to Man* (M. J. Schlesinger, M. Ashburner, and A. Tissieres, eds.), pp. 215–225, Cold Spring Harbor Laboratory, Cold Spring Harbor, New York.
33. Legon, S., Jackson, R. J., and Hunt, T., 1973, Control of protein synthesis in reticulocyte lysates by haemin, *Nature New Biol.* **241:**150–152.
34. Beuzard, R., and London, I. M., 1974, The effects of hemin and double-stranded RNA on α and β globin synthesis in reticulocyte and Krebs II ascites cell-free systems and the reversal of these effects on an initiation factor preparation, *Proc. Natl. Acad. Sci. USA* **71:**2863–2866.
35. Kaempfer, R., 1974, Identification of RNA-binding properties of an initiation factor capable of relieving translational inhibition induced by heme deprivation or double-stranded RNA, *Biochem. Biophys. Res. Commun.* **61:**591–597.
36. Clemens, M. J., Safer, B., Merrick, W. C., Anderson, W. F, and London, I. M., 1975, Initiation of protein synthesis in rabbit reticulocyte lysates by double-stranded RNA and oxidized glutathione: Indirect mode of action of polypeptide chain initiation, *Proc. Natl. Acad. Sci. USA* **72:**1286–1290.
37. Ernst, V., Levin, D. H., Ranu, R. S., and London, I. M., 1976, Control of protein synthesis in reticulocyte lysates: Effects of 3′:5′-cyclic AMP, ATP, and GTP on inhibitions induced by heme-deficiency, double-stranded RNA, and a reticulocyte translational inhibitor, *Proc. Natl. Acad. Sci. USA* **73:**1112–1116.
38. Legon, S., Brayley, A., Hunt, T., and Jackson, R. J., 1974, The effect of cyclic AMP and related compounds in the control of protein synthesis in reticulocyte lysates, *Biochem. Biophys. Res. Commun.* **56:**745–752.
39. Kramer, G., Cimadevilla, M., and Hardesty, B., 1976, Specificity of the protein kinase activity associated with the hemin-controlled repressor of rabbit reticulocyte, *Proc. Natl. Acad. Sci. USA* **73:**3078–3082.
40. Levin, D. H., Ranu, R. S., Ernst, V., and London, I. M., 1976, Regulation of protein synthesis in reticulocyte lysates: Phosphorylation of methionyl-$tRNA_f$ binding factor by protein kinase activity of the translational inhibitor isolated from heme-deficient lysates, *Proc. Natl. Acad. Sci. USA* **73:**3112–3116.

41. Gross, M., and Mendelewski, J., 1977, Additional evidence that the hemin-controlled translational repressor from rabbit reticulocytes is a protein kinase, *Biochem. Biophys. Res. Commun.* **74:**559–569.
42. Levin, D. H., Petryshyn, R., and London, I. M., 1980, Characterization of double-stranded-RNA-activated kinase that phosphorylates α subunit of eukaryotic initiation factor 2 (eIF-2α) in reticulocyte lysates, *Proc. Natl. Acad. Sci. USA* **77:**832–836.
43. Farrell, P. J., Hunt, T., and Jackson, R. J., 1978, Analysis of phosphorylation of protein synthesis initiation factor eIF-2 to two dimensional gel electrophoresis, *Eur. J. Biochem.* **89:**517–521.
44. Ernst, V., Levin, D. H., and London, I. M., 1979, *In situ* phosphorylation of the α subunit of eukaryotic initiation factor 2 in reticulocyte lysates inhibited by heme deficiency, double-stranded RNA, oxidized glutathione, or the heme-regulated protein kinase, *Proc. Natl. Acad. Sci. USA* **76:**2118–2122.
45. Leroux, A., and London, I. M., 1982, Regulation of protein synthesis by phosphorylation of eukaryotic initiation factor 2α in intact reticulocytes and reticulocyte lysates, *Proc. Natl. Acad. Sci. USA* **79:**2147–2151.
46. Scorsone, K. A., Paniers, R., Rowland, A. G., and Henshaw, E. C., 1987, Phosphorylation of eukaryotic initiation factor 2 during physiological stresses which affect protein synthesis, *J. Biol. Chem.* **262:**14538–14543.
47. Gross, M., and Rabinovitz, M., 1973, Control of globin synthesis in cell-free preparations of reticulocytes by formation of a translational repressor that is inactivated by hemin, *Proc. Natl. Acad. Sci. USA* **69:**1565–1568.
48. Ranu, R. S., and London, I. M., 1976, Regulation of protein synthesis in rabbit reticulocyte lysates: Purification and initial characterization of the cyclic 3′:5′-AMP independent protein kinase of the heme-regulated translational inhibitor, *Proc. Natl. Acad. Sci. USA* **73:**4349–4363.
49. Ranu, R. S., Levin, D. H., Delaunay, J., Ernst, V., and London, I. M., 1976, Regulation of protein synthesis in reticulocyte lysates: Characteristics of inhibition of protein synthesis by a translational inhibitor from heme-deficient lysates and its relationship to the initiation factor which binds Met-tRNA, *Proc. Natl. Acad. Sci. USA* **73:**2720–2724.
50. Mizuno, S., Fisher, J. E., and Rabinovitz, M., 1972, Hemin control of globin synthesis: Action of an inhibitor formed in the absence of hemin on the reticulocyte cell-free system and its reversal by a ribosomal factor, *Biochim. Biophys. Acta* **272:**638–650.
51. Trachsel, H., Ranu, R. S., and London, I. M., 1978, Regulation of protein synthesis in rabbit reticulocyte lysates: Purification and characterization of heme-reversible translational inhibitor, *Proc. Natl. Acad. Sci. USA* **75:**3654–3658.
52. Gross, M., and Mendelewski, J., 1978, An association between the formation of the hemin-controlled translational repressor and the phosphorylation of a 100,000 molecular weight protein, *Biochim. Biophys. Acta* **520:**650–663.
53. Chen, J.-J., Yang, J. M., Petryshyn, R., Kosower, N., and London, I. M., 1989, Disulfide bond formation in the regulation of eIF-2α kinase by heme, *J. Biol. Chem.* **264:**9559–9564.
54. Pal, J. K., Chen, J.-J., and London, I. M., 1991, Tissue distribution and immunoreactivity of heme-regulated eIF-2 alpha kinase determined by monoclonal antibodies, *Biochemistry* **30:**2555–2562.
55. Chen, J.-J., Pal, J. K., Petryshyn, R., Kuo, I., Yang, J. M., Throop, M. S., Gehrke, L., and London, I. M., 1991, Amino acid microsequencing of the internal tryptic peptides of heme-regulated eukaryotic initiation factor 2α subunit kinase: Homology to protein kinases, *Proc. Natl. Acad. Sci. USA* **88:**315–319.
56. Jackson, R. J., and Hunt, T., 1985, A novel approach to the isolation of rabbit reticulocyte heme-controlled eIF-2α protein kinase, *Biochim. Biophys. Acta* **826:**224–228.
57. Fagard, R., and London, I. M., 1981, Relationship between the phosphorylation and activity of the heme-regulated eIF-2α kinase, *Proc. Natl. Acad. Sci. USA* **78:**866–870.
58. Gross, M., and Redman, R., 1987, Effect of antibody to the hemin-controlled translational repressor in rabbit reticulocyte, *Biochim. Biophys. Acta* **908:**123–130.

59. London, I. M., Ernst, V., Fagard, R., Leroux, A., Levin, D., and Petryshyn, R., 1981, Regulation of protein synthesis by phosphorylation and heme, *Cold Spring Harbor Conf. Cell Proliferation* **8:**941–958.
60. Matts, R. L., Schatz, J. R., Hurst, R., and Kagen, R., 1991, Toxic heavy metal ions activate the heme-regulated eukaryotic initiation factor-2α kinase by inhibiting the capacity of hemin-supplemented reticulocyte lysates to reduce disulfide bonds, *J. Biol. Chem.* **266:**12695–12702.
61. Jackson, R. J., Campbell, E. A., Herbert, P., and Hunt, T., 1983, The preparation and properties of gel-filtered rabbit reticulocyte lysate protein synthesis systems, *Eur. J. Biochem.* **131:**289–301.
62. Jackson, R. J., Herbert, P., Campbell, E. A., and Hunt, T., 1983, The roles of sugar phosphates and thio-reducing systems in the control of reticulocyte protein synthesis, *Eur. J. Biochem.* **131:** 313–324.
63. Hunt, T., Herbert, P., Campbell, E. A., Delidakis, C., and Jackson, R. J., 1983, The use of affinity chromatography on 2′5′ ADP-Sepharose reveals a requirement for NADPH, thioredoxin and T thioredoxin reductase for the maintenance of high protein synthesis activity in rabbit reticulocyte lysates, *Eur. J. Biochem.* **131:**280–301.
64. Gross, M., 1978, Regulation of protein synthesis by hemin: Effect of dithiothreintol on the formation and activity of the hemin-controlled translational repressor, *Biochim. Biophys. Acta* **520:**642–649.
65. Rosenfeld, M., and Surgenor, D. M., 1950, Methemalbumin: Interaction between human serum albumin and ferriprotoporphyrin IX, *J. Biol. Chem.* **183:**663–677.
66. Beaven, G. H., Chen, S.-H., d'Albis, A., and Gratzer, W. B., 1974, A spectroscopic study of the maemin-human-serum-albumin system, *Eur. J. Biochem.* **41:**539–546.
67. Kudlicki, W., Fullilove, S., Read, R., Kramer, G., and Hardesty, B., 1987, Identification of spectrin-related peptides associated with the reticulocyte heme-controlled α subunit of eukaryotic translational initiation factor 2 kinase and of a M_r 95,000 peptide that appears to be the catalytic subunit, *J. Biol. Chem.* **262:**9695–9701.
68. Armstrong, R. N., and Kaiser, E. T., 1978, Sufhydryl group reactivity of adenosine 3′,5′-monophosphate dependent protein kinase from bovine heart: A probe of holoenzyme structure, *Biochemistry* **17:**2840–2845.
69. Bramson, H. M., Thomas, N., Matsueda, R., Nelson, N. C., Taylor, S. S., and Kaiser, E. T., 1982, Modification of the catalytic subunit of bovine heart CAMP-dependent protein kinase with affinity labels related to peptide substrates, *J. Bio. Chem.* **257:**10575–10581.
70. Ullrich, A., and Schlessing, J., 1990, Signal transduction by receptors with tyrosine kinase activity, *Cell* **61:**203–212.
71. Li, W., and Stanley, E. R., 1991, Role of dimerization and modification of the CSF-1 receptor in its activation and internalization during the CSF-1 response, *EMBO J.* **10:**277–288.
72. Li, W., and Schlessinger, J., 1991, Platelet-derived growth factor (PDGF)-induced disulfide-linked dimerization of PDGF receptor in living cells, *Mol. Cell. Biol.* **11:**3756–3761.
73. Bauskin, A. R., Alkalay, I., and Ben-Neriah, Y., 1991, Redox regulation of a protein tyrosine kinase in the endoplasmic reticulum, *Cell* **66:**685–696.
74. Hentze, M. W., Rouault, T. A., Harford, J. B., and Klausner, R. D., 1989, Oxidation–reduction and the molecular mechanism of a regulatory RNA–protein interaction, *Science* **244:**357–359.
75. Najita, L., and Sarnow, P., 1990, Oxidation–reduction sensitive interaction of a cellular 50-kDa protein with an RNA hairpin in the 5′ noncoding region of the poliovirus genome, *Proc. Natl. Acad. Sci. USA* **87:**5846–5850.
76. Abate, C., Patel, L., Rauscher III, F. J., and Curran, T., 1990, Redox regulation of FOS and Jun DNA-binding activity *in vitro*, **249:**1157–1161.
77. Ziegler, D. M., 1985, Role of reversible oxidation–reduction of enzyme thiols-disulfides in the metabolic regulation, *Annu. Rev. Biochem.* **54:**305–329.
78. Mizuno, S., 1975, Temperature sensitivity of protein synthesis initiation in the reticulocyte lysate system. Reduced formation of the 40 S ribosomal subunit Met-tRNA$_f$ complex at an elevated temperature, *Biochim. Biophys. Acta* **414:**273–282.

79. Bonanou-Tzedaki, S., Sohl, M. K., and Arnstein, H. R. V., 1981, Regulation of protein synthesis in reticulocyte lysates. Characterization of the inhibitor generated in the post-ribosomal supernatant by heating at 44°C, *Eur. J. Biochem.* **114:**69–77.
80. Mizuno, S., 1977, Temperature sensitivity of protein synthesis initiation. Inactivation of a ribosomal factor by an inhibitor formed at elevated temperature, *Arch. Biochem. Biophys.* **179:** 289–301.
81. Rose, D. W., Wettenhall, R. E. H., Kudlicki, W., Kramer, G., and Hardesty, B., 1987, The 90-kilodalton peptide of the heme-regulated eIF-2α kinase has sequence similarity with the 90-kilodalton heat shock protein, *Biochemistry* **26:**6583–6587.
82. Szyszka, R., Kramer, G., and Hardesty, B., 1989, The phosphorylation state of reticulocyte 90 kDa heat shock protein affects its ability to increase phosphorylation of peptide initiation factor 3α subunit by the heme-sensitive kinase, *Biochemistry* **28:**1435–1438.
83. Rose, D. W., Welch, W. J., Kramer, G., and Hardesty, B., 1989, Possible involvement of the 90 kDa heat shock protein in the regulation of protein synthesis, *J. Biol. Chem.* **264:**6239–6244.
84. Matts, R. L., and Hurst, R., 1989, Evidence for the association of the heme-regulated eIF-2α kinase with the 90 kDa heat shock protein in rabbit reticulocyte lysate *in situ*, *J. Biol. Chem.* **264:** 15542–15547.
85. Brugge, J. S., 1986, Interactions of the Rous sarcoma virus protein pp60^{src} with the cellular proteins pp50 and pp90, *Curr. Top. Microbiol. Immunol.* **123:**1–22.
86. Ziemiecki, A., Catelli, M.-G., Joab, I., and Moncharmont, B., 1986, Association of the heat shock protein hsp90 with steroid hormone receptors and tyrosine kinase oncogene products, *Biochem. Biophys. Res. Commun.* **138:**1298–1307.
87. Lindquist, S., and Craig, E. A., 1988, The heat-shock proteins, *Annu. Rev. Genet.* **22:**631–677.
88. Pain, V. M., and Henshaw, E. C., 1975, Initiation of protein synthesis in Ehrlich ascites tumour cells: Evidence of physiological variation in the association of methionyl-$tRNA_f$ with native 40-S ribosomal subunits *in vivo*, *Eur. J. Biochem.* **57:**335–342.
89. Pain, V. M., Lewis, J. A., Huvos, P., Henshaw, E. C., and Clemens, M. J., 1980, The effects of amino acid starvation on regulation of polypeptide chain initiation in Ehrlich ascites tumor cells, *J. Biol. Chem.* **255:**1486–1491.
90. Panniers, R., and Henshaw, E. C., 1984, Mechanism of inhibition of polypeptide chain initiation in heat-shocked Ehrlich ascites tumour cells, *Eur. J. Biochem.* **140:**209–214.
91. Flaim, K. E., Liao, W. S. K., Peavy, D. E., Taylor, J. M., and Jefferson, L. S., 1982, The role of amino acids in the regulation of protein synthesis in perfused rat liver, *J. Biol. Chem.* **257:**2939–2946.
92. Kelly, F. J., and Jefferson, L. S., 1985, Control of peptide-chain initiation in rat skeletal muscle, *J. Biol. Chem.* **260:**6677–6683.
93. Harmon, C. S., Proud, C. G., and Pain, V. M., 1984, Effects of starvation, diabetes and acute insulin treatment on the regulation of polypeptide-chain initiation in rat skeletal muscle, *Biochem. J.* **223:**687–696.
94. Duncan, R., and Hershey, J. W. B., 1985, Regulation of initiation factors during translational repression caused by serum depletion abundance, synthesis and turn-over rates, *J. Biol. Chem.* **260:**5486–5492.
95. Duncan, R., and Hershey, J. W. B., 1985, Regulation of initiation factors during translational repression caused by serum-deprivation covalent modification, *J. Biol. Chem.* **260:**5493–5497.
96. Duncan, R., and Hershey, J. W. B., 1984, Heat shock-induced translational alterations in HeLa cell: Initiation factor modifications and the inhibition of translation, *J. Biol. Chem.* **259:**11882–11889.
97. De Benedetti, A., and Baglioni, C., 1986, Activation of hemin-regulated initiation factor-2 kinase in heat-shocked HeLa cells, *J. Biol. Chem.* **261:**338–342.
98. Clemens, M. J., Pain, V. M., Henshaw, E. C., and London, I. M., 1976, Characterization of a macromolecular inhibitor of polypeptide chain initiation from Ehrlich ascites tumor cells, *Biochem. Biophys. Res. Commun.* **72:**768–775.

99. Ranu, R. S., 1980, Regulation of protein synthesis in eukaryotes by the protein kinases that phosphorylate initiation factor eIF-2: Evidence for a common mechanism of inhibition of protein synthesis, *FEBS Lett.* **112:**211–215.
100. Delaunay, J., Ranu, R. S., Levin, D. H., Ernst, V., and London, I. M., 1977, Characterization of a rat liver factor which inhibits initiation of protein synthesis in rabbit reticulocyte lysates, *Proc. Natl. Acad. Sci. USA* **74:**2264–2268.
101. Fagard, R., and Guguen-Guillouzo, C., 1983, The effect of hemin and of allyl isopropyl acetamide on protein synthesis in rat hepatocytes, *Biochem. Biophys. Res. Commun.* **114:** 612–619.
102. Hanks, S. K., Quinn, A. M., and Hunter, T., 1988, The protein kinase family: Conserved features and deducted phylogeny of the catalytic domains, *Science* **241:**42–52.
103. Chen, J.-J., Throop, M. S., Gehrke, L., Kuo, I., Pal, J. K., Brodsky, M., and London, I. M., 1991, Cloning of the cDNA of the heme-regulated eukaryotic initiation factor 2α (eIF-2α) kinase of rabbit reticulocytes: Homology to yeast GCN2 protein kinase and human double-stranded-RNA-dependent eIF-2α kinase, *Proc. Natl. Acad. Sci. USA* **88:**7729–7733.
104. Pathak, V. K., Schindler, D., and Hershey, J. W. B., 1988, Generation of a mutant form of protein synthesis initiation factor eIF-2 lacking the site of phosphorylation by eIF-2 kinases, *Mol. Cell. Biol.* **8:**993–995.
105. Pelletier, J., and Sonenberg, N., 1985, Insertion mutagenesis to increase secondary structure within the 5′ noncoding region of a eukaryotic mRNA reduces translational efficiency, *Cell* **40:**515–526.
106. Gallie, D. R., Sleat, D. E., Watts, J. W., Turner, P. C., and Wilson, T. M. A., 1987, A comparison of eukaryotic viral 5′-leader sequences as enhancers of mRNA expression *in vivo*, *Nucleic Acids Res.* **15:**8693–8711.
107. Gehrke, L., and Jobling, S. A., 1990, Untranslated leader sequences and enhanced messenger RNA translational efficiency, in: *NATO Conference on Regulation of Gene Expression, Series H: Cell Biology*, Volume 49 (J. E. G. McCarthy and M. Tuite, eds.) pp. 389–398, Springer-Verlag, Berlin.
108. Ramirez, M., Wek, R. C., and Hinnebusch, A. G., 1991, Ribosome association of GCN2 protein kinase, a translational activation of the GCN4 gene of *Saccharomyces cerevisiae*, *Mol. Cell. Biol.* **11:**3027–3036.
109. Kazlauskas, A., and Cooper, J. A., 1989, Autophosphorylation of the PDGF receptor in the kinase insert region regulates interactions with cellular proteins, *Cell* **58:**1121–1133.
110. Yu, J.-C., Heidaran, M. A. Pierce, J. H., Gutkind, J. S., Lombardi, D., Ruggiero, M., and Aaronson, S. A., 1991, Tyrosine mutation within the α platelet-derived growth factor receptor kinase insert domain abrogates receptor-associated phosphatidylinositol-3 kinase activity without affecting mitogenic or chemotactic signal transduction, *Mol. Cell. Biol.* **7:**3780–3785.
111. Reedjik, M., Liu, X., and Pawson, T., 1990, Interactions of phosphatidylinositol kinase, GTPase-activating protein (GAP), and GAP-associated proteins with the colony-stimulating factor 1 receptor, *Mol. Cell. Biol.* **10:**5601–5608.
112. Shurtleff, S. A., Downing, J. R., Rock, C. O., Hawkins, S. A., Roussel, M. F., and Sherr, C. J., 1990, Structural features of the colony-stimulating factor 1 receptor that affects its association with phosphatidylinositol 3-kinase, *EMBO J.* **9:**2415–2421.
113. Roussou, I., Thireos, G., and Hauge, B. M., 1988, Transcriptional–translational regulatory circuit in *Saccharomyces cerevisiae* which involves the GCN4 translational activator and the GCN2 protein kinase, *Mol. Cell. Biol.* **8:**2132–2139.
114. Wek, R. C., Jackson, B. M., and Hinnebusch, A. G., 1989, Juxtaposition of domains homologous to protein kinases and histidyl-tRNA synthetases in GCN2 protein suggests a mechanism for coupling GCN4 expression to amino acid availability, *Proc. Natl. Acad. Sci. USA* **86:**4579–4583.
115. Meurs, E., Chong, K., Galabru, J., Thomas, N. S. B., Kerr, I. M., Williams, B. R. G., and Hovanessian, A. G., 1990, Molecular cloning and characterization of human double-stranded RNA activated protein kinase induced by interferon, *Cell* **62:**379–390.
116. Knighton, D. R., Zheng, J., Ten Eyck, L. F., Ashford, V. A., Xuong, N.-H., Taylor, S., S., and

Sowadski, J. M., 1991, Crystal structure of the catalytic subunit of cyclic adenosine monophosphate-dependent protein kinase, *Science* **253:**407–414.
117. Knighton, D. R., Zheng, J., Ten Eyck, L. F., Xuong, N.-H., Taylor, S. S., and Sowadski, J. M., 1991, Structure of a peptide inhibitor bound to the catalytic subunit of cyclic adenosine monophosphate-dependent protein kinase, *Science* **253:**414–420.
118. Hinnebusch, A. G., 1988, Mechanisms of gene regulation in the general control of amino acid biosynthesis in *Saccharomyces cerevisiae*, *Microbiol. Rev.* **52:**248–273.
119. Thireos, G., Discoll-Penn, M., and Greer, H., 1984, 5′ untranslated sequences are required for the translational control of a yeast regulatory gene, *Proc. Natl. Acad. Sci. USA* **81:**5096–5100.
120. Hinnebusch, A. G., 1984, Evidence for translational regulation of the activation of general amino acid control in yeast, *Proc. Natl. Acad. Sci. USA* **81:**6442–6446.
121. Tzamarias, D., Roussou, I., and Thireos, G., 1989, Coupling of GCN4 mRNA translational activation with decreased rates of polypeptide chain initiation, *Cell* **57:**947–954.
122. Cigan, A. M., Pabich, E. K., Feng, L., and Donahue, T. F., 1989, Yeast translation initiation suppressor sui2 encodes the α subunit of eukaryotic initiation 2 and shares sequence identity with the human α subunit, *Proc. Natl. Acad. Sci. USA* **86:**2784–2788.
123. Ernst, H., Duncan, R. F., and Hershey, J. W. B., 1987, Cloning and sequencing of complementary DNAs encoding the α-subunit of translational initiation factor eIF-2. Characterization of the protein and its messenger RNA, *J. Biol. Chem.* **262:**1206–1212.
124. Gross, M., Redman, R., and Kaplansky, D. A., 1985, Evidence that the primary effect of phosphorylation of eIF2(a) in rabbit reticulocyte lysate is inhibition of the release of eIF2-GDP from 60S ribosomal subunits, *J. Biol. Chem.* **260:**9491–9500.
125. Thomas, N. S. B., Matts, R. L., Levin, D. H., and London, I. M., 1985, The 60 S ribosomal subunit as a carrier of eukaryotic initiation factor 2 and the site of reversing factor activity during protein synthesis, *J. Biol. Chem.* **260:**9860–9866.
126. Wek, R. C., Ramirez, M., Jackson, B. M., and Hinnebusch, A. G., 1990, Identification of positive-acting domains in GCN2 protein kinase required for translational activation of *GCN4* expression, *Mol. Cell. Biol.* **10:**2820–2831.
127. Rowlands, A. G., Montine, K. S., Henshaw, E. C., and Panniers, R., 1988, Physiological stresses inhibit guanine nucleotide exchange factor in Ehrlich cells, *Eur. J. Biochem.* **175:**93–99.
128. Clemens, M. J., Galpine, A., Austin, S. A., Panniers, R., Henshaw, E. C., Duncan, R., Hershey, J. W. B., and Pollard, J. W., 1987, Regulation of polypeptide chain initiation in Chinese hamster ovary cells with a temperature-sensitive leucyl-tRNA synthetase, *J. Biol. Chem.* **262:**767–771.
129. Pollard, J. W., Galpine, A. R., and Clemens, M. J., 1989, A novel role for aminoacyl-tRNA synthetases in the regulation of polypeptide chain initiation, *Eur. J. Biochem.* **182:**1–9.
130. Petryshyn, R., Chen, J.-J., and London, I. M., 1984, Growth-related expression of a double-stranded RNA-dependent protein kinase in 3T3 cells, *J. Biol. Chem.* **259:**14736–14742.
131. Petryshyn, R., Chen, J.-J., and London, I. M., 1988, Detection of activated double-stranded RNA-dependent protein kinase in 3T3-F442A cells, *Proc. Natl. Acad. Sci. USA* **85:**1427–1431.
132. Petryshyn, R., Rosa, F., Fagard, R., Levin, D. H., and London, I. M., 1984, Control of protein synthesis in human reticulocytes by heme-regulated and dsRNA dependent eIF-2α kinases, *Biochem. Biophys. Res. Commun.* **119:**891–898.
133. Sarre, T. F., Hermann, M., and Bader, M., 1989, Differential effect of hemin-controlled eIF-2α kinases from mouse erythroleukemia cells on protein synthesis, *Eur. J. Biochem.* **183:**137–143.
134. Sarre, T. F., 1989, Presence of haemin-controlled eIF-2α kinases in both undifferentiated and differentiating mouse erythroleukaemia cells, *Biochem. J.* **262:**569–574.
135. Yang, J. M., London, I. M., and Chen, J.-J., 1992, Effects of hemin and porphyrin compounds on intersubunit disulfide formation of heme-regulated eIF-2α kinase and the regulation of protein synthesis in reticulocyte lysates, *J. Biol. Chem.* **267:**20519–20524.
136. Dever, T. E., Feng, L., Wek, R. C., Cigar, A. M., Donahue, T. D., and Hinnebusch, A. G., 1992, Phosphorylation of initiation factor 2α by protein kinase GCN2 mediates gene-specific translational control of GCN4 in yeast, *Cell* **68:**585–596.

Chapter 18

Regulation of Reticulocyte eIF-2α Kinases by Phosphorylation

Gisela Kramer, Wieslaw Kudlicki, and Boyd Hardesty

1. INTRODUCTION

Rabbit reticulocytes have served as one of the model systems for studies on eukaryotic protein synthesis for more than three decades. Isolation and characterization of mammalian initiation factors are based to a major extent on results obtained with these cells. In 1975 it was found by Richard Jackson, Tim Hunt, and co-workers that reticulocyte initiation factor 2 (eIF-2) can be phosphorylated in its α subunit. Initial characterization of an apparently unique protein kinase followed.[1–4] This enzyme is active in reticulocytes under heme deficiency and leads to cessation of protein synthesis. Without knowing its identity, it had been described several years earlier as the causing agent for inhibition of protein synthesis; it was activated when the postribosomal supernatant was incubated without hemin. The name "heme-controlled repressor" (HCR) had been given to this agent.[5] A second protein kinase phosphorylating eIF-2α was found in reticulocytes; it is associated with polysomes and activated by incubation with low concentrations of double-stranded (ds) RNA and ATP.[1,6,7] Its relationship to the

GISELA KRAMER, WIESLAW KUDLICKI, AND BOYD HARDESTY • Department of Chemistry and Biochemistry, and Clayton Foundation Biochemical Institute, University of Texas at Austin, Austin, Texas 78712.

Translational Regulation of Gene Expression 2, edited by Joseph Ilan. Plenum Press, New York, 1993.

interferon-induced mammalian eIF-2α kinase described in other chapters of this book is obvious.

In this chapter we will review data showing that the phosphorylation state of the latter eIF-2α kinase directly affects its activity. The heme-controlled eIF-2α kinase is indirectly influenced by phosphorylation events. The phosphorylation state of proteins associated with it appears to affect the kinase's activity. We realize that the identity of this eIF-2α kinase is still controversial (compare Chapter 17 by Chen).

As each phosphorylation event is reversed by a counteracting phosphatase reaction, we will discuss which phosphatase appears to be involved in the dephosphorylation of eIF-2 and regulation of eIF-2α kinases and we will review what is known about the respective protein phosphatase in reticulocytes. In the end, we will present our working hypothesis concerning the association of mostly structural phosphoproteins with the heme-sensitive eIF-2α kinase and with both type 1 and type 2A phosphatases in a functional complex for translational control.

2. PHOSPHORYLATION/DEPHOSPHORYLATION OF eIF-2

2.1. eIF-2α Phosphorylation

Eukaryotic initiation factor 2, eIF-2, binds initiator transfer RNA (tRNA), Met-tRNA$_i$, in a GTP-dependent reaction and positions it properly on the ribosome in relation to the AUG initiation codon.[8] eIF-2 consists of three subunits, two of which (from human and rat) have been sequenced[9,10] and found to be of M_r 36,100 (α) and 38,400 (β), respectively. Genetic studies in yeast have implicated the latter subunit as a suppressor of initiator codon mutants;[11] 42% sequence identity exists between the yeast and human eIF-2β subunits.[11] In line with this relatively low evolutionary conservation of the eIF-2β sequence one may expect low sequence homology of the eIF-2α subunit. In fact, antibodies prepared against rabbit eIF-2 subunits do not recognize *Artemia* (brine shrimp) eIF-2.[12] Differences regarding eIF-2α phosphorylation and its effect have been found between lower eukaryotic and mammalian eIF-2. Wheat germ eIF-2 is not phosphorylated in the α subunit (which has a different electrophoretic mobility than its mammalian counterpart) by the reticulocyte heme-sensitive eIF-2α kinase (G. Kramer, W. Kudlicki, B. Hardesty, and J. Ravel, unpublished observations). *Artemia* eIF-2α is phosphorylated, but, unlike the reticulocyte factor, this eIF-2 is thereby not inhibited in partial reactions of peptide initiation.[12] Thus inhibition of protein synthesis through eIF-2α phosphorylation may be a characteristic of mammalian cells. The detailed mechanism by which eIF-2α phosphorylation leads to a block in peptide initiation is not totally clear; evidence has been presented that GDP bound to phosphorylated eIF-2 cannot be efficiently exchanged for GTP and that the guanine nucleotide exchange factor (GEF or eIF-2B) is sequestered in this complex (discussed in ref. 8).

The scientific history of mammalian eIF-2α phosphorylation is full of controversies, some of which have not been resolved. According to data from our laboratory, reticulocyte eIF-2α can be phosphorylated *in vitro* by the heme-sensitive kinase at two sites, serine-48 and serine-51.[13,14] Also, plasmid-expressed eIF-2α with serine-to-alanine mutations in either site has been shown to substitute for VA RNA (see Chapters 10 and 11) to antagonize the action of activated dsRNA-dependent eIF-2α kinase in COS-1 cells.[15] These data and other literature results on eIF-2α phosphorylation have been reviewed recently leading to the suggestion that the two sites may be phosphorylated sequentially;[16] alternatively, they may have different functions in translational control as discussed by Kaufman *et al.*[15] On the other hand, phosphorylation of only serine-51 may be sufficient to cause inhibition of globin synthesis in the reticulocyte lysate.[17]

It should be mentioned that *in vitro* eIF-2 is phosphorylated in its α subunit only when it is in its native state.[18] The isolated α subunit [the subunits can be separated by reversed-phase high-performance liquid chromatography (HPLC); see Fig. 1 of ref. 13] cannot serve as substrate. Requirement for native substrate is not unique for the eIF-2α kinases; we made similar observations with some substrates of casein kinase II.[19]

Most of the points discussed so far pertain to eIF-2α phosphorylation *in vitro* by the heme-controlled eIF-2α kinase. It should be emphasized, however, that increased phosphorylation of eIF-2α *in vivo* correlates with decreased protein synthesis. Elevated eIF-2α phosphorylation has been demonstrated for a variety of physiological stress situations under which translation is reduced in different mammalian cells (e.g., ref. 20 and literature cited therein). In virus-infected cells, dsRNA-dependent eIF-2α kinase can be activated and serve as a mechanism to reduce protein synthesis. Kaufman *et al.*[15] discuss the possibility that eIF-2α phosphorylation under certain conditions may affect translation of selected mRNAs, not global protein synthesis.

2.2. Reticulocyte Phosphatase Active with eIF-2(α-P)

Phosphatases that dephosphorylate phosphoserine and phosphothreonine residues in proteins are generally classified as type 1 or type 2 by criteria developed by Cohen and co-workers (reviewed in ref. 21). The major phosphatases regulating intracellular metabolic events are type 1 and type 2A enzymes.

eIF-2 phosphorylated in its α subunit by the heme-sensitive kinase, then reisolated, is efficiently dephosphorylated *in vitro* by the reticulocyte type 2A phosphatase.[22] This enzyme was isolated in its native undegraded three-subunit form. It appears to be a typical type 2A enzyme as found in other mammalian tissues (see the comparison given in Table V of ref. 22).

Type 2A phosphatases have been regarded as "soluble"[21] and are isolated from the postribosomal supernatant. We always found traces of "contaminating" polypeptides, however, that react with antispectrin antibodies in highly purified preparations of this phosphatase.[22,23] Does this point to intracellular interaction *in vivo* with the reticulocyte's membrane skeleton?

The reticulocyte type 2A native enzyme consists of a M_r 35,000 catalytic subunit and two noncatalytic subunits of about M_r 55,000 and 60,000, respectively. We have studied the role of the latter subunit and found that it has a regulatory function toward dephosphorylation by the catalytic subunit.[22]

Both the catalytic subunit and the M_r 60,000 polypeptide can be isolated under denaturing conditions from the native enzyme. They can be reconstituted to form a two-subunit type 2A phosphatase.[22,23] The M_r 60,000 peptide was found to have a stimulatory effect on the catalytic subunit with eIF-2(α-P) as substrate. Most dramatic is an increase in specific activity (given as nmole phosphate released from the substrate per min per mg protein of catalytic subunit) from 74 to 297 when the M_r 60,000 subunit is added to the reaction mixture. Experimental details are provided in ref. 22. The isolated M_r 60,000 polypeptide has an extended solution structure (axial ratio 10:1). After reconstitution with the catalytic subunit it confers this characteristic to the resulting two-subunit enzyme.[23]

Recently, Redpath and Proud[24] assayed phosphatase activity in diluted reticulocyte lysate using eIF-2(α-P) as substrate. Based on different assay conditions by which either type 1 (with an inhibitor 1 fragment) or type 2A (with 3 nM okadaic acid) were inhibited, these authors concluded that 70% of eIF-2(α-P) phosphatase in reticulocytes was the type 2A enzyme and most of the remaining activity was due to the type 1 enzyme. These data confirm our results[22,25] indicating that the type 2A phosphatase is the major enzyme dephosphorylating eIF-2(α-P); they also point to the fact that phosphatases are not as substrate-specific as protein kinases, especially those kinases that phosphorylate the α subunit of eIF-2.

3. THE dsRNA-DEPENDENT eIF-2α KINASE IS INACTIVE WHEN DEPHOSPHORYLATED

In responsive mammalian cells, interferon induces a dsRNA-dependent protein kinase that phosphorylates eIF-2α (see Chapter 8). This kinase is present in rabbit reticulocytes constitutively for unknown reasons. It was purified from these cells and partially characterized more than 10 years ago.[6,7] Activity was ascribed to a protein of about 68,000 daltons that was phosphorylated upon incubation with ATP in the presence of low concentrations of dsRNA. This eIF-2α kinase, in contrast to the heme-sensitive enzyme, sediments with polysomes during ultracentrifugation. It is not known how it is bound and whether it interacts directly with ribosomes.

Recently, the human dsRNA-dependent eIF-2α kinase has been cloned and sequenced.[26] The M_r 68,000 protein (by SDS–gel electrophoresis) was shown to contain 550 amino acids with all the conserved domains typical for serine/threonine kinases. In one of these domains the conserved Ala/Ser-Pro-Glu sequence is found, upstream of which the autophosphorylation site is assumed. Several serine and threonine residues are present in this 20-amino acid region. Multiple phosphorylation sites have been implied in studies with the enzyme from rabbit reticulocytes.[7]

dsRNA is required for the autophosphorylation reaction. It has been shown by Galabru and Hovannessian[27] that the human M_r 68,000 phosphoprotein remains active for phosphorylation of substrate (they used histones) after dsRNA had been separated from the kinase.

We demonstrated that a reticulocyte type 1 phosphatase dephosphorylated and thereby inactivated this enzyme.[28] The experiments leading to the latter result were carried out with the reticulocyte eIF-2α kinase. The enzyme fraction was first phosphorylated in the presence of poly(I)·poly(C). Half of the phosphorylated protein was incubated with the type 1 phosphatase in the presence of Mn^{2+}; the other half was incubated under the same conditions except that the phosphatase was omitted. Then both samples were chromatographed in parallel on small phosphocellulose columns which adsorbed the kinase but not dsRNA, ATP, and the phosphatase. After washing, the kinase fraction was eluted from both columns by increasing the salt concentration. Equal portions of either phosphorylated or dephosphorylated kinase were incubated in a phosphorylation assay with eIF-2 as substrate. Dephosphorylation reduced kinase activity to about 20% of the phosphorylated control. The dephosphorylated kinase, however, could be fully reactivated by including dsRNA [poly(I)·poly(C)] in the reaction mixture. Experimental details are given in ref. 28.

The phosphoprotein phosphatase used to dephosphorylate the activated dsRNA-dependent eIF-2α kinase was characterized as a type 1 enzyme. Its activity was inhibited by low concentrations of phosphatase inhibitor 2.[28] Type 1 phosphatases are recognized to occur in at least two forms, either as an inactive cytosolic complex consisting of the catalytic subunit plus inhibitor 2, or in "high-molecular-weight" structures where generally the catalytic subunit is bound to cellular particulate fractions through another protein.[21] Best characterized among the latter is the so-called G-subunit that ties the type 1 catalytic subunit to glycogen and regulates its activity. This feature of type 1 phosphatase will be discussed further in the last section of this chapter; the phosphatase used in the studies discussed above appears to copurify from the reticulocyte postribosomal supernatant with eIF-2α kinase activity until the two are separated by gel electrophoresis under nondenaturing conditions.[28] The native type 1 phosphatase requires Mn^{2+} for *in vitro* activity; about 2 mM was found to be the optimal concentration for the dephosphorylation of the human M_r 68,000 phosphoprotein, the dsRNA-activated eIF-2α kinase.[28]

4. THE HEME-SENSITIVE eIF-2α KINASE IS ASSOCIATED WITH PHOSPHOPROTEINS

4.1. The M_r 100,000 Phosphoprotein

Isolation and characterization of the heme-sensitive enzyme has been hindered by its extremely low concentration in reticulocytes (and other cells where there may be a similar enzyme; see ref. 29). Early results[1–4] indicated that incubation of purified reticulocyte eIF-2α kinase fractions with [γ-^{32}P]ATP led to phosphorylation of a polypeptide of M_r 100,000 or somewhat lower apparent molecular mass [analyzed by sodium dodecyl sulfate (SDS)–polyacrylamide gel electrophoresis and autoradiography]. This phosphoprotein was thought to be the kinase that was activated by autophosphorylation.[30] We have been able to isolate eIF-2α kinase fractions without this component.[31] Furthermore, we identified a protein of M_r 95,000 that is the only polypeptide in highly purified eIF-2α kinase preparations which detectably binds ATP.[31] These results have been challenged, however; Chen *et al.*[32] hold that the eIF-2α kinase is a homodimer with autophosphorylated $M_r \approx$ 92,000 subunits. Here we present some additional data to support our notion that it is not the M_r = 100,000 phosphoprotein, but a nonphosphorylated $M_r \approx$ 95,000 polypeptide that represents the catalytic subunit of the heme-sensitive eIF-2α kinase.

A second chromatographic procedure (besides the published one[31]) was developed by which we can obtain a kinase fraction devoid of the M_r 100,000 phosphopeptide but containing the M_r 95,000 catalytic subunit that is identified by affinity labeling with 8-azido-[α-^{32}P]ATP (Fig. 1). This eIF-2α kinase fraction can be separated from another one which does contain the M_r 100,000 phosphoprotein by chromatography on phosphocellulose provided the eIF-2α kinase fraction (purified through the polylysine–Sepharose step[31]) has been incubated with ATP/Mg^{2+}. Nonadsorbed protein (fraction 1) contains eIF-2α kinase with the M_r 100,000 phosphoprotein; adsorbed eIF-2α kinase that is eluted with higher KCl concentration (fraction 2) is deficient in the M_r 100,000 protein. Both fractions were further purified separately by polyacrylamide electrophoresis under nondenaturing conditions,[31] then analyzed for eIF-2α kinase activity and the presence of the M_r 95,000 (see below) and the M_r 100,000 polypeptides. The results (unpublished data) are given in Fig. 1.

As the type 2A phosphatase (described in Section 2.2) efficiently dephosphorylates the M_r 100,000 phosphoprotein (unpublished results), fraction 2 (apparently devoid of the M_r 100,000 phosphoprotein) was first incubated with this phosphatase, then with [γ-^{32}P]ATP. Still, no ^{32}P-phosphate was detected in the M_r 100,000 region of an SDS–polyacrylamide gel (unpublished data; cf. ref. 31).

Another line of evidence indicating that the M_r 100,000 phosphoprotein is not the catalytic subunit of the eIF-2α kinase stems from experiments involving limited proteolysis. As shown by the data presented in Fig. 2, the M_r 100,000

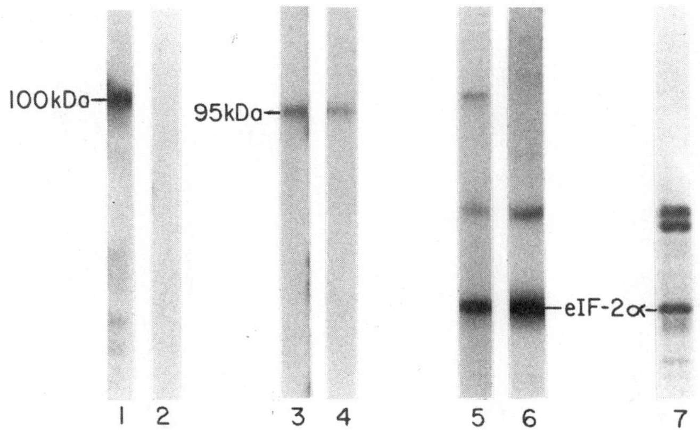

Figure 1. An eIF-2α kinase fraction devoid of the M_r 100,000 phosphoprotein eIF-2α kinase fractions 1 and 2 as defined in the text were assayed for M_r 100,000 phosphorylation (tracks 1 and 2), affinity-labeling with 8-azido-[α-^{32}P]ATP (tracks 3 and 4), and enzyme activity (tracks 5 and 6). After incubation, proteins in the samples were electrophoresed on SDS–polyacrylamide gels, then an autoradiogram was prepared which is shown. Results with fraction 1 are presented in tracks 1, 3, and 5, and those with fraction 2 in tracks 2, 4, and 6. In track 7, eIF-2 after gel electrophoresis and Coomassie staining is shown. Experimental procedures used are described in ref. 31.

phosphoprotein is degraded very rapidly (to peptides of low molecular mass—not indicated in the figure). However, eIF-2α kinase activity is not affected by the low concentrations of trypsin used in these experiments.

Further indication that the M_r 100,000 phosphoprotein is not the eIF-2α kinase comes from experiments involving polyproanthocyanidin (PPA).[19] This plant phenolic polymer interacts strongly with eIF-2 and a few other proteins. PPA seemingly inhibits eIF-2α phosphorylation by the heme-sensitive eIF-2α kinase (Fig. 1 of ref. 19). It also inhibits phosphorylation of the M_r 100,000 polypeptide; but it does not inhibit enzymatic activity of the eIF-2α kinase, in that a 14-amino acid peptide comprising the eIF-2α phosphorylation sites is phosphorylated even more strongly in the presence of 5 μM PPA than in its absence.[19] Taken together, the data indicate that the eIF-2α kinase appears to be active in the absence of the M_r 100,000 phosphoprotein, though under most isolation procedures the latter is associated with the kinase. In fact, identification and isolation of the heme-sensitive eIF-2α kinase is grossly facilitated by the M_r 100,000 protein, for which multiple (about five) phosphorylation sites have been estimated.[33]

4.2. The M_r 95,000 Catalytic Subunit

As indicated in Fig. 1, 8-azido-ATP affinity-labeling of the highly purified eIF-2α kinase fraction identifies a polypeptide of M_r 95,000 as the only ATP-

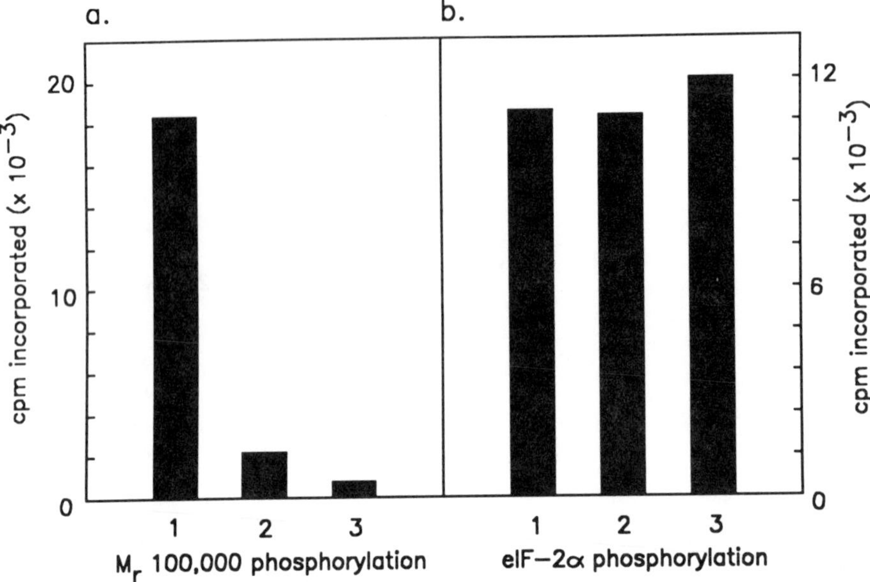

Figure 2. Higher protease sensitivity of the M_r 100,000 phosphoprotein than of the catalytic subunit of eIF-2. The eIF-2α kinase fraction 1 (Fig. 1) was phosphorylated with [γ-^{32}P]ATP, then incubated in the absence (bar 1) or presence of trypsin (100 ng/ml, bar 2; 300 ng/ml, bar 3). A tenfold excess of soybean trypsin inhibitor was added; then eIF-2α kinase activity was assayed. Samples were electrophoresed on SDS–polyacrylamide gels and radioactivity in the M_r 100,000 region of **(A)** the gel or **(B)** in eIF-2α was determined. Experimental procedures are given in ref. 31.

binding component in the enzyme fraction. Therefore we regard this component as the catalytic subunit.

This eIF-2α kinase has been called the heme-controlled repressor (see Section 1). It is activated in rabbit reticulocyte lysates under heme deficiency by an unknown mechanism. Direct binding of heme to the kinase[34] as well as hemin-aided disulfide bond formation[32] have been suggested as mechanisms by which the eIF-2α kinase is kept in an inactive form. An alternative hypothesis[35] involves the reversible association of the M_r 90,000 heat-shock protein hsp90 as a way of changing the activity of the eIF-2α kinase; this point will be discussed in more detail in the next section.

It appears that at least one effect of heme is directly on the kinase, in that highly purified preparations are inhibited for *in vitro* eIF-2α phosphorylation by the same low heme concentrations that prevent activation of the kinase in reticulocyte lysates.[36]

eIF-2α kinase activity that can be inhibited by heme has also been isolated from murine erythroleukemia cells.[29] On the other hand, monoclonal antibodies raised against the rabbit reticulocyte eIF-2α kinase did not identify cross-reacting polypeptides in any other tissue tested or in erythroid cells from other species, though eIF-2α kinase activity was detected in the latter sources.[37]

4.3. The M_r 90,000 Heat-Shock Protein

Highly purified preparations of the heme-sensitive eIF-2α kinase contain (besides the M_r 100,000 protein and the M_r 95,000 catalytic subunit) a relatively abundant M_r 90,000 polypeptide.[31,38] It can be partially separated from the kinase and, when added exogenously in apparently homogeneous form, increases eIF-2α phosphorylation by the kinase.[31] This M_r 90,000 polypeptide was subsequently shown to be the M_r 90,000 heat-shock protein hsp90.[39,40]

Hsp90 from different species, including mouse and human,[41,42] has been cloned and sequenced. It was found that there are two forms of hsp90 coded for by different genes.[41,43,44] Both forms (α and β) are constitutively present at 1–2% of cytoplasmic proteins in unstressed HeLa cells.[45,46] Lees-Miller and Anderson[46] determined a ratio of 39:61 (α:β) in exponentially growing HeLa cells. Under heat-shock conditions, expression of hsp90α is much more enhanced than that of hsp90β.[47] An obvious question is whether the two forms have a different function. In a study on nonactivated glucocorticoid receptor (which was shown previously[48] to contain hsp90 as a component) from WEHI-7 mouse thymoma cells, it was determined that approximately three-quarters of receptor-associated as well as of free hsp90 was present as the lower-molecular-weight (β) form.[49] Conversely, the predominant but not exclusive form of hsp90 associated with the rabbit reticulocyte heme-sensitive eIF-2α kinase appears to be the α species. This conclusion is reached by comparing the N-terminal sequence we published[39]—not appreciating the existence of two hsp90 forms—with N-terminal analyses done by Lees-Miller and Anderson.[50] The major α sequence was "contaminated" by the minor β sequence.[39] For (total, isolated) hsp90 from rabbit reticulocytes Lees-Miller and Anderson[50] determined a ratio of 3:1 (α:β) based on the different sequences at the N termini (Table I).

Hsp90 associated with the heme-sensitive rabbit reticulocyte eIF-2α kinase was identified after phosphorylation by casein kinase II (CKII) followed by trypsin digestion and subsequent sequencing of one of the isolated phosphopeptides that turned out to be identical to amino acids 49–62 of *Drosophila* hsp90.[39] Mammalian hsp90 sequences had not been published at the time of our analyses. Both serine and threonine residues were phosphorylated in this peptide[39] (Table I). Other sites that were phosphorylated by CKII *in vitro* were shown also to be phosphorylated *in vivo* in HeLa cells.[46] In addition, the α form of hsp90 can be phosphorylated *in vitro* on two threonine residues close to the N-terminus by a dsDNA-dependent kinase.[50] We found that hsp90 can be phosphorylated *in vitro* by cAMP-dependent kinase.[51] Not surprisingly, up to ten hsp90 species were resolved by high-resolution two-dimensional gel electrophoresis.[45] A list (prob-

Table I. Identified and Potential Phosphorylation Sites on hsp90[a]

	1			a		a					11				
α	P	E	E	T	Q	T	Q	D	Q	P	M	E	E	E	E
β	P	E	E	—	—	V	H	H	G	—	—	—	E	E	E

	60		c		c			c				c		73	
α} β}	Y	E	S	L	T	D	P	S	K	L	D	S	G	K	E

	206				d																226				b							
α	V	K	K	H	S	Q	F	I	G	Y	P	I	I	L	F	V	E	K	E	R	D	K	E	V	S	D	D	E	A	E	E	K
β														T	L						E			T								

	238																								b							269
α	E	E	K	E	E	E	K	E	K	E	E	K	E	S	D	D	K	P	E	I	E	D	V	G	S	D	E	E	E	E	E	K
β					G				E		D			D	E	E			K										D	D	S	G

	456				d										e						e				480			
α	R	K	K	L	S	E	L	L	R	Y	Y	T	S	A	S	G	D	E	M	V	S	L	K	D	Y	C	T	R
β		R	R								H			Q						T						V	S	

[a]Selected sequences are given for the two mouse hsp90 forms.[44] Except for peptide I, where the full sequences of α and β are shown, βsequences are identical with α sequences unless indicated (α and β designate the larger and smaller hsp90 forms, respectively). Arabic numbers indicate the position of the amino acid in the hsp90α sequence. Phosphorylation sites are shown by lowercase letters: a = sites for dsDNA-dependent kinase[50]; b = *in vivo* sites (for CKII)[46]; c = *in vitro* sites for CKII[39]; d = potential sites for cAMP-dependent kinase[52]; e = potential CKII sites.[53]

ably incomplete) of known and potential phosphorylation sites in both forms of hsp90 is given in Table I.

Sequence requirements as qualification for casein kinase II phosphorylation sites have been defined by Kuenzel *et al.*[53] In the sequences shown in Table I, *in vivo* identified sites (most likely CKII sites) are indicated by the letter b. At least one serine and Thr-64 in the tryptic peptide spanning residues 60–73 of hsp90α were found to be CKII sites *in vitro*.[39] (In the sequence shown, Glu-74 is included to indicate that Ser-71 might be a phosphorylation site; however, the finding that trypsin hydrolyzed the bond between Lys-73 and Glu-74, but not the one after Lys-68, may indicate that Ser-67 but not Ser-71 had been phosphorylated.) In fact, according to the requirements for CKII phosphorylation,[53] Ser-67 has to be phosphorylated for Thr-64 to be recognized by CKII. Another serine residue (Ser-62) in this tryptic peptide may also serve as a CKII site. Two possible phosphorylation sites for the cAMP-dependent kinase are indicated by d in Table I. They are the only serine residues in the hsp90 sequences that obey the criteria developed by Krebs and co-workers[52] as recognition sites for this kinase. Either of them is in proximity to CKII sites, so that they could influence the phosphorylatability of the latter as discussed below.

The phosphorlyation state of hsp90 appears to be important for its interaction with the eIF-2α kinase. This was shown by Szyszka *et al.*,[51] who preincubated hsp90 (separated from the eIF-2α kinase preparation) with a type 1 reticulocyte phosphatase and thereby abolished the stimulatory effect of hsp90 on the kinase. Rephosphorylation by CKII but not by the cAMP-dependent kinase restored the biological activity of hsp90. These data are summarized in Fig. 3. Interestingly, preincubation by the cAMP-dependent kinase of the dephosphorylated hsp90 prevented its reactivation by CKII (unpublished data).

The amount of hsp90 associated with the heme-sensitive eIF-2α kinase is likely to be only a fraction of the reticulocyte hsp90. Other roles for hsp90 (in other cells) have been reported (reviewed in refs. 54 and 55). An interesting observation was made by Matts and Hurst.[35] They were able to immunoprecipitate eIF-2α kinase with monoclonal anti-hsp90 antibodies from reticulocyte lysates, but only from those incubated in the presence of hemin, not from heme-deficient lysates. The specific monoclonal antibody used in these experiments recognized bound and free hsp90, whereas another one would only react with free hsp90 in the reticulocyte lysate and consequently not coimmunoprecipitate eIF-2α kinase. Matts and Hurst[35] presented evidence for association of hsp90 with the eIF-2α kinase, for binding of hsp90 to other proteins, and for free hsp90 under certain

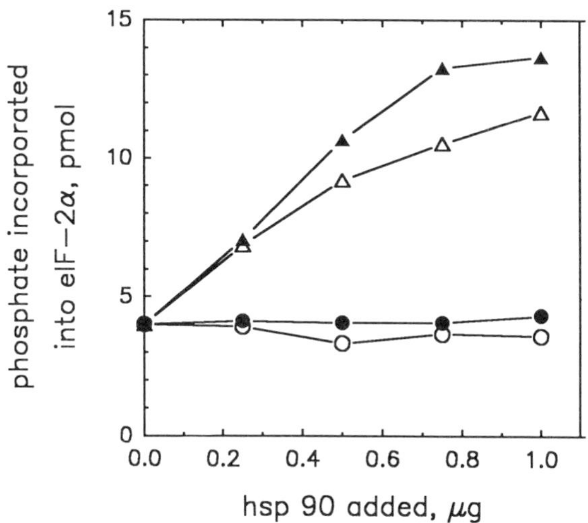

Figure 3. Stimulation of eIF-2α kinase activity by hsp90 depends on the phosphorylation state of the latter Hsp90 was preincubated with Mn^{2+} in the absence or presence of type 1 phosphatase; ethyleneglycol-bis (β-aminoethyl ether) N,N,N′,N′-tetraacetic acid (EGTA) was added to chelate Mn^{2+}; then hsp90 was incubated under phosphorylation conditions in the absence or presence of either CKII or cAMP-dependent kinase; this was followed by 5 min of incubation at 70°C to inactivate the enzymes.[51] Finally, eIF-2α phosphorylation was carried out with [γ-^{32}P]ATP in the absence or presence of the indicated amounts of preheated hsp90. This assay and quantification of ^{32}P-phosphate incorporated into eIF-2α were carried out as detailed in ref. 31. Results are for native hsp90 (closed triangles), dephosphorylated hsp90 (closed circles), hsp90 rephosphorylated by cAMP-dependent kinase (open circles), and hsp90 rephosphorylated by CKII (open triangles).

conditions, especially when the lysate had been treated with *N*-ethylmaleimide to activate rapidly the heme-sensitive eIF-2α kinase. Furthermore, these authors obtained some data indicating reversible association of hsp90 with the eIF-2α kinase. They developed the hypothesis that hsp90 when bound to the eIF-2α kinase keeps the enzyme in an inactive state just as hsp90 does when associated with steroid receptors (reviewed in ref. 55).

4.4. High-Molecular-Weight Functional Complex Containing the Heme-Sensitive eIF-2α Kinase and Phosphatase Activity

More and more data become available indicating that functional complexes exist in which different proteins, enzymes, and sometimes their regulatory factors are held together to carry out metabolic processes in the cell in an organized manner. Several years ago Srere[56] reviewed evidence for complexes of sequential metabolic enzymes. Functional units consisting of polysomes together with all components necessary for protein synthesis have been discussed and found to exist.[57,58] Mammalian mitochondrial pyruvate dehydrogenase complex is an extensively studied example (reviewed in ref. 59). This complex also contains a specific protein kinase and phosphatase that control the activity of the dehydrogenase.[59] Progress in understanding the structure of functional complexes may be slow because proteins in some of these complexes are not held together as rigidly as in the pyruvate dehydrogenase complex; also, protease activity may pose a problem in that large structural proteins are degraded during isolation procedures.[60]

Here we present some evidence to support our working hypothesis that the heme-sensitive eIF-2α kinase interacts with other proteins to form a functional complex. The following observations form the basis for our hypothesis. First, during the isolation procedure several proteins (listed in Table II) appear to

Table II. Proteins That May Directly or Indirectly Interact with the Heme-Sensitive eIF-2α Kinase

Name	Antibodies	Phosphoprotein	Structural protein
M_r 100,000	—	Yes	—
Hsp90	Yes	Yes	Yes
Phosphatase PP1-C	Yes[a]	—	No
Phosphatase type 2A catalytic subunit	Yes[b]	—	No
G subunit (R_G)	Yes[c]	Yes	Yes
Regulin	Yes	Yes	Yes

[a]Monoclonal antibodies were kindly provided by Dr. J. Vandenheede, Leuven, Belgium.
[b]Monoclonal antibodies raised against the bovine heart enzyme[61] were kindly provided by Dr. M. Mumby, Dallas, Texas.
[c]Antibodies against the N-terminal peptide of R_G (ref. 62) were kindly provided by Dr. A. DePaoli-Roach, Indianapolis, Indiana.

copurify with the eIF-2α kinase from different matrices, even during gradient elution from polylysine–Sepharose where relatively high salt concentrations are required to desorb this fraction (see Fig. 1 in ref. 31).

Mainly by using antibodies, we identified the components listed in Table II in the fraction having highest eIF-2α kinase activity. Surprisingly, both type 1 and type 2A phosphatases were detected to elute exactly where eIF-2α kinase was found. Monoclonal antibodies against either catalytic subunit aided in the identification of these proteins. These monoclonal antibodies do not cross-react. The bulk of type 1 phosphatase (mainly as the inactive complex with inhibitor 2) elutes at a much lower salt concentration from polylysine–Sepharose. The majority of type 2A phosphatase was separated from eIF-2α kinase at an early step in the purification procedure.[22,31] We do not know whether the other subunits of this phosphatase are present and how the phosphatase (or for that matter all other proteins in the "complex") interacts. We have evidence for the presence of the "G-subunit,"[21] the protein that binds PP1-C and tags it to intracellular structures. The G-subunit (as a degradation product of $M_r \approx 85{,}000$) was detected by specific antibodies raised against the N terminus of this protein[62] and kindly provided by Dr. A. DePaoli-Roach.

The second argument for the presence of a complex containing eIF-2α kinase plus the phosphatases stems from immunoprecipitation experiments. Preliminary data are presented in Fig. 4. They indicate that one line of monoclonal antibodies (Bc 2-5) is able to immunoprecipitate both kinase and phosphatase activity. For the data shown in Fig. 4, the polylysine–Sepharose fraction mentioned above in relation to Table II was treated with either antibodies Bc 2-5 or commercial mouse IgG, then incubated with insoluble protein A. After centrifugation, aliquots of each supernatant were assayed for phosphatase[63] or eIF-2α kinase activity.[31]

If eIF-2α kinase copurifies with phosphatases and both activities can be immunoprecipitated simultaneously; an obvious question is then how these proteins are held together. As indicated in Table II, at least two structural proteins appear to be present in this "complex." Hsp90 has an extended solution structure with an axial ratio of 21:1.[39] Evidence for its association with this eIF-2α kinase has been discussed in Section 4.2. Regulin is another structural protein we studied previously.[61] It has an extended solution structure and a tendency to self-aggregate in low salt concentrations;[61] also, it was found to bind heme.[65] Though regulin appears to be an abundant protein in reticulocytes,[61] we have not been able to detect it in any other tissues or species by using antiregulin monoclonal antibodies. This is reminiscent of the tissue distribution found for the eIF-2α kinase.[37]

Several of the proteins listed in Table II are phosphoproteins. As discussed in Section 4.2, the phosphorylation state of hsp90 is important for its interaction with the eIF-2α kinase. Type 1 phosphatase appears to be the enzyme that changes the biological activity of hsp90. On the other hand, type 2A phosphatase efficiently dephosphorylates the M_r 100,000 phosphoprotein. We do not know whether this dephosphorylation affects the eIF-2α kinase activity. Also, we have no data on

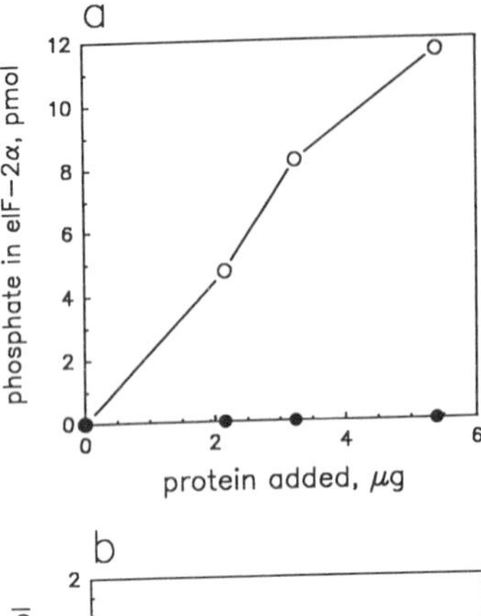

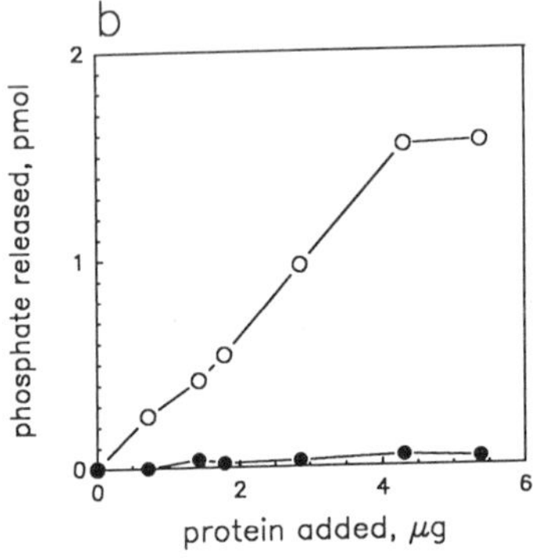

Figure 4. Coimmunoprecipitation of eIF-2α kinase and phosphatase activity Immunoprecipitation was carried out as outlined in the text using monoclonal antibodies (BC 2-5) or mouse IgG (Sigma) following the procedure of DePaoli-Roach and Lee.[64] **(A)** Remaining eIF-2α kinase determined as described;[31] **(B)** phosphatase activity depending on the amount of supernatant protein added. The assay was carried out as detailed previously,[63] using [^{32}P]histones as substrate.

whether casein kinase II is the only enzyme phosphorylating regulin[19] and which phosphatase is active with phosphorylated regulin. The presence of both type 1 and type 2A phosphatases in the eIF-2α kinase complex, however, may be of physiological importance.

5. CONCLUSION

After an early period of controversy, it is established now that eIF-2α phosphorylation causes a drastic decrease in protein synthesis. The activity of the dsRNA-dependent eIF-2α kinase appears to be controlled by autophosphorylation and counteracting dephosphorylation. Phosphorylation/dephosphorylation events also influence the activity of the heme-controlled eIF-2α kinase. In this case, however, clouds still prevent a clear insight into a complicated regulatory mechanism.

ACKNOWLEDGMENTS. Recent research done by the authors has been funded by grants to B.H. from the Foundation for Research and the Texas Advanced

Technology Program. We thank Dr. O. W. Odom for critically reading and Lisa Chronis for typing the manuscript.

REFERENCES

1. Farrell, P. J., Balkow, K., Hunt, T., Jackson, R., and Trachsel, H., 1977, Phosphorylation of initiation factor eIF-2 and the control of reticulocyte protein synthesis, *Cell* **11:**187–200.
2. Kramer, G., Cimadevilla, M., and Hardesty, B., 1976, Specificity of the protein kinase activity associated with the hemin-controlled repressor of rabbit reticulocyte, *Proc. Natl. Acad. Sci. USA* **73:**3078–3082.
3. Levin, D. H., Ranu, R. S., Ernst, V., and London, I. M., 1976, Regulation of protein synthesis in reticulocyte lysates: Phosphorylation of methionyl-$tRNA_f$ binding factor by protein kinase activity of translational inhibitor isolated from heme-deficient lysates, *Proc. Natl. Acad. Sci. USA* **73:** 3112–3116.
4. Gross, M., and Mendelewski, J., 1977, Additional evidence that the hemin-controlled repressor from rabbit reticulocytes is a protein kinase, *Biochem. Biophys. Res. Commun.* **74:**559–569.
5. Gross, M., and Rabinovitz, M., 1972, Control of globin synthesis in cell-free preparations of reticulocytes by formation of a translational repressor that is inactivated by hemin, *Proc. Natl. Acad. Sci. USA* **69:**1565–1568.
6. Petryshyn, R., Levin, D. H., and London, I. M., 1980, Purification and properties of the double-stranded RNA-activated eukaryotic RNA dependent protein kinase from reticulocyte lysate, *Biochem. Biophys. Res. Commun.* **94:**1190–1198.
7. Grosfeld, H., and Ochoa, S., 1980, Purification and properties of the double-stranded RNA-activated eukaryotic initiation factor 2 kinase from rabbit reticulocytes, *Proc. Natl. Acad. Sci. USA* **77:**6526–6530.
8. Pain, V. M., 1986, Initiation of protein synthesis in mammalian cells, *Biochem. J.* **235:**625–637.
9. Ernst, H., Duncan, R., and Hershey, J. W. B., 1987, Cloning and sequencing of complementary DNAs encoding the α-subunit of translational initiation factor eIF-2, *J. Biol. Chem.* **262:**1206–1212.
10. Pathak, V. K., Nielsen, P. J., Trachsel, H., and Hershey, J. W. B., 1988, Structure of the β subunit of translational initiation factor eIF-2, *Cell* **54:**633–639.
11. Donahue, T. F., Cigan, A. M., Pabich, E. K., and Valavicius, B. C., 1988, Mutations at a Zn(II) finger motif in the yeast eIF-2β gene after ribosomal start-site selection during the scanning process, *Cell* **54:**621–632.
12. Mehta, H. B., Dholakia, J. N., Roth, W. W., Parakh, B. S., Montelaro, R. C., Woodley, C. L., and Wahba, A. J., 1986, Structural studies on the eukaryotic chain initiation factor 2 from rabbit reticulocytes and brine shrimp *Artemia* embryos. Phosphorylation by the heme-controlled repressor and casein kinase II, *J. Biol. Chem.* **261:**6705–6711.
13. Wettenhall, R. E. H., Kudlicki, W., Kramer, G., and Hardesty, B., 1986, The NH_2-terminal sequence of the α and γ subunits of eukaryotic initiation factor 2 and the phosphorylation site for the heme-regulated eIF-2α kinase, *J. Biol. Chem.* **261:**12444–12447.
14. Kudlicki, W., Wettenhall, R. E. H., Kemp, B. E., Szyszka, R., Kramer, G., and Hardesty, B., 1987, Evidence for a second phosphorylation site on eIF-2α from rabbit reticulocytes, *FEBS Lett.* **215:**16–20.
15. Kaufman, R. J., Davies, M. V., Pathak, V. K., and Hershey, J. W. B., 1989, The phosphorylation state of eucaryotic initiation factor 2 alters translation efficiency of specific mRNAs, *Mol. Cell. Biol.* **9:**946–958.
16. Kramer, G., 1990, Two phosphorylation sites on eIF-2α, *FEBS Lett.* **267:**181–182.
17. Price, N. T., Welsh, G. I., and Proud, C. G., 1991, Phosphorylation of only serine-51 in protein

synthesis initiation factor-2 is associated with inhibition of peptide-chain initiation in reticulocyte lysates, *Biochem. Biophys. Res. Commun.* **176:**993–999.

18. Kramer, G., and Hardesty, B., 1981, Phosphorylation reactions that influence the activity of eIF-2, *Curr. Top. Cell. Regulat.* **20:**185–203.
19. Kudlicki, W., Picking, W. D., Kramer, G., Hardesty, B., Smailov, S. K., Mukhamedzhanov, B. G., Lee, A. V., and Iskakov, B. K., 1991, Eukaryotic protein synthesis initiation factor 2. A target for inactivation by proanthocyanidin, *Eur. J. Biochem.* **197:**623–629.
20. Scorsone, K. A., Panniers, R., Rowlands, A. G., and Henshaw, E. C., 1987, Phosphorylation of eukaryotic initiation factor 2 during physiological stresses which affect protein synthesis, *J. Biol. Chem.* **262:**14583–14543.
21. Cohen, P., 1989, The structure and regulation of protein phosphatases, *Annu. Rev. Biochem.* **58:** 453–508.
22. Chen, S. C., Kramer, G., and Hardesty, B., 1989, Isolation and partial characterization of an M_r 60,000 subunit of a type 2A phosphatase from rabbit reticulocytes, *J. Biol. Chem.* **264:**7267–7275.
23. Kramer, G., Chen, S.-C., Wollny, E., and Hardesty, B., 1987, Characterization of a high molecular weight phosphatase from rabbit reticulocytes, *Adv. Prot. Phosphatases* **4:**269–292.
24. Redpath, N. T., and Proud, C. G., 1990, Activity of protein phosphatases against initiation factor-2 and elongation factor-2, *Biochem J.* **272:**175–180.
25. Kramer, G., Chen, S.-C., Szyszka, R., and Hardesty, B., 1989, Reticulocyte phosphoprotein phosphatases active in regulating eIF-2α phosphorylation, *Adv. Prot. Phosphatases* **5:**611–633.
26. Meurs, E., Chong, K., Galabru, J., Thomas, N. S. B., Kerr, I. M., Williams, B. R. G.,and Hovanessian, A. G., 1990, Molecular cloning and characterization of the human double-stranded RNA-activated protein kinase induced by interferon, *Cell* **62:**379–390.
27. Galabru, J., and Hovanessian, A., 1987, Autophosphorylation of the protein kinase dependent on double-stranded RNA, *J. Biol. Chem.* **262:**15538–15544.
28. Szyszka, R., Kudlicki, W., Kramer, G., Hardesty, B., Galabru, J., and Hovanessian, A., 1989, A type 1 phosphoprotein phosphatase active with phosphorylated $M_r = 68{,}000$ initiation factor 2 kinase, *J. Biol. Chem.* **264:**3827–3831.
29. Sarre, T. F., Hermann, M., and Bader, M., 1989, Differential effect of hemin-controlled eIF-2α kinases from mouse erythroleukemia cells on protein synthesis, *Eur. J. Biochem.* **183:**137–143.
30. Trachsel, H., Ranu, R. S., and London, I. M., 1978, Regulation of protein synthesis in rabbit reticulocyte lysates: Purification and characterization of heme-reversible translational inhibitor, *Proc. Natl. Acad. Sci. USA* **75:**3654–3658.
31. Kudlicki, W., Fullilove, S., Read, R., Kramer, G., and Hardesty, B., 1987, Identification of spectrin-related peptides associated with the reticulocyte heme-controlled α subunit of eukaryotic translational initiation factor 2 kinase and of a M_r 95,000 peptide that appears to be the catalytic subunit, *J. Biol. Chem.* **262:**9695–9701.
32. Chen, J.-J., Yang, J. M., Petryshyn, R., Kosower, N., and London, I. M., 1989, Disulfide bond formation in the regulation of eIF-2α kinase by heme, *J. Biol. Chem.* **264:**9559–9564.
33. Gross, M., 1978, Regulation of protein synthesis by hemin. An association between the formation of the hemin-controlled translational repressor and the phosphorylation of a 100,000 molecular weight protein, *Biochim. Biophys. Acta* **520:**650–663.
34. Gross, M., 1974, Regulation of protein synthesis by hemin. Regulation by hemin of the formation and inactivation of a translational repressor of globin synthesis in rabbit reticulocyte lysates, *Biochim. Biophys. Acta* **340:**484–497.
35. Matts, R. L., and Hurst, R., 1989, Evidence for the association of the heme-regulated eIF-2α kinase with the 90-kda heat shock protein in rabbit reticulocyte lysate *in situ*, *J. Biol. Chem.* **264:** 15542–15547.
36. Hardesty, B., Kramer, G., Kudlicki, W., Chen, S.-C., Rose, D., Zardeneta, G., and Fullilove, S., 1985, Heme-controlled eIF-2α kinase and protein phosphatases in reticulocytes: Interaction with components of the membrane skeleton, *Adv. Prot. Phosphatases* **1:**235–257.

37. Pal, J. K., Chen, J.-J., and London, I. M., 1991, Tissue distribution and immunoreactivity of heme-regulated eIF-2α kinase determined by monoclonal antibodies, *Biochemistry* **30:**2555–2562.
38. Wallis, M. H., Kramer, G., and Hardesty, B., 1980, Partial purification and characterization of a 90,000-dalton peptide involved in activation of the eIF-2α protein kinase of the hemin-controlled translational repressor, *Biochemistry* **19:**798–804.
39. Rose, D. W., Wettenhall, R. E. H., Kudlicki, W., Kramer, G., and Hardesty, B., 1987, The 90-kilodalton peptide of the heme-regulated eIF-2α kinase has sequence similarity with the 90-kilodalton heat shock protein, *Biochemistry* **26:**6583–6587.
40. Rose, D. W., Welch, W. J., Kramer, G., and Hardesty, B., 1989, Possible involvement of the 90-kda heat shock protein in the regulation of protein synthesis, *J. Biol. Chem.* **264:**6239–6244.
41. Moore, S. K., Kozak, C., Robinson, E. A., Ullrich, S. J., and Appella, E., 1987, Cloning and nucleotide sequence of the murine hsp 84 cDNA and chromosome assignment of related sequences, *Gene* **56:**29–40.
42. Rebbe, N. F., Ware, J., Bertina, R. M., Modrich, P., and Stafford, D. W., 1987, Nucleotide sequence of a cDNA for a member of the human 90-kda heat shock protein family, *Gene* **53:** 235–245.
43. Simon, M. C., Kitchener, K., Kao, H.-T., Hickey, E., Weber, L., Voellmy, R., Heintz, N., and Nevins, J. R., 1987, Selective induction of human heat shock gene transcription by the adenovirus E1A gene products, including the 12S E1A product, *Mol. Cell. Biol.* **7:**2884–2890.
44. Moore, S. K., Kozak, C., Robinson, E. A., Ullrich, S. J., and Appella, E., 1989, Murine 86- and 84-kda heat shock proteins, cDNA sequences, chromosome assignments, and evolutionary origins, *J. Biol. Chem.* **264:**5343–5351.
45. Welch, W. J., and Feramisco, J. R., 1982, Purification of the major mammalian heat shock proteins, *J. Biol. Chem.* **257:**14949–14959.
46. Lees-Miller, S. P., and Anderson, C., 1989, Two human 90-kda heat shock proteins are phosphorylated *in vivo* at conserved serines that are phosphorylated *in vitro* by casein kinase II, *J. Biol. Chem.* **264:**2431–2437.
47. Barnier, J. V., Bensaude, O., Morange, M., and Babinet, C., 1987, Mouse 89 kd heat shock protein. Two polypeptides with distinct developmental regulation, *Exp. Cell Res.* **170:**186–194.
48. Housely, P. R., Sanchez, E. R., Westphal, H. M., Beato, M., and Pratt, W. B., 1985, The molybdate-stabilized L-cell glucocorticoid receptor isolated by affinity chromatography or with a monoclonal antibody is associated with a 90-92-kda nonsteroid-binding phosphoprotein, *J. Biol. Chem.* **260:**13810–13817.
49. Mendel, D. B., and Orti, E., 1988, Isoform composition and stoichiometry of the ~90-kda heat shock protein associated with glucocorticoid receptors, *J. Biol. Chem.* **263:**6695–6702.
50. Lees-Miller, S. P., and Anderson, C., 1989, The human double-stranded DNA-activated protein kinase phosphorylates the 90-kda heat-shock protein, hsp 90α at two NH_2-terminal threonine residues, *J. Biol. Chem.* **264:**17275–17280.
51. Szyszka, R., Kramer, G., and Hardesty, B., 1989, The phosphorylation state of the reticulocyte 90-kda heat shock protein affects its ability to increase phosphorylation of peptide initiation factor 2α subunit by the heme-sensitive kinase, *Biochemistry* **28:**1435–1438.
52. Krebs, E. G., and Beavo, J. A., 1979, Phosphorylation–dephosphorylation of enzymes, *Annu. Rev. Biochem.* **48:**923–959.
53. Kuenzel, E. H., Mulligan, J. A., Sommercorn, J., and Krebs, E. G., 1987, Substrate specificity determinants for casein kinase II as deduced from studies with synthetic peptides, *J. Biol. Chem.* **262:**9136–9140.
54. Lindquist, S., 1986, The heat-shock response, *Annu. Rev. Biochem.* **55:**1151–1191.
55. Hardesty, B., and Kramer, G., 1990, The 90,000 dalton heat shock protein, a lot of smoke but no function as yet, *Biochem. Cell. Biol.* **67:**749–750.
56. Srere, P., 1987, Complexes of sequential metabolic enzymes, *Annu. Rev. Biochem.* **56:**89–124.
57. Ryazonov, A., 1989, Organization of soluble enzymes in the cell. Relay at the surface, *FEBS Lett.* **237:**1–3.

58. Spirin, A. S., Baranov, V. I., Ryabova, L. A., Ovovdov, S. Y., and Alakhov, Y. B., 1988, A continuous cell-free translation system capable of producing polypeptides in high yield, *Science* **242:**1162–1164.
59. Reed, L. J., and Hackert, M., 1990, Structure–function relationships in dihydrolipoaminde acyltransferases, *J. Biol. Chem.* **265:**8971–8974.
60. Zardeneta, G., Kramer, G., and Hardesty, B., 1988, Quantification and characterization of regulin, a M_r-230,000 highly elongated protein of rabbit reticulocytes, *Eur. J. Biochem.* **178:** 267–276.
61. Mumby, M. C., Green, D. D., and Russell, K. L., 1985, Structural characterization of cardiac protein phosphatase with a monoclonal antibody, *J. Biol. Chem.* **260:**13763–13770.
62. Tang, P. M., Bondor, J. A., Swiderek, K., and DePaoli-Roach, A. A., 1991, Molecular cloning and expression of the regulatory (R_{GI}) subunit of the glycogen-associated protein phosphatase, *J. Biol. Chem.* **266:**15782–15789.
63. Wollny, E., Watkins, K., Kramer, G., and Hardesty, B., 1984, Purification to homogeneity and partial characterization of a 56,000-dalton protein phosphatase from rabbit reticulocytes, *J. Biol. Chem.* **259:**2484–2492.
64. DePaoli-Roach, A. A., and Lee, F. T., 1985, Phosphoprotein phosphatase inhibitor 2 is phosphorylated at both serine and threonine residues in mouse diaphragm, *FEBS Lett.* **183:**423–429.
65. Zardeneta, G., 1988, Purification and characterization of regulin, Ph.D. Dissertation, University of Texas at Austin, Austin, Texas.

Chapter 19

Initiation Mechanisms Used in the Translation of Bicistronic mRNAs

William C. Merrick and Donald D. Anthony

1. INTRODUCTION

The mechanism and control of eukaryotic protein synthesis have been the topic of many review articles in the past few years.[1–15] While some have dealt in fine detail with parts of the process, others have focused more on control than mechanism. In large part, all of these reviews have dealt with what is considered "normal" eukaryotic protein synthesis. This begins with a messenger RNA (mRNA) with a 5′ terminal m^7G cap structure, usually a 5′ untranslated region of 30–150 bases followed by a coding region which begins with the Met-transfer RNA: ($tRNA_i$) initiating AUG codon in the following favored context: A/GXXAUGG. The mRNA usually contains several hundred bases 3′ to the coding region and is followed by a poly(A) tail, where the poly(A) length is 100–150 bases. A generalized pathway describing the sequential utilization of translation factors has been proposed to account for the binding of eukaryotic mRNAs to ribosomes.[1] In this pathway, eIF-2 directs the binding of Met-$tRNA_i$ to the 40S subunit prior to the binding of mRNA. Binding of the mRNA to the 40S subunit is achieved by

---

William C. Merrick and Donald D. Anthony • Department of Biochemistry, School of Medicine, Case Western Reserve University, Cleveland, Ohio 44106-4935.

Translational Regulation of Gene Expression 2, edited by Joseph Ilan. Plenum Press, New York, 1993.

interaction with the three-subunit factor eIF-4F, which specifically recognizes the m^7G cap structure. Subsequently, eIF-4A and eIF-4B, in the presence of ATP, serve to unwind the mRNA and allow for its attachment to the 40S subunit. A correct match of the initiator tRNA with the correct AUG start code word appears to be accomplished by the ATP-dependent process termed scanning,[7,15] whereby the 40S subunit moves in a 5′ to 3′ direction in search of the initiating AUG. Current reports[16–18] support the hypothesis that recognition of the initiating AUG codon is effected via the anticodon of the initiator tRNA with the AUG codon.

The above statements appear to apply to almost all normal eukaryotic mRNAs. There are a number of viral or synthetic mRNAs, however, which appear to utilize two alternate forms of initiation, reinitiation and internal initiation, which are represented schematically in Fig. 1. For the process of reinitiation, a normal initiation event takes place, but at the termination step, the 40S subunit is not released from the mRNA. Rather, the 40S subunit appears to be able to

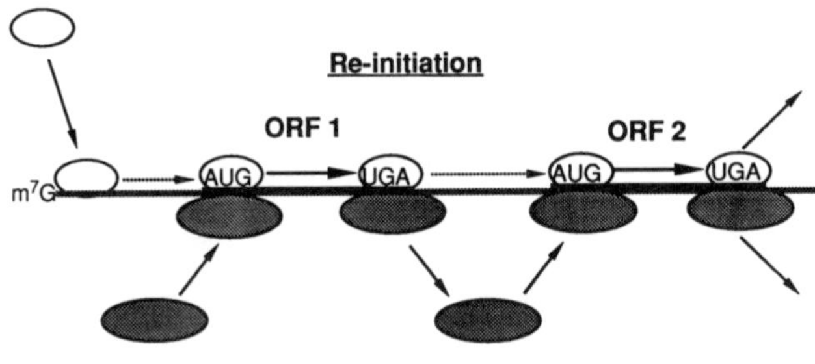

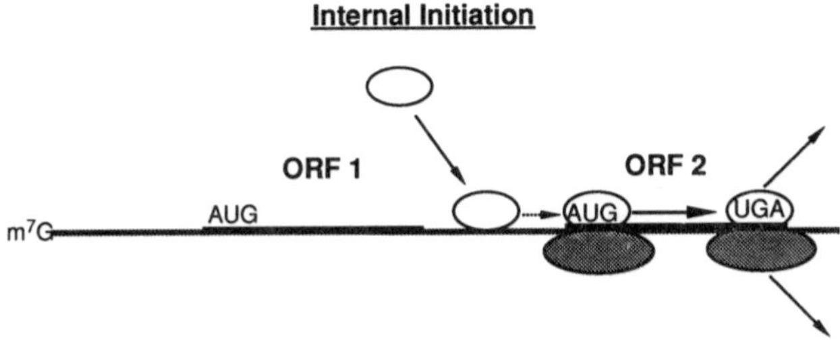

Figure 1. Alternate initiation of translation mechanisms. A schematic of the two rare initiation events in eukaryotic systems, reinitiation (top) and internal initiation (bottom). The coding region begins with the AUG initiation codon and ends with the UGA stop codon. The arrows with dashed lines indicate those portions of the mRNA where "scanning" is thought to occur. The open ellipse represents the 40S subunit, while the shaded ellipse represents the 60S subunit (see text for details).

scan the mRNA and acquire a fresh initiator tRNA as the ternary complex (eIF-2·GTP·Met-tRNA$_i$). The best studies on this process have been from work done on GCN4 expression in yeast.[19–24]

The second rare initiation event is termed internal initiation. In this process, a 40S subunit with accompanying translation factors and Met-tRNA$_i$ binds to an internal portion of the mRNA, not at the m^{7}G cap structure (if it is present). This binding may be directly to the AUG codon or may be to the 5′ side of the AUG start codon followed by scanning to locate the AUG codon. The best-studied examples of internal initiation have come from work with the piconaviruses, especially encephalomyocarditis (EMC) virus and poliovirus.[25–29]

In an attempt to understand these alternate processes, we have constructed several bicistronic mRNAs.[30] By the use of these mRNAs, it has been possible to examine, in part, the translation factor requirements for the translation of these mRNAs. As will be seen shortly, we were not able to find the simple answers we sought, but we are now able to better design the controls for such studies by ourselves and others. Second, it has become apparent to us that trying to design useful mRNAs is not as straightforward as anticipated. The reason for this, besides our limited mental capacities, is not clear.

2. RATIONALE: WHY BICISTRONIC mRNAs?

The necessity for a bicistronic mRNA to study reinitiation is obvious. For the studies on internal initiation, the use of bicistronic mRNAs is less obvious. If an internal initiation event occurs, how does one distinguish between binding which occurs at the 5′ end of the mRNA that is not dependent on an m^{7}G cap structure from a binding where the 40S subunit binds well removed from the 5′ end? The idea is the introduction of a "reporter" to indicate the passage of the ribosome. In this instance, the reporter is the protein which is encoded by the first or upstream open reading frame so that it is the second or downstream open reading frame which is then the experimental test system.

The system chosen for the translational studies of bicistronic mRNAs was the reticulocyte lysate, as this system translates mRNAs at the *in vivo* rate. The mRNAs constructed were chosen to be portions of the α-globin mRNA coding region so that we would avoid complications due to codon usage. As a control bicistronic mRNA, the TK/CAT and TK/P2CAT constructs of Pelletier and Sonenberg were also used, as these constructs had been tested both *in vitro* and *in vivo* for internal initiation.[25]

Having selected appropriate reading frames, the next question was, what type of intercistronic spacer should be chosen? Given that eIF-4F is inactivated during poliovirus infection and that poliovirus mRNA is initiated internally,[1,25] it was anticipated that internal initiation might predominantly use the other two mRNA-specific initiation factors, eIF-4A and eIF-4B. Model studies indicated that these proteins interacted best with single-stranded mRNAs.[31,32] Therefore an alternating sequence of U and C for 30 bases was used, as most studies have suggested that a ribosome covers approximately 20–30 bases of an mRNA. The selection of an intercistronic spacer for reinitiation was complicated by a lack of knowledge of the process, although an optimal distance of about 75 nucleotides has been suggested.[33] A diagrammatic representation of the different mRNAs tested is presented in Fig. 2.[30]

3. THE TEST SYSTEM

Having generated these different mRNAs by *in vitro* transcription using SP6 RNA polymerase, how would one determine whether a given coding region was translated by the normal, m^7G cap-dependent process or by the rare alternate schemes, reinitiation or internal initiation? The analysis is represented in Table I and depends on the observation that m^7GTP, an analog of the cap structure of eukaryotic mRNAs, effectively inhibits normal translation. Thus, if translation of the second open reading frame (ORF2) is not inhibited by m^7GTP, the inference is that this reading frame is using internal initiation. On the other hand, if the translation of ORF2 is sensitive to the addition of m^7GTP, then it is inferred that initiation of the second reading frame is achieved by reinitiation. It should be noted, however, that as a formal possibility, this same result might be observed if it was necessary, for structural reasons, to translate ORF1 in order to make ORF2 available for internal initiation. This complication referred to as "coupled internal initiation," derives its name in part from the many prokaryotic examples of coupled translation.

4. REALITY: THE TRANSLATION RESULTS

The initial analysis of the synthetic constructs in the presence or absence of m^7GTP is presented in Table II. The necessity for the numerous constructs used becomes apparent, as the "expected" results were not always obtained. The results with the capped TK/CAT and TK/P2CAT transcripts were as expected, with enhanced translation of the CAT coding region when the poliovirus leader P2 segment was present in the intercistronic region, consistent with the suggestion that this segment facilitates internal initiation.[25] The results with the bicistronic globin constructs were less predictable. The 103F transcript displayed reasonable translation from ORF2. Addition of the $(UC)_{15}$ sequence to the intercistronic region (103G) led to a decrease in the translation of ORF2 rather than the expected stimulation. The two transcripts with the shortest intercistronic region (103H and 103I) also yielded relatively high levels of ORF2 translation; the inhibition of ORF2 translation in these latter mRNAs by m^7GTP is consistent with the

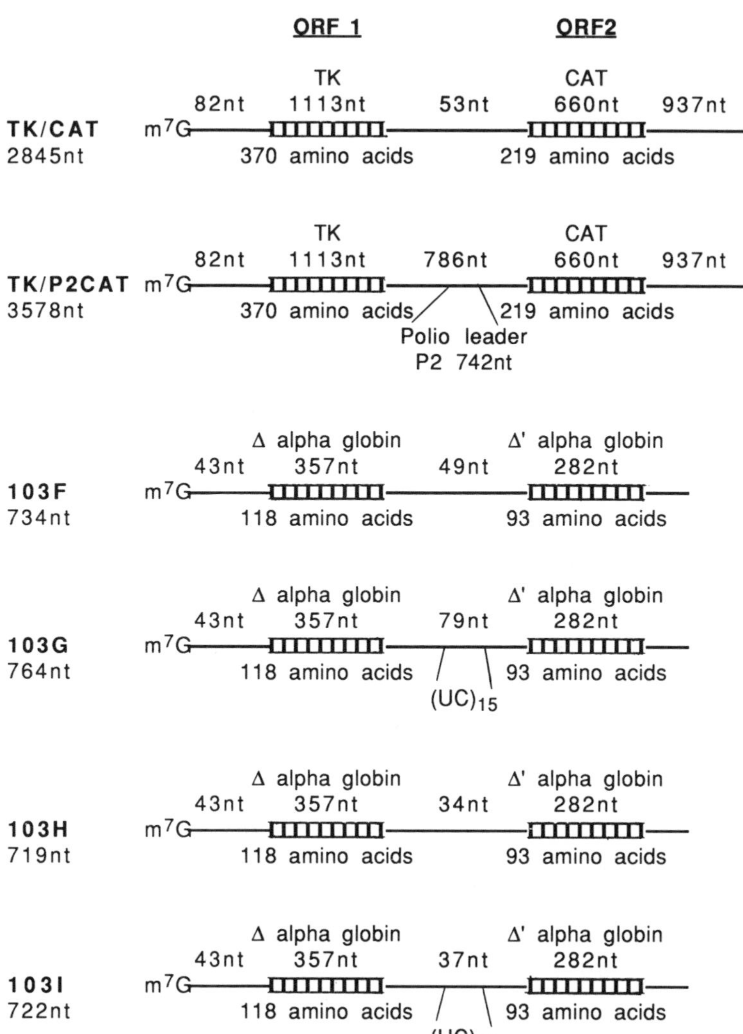

Figure 2. Bicistronic mRNA constructs. Schematic representation of the six bicistronic mRNAs tested in this report. The exact sequence and details of construction have been reported.[30] The 43 nucleotides to the 5′ side of Δ α-globin make up the authentic 5′ untranslated region of the α-globin mRNA. Each of the α-globin coding regions begins at the usual AUG start codon, but then is truncated to produce a smaller polypeptide which can be readily resolved by sodium dodecyl sulfate gel electrophoresis. TK indicates the coding region for the enzyme thymidine kinase and CAT indicates the coding region for chloramphenicol acetyltransferase.

Table I. Expected Translation Results[a]

m^{7}G — 5′-UTR — Initiation mechanism	ORF1 Peptide 1	ORF2 Peptide 2	3′-UTR — (A)$_n$
Normal initiation	Yes	No	
+ m^{7}GTP	No	No	
Internal initiation	Yes	Yes	
+ m^{7}GTP	No	Yes	
Reinitiation	Yes	Yes	
+ m^{7}GTP	No	No	
Coupled internal initiation	Yes	Yes	
+ m^{7}GTP	No	No	

[a]The expected ideal translation results with synthetic bicistronic mRNAs that contain nonoverlapping ORFs are represented here (yes: translation occurs; no: translation does not occur) for each particular initiation mechanism of translation. The expected result is given for the case of translation with and without m^{7}GTP addition.

Table II. The m^{7}GTP Sensitivity of Capped and Uncapped mRNAs[a]

	Capped mRNA		Uncapped mRNA		
mRNA	ORF1	ORF2	ORF1		ORF2
TK/CAT	100	5	100	(30)	55
+ m^{7}GTP	40	10	110		55
TK/P2CAT	100	10	100	(30)	180
+ m^{7}GTP	30	25	110		370
103F	100	35	100	(10)	130
+ m^{7}GTP	40	50	220		220
103G	100	10	100	(10)	740
+ m^{7}GTP	20	15	150		670
103H	100	30	100	(10)	60
+ m^{7}GTP	30	15	280		130
103I	100	40	100	(10)	170
+ m^{7}GTP	30	20	330		330

[a]Various mRNA transcripts were obtained by use of SP6 polymerase and plasmids as detailed in ref. 30. These transcripts were translated in nuclease-treated reticulocyte lysate. Quantitation of translation products was achieved by SDS gel electrophoresis followed by autoradiography and finally laser densitometry of the appropriate bands on X-ray film (see ref. 30 for full details). The translation efficiency for each mRNA is standardized to the ORF1 translation efficiency when the mRNA alone was added to the reaction (underlined values). Values in parentheses represent the translation efficiency of ORF1 with the uncapped mRNA relative to the capped mRNA. Each value represented is the average of three or more experiments. Variation from experiment to experiment was less than 10% of the value represented.

mechanism of reinitiation or coupled internal initiation (see Table I). An observation not expected was that for several of the transcripts, the level of ORF2 translation was stimulated by the addition of m^7GTP, while the translation of ORF1 in all mRNAs was inhibited by m^7GTP addition as expected.

As an independent assessment of the cap dependence of translation, the same transcripts were prepared without the m^7G cap structure. As expected, the translation efficiency of ORF1 was approximately the same as that for its capped equivalent in the presence of added m^7GTP. For the two transcripts which appeared capable of reinitiation or coupled-internal initiation (103H and 103I), the translation efficiency of ORF2 was similar to that of the capped transcript in the presence of m^7GTP ($60\% \times 0.1 = 6\%$ versus 15%; $170\% \times 0.1 = 17\%$ versus 20%). An unexpected observation of these studies with uncapped mRNAs was that the addition of m^7GTP often stimulated the translation of both ORF1 and ORF2. There are two simple interpretations of this observation, both of which require that eIF-4F stimulate internal initiation. The first is that capped mRNA fragments resulting from the micrococcal nuclease treatment of the lysate are able to serve as competitive inhibitors of the cap structure for eIF-4F more efficiently than m^7GTP. In the presence of excess m^7GTP, the fragments become displaced and the eIF-4F–m^7GTP complex is capable of stimulating internal initiation. The second interpretation would be that the eIF-4F–m^7GTP complex is more active than free eIF-4F in stimulating internal initiation, although RNA-dependent ATPase assays have not shown such a stimulation.

When all the bicistronic constructs are compared, ORF2 translation is m^7GTP-sensitive (translated via a cap-dependent mechanism consistent with reinitiation or coupled internal initiation) in 103H and 103I, while ORF2 translation in constructs 103F, 103G, TK/CAT, and TK/P2CAT is not m^7GTP-sensitive (translated in a cap-independent fashion consistent with internal initiation: Table II). A potassium acetate concentration curve (70–200 mM) was performed to assess any variation in the m^7GTP sensitivity of ORF2 translation. At all potassium acetate concentrations tested ORF2 translation was via a cap-dependent mechanism for 103H and 103I (data not shown). In contrast, TK/CAT and TK/P2CAT displayed cap-independent translation of ORF2 under all salt conditions. Somewhat different, however, was the finding that 103F and 103G displayed a cap-dependent ORF2 translation at a salt concentration of 120 mM or higher (110 mM potassium acetate is the standard assay salt in this study), while ORF2 was translated via a cap-independent mechanism at 110 mM potassium acetate and lower salt concentrations. A simple interpretation of these results would be that at increased ionic strength, the mRNA is capable of undergoing more extensive folding such that the intercistronic region is no longer available/suitable for internal initiation. Given the observation, however, that our ability to predict the outcome of the translation of the different mRNAs has proven suspect, there may easily be a more complicated reason for the observed salt influence on initiation mechanism selection.

The effect of intercistronic length or sequence was not as expected. The presence of the poly(UC) region (103G and 103I) did not enhance internal initiation in the capped transcripts, although it had an enormous effect in the uncapped 103G transcript, yielding four to seven times more protein from ORF2 than ORF1, and about fivefold more protein than from ORF2 in the capped 103G transcript when translated in the presence of m^7GTP. Comparison of any of the four globin transcripts for the effect of intercistronic length, especially as related to reinitiation, is not possible from these experiments due to the contribution of internal initiation, which is dominant in the 103F and 103G mRNAs. Second, the use of the uncapped mRNAs would appear to indicate that, for the 103H and 103I constructs, appreciable internal initiation (as well as reinitiation) is occurring. This can be seen as the ratio of ORF1 to ORF2 (3:1) for the capped transcripts and for the uncapped transcripts (1–2:1). Although conclusions about requirements in an mRNA for reinitiation are not possible from these experiments, the suggestion is made that studies which attempt to evaluate the mRNA structural requirements for reinitiation should include some measure of internal initiation (for *in vivo* studies, the influence of poliovirus infection; for *in vitro* studies the effect of m^7GTP addition and the translation of the equivalent uncapped transcript).

5. INFLUENCE OF INITIATION FACTORS ON BICISTRONIC mRNA TRANSLATION

Initial attempts at determining initiation factor involvement in the atypical mechanisms of internal initiation, reinitiation, and/or coupled internal initiation have been initiation factor addition assays with the same rabbit reticulocyte lysate translation system. Because each synthetic construct contained two individual ORFs, the effect that the addition of an initiation factor (eIF-4A, eIF-4B, and/or eIF-4F) had on each ORF could easily be determined. If our initial predictions were valid, based upon the poliovirus system and model assays, then it is expected that the addition of eIF-4A and/or eIF-4B might favor internal initiation, while the addition of eIF-4F should favor normal, cap-dependent initiation. The results in Tables III and IV prove once again that our simple predictions were at least partially incorrect. In part this may reflect the fact that the amount of mRNA added (0.1 μg) was in the linear range (as judged by hot trichloroacetic acid precipitation of radioactivity) and thus sufficient levels of translation factors might be present in the lysate. Second, it is possible that one or more factors may be in sufficent excess that no effect would be seen. Of the factors added, eIF-4A would be a candidate for factor excess, as it has been reported to be present at tenfold higher concentrations than eIF-4B or eIF-4F.[34,35]

Of all the constructs tested, only ORF2 of 103G showed a significant stimulation with added eIF-4A, one of the two mRNAs with the $(UC)_{15}$ sequence in the intercistronic region. In contrast, eIF-4B addition stimulated translation

Table III. Factor Additions with Capped mRNAs[a]

Condition	TK/CAT	TK/P2CAT	103F	103G	103H	103I
mRNA	*100*	*100*	*100*	*100*	*100*	*100*
	5	**10**	**35**	**10**	**30**	**40**
+ m⁷GTP	40	30	40	20	30	30
	5	**25**	**50**	**15**	**15**	**20**
+ eIF-4A	85	120	80	110	120	110
	15	**15**	**45**	**40**	**40**	**30**
+ eIF-4B	70	140	85	150	130	100
	5	**25**	**70**	**100**	**85**	**35**
+ eIF-4F	65	150	120	120	180	100
	5	**45**	**100**	**140**	**60**	**65**
+ eIF-4A/4B	—	120	150	—	—	—
	—	**25**	**95**	—	—	—
+ eIF-4B/4F	—	120	—	—	—	—
	—	**50**	—	—	—	—
+ eIF-4A/4B/4F	—	190	120	—	180	120
	—	**95**	**350**	—	**170**	**90**

[a]Translation reactions were performed and the protein products were analyzed as described in Table II and in ref. 30. Where indicated, 1 μg of each protein factor was included. Translation efficiency under varying conditions for each mRNA was standardized to the translation efficiency of ORF1 when the mRNA alone was added to the reaction (italized values of 100). The upper value in each set represents the ORF1 translation efficiency and the lower (bold) value represents the ORF2 translation efficiency. Each value represented is the average of three or more experiments. Variation from one experiment to the next was less than 10% of the value represented.

of several mRNAs. In this instance, modest stimulation was seen for ORF1 in three constructs, while ORF2 showed significant stimulation in four constructs. The stimulation of ORF1 cannot be simply explained, as the identical 5′ untranslated region (UTR) of a similar construct did not show stimulation (i.e., compare TK/CAT to TK/P2CAT or the four α-globin mRNA bicistronic mRNAs). Those mRNAs which did show stimulation of ORF2 translation generally also showed enhanced translation with m⁷GTP addition and thus would seem to be initiated by an internal initiation mechanism; the exception is 103I.

The addition of eIF-4F generally stimulated translation of ORF1 as expected. The unexpected finding was that eIF-4F stimulated the translation of all the ORF2s except in the TK/CAT mRNA (note: this mRNA uniquely showed inhibition of translation of ORF1 with all factor additions; the significance of this is not clear). For several of the mRNAs, the translation of ORF2 was about the same as ORF1. The same trend was noted when all three mRNA-specific initiation factors were added (note that in the translation of 103F, ORF2 translation is now about threefold greater than ORF1 translation).

While simple mechanistic conclusions could not be reached from the factor addition experiments, two general conclusions were evident. The first is that eIF-4F facilitates internal initiation. This, in spite of the long history of eIF-4F

Table IV. Factor Additions with Uncapped mRNAs[a]

Condition	TK/CAT	TK/P2CAT	103F	103G	103H	103I
mRNA	*100*	*100*	*100*	*100*	*100*	*100*
	55	**180**	**130**	**740**	**60**	**170**
+ m^{7}GTP	110	110	220	150	280	330
	55	**370**	**220**	**670**	**130**	**330**
+ eIF-4A	—	130	100	85	—	—
	—	**140**	**180**	**670**	—	—
+ eIF-4B	—	210	250	500	—	—
	—	**210**	**270**	**1900**	—	—
+ eIF-4F	—	190	520	470	—	—
	—	**250**	**280**	**2300**	—	—
+ eIF-4A/4B	180	190	—	—	—	—
	90	**300**	—	—	—	—
+ eIF-4A/4B/4F	—	—	1200	1300	—	—
	—	—	**970**	**4300**	—	—

[a]Values represented in this table were derived as was done for Table III. Translation efficiency under varying conditions for each mRNA was standardized to the translation efficiency of ORF1 when the mRNA alone was added to the reaction (italized values of 100). The upper value in each set represents the ORF1 translation efficiency and the lower value (bold) represents the ORF2 translation efficiency. See Table II for comparison of capped versus uncapped mRNA translation efficiency.

functioning as the cap-binding protein and facilitating normal, cap-dependent translation. The second conclusion or inference was that there might be a competition between the two "initiation sites" for the translation factors, the m^{7}G cap and some sequence or structure that allows for internal initiation. To better examine this second conclusion, the same mRNA transcripts were made, but without an m^{7}G cap (Table IV). While the 100% value for ORF1 is less in absolute terms relative to the equivalent capped transcript (see Table II), it serves as a yardstick to measure the relative translation of ORF2. As is immediately obvious, the level of ORF2 translation is generally similar (50–200% relative to ORF1). The unique exception is the 103G mRNA, where there is considerably more translation of ORF2 (approximately four- to sixfold greater). The second observation is that the factor additions appear to show the same trend in the stimulation of translation for both ORF1 and ORF2. Thus it would seem that the translation of an uncapped mRNA is likely to be mechanistically similar to the process for internal initiation and that the existence of a free 5′ end is not influential in the efficiency or mechanism of translation.

6. DISCUSSION

The conclusions reached from the above studies have given some insight into the function/role of translation initiation factors in the three different modes of translation: normal initiation (m^7G cap-dependent); internal initiation (cap-independent); and reinitiation. While most of the data relate to internal initiation, one general principle has emerged which is that monocistronic m^7G capped mRNAs have potential for both normal and internal initiation and that polycistronic mRNAs have the potential for all three types of initiation. In this regard, the effect of ionic strength on whether 103F or 103G transcripts were translated by internal initiation (low ionic strength) or reinitiation (high ionic strength) most likely reflects differences in the mRNA structure (or proteins perhaps associated with the mRNA), as the linear sequence is the same.

A possible disappointing feature to these studies was our inability to predict accurately what intercistronic sequence might best suit internal initiation or reinitiation. In part this may easily reflect that not all of the factors necessary for initiation have been identified. For the process or reinitiation and normal initiation, a major element has not been defined biochemically and that is the process of mRNA scanning. It is not clear whether one or more translation factors is required for this process or whether this is an inherent property of the ribosome. If a translation factor is involved, it is not clear if there is a competition between complexes for this factor or what might be the best "substrate" or target to bind this protein. For internal initiation, several additional cellular proteins have been identified which seem to recognize specific sequences or structures within an mRNA (see below). This is in contrast, however, to what appears to be an inherent ability of all mRNAs (especially as uncapped mRNAs) to undergo a cap-independent or internal initiation event, although with reduced efficiency.

In this regard, our ability to predict that an unstructured RNA sequence would facilitate internal initiation was not very good. The insertion of the $(UC)_{15}$ linker in the intercistronic region of construct 103G slightly favored internal initiation, but less so than the parent lacking this structure (103F). Yet when tested as an uncapped mRNA, this sequence was quite effective and the translation of ORF2 was 74% of ORF1 in the capped mRNA. The parent construct, 103F, did not do as well in the translation of ORF2 in the uncapped mRNA, which was only 13% of ORF1 translation in the capped mRNA. As noted earlier, this apparent inconsistency appears to be the result of competition between the different initiation sites for the translation factors heavily favoring the m^7G cap. Thus, if only the uncapped mRNAs are examined, it is evident that the poliovirus P2 sequence element and the $(UC)_{15}$ linker both enhanced internal initiation approximately three- to fivefold. Clearly, to be able to make more accurate guesses as to what might be an optimal sequence for internal initiation, we need to be more knowledgeable about the possible factors involved and the effects of competition for factors on our "expected" results.

The key element in mRNA translation is the m^7G cap. Based upon these and numerous other studies, this element provides for the most efficient attraction of the translation factors necessary for the attachment of an mRNA to the 40S subunit. This is perhaps most simply visualized in an mRNA competition assay, where the ability to translate via an internal mechanism is lost as the concentration of the capped mRNA increases.[30] The general inference is that the attachment of eIF-4F to an RNA enhances the ability of eIF-4A and eIF-4B to bind to and unwind that mRNA to allow its subsequent binding to the 40S subunit.

By this token, all internal initiation events would be anticipated to be weak and normally noncompetitive with respect to most cellular mRNAs. So how might such mRNAs escape this fate? In one strategy, poliovirus infection causes the loss of eIF-4F activity necessary for cap-dependent translation and consequently this would be expected to enhance the possibility of internal initiation.[1] This effect would appear to have an equivalent cellular response in that phosphorylation of eIF-4F (especially the 24,000-dalton subunit) is necessary for biologic activity.[5,36–40] Thus, circumstances which yield eIF-4F in a reduced state of phosphorylation would tend to favor enhanced translation via internal initiation, not because this process has become more efficient, but rather because there has been a loss of the highly competitive sequestration of translation factors to the 5′ m^7G cap structure of mRNA.

At the time when these studies were in progress, several laboratories identified specific regions in picornavirus mRNAs (which naturally lack an m^7G cap) that were necessary to facilitate internal initiation.[25,29] In addition, specific cellular proteins were identified which bind to these regions and are therefore thought to enhance the translation of these mRNAs.[26–29,41] While these studies are likely to provide more definitive insight into the mechanistic details of the internal initiation of these viral mRNAs, there has been no proven role for such proteins in the translation of cellular mRNAs. There has been only a single report of a cellular mRNA being translated as a result of internal initiation, the immunoglobulin heavy-chain binding protein.[42,43] It is anticipated that additional cellular mRNAs will be identified which also use an internal initiation mechanism.

ACKNOWLEDGMENTS. The authors thank Dr. Lee Gehrke (MIT) for supplying the α-globin construct and many helpful suggestion, Dr. Nahum Sonenberg (McGill University) for supplying the TK/CAT and TK/P2CAT constructs, and Dr. Thomas Dever for many helpful discussions. We also thank Toni L. Bodnar for her excellent editorial assistance in the preparation of this manuscript. This work was supported in part by National Institutes of Health grant GM-26796 and training grant T32-GM07520.

REFERENCES

1. Merrick, W. C., 1992, *Microbiol. Rev.* **5**:291–315.
2. Bielka, H., 1985, *Prog. Nucleic Acid. Res. Mol. Biol.* **32**:267–289.
3. Duncan, R. F., 1993, *Prog. Nucleic Acid Res. Mol. Biol.*, in press.
4. Edery, I., Pelletier, J., and Sonenberg, N., 1987, In: *Translational Regulation of Gene Expression* (J. Ilan, ed.), pp. 335–366, Plenum Press, New York.
5. Hershey, J. W. B., 1991, *Annu. Rev. Biochem.* **60**:717–755.

6. Kaufman, R. J., 1990, *Genet. Eng.* **12:**243–273.
7. Kozak, M., 1989, *J. Cell Biol.* **108:**229–241.
8. Müeller, P. P., and Trachsel, H., 1990, *Eur. J. Biochem.* **191:**257–261.
9. Nygård, O., and Nilsson, L., 1990, *Eur. J. Biochem.* **191:**1–17.
10. Pain, V. M., 1986, *Biochem. J.* **235:**625–637.
11. Rhoads, R. E., Hiremath, L. S., Rychlik, W., Gardner, P. R., and Morgan, J. L., 1985, In: *Nuclear Envelope Structure and RNA Maturation*, pp. 427–464, Alan R. Liss, New York.
12. Ryanzanov, A. G., Rudkin, B. B., and Spirin, A. S., 1991, *FEBS Lett.* **285:**170–175.
13. Safer, B., 1989, *Eur. J. Biochem.* **186:**1–3.
14. Sonenberg, N., 1988, *Prog. Nucleic Acid Res. Mol. Biol.* **35:**173–207.
15. Kozak, M., 1980, *Cell* **22:**459–467.
16. Cigan, A. M., Feng, L., and Donahue, T. F., 1988, *Science* **242:**93–97.
17. Cigan, A. M., Pabich, E. K., Feng, L., and Donahue, T. F., 1989, *Proc. Natl. Acad. Sci. USA* **86:** 2784–2788.
18. Donahue, T. F., Cigan, A. M., Pabich, E. K., and Valavicius, B. C., 1988, *Cell* **54:**621–632.
19. Abastado, J. P., Miller, P. F., Jackson, B. M., and Hinnebusch, A. G., 1991, *Mol. Cell. Biol.* **11:** 486–496.
20. Dever, T. E., Feng, L., Wek, R. C., Cigan, A. M., Donahue, T. F., and Hinnebusch, A. G., 1992, *Cell* **68:**585–595.
21. Müeller, P. P., and Hinnebusch, A. G., 1986, *Cell* **45:**201–207.
22. Krupitza, G., and Thireos, G., 1990, *Mol. Cell. Biol.* **10:**4375–4378.
23. Roussou, I., Thireos, G., and Hauge, B. M., 1988, *Mol. Cell. Biol.* **8:**2132–2139.
24. Tzamarias, D., Alexandraki, D., and Thireos, G., 1986, *Proc. Natl. Acad. Sci. USA* **83:**4849–4853.
25. Pelletier, J., and Sonenberg, N., 1988, *Nature* **334:**320–325.
26. Borovjagin, A. V., Evstafieva, A. G., Ugarova, T. Y., and Shatsky, I. N., 1990, *FEBS Lett.* **261:** 237–240.
27. Del Angle, R. M., Papvassiliou, A. G., Fernández-Tomás, C., Silverstein, S. J., and Racaniello, V. R., 1989, *Proc. Natl. Acad. Sci. USA* **86:**8299–8303.
28. Meerovitch, K., Pelletier, J., and Sonenberg, N., 1989, *Genes Dev.* **3:**1026–1034.
29. Jang, S. K., and Wimmer, E., 1990, *Genes Dev.* **4:**1560–1572.
30. Anthony, Jr., D. D., and Merrick, W. C., 1991, *J. Biol. Chem.* **266:**10218–10226.
31. Grifo, J. A., Abramson, R. D., Satler, C. A., and Merrick, W. C., 1984, *J. Biol. Chem.* **259:**8648–8654.
32. Abramson, R. D., Dever, T. E., and Merrick, W. C., 1988, *J. Biol. Chem.* **263:**6016–6019.
33. Kozak, M., 1987, *Mol. Cell. Biol.* **7:**3438–3445.
34. Duncan, R., Milburn, S. C., and Hershey, J. W. B., 1987, *J. Biol. Chem.* **262:**380–388.
35. Browning, K. S., Humphreys, J., Hobbs, W., Smith, G. B., and Ravel, J. M., 1990, *J. Biol. Chem.* **265:**17967–17973.
36. Morely, S. J., Dever, T. E., Etchison, D. E., and Traugh, J. A., 1991, *J. Biol. Chem.* **266:**4669–4672.
37. Duncan, R., and Hershey, J. W. B., 1985, *J. Biol. Chem.* **260:**5493–5497.
38. Hershey, J. W. B., 1989, *J. Biol. Chem.* **264:**20823–20826.
39. Morley, S. J., and Traugh, J. A., 1990, *J. Biol. Chem.* **265:**10611–10616.
40. Josh-Barve, S., Rychlik, W., and Rhoads, R. E., 1990, *J. Biol. Chem.* **265:**2979–2983.
41. Borovjagin, A. V., Ezrokhi, M. V., Rostapshov, V. M., Ugarova, T. Y., Bystrova, T. F., and Shatsky, I. N., 1991, *Nucleic Acids Res.* **19:**4999–5005.
42. Sarnow, P., 1989, *Proc. Natl. Acad. Sci. USA* **86:**5795–5799.
43. Macejak, D. G., and Sarnow, P., 1991, *Nature* **353:**90–94.

Chapter 20

Protein Synthesis Initiation in Animal Cells

Mechanism of Ternary and Met-tRNA$_f$·40S·mRNA Complex Formation and the Regulatory Role of an eIF-2-Associated 67-kDa Polypeptide

Naba K. Gupta, Bansidhar Datta, Manas K. Ray, and Ananda L. Roy

1. INTRODUCTION

In a chapter published in the predecessor of this volume we reviewed our work and also discussed several controversies in mammalian peptide chain initiation research.[1] During the past several years we have sought to resolve some of these controversies and reach agreements. A brief description of these controversies and their current status follows.

1. *Mechanism*. The first step in peptide chain initiation is the formation of a ternary complex between the eukaryotic peptide chain initiation factor 2 (eIF-2), Met-transfer RNA (tRNA)$_f$, and GTP (Met-tRNA$_f$·eIF-2·GTP). The next step is the transfer of Met-tRNA$_f$ from the ternary complex to 40S ribosomes and

Naba K. Gupta, Bansidhar Datta, Manas K. Ray, and Ananda L. Roy • Department of Chemistry, University of Nebraska, Lincoln, Nebraska 68588-0304.

Translational Regulation of Gene Expression 2, edited by Joseph Ilan. Plenum Press, New York, 1993.

formation of Met-tRNA$_f$·40S·mRNA complex. Two distinctly different views were reported regarding the characteristics and factor requirements for ternary and Met-tRNA$_f$·40S·mRNA complex formation. According to one view, described in the Staehelin–Anderson–Hershey model, eIF-2 alone forms a near-stoichiometric amount of ternary complex in the presence of Mg^{2+} and the complex thus formed binds almost quantitatively to 40S ribosomes in the absence of messenger RNA (mRNA). On the other hand, we have reported that eIF-2 requires multiple ancillary protein factors for ternary complex formation in the presence of Mg^{2+}, and Met-tRNA$_f$ binding to 40S ribosomes is almost totally dependent on added mRNA (for details, see refs. 1–4). Recently we collaborated with the groups of Hershey and Merricks to resolve these differences. All three laboratories independently analyzed the factor preparations from the other laboratories and agreed on the requirements of three different protein factors, eIF-2B, eIF-3, and Co-eIF-2A, for a near-stoichiometric amount of ternary complex formation by eIF-2. These three laboratories jointly proposed a mechanism for ternary complex formation.[5]

2. *Regulation*. An important regulatory mechanism involves one or more eIF-2 kinases which are believed to remain in latent forms in animal cells and are activated under certain physiological conditions to inhibit protein synthesis. In reticulocyte lysate one of these eIF-2 kinases (HRI) is activated during heme deficiency and another eIF-2 kinase (dsI) is activated in the presence of double-stranded RNA and ATP. Several laboratories previously reported that protein synthesis inhibition in heme-deficient reticulocyte lysate was reversed by addition of either eIF-2 or a reticulocyte cell supernatant factor, eIF-2B (for details, see ref. 1). We recently discovered that a 67-kDa polypeptide present in both partially purified eIF-2 and eIF-2B preparations was necessary for reversal of protein synthesis inhibition in heme-deficient reticulocyte lysate.[6–10] This 67-kDa polypeptide (p67) protects the eIF-2 α-subunit from eIF-2-kinase(s)-catalyzed phosphorylation and promotes protein synthesis in the presence of active eIF-2 kinases. We also provided evidence that all animal cells, including heme-supplemented reticulocyte lysate, contain active eIF-2 kinase(s).[10] The activity of the eIF-2 kinase(s), however, may or may not be evident, depending on the variable presence of p67 in the system. Under certain physiological conditions, such as during heme deficiency in reticulocyte lysate and also during nutritional deprivation, p67 is degraded, thus allowing eIF-2 kinase(s) to phosphorylate eIF-2 and inhibit protein synthesis. Our work thus suggests that protein synthesis in animal cells is not regulated by the reversible activation–inactivation of eIF-2 kinase(s) as was previously believed, but by the variable presence of p67.

In this chapter we will describe our recent work on the mechanism of the basic steps of peptide chain initiation leading to Met-tRNA$_f$·40S·mRNA complex formation and the roles of p67 in the regulation of protein synthesis in animal cells.

2. MECHANISMS OF TERNARY (Met-tRNA$_f$·eIF-2·GTP) AND Met-tRNA$_f$·40S·mRNA COMPLEX FORMATION

2.1. Background

As noted earlier, there were two distinctly different views regarding the characteristics and factor requirements for ternary and Met-tRNA$_f$·40S·mRNA complex formation.[1] According to the Staehelin–Anderson–Hershey model,[4,11–13] purified eIF-2 alone forms a near stoichiometric amount of ternary complex and the complex thus formed binds almost quantitatively to 40S ribosomes in the absence of mRNA. Several protein factors, called eIF-4 group proteins, bind to the cap site in mRNA and subsequently transfer the bound mRNA to Met-tRNA$_f$·40S ribosomes. On the other hand, our laboratory reported the requirement of multiple protein factors for ternary complex formation and requirement of mRNA for Met-tRNA$_f$ binding to 40S ribosomes.[1,3,14] According to our model, eIF-2 is isolated as a binary complex, eIF-2·GDP. In the presence of Mg^{2+}, GDP remains tightly bound to eIF-2 and prevents ternary complex formation. A high-molecular-weight protein complex Co-eIF-2C promotes GDP displacement from eIF-2·GDP (eIF-2B activity) and also stimulates ternary complex formation by eIF-2 (now identified as eIF-3 activity, see Section 2.2.4). The ternary complex formed in the presence of Co-eIF-2C, however, is unstable to natural mRNA and requires an additional protein factor, Co-eIF-2A, for stabilization. The ternary complex formed in the presence of Co-eIF-2A and Co-eIF-2C actively transfers Met-tRNA$_f$ to 40S ribosomes and such a transfer reaction is almost entirely dependent on the presence of a mRNA.

To resolve these differences, we recently collaborated with the Hershey and Merrick groups. All three laboratories independently analyzed the factor preparations from other laboratories and agreed on the requirements of three protein factors, eIF-2B, eIF-3, and Co-eIF-2A, for a near stoichiometric amount of ternary complex formation in the presence of physiological concentrations of Mg^{2+} and natural mRNA.[5] In this collaborative work our previously used Co-eIF-2C protein complex was replaced by two purified factor preparations containing Co-eIF-2C component activities, eIF-2B and eIF-3.[5,14] Based on the results of these studies these three groups jointly proposed a mechanism of ternary complex formation by eIF-2.[5]

In this section we will present the results of this collaborative work and also our related results on ternary and Met-tRNA$_f$·40S·mRNA complex formation.

2.2. Mechanism of Ternary Complex Formation by eIF-2

2.2.1. Purification of eIF-2 and eIF-2 Ancillary Protein Factors

Three ancillary protein factors, eIF-2B, eIF-3, and Co-eIF-2A, were required for ternary complex formation by eIF-2. In our collaborative work with the

Hershey and Merrick groups, all three laboratories purified eIF2 according to previously described procedures.[13–15] eIF-3 was purified in Merrick's laboratory[13] as well as in Hershey's laboratory.[15] Co-eIF-2A and eIF-2B were prepared in our laboratory following the previously published procedures.[14,16] All three laboratories used these factor preparations and the results obtained, using similar preparations from different laboratories, were the same.

Figure 1 shows the sodium dodecyl sulfate–polyacrylamide gel electrophoresis (SDS–PAGE) of the eIF-2, Co-eIF-2A, and eIF-3 preparations used in this work. Both eIF-2 preparations from Merrick's laboratory (lane 1) and our laboratory (lane 2) contained, in addition to the usual three subunits, the 67-kDa polypeptide (p67). Co-eIF-2A gave a single polypeptide band (94 kDa) and eIF-3 gave multiple polypeptide bands, including the 180-kDa polypeptide (p180).

2.2.2. Requirements of Ancillary Protein Factors for Ternary Complex Formation by eIF-2

We previously reported that eIF-2 alone formed very little ternary complex in the absence of Mg^{2+} and such complex formation was almost completely inhibited in the presence of Mg^{2+} and natural mRNA.[1,14] The results presented in Fig. 2 are in agreement with these reports.

At the lowest concentration (0.2 μg; approximately 1 pmole assuming a molecular weight of 200 kDa for eIF-2 containing p67) eIF-2 formed negligible ternary complex. In the absence of Mg^{2+}, ternary complex formation by eIF-2 followed a sigmoidal curve. At a higher concentration, 5 pmole input eIF-2 (1 μg) bound approximately 0.8 pmole Met-tRNA$_f$, i.e., 16% of the input eIF-2 molecules formed ternary complex. The ternary complex formation by eIF-2 alone, however, was almost completely inhibited by addition of globin mRNA and Mg^{2+}.

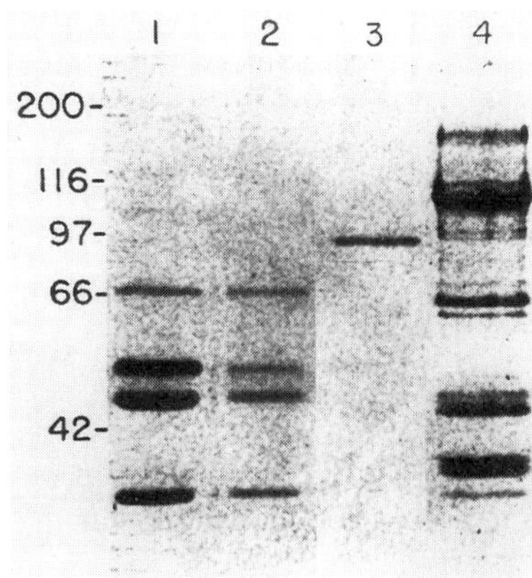

Figure 1. SDS–PAGE of different factor preparations. Lane 1, eIF-2 from Merrick;[13] lane 2, eIF-2 from Roy *et al.*[14]; lane 3, Co-eIF-2A from Roy *et al.*[14]; lane 4; eIF-3 from Merrick.[13] Data were obtained from Gupta *et al.*[5]

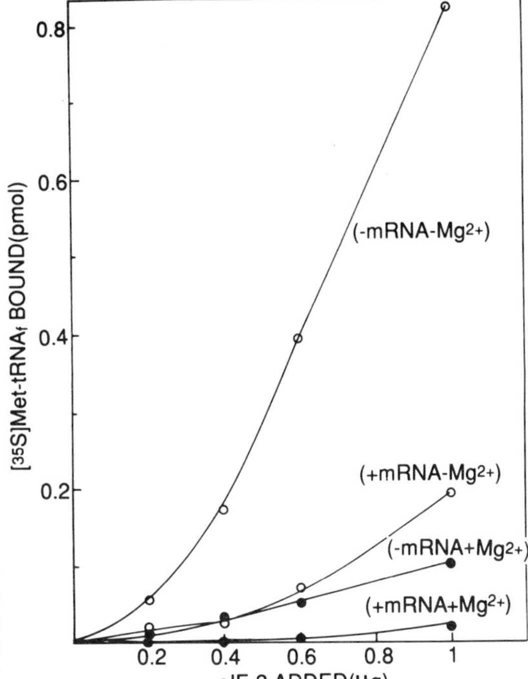

Figure 2. Characteristics of ternary complex formation by eIF-2. Where indicated, 1 mM Mg^{2+} and 0.5 μg globin mRNA were added. eIF-2 from Merrick[13] was used. Data were obtained from Gupta *et al.*[5]

The results described in Fig. 3 show that both Co-eIF-2A and eIF-3 strongly stimulated ternary complex formation by eIF-2. In the presence of either excess Co-eIF-2A or eIF-3, 0.2 μg (1 pmole) eIF-2 formed approximately 0.7 pmole ternary complex, i.e., 70% of the input eIF-2 molecules bound Met-tRNA$_f$. As reported previously,[1,14] however, the ternary complex formed in the presence of Co-eIF-2A was mostly stable in the presence of mRNA, whereas the ternary complex formed in the presence of eIF-3 was unstable to mRNA.

In the presence of Mg^{2+}, eIF-2 becomes inactive and dependent on eIF-2B for ternary complex formation. The results presented in Fig. 4 describe ternary complex formation by eIF-2 in the presence of 1 mM Mg^{2+} and additional protein factors. As shown in Fig. 4, eIF-2 alone, or eIF-2 in the presence of either Co-eIF-2A or eIF-3, formed very little ternary complex in the presence of Mg^{2+}. At the catalytic concentration used, eIF-2B alone did not stimulate ternary complex formation by eIF-2, but when added in the presence of either Co-eIF-2A or eIF-3, significantly stimulated ternary complex formation by eIF-2.

The results presented in Table I show the requirement of all three protein factors, eIF-2B, eIF-3, and Co-eIF-2A, for the near stoichiometric amount of ternary complex formation in the presence of Mg^{2+}. When all three protein factors

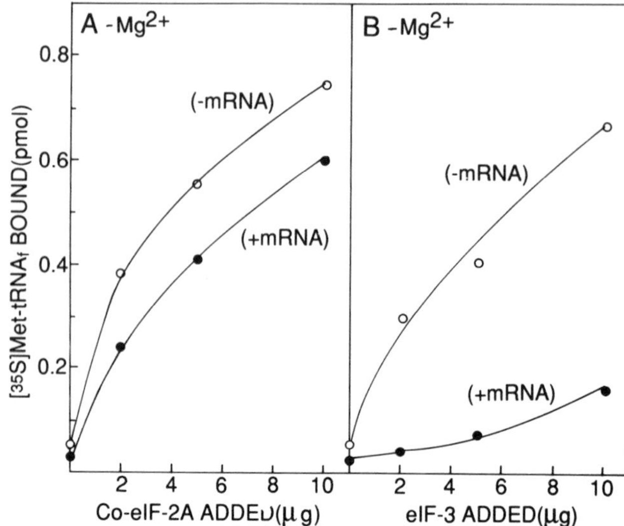

Figure 3. Ternary complex formation by eIF-2 in the presence of Co-eIF-2A and eIF-3. Ternary complex formation was analyzed in the absence of Mg^{2+}. Where indicated **(A)** Co-eIF-2A and **(B)** eIF-3 were added. eIF-2 from Merrick[13] was used. Data were obtained from Gupta *et al.*[5]

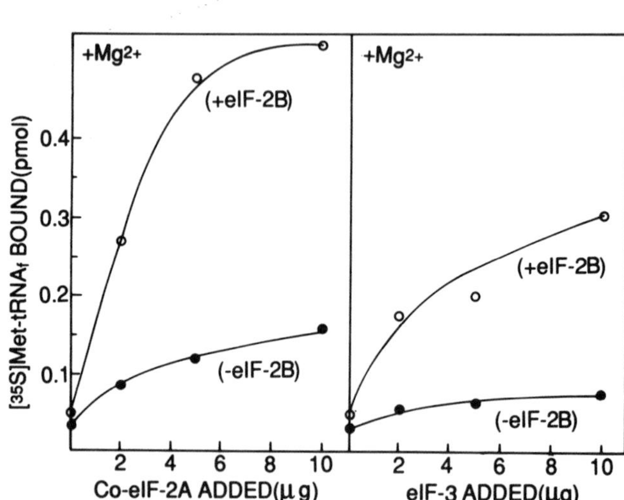

Figure 4. Ternary complex formation by eIF-2 in the presence of eIF-2B, Co-eIF-2A, and eIF-3. Ternary complex formation was analyzed in the presence of 1 mM Mg^{2+}. Where indicated 1 μg of eIF-2B, Co-eIF-2A, and eIF-3 were added. eIF-2 from Roy *et al.*[14] was used. Data were obtained from Gupta *et al.*[5]

Table I. Ternary Complex Formation by eIF-2 in the Presence of Different Factors, mRNA, and Mg^{2+}[a]

Factors added	Ternary complex formed (pmole)	
	−mRNA	+mRNA
None	0.07	0.04
eIF-2B	0.05	0.04
Co-eIF-2A	0.10	0.09
eIF-3	0.05	0.04
eIF-2B + eIF-3	0.30	0.05
eIF-2B + Co-eIF-2A	0.44	0.32
eIF-2B + Co-eIF-2A + eIF-3	0.72	0.58

[a]Standard millipore filtration conditions were used in the presence of 1 mM Mg^{2+}. Concentrations of protein factors used were: eIF-2, 0.15 μg; eIF-2B, 1 μg; Co-eIF-2A, 8 μg; and eIF-3, 10 μg. Where indicated, the reactions also contained 2 μg globin mRNA.

were present 0.15 μg or 0.75 pmole of input eIF-2 formed 0.72 pmole of ternary complex and the complex was significantly stable to mRNA.

These results show that three distinct protein factors, eIF-2B, eIF-3, and Co-eIF-2A, are required for ternary complex formation by eIF-2 in the presence of Mg^{2+} and natural mRNA. A brief description of these factor activities follows.

1. eIF-2B. This activity is presumably required for GDP displacement from eIF-2·GDP during ternary complex formation in the presence of Mg^{2+}. It has been previously reported that approximately 40–60% of the isolated eIF-2 molecules contain bound GDP.[17,18] As shown in Figs. 2 and 4, however, almost all the eIF-2 molecules (with or without bound GDP) remained in an inactive form in the presence of 1 mM Mg^{2+} and ternary complex formation was totally dependent on the presence of eIF-2B. These results are in agreement with our past postulation[1,18] that all of the eIF-2 molecules assume an inactive conformation in the presence of Mg^{2+} and eIF-2B restores the active conformation of the eIF-2 molecules. In this active conformation, GDP, from eIF-2·GDP, can be displaced and ternary complex can be formed in the presence of eIF-3 and/or Co-eIF-2A.

2. eIF-3. The precise mechanism of eIF-3 stimulation of eIF-2 activity is not clear. Previous work suggested that eIF-3 stabilizes the ternary complex bound on 40S ribosomes.[19] The current work demonstrates that eIF-3 is indeed an eIF-2 ancillary protein factor; eIF-3 stimulates ternary complex formation by eIF-2 and also possibly stabilizes the complex. The results of our preliminary experiments indicate that the 180-kDa polypeptide (p180) component in eIF-3 is responsible for eIF-3 stimulation of eIF-2 activity.[20]

3. Co-eIF-2A. Ternary complex formed with eIF-2B + eIF-3 was unstable in the presence of natural mRNA. Addition of Co-eIF-2A caused further stimulation

of ternary complex formation and the complex was stable to physiological concentrations of a natural mRNA. Co-eIF-2A is thus required for stabilization of the ternary complex in the presence of natural mRNAs.

2.2.3. A Proposed Mechanism for Ternary Complex Formation

Based on the results presented above, all three groups (Gupta, Hershey, and Merrick) proposed a tentative model for ternary complex formation by eIF-2 in the presence of eIF-2B, eIF-3, and Co-eIF-2A (Fig. 5).[5] In the presence of Mg^{2+}, eIF-2B at catalytic concentrations promotes GDP displacement from eIF-2·GDP and facilitates ternary complex formation. This ternary complex formation is significantly stimulated by eIF-3. The ternary complex formed in the presence of eIF-2B + eIF-3, however, is unstable to natural mRNA. Co-eIF-2A is required to further stimulate ternary complex formation by eIF-2 and subsequently to stabilize the complex in the presence of natural mRNA.

2.3. Mechanisms of Met-tRNA$_f$·40S·mRNA Complex Formation

2.3.1. Characteristics and Factor Requirement for Met-tRNA$_f$·40S·mRNA Complex Formation

As has been mentioned, in our earlier work we used eIF-2, Co-eIF-2A, and Co-eIF-2C protein complex for ternary and Met-tRNA$_f$·40S·mRNA complex formation. This Co-eIF-2C protein complex contains similar polypeptide components as eIF-3 and also contains eIF-2B activity. We reported[14] that the ternary complex formed in the presence of eIF-2 + Co-eIF-2C efficiently transferred Met-

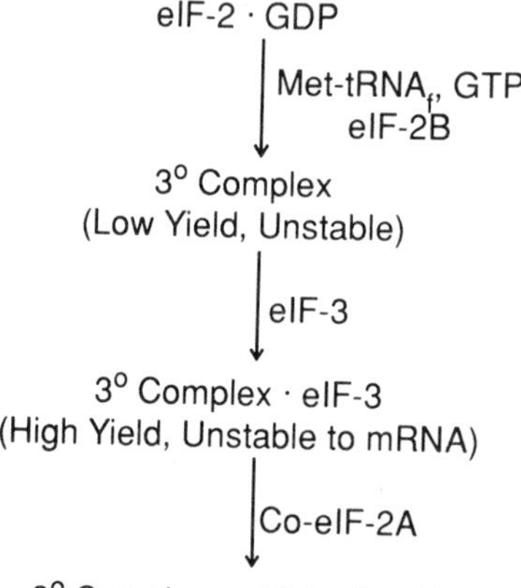

Figure 5. A tentative proposed model for ternary complex formation by eIF-2. From Gupta *et al.*[5]

tRNA$_f$ to 40S ribosomes in the presence of AUG codon, and Co-eIF-2A, besides eIF-2 + Co-eIF-2C, was necessary for a similar Met-tRNA$_f$ transfer to 40S ribosomes in the presence of a natural mRNA. The results of a typical experiment are shown in Table II. In these experiments Met-tRNA$_f$ binding to 40S ribosomes was assayed by density gradient centrifugation followed by Millipore filtration. Met-tRNA$_f$ binding to 40S ribosomes, under the assay condition, was completely dependent on the presence of either AUG codon or a natural mRNA. Control experiments without AUG codon or globin mRNA showed negligible binding in each case. Four different natural mRNAs, both capped (globin and brome mosaic virus (BMV) RNA) and uncapped (cowpea mosaic virus (CPMV) RNA and poliovirus RNA), were used in these experiments. As shown in Table II, only two factors, eIF-2 and Co-eIF-2C, were required for efficient Met-tRNA$_f$ binding to 40S ribosomes in the presence of AUG codon. The addition of Co-eIF-2A had no significant effect on such binding activity. The same factor combination, however, namely eIF-2 + Co-eIF-2C, was almost completely inactive in promoting a natural mRNA-dependent Met-tRNA$_f$·40S complex formation. Such complex formation required Co-eIF-2A, in addition to eIF-2 + Co-eIF-2C. All the natural mRNAs used in this experiment actively stimulated (three- to fourfold) Met-tRNA$_f$ binding to 40S ribosomes.

It should be noted, however, that for stable Met-tRNA$_f$·40S·mRNA complex formation it was necessary to use a nonhydrolyzable GTP analog, such as 5′ guanylyl imidodiphosphate (GMP)-PNP or GTP + nucleosiside 5′-diphosphate kinase (NDK) + ATP, to prevent GTP hydrolysis in the ternary complex and release of eIF-2 from 40S ribosomes after Met-tRNA$_f$·40S complex formation. We believe that Met-tRNA$_f$·40S complex formed with a natural mRNA under *in vitro* conditions is unstable and dissociates. In the presence of NDK + ATP or GMP-

Table II. Factor Requirements for Met-tRNA$_f$·40S Ribsosome Complex Formation with AUG Codon and Different Natural mRNAs[a]

	[^{35}S]Met-tRNA$_f$ bound to 40S ribosomes (pmole)	
mRNA added	−Co-eIF-2A, +Co-eIF-2C	+Co-eIF-2A, +Co-eIF-2C
None	0.27	0.27
AUG	1.44	1.44
Globin	0.29	1.16
CPMV RNA	0.25	0.84
BMV RNA	0.26	0.80
Poliovirus RNA	0.27	0.80

[a]Met-tRNA$_f$·40S complex formation was assayed by sucrose density gradient centrifugation followed by Millipore filtration of the gradient fractions. When indicated 4 μg of globin mRNA, 10 μg of CPMV RNA, 10 μg of BMV RNA, and 8 μg of polioviral RNA were added. Concentrations of protein factors used were as follows: eIF-2 0.4 μg; Co-eIF-2A, 8 μg; and Co-eIF-2C, 12 μg. Data obtained from Roy *et al.*[14]

PNP (in place of GTP), the intact ternary complex remains stably bound to 40S ribosomes and confers added stability to the Met-$tRNA_f$·40S initiation complex formed with a natural mRNA.

2.3.2. A Proposed Mechanism for Met-$tRNA_f$·40S·mRNA Complex Formation

In Section 2.2.3 we described a mechanism for ternary complex formation jointly proposed by three independent groups,[5] Gupta, Hershey, and Merrick. Three protein factors, eIF-2B, eIF-3, and Co-eIF-2A, were required for a near stoichiometric amount of ternary complex formation in the presence of Mg^{2+} and natural mRNA. In our previous work[14] we used Co-eIF-2C protein complex, which, as mentioned earlier, is a mixture of eIF-2B and eIF-3. We observed that the ternary complex formed using eIF-2 + Co-eIF-2C + Co-eIF-2A actively transferred Met-$tRNA_f$ to 40S ribosomes and such a transfer reaction was almost completely dependent on the presence of a natural mRNA. There are also indications that eIF-2 is released after the ternary complex is bound to 40S ribosomes. We propose that all the protein factors, eIF-2, eIF-2B, eIF-3, and Co-eIF-2A, are released and recycled after Met-$tRNA_f$ is bound to 40S ribosomes and Met-$tRNA_f$·40S·mRNA complex is formed. Our proposed mechanism for the early steps of mammalian peptide chain initiation leading to Met-$tRNA_f$·40S·mRNA complex formation is shown diagrammatically in Fig. 6.

2.4. Our Views Regarding Different Peptide Chain Initiation Models for Ternary and Met-$tRNA_f$·40S·mRNA Complex Formation

2.4.1. Characteristics of the Peptide Chain Initiation Factors

There is now agreement that eIF-2 alone is almost totally inactive in the presence of Mg^{2+} and requires three ancillary protein factors for ternary complex

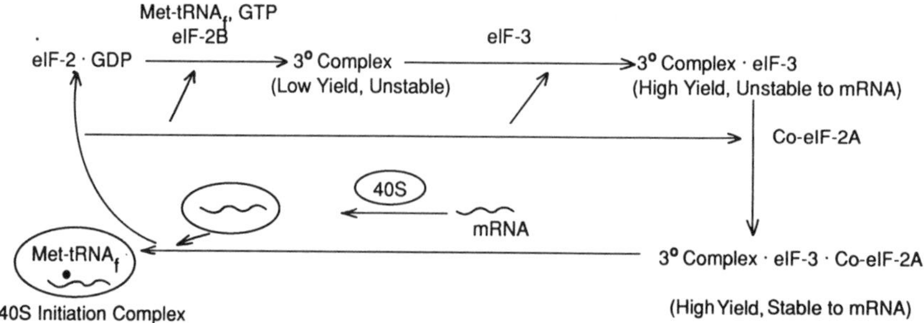

Figure 6. A tentative proposed model for Met-$tRNA_f$·40S·mRNA complex formation.

formation. In our laboratory we observed that the ternary complex formed in the presence of these protein factors, eIF-2 + Co-eIF-2C (eIF-2B + eIF-3) + Co-eIF-2A, efficiently transferred Met-tRNA$_f$ to 40S ribosomes and such a transfer reaction was almost entirely dependent on the presence of natural mRNAs. Therefore, according to our model, only four purified protein factors, eIF-2, eIF-2B, eIF-3, and Co-eIF-2A, are required for stable Met-tRNA$_f$·40S·mRNA complex formation.

According to the Staehelin–Anderson–Hershey model,[4,11–13] several additional protein factors, such as eIF-1 and several eIF-4 group proteins (4A, 4B, 4C, 4D, 4E, and 4F), are also required for Met-tRNA$_f$·40S·mRNA complex formation. Originally these protein factors were characterized by their requirements for Met-tRNA$_f$·40S·mRNA complex formation.[4,11–13] As suggested in our earlier papers,[1,3,5] however, these experiments were carried out under unphysiological conditions and are not reproducible.[1,3] Our recent collaborative work with the Hershey and Merrick groups is in agreement with this suggestion (for an explanation, see ref. 5). Therefore, at present there is no direct evidence that these protein factors (eIF-1 and eIF-4 group proteins) are peptide chain initiation factors.

Numerous laboratories have accepted the Staehelin–Anderson–Hershey model, however, and have worked extensively on several properties of these protein factors, which include covalent modifications,[21] RNA helicase activities,[22] and also cross-linking to the cap sites in mRNAs.[23] As analyzed in our previous publications,[1,3,5] however, these studies do not provide any evidence for the peptide chain initiation roles of these protein factors. It will now be imperative that the requirements of these protein factors be reexamined using physiologically significant partial reactions for ternary and Met-tRNA$_f$·40S·mRNA complex formation. Without this information, study of the peripheral properties of these protein factors will only provide information regarding these protein factors and may not have relevance to the actual process of peptide chain initiation.

We must emphasize the following points regarding the requirements of peptide chain initiation factors for Met-tRNA$_f$·40S·mRNA complex formation.

1. Our work indicates that only four protein factor preparations, eIF-2, eIF-2B, Co-eIF-2A, and eIF-3, are required for Met-tRNA$_f$·40S·mRNA complex formation. Two of these protein factors, eIF-2B and eIF-3, are high-molecular-weight protein complexes. The active polypeptide component(s) in these protein complexes have not yet been identified. Our work has indicated that the 180-kDa polypeptide component in eIF-3 may be necessary for ternary complex formation.[5,20] This protein complex also contains a 50-kDa polypeptide component which is presumably the same as eIF-4A.[14] Also, this protein complex may contain trace amount of other polypeptides, such as eIF-4E,[24] not visible by SDS–PAGE, but necessary for Met-tRNA$_f$·40S·mRNA complex formation.

2. Requirement of a specific protein factor in some cases may depend on stringent experimental conditions which may include the use of very low concen-

trations of the other protein factors. For example, in *Escherichia coli* protein synthesis the requirement of the peptide chain initiation factor IF-1 for fMet-tRNA$_f$·40S·AUG complex formation could only be demonstrated in the presence of suboptimal concentrations of the other protein factor IF-3.[25] In our experiments we may have used saturating concentrations of the protein factors and have thus missed the requirements of additional unidentified proteins. Also, the requirement of some protein factor(s), such as cap-binding protein(s), may only be demonstrated under some specific salt and Mg^{2+} concentrations, not used in our assays.

Further work will therefore be necessary to gain a better understanding of the requirements of different protein factors for Met-tRNA$_f$·40S·mRNA complex formation.

2.4.2. mRNA and ATP Requirement for Met-tRNA$_f$·40S Complex Formation

According to our proposed model, Met-tRNA$_f$ binding to the small ribosomal subunit (40S) in mammalian cells, as in prokaryotes, is almost totally dependent on the presence of mRNA.[14] Met-tRNA$_f$·40S·mRNA complex is formed by direct interaction between the anticodon in Met-tRNA$_f$ and the initiation codon in ribosome-bound mRNA. There is no requirement for ATP in this complex formation.[14]

According to the Staehelin–Anderson–Hershey model,[4] several eIF-4 group proteins bind to the cap site in mRNA in an ATP-dependent manner and subsequently promote binding of the cap site in mRNA to 40S ribosomes. In agreement with Kozak's scanning model,[26] the Staehelin–Anderson–Hershey model then proposes[4] that the 40S ribosomes initially bound to the cap site in mRNA migrate along the 5′-untranslated region driven by ATP hydrolysis until they reach the initiation site in mRNA.

Our views are as follows: (1) mRNA scanning involving several eIF-4 group proteins, as described in the Staehelin–Anderson–Hershey model, is not based on any experimental evidence. (2) In almost 95% of the eukaryotic mRNAs the 5′-most AUG codon acts as the initiation codon. In several cases, however, such as in poliovirus RNA, the initiation site is at a considerable distance from the 5′ end. In such cases there is convincing evidence for direct entry of the 40S ribosomes to the internal initiation site, reviewed in ref. 27. (3) Our experimental results suggest that Met-tRNA$_f$ binds directly to the initiation site in mRNA in the Met-tRNA$_f$·40S complex in an ATP-independent manner.

Finally, we should emphasize that a critical problem in studies of mRNA–ribosome interaction is the lack of suitable assay methods for such studies. In prokaryotes Steitz elegantly performed similar studies many years ago by directly determining the nucleotide sequences in mRNAs involved in f-Met-tRNA$_f$·30S·mRNA complex formation.[28] Unfortunately, a similar technique has not been used in eukaryotic studies, as a similar Met-tRNA$_f$·40S·mRNA complex, described in the widely accepted Staehelin–Anderson–Hershey model, is not experimentally reproducible. We believe that the experimental procedures developed in our laboratory and described in this chapter for the formation of a stable Met-tRNA$_f$·40S·mRNA complex will be useful to provide answers to these critical questions of the mechanism of mRNA–40S ribosome interaction.

3. REGULATORY ROLE OF AN eIF-2-ASSOCIATED 67-kDa POLYPEPTIDE

3.1. Background

Animal cells contain one or more protein synthesis inhibitors, such as HRI (heme-regulated inhibitor) and dsI (double-stranded RNA activated inhibitor). These inhibitors are also eIF-2 kinases and phosphorylate specifically the eIF-2 α-subunit. Protein synthesis becomes inhibited as phosphorylated eIF-2 [eIF-2α(P)·GDP] does not form a ternary complex. It is generally believed that these inhibitors remain in inactive forms and are activated under certain physiological conditions to inhibit protein synthesis. In reticulocyte lysate one of these inhibitors, HRI, becomes activated in the absence of heme, and another inhibitor, dsI, is activated in the presence of double-stranded RNA and ATP.[1,4] Numerous reports now indicate that animal cells widely use this eIF-2 α-subunit phosphorylation mechanism to regulate protein synthesis under different physiological conditions, which include nutritional deprivation,[29,30] heat shock,[30–32] and virus infection.[33–40] There are also indications that protein synthesis during several virus infections may also be regulated both by eIF-2 kinases and eIF-2 kinase inhibitors. Three different eIF-2 kinase inhibitors have been characterized: (1) A high level of a small RNA transcript called VAI RNA inhibits the activation of eIF-2 kinase in adenovirus-infected cells.[34,35] (2) A protein factor synthesized in vaccinia virus-infected cells inhibits the activation of an eIF-2 kinase.[37,38] (3) A protein factor synthesized in poliovirus-infected cells inhibits the activity of an already activated eIF-2 kinase.[39,40]

Several laboratories have reported that protein synthesis inhibition in heme-deficient reticulocyte lysate can be reversed by addition of either eIF-2[41–47] or a reticulocyte cell supernatant factor, which we term RF (reversal factor).[46,47] Several laboratories have also claimed that the eIF-2B activity in the RF preparation is responsible for this reversal activity.[46,47]

Since 1978 we have reported in a series of papers that: (1) Homogeneous eIF-2 preparations do not reverse protein synthesis inhibition in heme-deficient reticulocyte lysates.[48] (2) Partially purified eIF-2 preparations, which reverse protein synthesis inhibition in heme-deficient reticulocyte lysates, also reverse HRI inhibition of ternary complex formation.[49] (3) This eIF-2B preparation also contains free eIF-2 α-subunit and this subunit cannot be phosphorylated by HRI.[50] (4) The reversal activity of eIF-2B preparation can be directly correlated to the presence of the unphosphorylated eIF-2 α-subunit and not to the eIF-2B activity.[51]

In 1988 we reported[6] that a partially purified eIF-2 preparation, which reverses protein synthesis inhibition in heme-deficient reticulocyte lysate, con tains an extra 67-kDa polypeptide (p67) besides the usual three eIF-2 subunits (α, β, and γ). Similarly, an active RF preparation contains p67 and also free eIF-2 α-subunit. We provided evidence that this p67 and also the eIF-2 α-subunit are necessary for protein synthesis inhibition reversal, and p67 in both eIF-2 and eIF-2B preparations protects eIF-2 α-subunit from eIF-2-kinase-catalyzed phosphorylation.

In this section we will describe significant results related to the characteristics and roles of p67 in the regulation of protein synthesis in animal cells.

3.2. Isolation and Characterization of p67

3.2.1. Isolation of p67

p67 is present in both partially purified reticulocyte eIF-2 preparation and also in the reticulocyte cell supernatant factor RF.[6] This polypeptide copurifies with eIF-2 and can be isolated in association with a near homogeneous eIF-2 preparation containing the standard three-subunit eIF-2. Upon further purification free p67 and eIF-2 containing the three subunits can be obtained.

Figure 7 describes the SDS–PAGE of a partially purified eIF-2 preparation (fraction IV) containing p67 (lane 1), further purified three-subunit eIF-2 (fraction V) (lane 2), and free p67 (lane 3).

3.2.2. Requirement of p67 for Reversal of Protein Synthesis Inhibition in Heme-Deficient Reticulocyte Lysate

We studied the activities of the above purified protein factors to reverse protein synthesis inhibition in heme-deficient reticulocyte lysate[6] (Fig. 8). Frac-

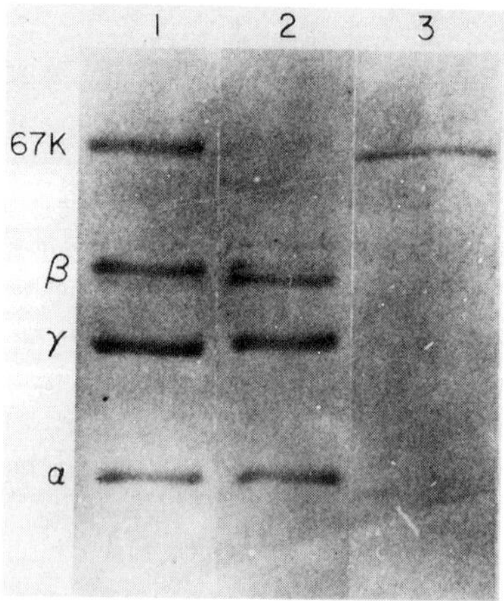

Figure 7. SDS–PAGE of different factor preparations. Lane 1, fraction IV eIF-2; lane 2, fraction V eIF-2; lane 3, isolated p67. Data were obtained from Datta *et al.*[6]

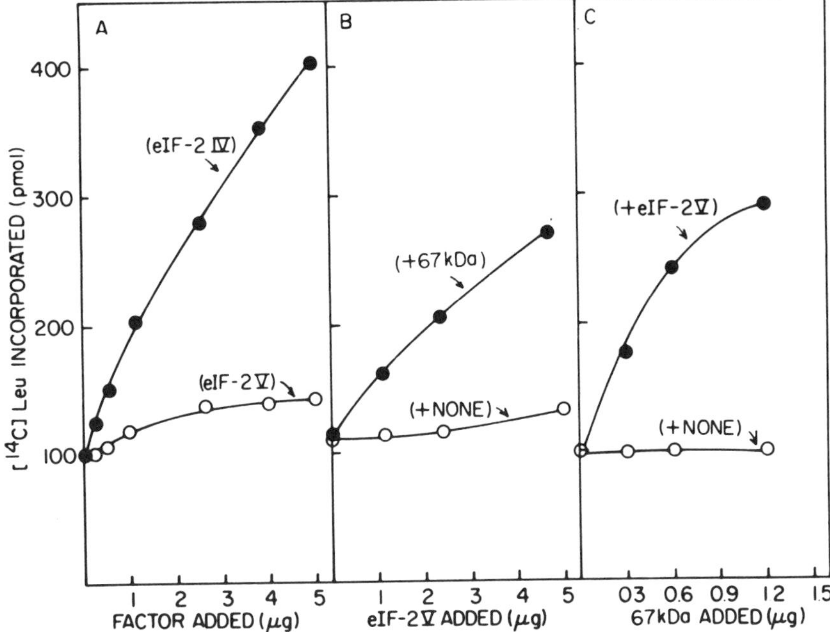

Figure 8. Protein synthesis inhibition reversal activity of fraction IV eIF-2, fraction V eIF-2, and p67. Data were obtained from Datta *et al.*[6]

tion IV eIF-2 efficiently reversed protein synthesis inhibition in heme-deficient reticulocyte lysate and fraction V eIF-2 was completely inactive (Fig. 8A). Also p67 alone was almost completely inactive (Fig. 8C). The reversal activity could be reconstituted, however, by combining fraction V eIF-2 and the isolated p67 (Figs. 8B and 8C), indicating that both the three-subunit eIF-2 and p67 are essential for protein synthesis.

3.2.3. p67 Protects eIF-2 α-subunit from eIF-2-Kinase(s)-Catalyzed Phosphorylation

The results presented in Fig. 9 show that p67 protects eIF-2 α-subunit from eIF-2-kinase-catalyzed phosphorylation. In this experiment eIF-2 (IV), eIF-2 (V), and eIF-2 (V) + equimolar amount of p67 were incubated in the presence of HRI and [γ-^{32}P]ATP. The reaction products were analyzed by SDS–PAGE followed by autoradiography. As shown in Fig. 9, the α-subunit in the three-subunit eIF-2 was extensively phosphorylated by HRI (lane 2). Under similar conditions the α-subunit in eIF-2 containing p67 was very poorly phosphorylated (lane 1). Also, the addition of p67 to the three-subunit eIF-2 strongly inhibited eIF-2 α-subunit phosphorylation (lane 3). Results of several control experiments indicated that

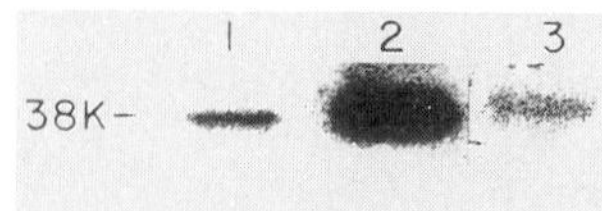

Figure 9. Phosphorylation of eIF-2 α-subunit (38 kDa) with ATP and HRI in the presence and absence of p67. Phosphorylation of eIF-2 α-subunit using HRI and [γ-^{32}P]ATP was carried out as described.[6,7] Lane 1, fraction IV eIF-2; lane 2, fraction V eIF-2; lane 3, fraction V eIF-2 + p67. Data were obtained from Datta *et al.*[7]

p67 is not an eIF-2 phosphatase and also does not inactivate eIF-2 kinase activity in HRI. These results thus suggest that p67 protects the eIF-2 α-subunit from eIF-2-kinase-catalyzed phosphorylation.

3.2.4. p67 Contains Multiple O-Linked Glc *N*-Acetylglucosamine (GlcNAc) Residues and These GlcNAc Residues on p67 May Be Necessary for p67 Activity to Protect eIF-2 α-Subunit from eIF-2-Kinase-Catalyzed Phosphorylation

3.2.4a. Detection of the GlcNAc Moieties on p67. Lectins such as wheat germ agglutinin (WGA) and concanavalin A (ConA) have been used widely to detect glycosyl residues on proteins.[8] WGA binds specifically to GlcNAc and sialic acid residues and ConA binds specifically to glucosyl and mannosyl residues. The result presented in Fig. 10 show that p67 binds specifically to WGA and that this binding is prevented by 0.2 M GlcNAc. In this experiment purified p67 was subjected to SDS–PAGE and was then transferred to nitrocellulose filters. Different lanes in the nitrocellulose filters containing p67 were cut into four parts. One part was used in standard immunoblot assay using p67 monoclonal antibodies (lane 1). The other parts were incubated separately with either biotinylated WGA (lanes 2 and 3) or biotinylated ConA (lane 4). Bound lectins were then visualized

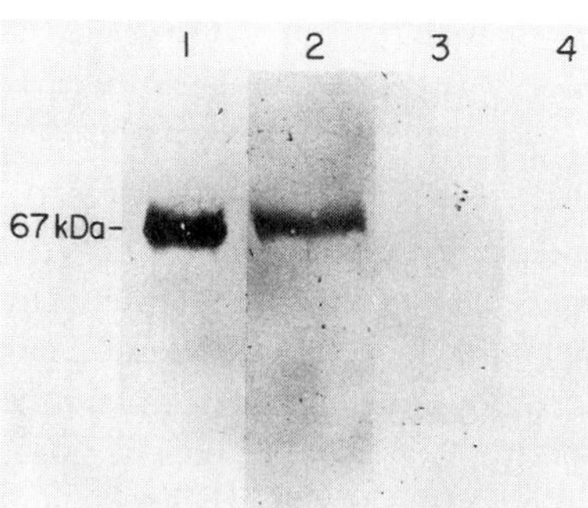

Figure 10. Detection of p67 by WGA. See text for details. Data were obtained from Datta *et al.*[8]

following treatment of the filters with avidin-conjugated horseradish peroxidase. Strong binding of WGA to p67 was apparent (lane 2), and this binding was completely inhibited in the presence of 0.2 m GlcNAc (lane 3), a competitive inhibitor of WGA. No detectable binding of ConA to p67 was observed (lane 4).

3.2.4b. Characterization of p67 As a Multiple O-Linked GlcNAc-Containing Protein. We used several standard experimental protocols, originally described by Hart and co-workers,[52,53] to characterize GlcNAc-containing O-linked glycoprotein. p67 was labeled with [^{3}H]galactose using UDP-[^{3}H]galactose in the presence of galctosyl transferase. Approximately 12[^{3}H]galactose molecules were incorporated into p67, indicating 12 GlcNAc residues on p67. Furthermore, [^{3}H]galactose-labeled p67 was resistant to Endo-β-*N*-acetylglucosaminidase F (Endo-F), but released [^{3}H]Galβ1 → 4 GlcNAcitol after treatment with mild alkali under reducing conditions.

3.2.4c. Roles of the GlcNAc Moieties in p67 Activity to Protect eIF-2 α-Subunit from eIF-2-Kinase-Catalyzed Phosphorylation. We previously observed that WGA, which binds specifically to the GlcNAc moieties on p67, inhibits p67 activity to protect eIF-2 α-subunit from eIF-2-kinase-catalyzed phosphorylation.[8] Based on this observation we postulated that p67 used these GlcNAc residues to bind to the eIF-2 α-subunit phosphorylation site and thus protects this site from eIF-2-kinase-catalyzed phosphorylation. The possibility that WGA binding sterically hindered p67–eIF-2 interaction, however, was not ruled out. Recently we studied[54] this problem using p67 polyclonal antibodies and four different p67 monoclonal antibodies (D_1, D_2, D_3, and D_4) each specific for the glycosyl residues on p67. We observed that the polyclonal antibodies and only one monoclonal antibody (D_1) strongly inhibited protein synthesis in reticulocyte lysate and also inhibited p67 activity to protect eIF-2 α-subunit from eIF-2-kinase(s)-catalyzed phosphorylation. These observations thus suggest that a specific glycosyl residue on p67 is necessary for p67 activity to protect eIF-2 α-subunit from eIF-2-kinase(s)-catalyzed phosphorylation.

3.3. Roles of p67 and eIF-2 Kinase(s) in the Regulation of Protein Synthesis Studied Using Several Cell-Free Lysates and an Animal Cell in Culture

3.3.1. Introduction

It is now generally believed that protein synthesis in animal cells is regulated by reversible activation–inactivation of eIF-2 kinase(s). Under conditions of protein synthesis these eIF-2 kinase(s) remain in inactive forms. Under certain other physiological conditions, such as during heme deficiency and nutritional deprivation, the eIF-2 kinases are activated and inhibit protein synthesis. In most of these studies the activities of the eIF-2 kinase(s) in cell-free lysates were measured by their abilities to phosphorylate eIF-2 α-subunit. Lack of this phosphorylation was used as a measure of the inactive forms of the eIF-2 kinase(s). Our recent work on p67, however, as described in this chapter, would suggest that the presence of p67 in the cell lysate would lead to inhibition of eIF-2 α-subunit phosphorylation, even in the presence of activated eIF-2 kinase(s). It was therefore necessary to measure both p67 and eIF-2 kinase activities to ascertain the role of each factor(s) in regulation of protein synthesis.

3.3.2. Assays for eIF-2 Kinase(s) and p67 Activity in a Mixture

To assay both eIF-2 kinase(s) and p67 activities in a mixture we developed an experimental procedure which measures eIF-2 α-subunit phosphorylation in the mixture and also after inhibition of p67 activity by preincubation with either p67 antibodies or WGA. As mentioned earlier,[8] WGA binds to the glycosyl residues on p67 and interferes with the p67 activity to protect eIF-2 α-subunit from eIF-2-kinase-catalyzed phosphorylation.

The characteristics of this assay method are shown in Fig. 11. As before, HRI efficiently phosphorylated three-subunit eIF-2 (lane 1) and addition of p67 protected eIF-2 α-subunit from the phosphorylation reaction (lane 2). The p67 activity to protect eIF-2 α-subunit was almost completely inhibited, however, by preincubation of the eIF-2 + p67 mixture either with p67-antibodies (lane 3) or with WGA (lane 6).

Under similar conditions preincubation of the eIF-2 + p67 mixture with either eIF-2 α-subunit or β-subunit antibodies had no significant effects (lanes 4 and 5). None of the antibodies (p67, eIF-2 α- and β-subunits), when added alone in the absence of p67, inhibited HRI-catalyzed phosphorylation of eIF-2 α-subunit.

This experimental procedure therefore provides an assay for eIF-2 kinase activity in the presence of p67; eIF-2 kinase activity to phosphorylate eIF-2 α-subunit became evident as p67 was removed by preincubation with p67 antibodies.

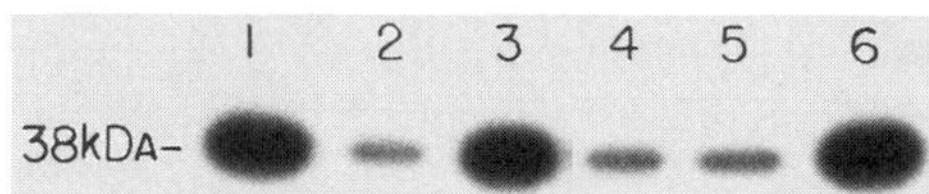

Figure 11. Assay for eIF-2 kinases and p67 activities in a mixture. Phosphorylation of eIF-2 α-subunit was carried out in the presence of HRI, $\gamma[^{32}P]$ATP, and, where indicated, p67 and different antibodies and WGA. Lane 1, eIF-2 alone; lane 2, eIF-2 + p67; lane 3, eIF-2 + p67 + p67 antibodies; lane 4, eIF-2 + p67 + eIF-2 + α-subunit antibodies; lane 5, eIF-2 + p67 + eIF-2 β-subunit antibodies; lane 6, eIF-2 + p67 + WGA. Data were obtained from Ray *et al.*[10]

3.3.3. Studies Using Animal Cell Lysates

We used the above assay method and studied eIF-2 kinases and p67 activities using four different cell lysates: heme-deficient rabbit reticulocyte lysate, heme-supplemented rabbit reticulocyte lysate, and cell lysates prepared from rat liver and rat brain (Fig. 12). In the absence of p67 antibodies significant eIF-2 α-subunit phosphorylation was observed only with heme-deficient reticulocyte lysate. In contrast, the heme-supplemented reticulocyte lysate, and also the lysates from liver and brain, showed very little or no phosphorylation. When the same cell lysates were preincubated with p67 antibodies, however, significant phosphorylation of eIF-2 α-subunit was observed with all the cell lysates. Also, the extent of phosphorylation with both heme-deficient and heme-supplemented reticulocyte lysate was comparable. These results thus show that both eIF-2 kinase(s) and p67 are present in active forms in all of the cell lysates, and p67 inhibits eIF-2 α-subunit phosphorylation by active eIF-2 kinase(s). Removal of p67 by p67 antibodies facilitates eIF-2-kinase(s)-catalyzed phosphorylation of the eIF-2 α-subunit.

The results presented in Fig. 13 show that p67 is indeed present in heme-supplemented reticulocyte lysate, and this polypeptide is partially deglycosylated in heme-deficient reticulocyte lysate. For these experiments the reticulocyte lysates were preincubated with p67 monoclonal (panel A) or polyclonal (panel B)

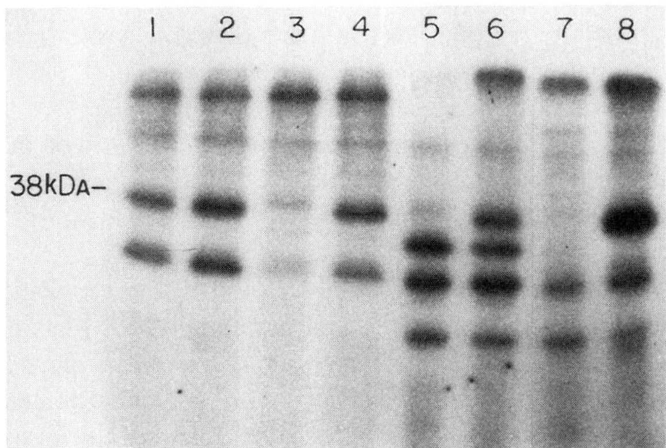

Figure 12. Analysis of eIF-2 kinases and p67 in different cell lysates. The assay procedure was the same as described in Fig. 11. Lane 1, heme-deficient reticulocyte lysate; lane 2, heme-deficient reticulocyte lysate preincubated with p67 antibodies; lane 3, heme-supplemented reticulocyte lysate; lane 4, heme-supplemented reticulocyte lysate preincubated with p67 antibodies; lane 5, lysate from rat brain; lane 6, lysate from rat brain preincubated with p67 antibodies; lane 7, lysate from rat liver; lane 8, lysate from rat liver preincubated with p67 antibodies. Data were obtained from Ray *et al.*[10]

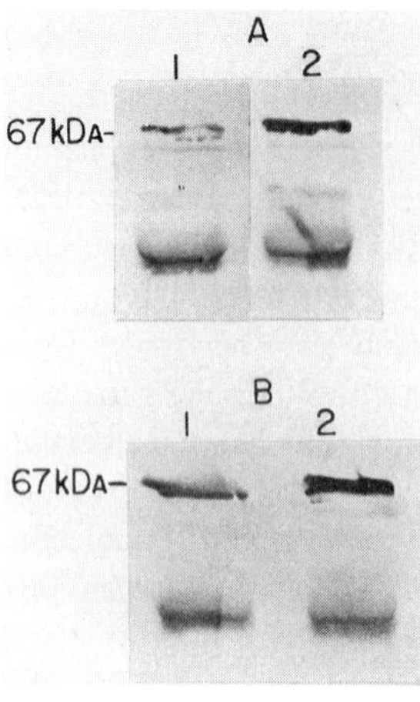

Figure 13. Analysis of p67 in heme-deficient and in heme-supplemented reticulocyte lysate. See text for details. **(A)** p67 in the lysate was immunoprecipitated using p67 monoclonal antibodies and protein A agarose and was subsequently analyzed by immunoblotting using p67 polyclonal antibodies: lane 1, heme-deficient lysate; lane 2, heme-supplemented lysate. **(B)** p67 in the lysates was immunoprecipitated using p67 polyclonal antibodies and protein A agarose and was subsequently analyzed by immunoblotting using p67 polyclonal antibodies: lane 1; heme-deficient lysate; lane 2, heme-supplemented lysate. The lower-molecular-weight band in each figure represents the heavy chain of IgG. Data were obtained from Ray *et al.*[10]

antibodies and the antigen–antibody complexes were precipitated with protein A agarose.[55] The precipitates were then assayed by immunoblot procedure using p67 polyclonal antibodies. As shown in Fig. 3, p67 is present in heme-supplemented reticulocyte lysate (panels A and B, lane 2) and also in heme-deficient reticulocyte lysate (panels A and B, lane 1). The level of this polypeptide, however, in heme-deficient reticulocyte lysate was significantly lower (panels A and B, lane 1). In experiments described in panel A, p67 present in the lysates was immunoprecipitated using p67 monoclonal antibodies which reacted only with glycosylated p67. As shown in panel A, the p67 level in heme-deficient lysate (lane 1) was less than 20% of that present in heme-supplemented lysate (lane 2) indicating extensive deglycosylation of p67 in heme-deficient lysate.

These results thus provide evidence that p67 is present in both heme-supplemented and heme-deficient reticulocyte lysate. In heme-deficient lysate, p67 is partially deglycosylated. This allows eIF-2 kinases to phosphorylate eIF-2 and thus inhibit protein synthesis.

Using a similar immunoblot experiment, we also observed that p67 is present in glycosylated form in both rat liver and rat brain extracts (data not shown).

It should be noted, however, that the results presented in Fig. 12 indicate that p67 also inhibits phosphorylation of one or more additional proteins present in reticulocyte lysate ($\approx$30 kDa) and in lysates from rat brain and rat liver ($\approx$65 and 105 kDa). In separate experiments we observed that p67 protects specifically eIF-2 α-subunit from both HRI- and dsI-catalyzed phosphorylation, but does not inhibit dsI-catalyzed phosphorylation of histones or casein kinase-catalyzed phosphorylation of eIF-2 β-subunit (data not shown). The results presented in Fig. 12 may indicate that under *in vitro* conditions p67 may use its glycosyl residues to bind to these proteins and thus inhibit their phosphorylation. On the other hand, the possibility that p67 may inhibit phosphorylation of a select group of closely related proteins cannot be ruled out. Abundant evidence reported previously and also presented in this chapter, however, suggests that a major function of p67 is to protect eIF-2 α-subunit from eIF-2-kinase(s)-catalyzed phosphorylation and thus to promote protein synthesis in the presence of active eIF-2 kinases.

3.3.4. Studies Using a Tumor Hepatoma Cell Line (KRC-7)

We used a tumor hepatoma cell line (KRC-7)[56] and analyzed the roles of p67 and eIF-2 kinase(s) in regulating protein synthesis under different growth conditions. We grew the cells to confluency, then serum-depleted the medium for 3–4 days, and finally stimulated the serum-starved cells with a mitogen, namely tumor-promoting phorbol ester (TPA). We measured protein synthesis activities, the levels of different polypeptides, and also activation states of eIF-2 kinase(s) at different intervals during confluency, after serum starvation, and after TPA addition. Protein synthesis rate was measured by incorporation of [^{35}S]methionine into cellular proteins following the procedure as described.[7] The protein synthesis rate was maximum in confluent cells and was reduced to 30% of the maximum level after 100 hr of serum starvation. Upon TPA addition the serum-starved cells regained 70% of the original protein synthesis activity of the confluent cells within 4 hr.

3.3.4a. Levels of Different Polypeptides. Immunoblot analysis was used to measure levels of different polypeptides (p67, dsI, eIF-2 β- and α-subunit) at different intervals after serum starvation and after TPA addition (Fig. 14). p67 levels were measured using both monoclonal (upper panel) and polyclonal (lower

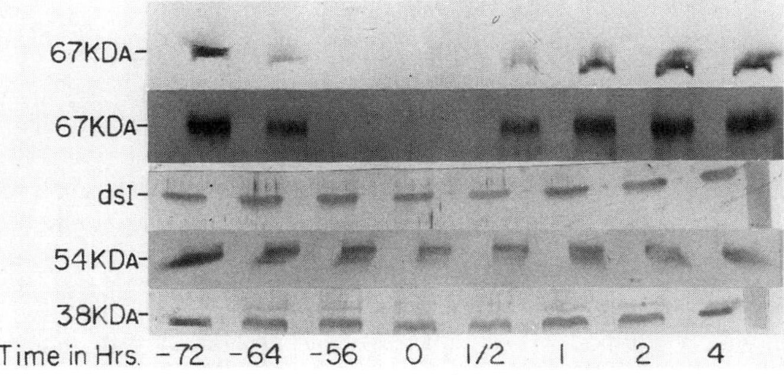

Figure 14. Immunoblot analysis of p67, dsI, eIF-2 β-subunit (p54), and eIF-2 α-subunit in tumor hepatoma cells (KRC-7); confluent, quiescent, and TPA-induced. p67 was analyzed using p67 monoclonal (upper panel) and polyclonal antibodies (lower panel). Data were obtained from Ray *et al.*[10]

panel) antibodies. dsI and eIF-2 β- and α-subunit levels were measured using polyclonal antibodies.

As shown in Fig. 14, levels of eIF-2 subunits (β and α), and, more importantly, the level of dsI, remained essentially unchanged in confluent (72 hr), serum-starved, and TPA-stimulated cells. On the other hand, the p67 level changed dramatically. This polypeptide was prominent in confluent cells, but disappeared rapidly after serum starvation. It again became prominent after TPA addition. A significant observation is that p67 level decreased faster after serum depletion and appeared more slowly after TPA addition when measured using monoclonal antibodies (upper panel) than when measured using polyclonal antibodies (lower panel). These results, in agreement with our earlier observation with heme-deficient reticulocyte lysates (Fig. 13), may suggest that the first step after serum depletion is deglycosylation of p67 and deglycosylated p67 is subsequently degraded.

3.3.4b. eIF-2 Kinase Activity. The results presented in Fig. 14 show that dsI polypeptide level remains essentially unchanged under different growth conditions. It is known, however, that the eIF-2 kinases such as HRI and dsI may remain in either active or inactive forms. We have therefore analyzed eIF-2 kinase activities by analyzing eIF-2 α-subunit phosphorylation of endogenous eIF-2 using cell extracts at different intervals following the assay method described in Fig. 11. As shown in Fig. 15, eIF-2 α-subunit phosphorylation was low with extracts from confluent and TPA-induced cells, but was significantly increased when a similar extract from quiescent cell was used (panel A). As described in Fig. 11, however, when the same cell extracts were first preincubated with p67 polyclonal antibodies to remove endogenous p67 (panel B), or with WGA to inhibit eIF-2–p67 interaction (panel C), the extent of eIF-2 α-subunit phosphorylation using all the cell extracts was significantly increased and was nearly similar in all cases.

3.4. Our Views Regarding the Roles of p67 and eIF-2 Kinases in Regulation of Protein Synthesis

The results described in this chapter indicate that all animal cells contain one or more eIF-2 kinases in the active form. The activities of these eIF-2-kinases may or may not be evident, depending on the variable presence of p67. The p67 protects eIF-2 α-subunit from eIF-2-kinase(s)-catalyzed phosphorylation and this promotes protein synthesis in the presence of active eIF-2 kinase(s).

Two widely studied eIF-2 kinases are HRI and dsI. In heme-deficient reticulocyte lysates only HRI is active, as dsI requires double-stranded RNA and ATP for activation. We have provided evidence that in both heme-supplemented and heme-deficient lysates HRI is present in active form and possibly in equal amounts. In heme-supplemented lysate this HRI activity is not evident, due to the

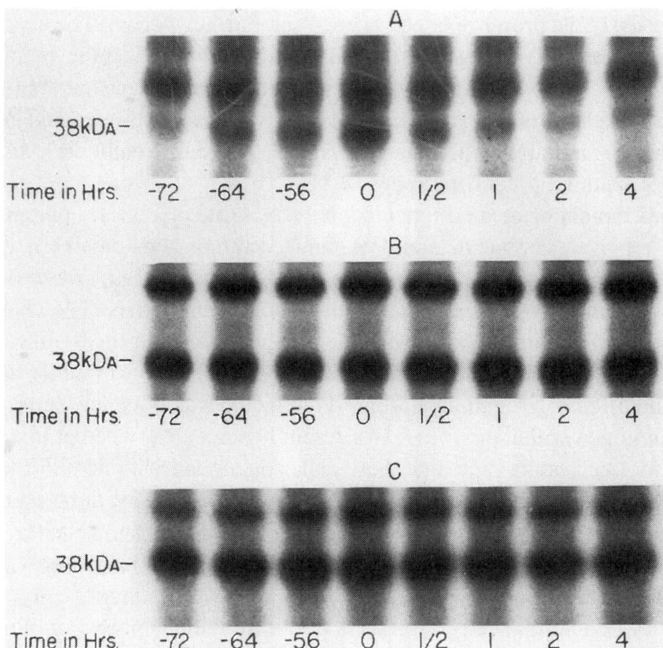

Figure 15. eIF-2 α-subunit phosphorylation using KRC-7 cell lysate, using different growth conditions. Phosphorylation of eIF-2 α-subunit in the cell lysate by endogenous eIF-2 kinase(s) was carried out using γ[^{32}P]ATP following the procedure described in Fig. 11. **(A)** The cell lysates alone; **(B)** the cell lysates preincubated with p67 polyclonal antibody; **(C)** the cell lysates preincubated with wheat germ agglutinin (WGA).

presence of p67, whereas in heme-deficient lysates p67 appear to be degraded, thus allowing HRI to actively phosphorylate eIF-2 α-subunit. Removal of p67 from both heme-supplemented and heme-deficient reticulocyte lysates using p67 antibodies gave almost similar extents of eIF-2 α-subunit phosphorylation. This indicates that the eIF-2 kinase in active form may be present at the same level in both systems. These results would suggest that eIF-2 α-subunit phosphorylation observed in heme-deficient reticulocyte lysate is not due to activation of HRI as is widely believed, but is due to degradation of p67. We also provide evidence that p67 degradation involves partial deglycosylation in heme-deficient lysates. It may therefore be assumed that hemin, present in heme-supplemented lysate, prevents p67 deglycosylation, possibly by inhibiting the deglycosylating enzyme(s).

Reticulocyte lysates also contain another eIF-2 kinase, namely dsI in inactive form, and this eIF-2 kinase is activated in the presence of double-stranded RNA and ATP. Activated dsI inhibits protein synthesis even in the presence of hemin and presumably in the presence of p67. The mechanism of eIF-2 α-subunit phosphorylation by dsI in the presence of p67 is not clear. Further work will be necessary to examine p67 activity during dsI inhibition of protein synthesis in heme-supplemented reticulocyte lysate. It may be possible, however, that dsI is formed in relatively large excess in the presence of double-stranded RNA and ATP, and at a very high concentration dsI may phosphorylate eIF-2 α-subunit even in the presence of equimolar concentrations of p67.

Like heme-supplemented reticulocyte lysates, the cell lysates prepared from rat brain, rat liver, confluent, and TPA-induced tumor hepatoma cells (KRC-7) contain both p67 and eIF-2 kinase(s) in active forms and eIF-2 kinase(s) activity becomes evident only after removal of p67 by p67 antibodies. The characteristics of these eIF-2 kinases, and whether they are HRI-like or dsI-like, are not known. dsI polypeptide is present in a nearly unchanged level in tumor hepatoma cells under different growth conditions. This includes confluency, serum starvation, and mitogen stimulation (Fig. 14). Again it is not clear whether this dsI is in active form. Our results show that these cells contain at least one eIF-2 kinase in active form. The identity of this eIF-2 kinase with dsI remains to be established. As noted earlier, dsI exists in inactive form in the cells and requires double-stranded RNA and ATP for activation. The cell extracts used in our study were not treated with double-stranded RNA and ATP. Also, our present work suggests that the HRI may be a normal cell constituent, as this inhibitor is present in active form and in nearly equal amounts in both heme-supplemented and heme-deficient reticulocyte lysates. Possibly a HRI-like eIF-2 kinase is present in active form in all animal cells and dsI activation serves specialized functions under certain physiological conditions. As postulated earlier, this inhibitor (dsI) may be formed in large excess upon activation under certain physiological conditions, such as virus infections, and may inhibit protein synthesis in the presence of normal p67 concentrations.

The roles and requirements for p67 in KRC-7 cells under different growth conditions are clearly evident. This polypeptide is present in high concentrations in confluent cells and disappears rapidly from the cells after serum depletion with accompanying decrease in the protein synthesis rate in the cells. Also, this polypeptide reappears soon after TPA addition to the serum-starved resting cells as the cells regain protein synthesis activity. Thus, there is a clear correlation between the p67 level and the protein synthesis activity of the cells, indicating a direct involvement of p67 in protein synthesis. Furthermore, a comparison of the p67 levels studied using monoclonal (Fig. 14, upper panel) and polyclonal (Fig. 14, lower panel) antibodies reveals (in agreement with our observation with the heme-deficient reticulocyte lysates, Fig. 13) that the initial event in p67 degradation involves deglycosylation of p67 and deglycosylated p67 is unstable and subsequently degraded. These results, therefore, establish that: (1) p67 is necessary to protect eIF-2 α-subunit from eIF-2-kinase(s)-catalyzed phosphorylation and thus to promote protein synthesis in the presence of active eIF-2 kinase(s). (2) At least one eIF-2 kinase is present in all of the cells in active form under different growth conditions. (3) p67 degrades rapidly in serum-depleted cells, thus allowing eIF-2 kinase(s) to phosphorylate eIF-2 and inhibit protein synthesis. (4) The initial event of p67 degradation involves deglycosylation of the p67 polypeptide.

An important aspect of p67 regulation is that this is the only polypeptide presently studied which is both degradable and also inducible. The levels of this polypeptide correlate directly with the protein synthesis activities of the cells. As evident in our study, the initial event for degradation involves deglycosylation of p67. The deglycosylated p67 is unstable and subsequently degrades. Again, addition of a mitogen, such as TPA, induces an increased appearance of p67, presumably by transcription activation of the p67 gene. Thus, the p67 level in the cell is regulated both transcriptionally and also at the posttranscription level. This factor is essential for protein synthesis and plays a critical role in protein synthesis regulation.

Finally, the differential expression of p67 during different developmental stages may be responsible for preferential translation of different mRNAs under various physiological conditions, including viral infection. A plausible mechanism of "shutoff" of host protein synthesis and preferential translation of viral mRNAs during virus infection of animal cells may be postulated as follows: During the early phase of viral infection an eIF-2 kinase, such as dsI, is activated, leading to eIF-2 α-subunit phosphorylation and inhibition of host protein synthesis. In the absence of mRNA translation the host mRNAs available at the early stages of viral infection may decay and the viruses use the host machinery for its own gene expression. During the late phase of viral infection, activation of the p67 gene may lead to increased synthesis of p67, leading to inhibition of eIF-2 α-subunit phosphorylation and resumption of translation of viral mRNAs available at that stage. Interestingly, if this postulated mechanism of viral infection proves correct and viral infection depends on "shutoff" of host protein synthesis, the increased availability of p67 by the expression of the transfected cloned p67 gene at the initial stage of viral infection may prevent "shutoff" of host protein synthesis and consequent viral production.

ACKNOWLEDGMENTS. This investigation was supported by a National Institutes of Health Grant GM 22079 and a Nebraska State Grant for Cancer and Smoking Disease (91-21). The authors thank Dr. Nahum Sonenberg for critical reading of the manuscript. The authors also gratefully recognize the assistance of Janelle Jones in the preparation of this manuscript.

REFERENCES

1. Gupta, N. K., Ahmad, M. F., Chakrabarti, D., and Nasrin, N., 1987, In: *Translation Regulation in Gene Expression*, (J. Ilan, ed.), pp. 287–334, Plenum Press, New York.

2. Proud, C. G., 1986, *Trends Biochem. Sci.* **11:**73–77.
3. Gupta, N. K., 1987, *Trends Biochem. Sci.* **12:**15–18.
4. Hershey, J. W. B., 1991, *Annu. Rev. Biochem.* **60:**717–755.
5. Gupta, N. K., Roy, A. L., Nag, M. K., Kinzy, T. G., MacMillan, S., Hileman, R. E., Dever, T. E., Wu, S., Merrick, W. C., and Hershey, J. W. B., 1990, In: *Post-Transcriptional Control of Gene Expression* (J. E. G. McCarthy and M. F. Tuite, eds.), pp. 521–526, Springer-Verlag, Berlin.
6. Datta, B., Chakrabarti, D., Roy, A. L., and Gupta, N. K., 1988, *Proc. Natl. Acad. Sci. USA* **85:** 3324–3328.
7. Datta, B., Ray, M. K., Chakrabarti, D., and Gupta, N. K., 1988, *Indian J. Biochem.* **25:**478–482.
8. Datta, B., Ray, M. K., Chakrabarti, D., Wylie, D., and Gupta, N. K., 1989, *J. Biol. Chem.* **264:** 20620–20624.
9. Gupta, N. K., Datta, B., Roy, A. L., and Ray, M. K., 1990, In: *Post-Transcriptional Control of Gene Expression* (J. E. G. McCarthy and M. F. Tuite, eds.), pp. 511–520, Springer-Verlag, Berlin.
10. Ray, M., Datta, B., Chakraborty, A., Chattopadhyay, A., Meza-Keuten, S., and Gupta, N. K., 1992, *Proc. Natl. Acad. Sci. USA* **89:**539–543.
11. Schreier, M. H., Erni, B., and Staehelin, T., 1977, *J. Mol. Biol.* **116:**727.
12. Benne, R., and Hershey, J. W. B., 1978, *J. Biol. Chem.* **253:**3078.
13. Merrick, W. C., 1979, *Meth. Enzymol.* **60:**101–108.
14. Roy, A. L., Chakrabarti, D., Datta, B., Hileman, R. E., and Gupta, N. K., 1988, *Biochemistry* **27:** 8203–8209.
15. Benne, R., Brown-Luedi, M. L., and Hershey, J. W. B., 1979, *Meth. Enzym.* **60:**15–35.
16. Grace, M., Bagchi, M., Ahmad, F., Yeager, T., Olson, C., Chakravarty, I., Nasrin, N., and Gupta, N. K., 1984, *Proc. Natl. Acad. Sci. USA* **81:**5379–5383.
17. Siekierka, J., Manne, V., Mauser, L., and Ochoa, S., 1983, *Proc. Natl. Acad. Sci. USA* **80:**1232–1235.
18. Bagchi, M. K., Chakravarty, I., Ahmad, F., Nasrin, N., Banerjee, A., Olson, C., and Gupta, N. K., 1985, *J. Biol. Chem.* **260:**6950–6954.
19. Peterson, D. T., Merrick, W. C., and Safer, B., 1979, *J. Biol. Chem.* **254:**2509–2516.
20. Roy, A. L., 1989, Ph.D. Thesis, University of Nebraska, Lincoln, Nebraska.
21. Hershey, J. W. B., 1989, *J. Biol. Chem.* **264:**20823–20826.
22. Rozen, F., Edery, I., Meerovitch, K., Dever, T. E., Merrick, W. C., and Sonenberg, N., 1990, *Mol. Cell Biol.* **10:**1134–1144.
23. Sonenberg, N., 1987, *Adv. Virus Res.* **33:**175–204.
24. Sonenberg, N., Trachsel, H., Hecht, S., and Shatkin, A. J., 1980, *Nature* **285:**331–333.
25. Wahba, A., Iwasaki, K., Miller, M. J., Sabol, S., Sillero, M. A. G., and Vasquez, C., 1989, *Cold Spring Harbor Symp. Quant. Biol.* **36:**291–299.
26. Kozak, M., 1989, *J. Cell Biol.* **108:**229–241.
27. Sonenberg, N., 1988, *Prog. Nucleic Acid Res. Mol. Biol.* **35:**173–207.
28. Steiz, J. A., 1969, *Cold Spring Harbor Symp. Quant. Biol.* **36:**621–633.
29. Duncan, R., and Hershey, J. W. B., 1985, *J. Biol. Chem.* **260:**5493–5497.
30. Scorsone, K. A., Panniers, R., Rowlands, A. G., and Henshaw, E., 1987, *J. Biol. Chem.* **262:** 14538–14543.
31. Duncan, R., and Hershey, J. W. B., 1984, *J. Biol. Chem.* **259:**11882–11884.
32. Benedetti, A. D., and Baglioni, C., 1987, *J. Biol. Chem.* **262:**338–342.
33. Reichel, P. A., Merrick, W. C., Siekierka, J., and Mathews, M. B., 1985, *Nature* **313:**196–200.
34. Schneider, R. J., Safer, B., Munemitsu, S. M., Samuel, C. E., and Shenk, T., 1985, *Proc. Natl. Acad. Sci. USA* **82:**4321–4325.
35. Siekierka, J., Mariano, T., Reichel, P. A., and Mathews, M. B., 1985, *Proc. Natl. Acad. Sci. USA* **82:**1259–1263.
36. Whitaker-Dowling, P. A., and Younger, J. S., 1984, *Virology* **137:**171–181.
37. Rice, A. P., and Kerr, I. M., 1984, *J. Virol.* **50:**229–236.

38. Akkaraju, G. R., Whitaker-Dowling, P., Younger, J. S., and Jagus, R., 1989, *J. Biol. Chem.* **264:** 10321–10325.
39. Ransoni, L. J., and Dasgupta, A., 1987, *J. Virol.* **61:**1781–1787.
40. Ransoni, L. J., and Dasgupta, A., 1988, *J. Virol.* **62:**3551–3588.
41. Clemens, M. J., Henshaw, E. W., Rahamimoff, H., and London, I. M., 1974, *Proc. Natl. Acad. Sci. USA* **71:**2946–2950.
42. Kaempfer, R., 1974, *Biochem. Biophys. Res. Commun.* **61:**591–597.
43. Clemens, M. J., 1976, *Eur. J. Biochem.* **66:**413–422.
44. Amesz, H., Gouman, H., Haubrich-Morree, T., Voorma, H. O., and Benne, R., 1979, *Eur. J. Biochim.* **98:**513–520.
45. Siekierka, J., Mitusi, K., and Ochoa, S., 1981, *Proc. Natl. Acad. Sci. USA* **78:**220–223.
46. Konieczny, A., and Safer, B, 1983, *J. Biol. Chem.* **258:**3402–3408.
47. Shaun, N., Matts, R. L., Petryshyn, R., and London, I. R., 1984, *Proc. Natl. Acad. Sci. USA* **81:** 6998–7002.
48. Ralston, R. D., Das, A., Dasgupta, A., Roy, R., Palmieri, S., and Gupta, N. K., 1978, *Proc. Natl. Acad. Sci. USA* **75:**4858–4862.
49. Ralston, R., Das, A., Grace, M., Das, H. K., and Gupta, N. K., 1979, *Proc. Natl. Acad. Sci. USA* **76:**5490–5494.
50. Grace, M., Ralston, R. O., Banerjee, A. C., and Gupta, N. K., 1982, *Proc. Natl. Acad. Sci. USA* **79:**6517–6521.
51. Grace, M., Bagchi, M., Ahmad, F., Yeager, T., Olson, C., Chakravarty, I., Nasrin, N., Banerjee, A., and Gupta, N. K., 1984, *Proc. Natl. Acad. Sci. USA* **81:**5379–5381.
52. Holt, G. D., and Hart, G. W., 1986, *J. Biol. Chem.* **261:**8049–8057.
53. Holt, G. D., Haltiwanger, R. S., Torres, C. R., and Hart, G. W., 1987, *J. Biol. Chem.* **262:**14847–14850.
54. Chakraborty, A., Ray, M. K., Datta, B., Chakrabarti, D., Wylie, D., and Gupta, N. K., 1992, Annual Meeting of the Fed. Am. Soc. Biochem. and Mol. Biol. (Abstract), *FASEB J.* **6:**A453.
55. Reed, R., Griffith, J., and Maniatis, T., 1988, *Cell* **52:**949–969.
56. Trevillyan, J. M., Kulkarni, R. K., and Byers, C. V., 1984, *J. Biol. Chem.* **259:**897–902.

Chapter 21

Phosphorylation of Elongation Factor 2

A Mechanism to Shut Off Protein Synthesis for Reprogramming Gene Expression

Alexey G. Ryazanov and Alexander S. Spirin

1. INTRODUCTION

In 1987 the phosphorylation of elongation factor 2 (eEF-2) was discovered.[1] At that time it seemed as though this was merely a demonstration that one more translation factor could be phosphorylated. It now appears, however, that eEF-2 phosphorylation is involved in the regulation of the cell cycle, cell differentiation, and many other processes. Here we summarize what is currently known about eEF-2 phosphorylation and its physiological role.

2. DISCOVERY OF eEF-2 PHOSPHORYLATION

In 1983 Palfrey reported that incubation of cytoplasmic extracts from different rat tissues with [γ-^{32}P]ATP resulted in an intensive labeling of a soluble 100-

Alexey G. Ryazanov • Institute of Protein Research, Academy of Sciences of Russia, 142292, Pushchino, Moscow Region, Russia; and Department of Pharmacology, UMDNJ–Robert Wood Johnson Medical School, Piscataway, New Jersey 08854. Alexander S. Spirin • Institute of Protein Research, Academy of Sciences of Russia, 142292, Pushchino, Moscow Region, Russia.

Translational Regulation of Gene Expression 2, edited by Joseph Ilan. Plenum Press, New York, 1993.

kDa polypeptide.[2] Phosphorylation of this polypeptide was Ca^{2+}/calmodulin-dependent.[2] Later, Nairn *et al.*[3] found that the protein kinase responsible for this phosphorylation differed from all previously characterized Ca^{2+}/calmodulin-dependent protein kinases. The novel kinase was called Ca^{2+}/calmodulin-dependent protein kinase III.[3] Shortly thereafter the 100-kDa polypeptide was identified as eEF-2.[1]

This phosphorylation has several striking features. It is the fastest and most prominent phosphorylation in mammalian cell extracts.[2,3] Furthermore, in many cases the phosphorylation of eEF-2 represents the only Ca^{2+}-sensitive phosphorylation in the cell extract (see Fig. 1). The exceptional features of eEF-2 phosphorylation prompted the investigation of this phenomenon in more detail and the characterization of the protein kinase responsible for eEF-2 phosphorylation.

3. eEF-2 KINASE: A NOVEL Ca^{2+}/CALMODULIN-DEPENDENT PROTEIN KINASE

The eEF-2 kinase deserves special attention. The kinase is strictly Ca^{2+}/calmodulin-dependent and differs in its properties from four other well-character-

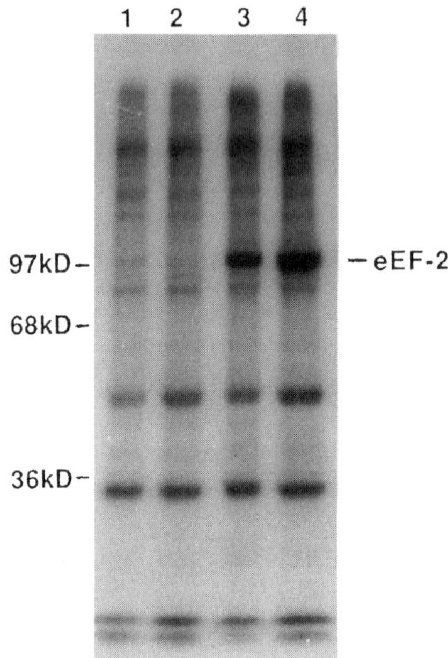

Figure 1. Phosphorylation of eEF-2 in extracts from the gastropod mollusk *Patella vulgata* oocytes. The figure represents an autoradiograph of the gel after SDS–electrophoresis of oocyte extracts incubated with [γ-^{32}P]ATP. Purified rabbit reticulocyte eEF-2 was added to the incubation mixture in lanes 2 and 4 and 100 μM $CaCl_2$ was added to the incubation mixture in lanes 3 and 4.

ized Ca^{2+}/calmodulin-dependent protein kinases, namely Ca^{2+}/calmodulin-dependent protein kinases I and II, phosphorylase kinase, and myosin light-chain kinase.[1–4] The eEF-2 kinase has been partially purified from different mammalian tissues. Its molecular mass, according to gel filtration, has been estimated to be about 140 kDa.[3,4] It is probably a monomeric protein, since upon purification from rabbit reticulocyte lysate it is copurified with a 105- to 110-kDa polypeptide (A. G. Ryazanov, unpublished observations). Up to three threonine residues (Thr-53, Thr-56, and Thr-58) can be phosphorylated in the eEF-2 molecule during incubation of partially purified eEF-2 kinase with eEF-2 (see Fig. 2).[5] The site of phosphorylation in eEF-2 in the crude reticulocyte lysate is Thr-56 and, to a small extent, Thr-58.[6] In the case of *in vivo* phosphorylation only one site, most probably Thr-56, has been revealed.[7,8]

It is likely that the eEF-2 kinase is present in all eukaryotic organisms. It has been recently identified in oocytes and eggs of different marine invertebrates, such as echinoderms, mollusks, and annelids (H. Abdelmajid, P. Guerrier, and A. G. Ryazanov, unpublished observations), in neurons of mollusks (R. Gillette and A. G. Ryazanov unpublished observations), and in yeasts.[9]

4. DO OTHER PROTEIN KINASES PHOSPHORYLATE eEF-2?

Tests of a number of well-characterized protein kinases other than eEF-2 kinase show that they are unable to phosphorylate eEF-2.[3,10] In eEF-2 immunoprecipitates of extracts from human fibroblasts metabolically labeled with ^{32}P, however, a small amount of phosphoserine was found.[7] Moreover, it was reported recently that *in vitro* phosphorylated eEF-2 from chick embryo contains, together with phosphothreonine, a small amount of phosphoserine and phosphotyrosine.[11]

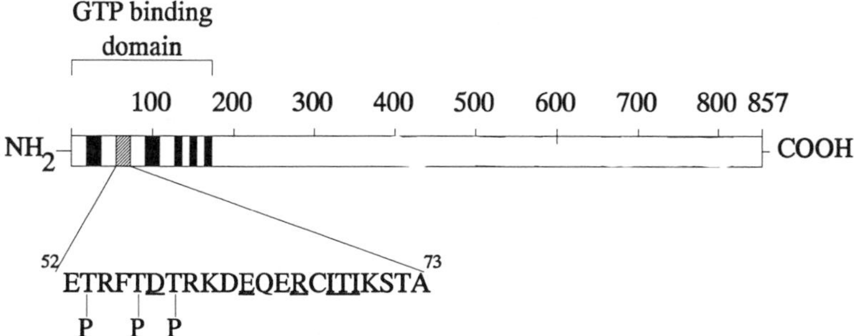

Figure 2. Schematic representation of the structure of eEF-2 and location of phosphorylation sites. The hatched box indicates the homologous region of eEF-2 with other elongation factors. Amino acids identical in all elongation factors are underlined. Three threonine residues which can be phosphorylated by eEF-2 kinase are located in this region. The black boxes correspond to the homologous regions of eEF-2 with other GTP-binding proteins.

These results suggest that other protein kinases capable of phosphorylating eEF-2 may exist, but the extent of eEF-2 phosphorylation by those protein kinases is insignificant in comparison with the phosphorylation of threonines catalyzed by the eEF-2 kinase.

5. eEF-2 PHOSPHATASE: TYPE 2A PROTEIN PHOSPHATASE

Type 2A protein phosphatase is the main phosphatase responsible for the dephosphorylation of eEF-2. This is supported by two kinds of experiments. First, the phosphatase inhibitor okadaic acid strongly inhibits the eEF-2 dephosphorylation at the concentration at which it affects only type 2A protein phosphatase.[12–14] Second, purified phosphatase 2A rapidly dephosphorylates phosphorylated eEF-2.[13,15] Interestingly, this phosphatase exhibits a certain specificity toward eEF-2: upon addition of purified type 2A phosphatase to a mixture of different phosphorylated proteins, eEF-2 is dephosphorylated more rapidly than any other protein.[13]

6. EFFECTORS OF eEF-2 PHOSPHORYLATION OTHER THAN Ca^{2+} AND CALMODULIN

It is clear now that Ca^{2+} is not the only second messenger capable of regulating eEF-2 phosphorylation. It has been shown in the case of extracts from mollusk neurons that the ability of the extracts to phosphorylate eEF-2 drops severalfold after an increase of pH from 7.0 to 7.5[16] (also R. Gillette and A. G. Ryazanov, unpublished results). In rabbit reticulocyte lysate, oxidized glutathione (GSSG) was shown to stimulate drastically the phosphorylation of the 95-kDa polypeptide,[17] which was later identified as eEF-2.[1] Also, in reticulocyte lysate it was shown that a high concentration of cAMP (1–10 mM) can induce dephosphorylation of eEF-2.[18] This effect seems to be due to direct inhibition of the eEF-2 kinase by cAMP.[19] The physiological relevance of the effect of cAMP, however, is unclear, because the concentrations of cAMP that are inhibitory for eEF-2 kinase are much higher than the concentration which can be found in the cytoplasm *in vivo*. An interesting effect has been recently observed with Mn^{2+}, which is found to stimulate specifically the eEF-2 phosphorylation in rat pancreas extract.[20] There is an indication that the eEF-2 kinase can be regulated by the heat-shock protein hsp90.[21] Incidentally, hsp90 is always copurified with the eEF-2 kinase[21] (also A. G. Ryazanov, unpublished observations).

Another possible way of regulating eEF-2 phosphorylation is through phosphorylation of the eEF-2 kinase itself. It has been reported recently that incubation of rat liver extract with alkaline phosphatase completely inactivates eEF-2 kinase while subsequent incubation of the inactive kinase with the catalytic subunit of the cAMP-dependent protein kinase restores the eEF-2 kinase activity.[21] Thus, the eEF-2 kinase itself must be phosphorylated to be active.

7. eEF-2 BECOMES INACTIVE AFTER PHOSPHORYLATION

One of the most important questions concerning any type of protein phosphorylation is the effect of the phosphorylation on the biological activity of the respective protein. The first experiments revealed that phosphorylation of eEF-2 by eEF-2 kinase has a strong inhibitory effect on this activity.[15,22,23] Phosphorylated eEF-2 purified from rabbit reticulocyte lysate[22] and eEF-2 phosphorylated *in vitro* by partially purified eEF-2 kinase[15,23] is found inactive in a poly(U)-directed cell-free translation system.

Phosphorylation of just a single threonine residue, namely Thr-56 (the same that can be phosphorylated *in vivo*), has been found to be sufficient to inactivate eEF-2 (C. G. Proud and A. G. Ryazanov, unpublished observations). This threonine residue is located in the so-called "effector domain,"[24] a protein sequence which has a strong homology among different elongation factors, but which has no homology with corresponding regions in other GTP-binding proteins (see Fig. 2). The effector domain is most likely involved in the interaction of eEF-2 with the ribosome. At the same time it has been found that phosphorylated eEF-2 can form complexes with GTP and ribosomes, but is unable to catalyze the translocation reaction.[25] The explanation can be that either the binding of phosphorylated eEF-2 to the ribosome is not strong enough to induce translocation (it is in fact weaker in comparison with that of nonphosphorylated eEF-2[26]) or that the binding is topographically incorrect.

Correlation between the degree of eEF-2 phosphorylation and the inhibition of protein synthesis has also been demonstrated in a more natural cell-free translation system based on reticulocyte lysate. Addition of millimolar concentrations of cAMP was found to stimulate both the protein synthesis and the dephosphorylation of eEF-2,[18] while the phosphatase inhibitor okadaic acid induced a strong increase in the phosphorylation of eEF-2 and inhibited protein synthesis.[12] It was also found that phosphorylated eEF-2 was unable to restore protein synthesis upon addition to the eEF-2-depleted rabbit reticulocyte translation system, whereas addition of the unphosphorylated form could.[26]

8. DOES Ca^{2+} AFFECT TRANSLATION *IN VIVO* THROUGH EF-2 PHOSPHORYLATION?

In experiments with intact rabbit reticulocytes it has been found that the elevation of intracellular Ca^{2+} concentration after treatment of the reticulocytes with the Ca^{2+} ionophore A23187 results in strong inhibition of the elongation.[27]

This is consistent with the fact that the elevation of Ca^{2+} in the cytoplasm affects eEF-2 phosphorylation (see the next section).

The effect of calcium ionophores on protein synthesis in nucleated cells is more complex. In cells with developed endoplasmic reticulum, calcium ionophores, besides transiently increasing cytoplasmic Ca^{2+} concentration, also induce long-lasting depletion of Ca^{2+} stores in the endoplasmic reticulum. This depletion has a strong inhibitory effect on overall protein synthesis at the level of initiation (reviewed in ref. 28). Thus, ionophores and natural hormones which induce Ca^{2+} mobilization from the endoplasmic reticulum may affect translation in two ways: depletion of Ca^{2+} in the endoplasmic reticulum through unknown mechanisms results in the inhibition of initiation, while simultaneous increase of Ca^{2+} concentration in the cytoplasm results in the inhibition of elongation.

It should be mentioned that the elevation of Ca^{2+} concentration in the cytoplasm, as a rule, is of a transient nature; in most cases it lasts for several seconds and rarely for a few minutes. Consequently, one can expect that the phosphorylation of eEF-2 *in vivo* and the resultant inhibition of protein synthesis should also be transient (about several minutes).

9. WHEN DOES PHOSPHORYLATION OF eEF-2 OCCUR *IN VIVO*?

Examples of eEF-2 phosphorylation *in vivo* are listed in Table I. These cases include the transition of quiescent cells into proliferative state, mitosis, the activation of neurons, and the activation of several types of secretory cells. As expected, eEF-2 becomes phosphorylated *in vivo* in situations when a transient increase of Ca^{2+} concentration in the cytoplasm takes place. In all cases, phosphorylation of eEF-2 was found to be transient and continued for only a few minutes, which is consistent with the transient nature of Ca^{2+} increase in the cytoplasm. Does this mean that in all the examples a transient inhibition of protein synthesis is observed? If so, what is the role of the protein synthesis inhibition? To answer these questions, some examples will be discussed in more detail.

10. IS THE TRANSIENT INHIBITION OF PROTEIN SYNTHESIS REQUIRED FOR TRANSITION OF QUIESCENT CELLS INTO PROLIFERATIVE STATE?

The first case in which *in vivo* phosphorylation of eEF-2 was described was mitogenic stimulation of quiescent human fibroblasts.[7] The treatment of these cells with serum, bradykinin, vasopressin, or epidermal growth factor resulted in transient phosphorylation of eEF-2. Phosphorylation of eEF-2 increased almost immediately after mitogen treatment, reached the maximum after about 0.5–1 min, and then declined back within 5 min. This result is not unexpected, since a

Table I. Phosphorylation of eEF-2 *in Vivo*

Cell type	Process during which phosphorylation of eEF-2 occurs	Inductors of eEF-2 phosphorylation	Ref.
Quiescent human fibroblasts, HSWP (foreskin-derived), WI38 (lung-derived)	G_0–G_1 transition	Serum, bradykinin, vasopressin, epidermal growth factor, ionophore A23187	7
Quiescent mouse fibroblasts NIH 3T3 (HIR 3.5)	G_0–G_1 transition	Serum, insulin	29
Transformed human amnion cells (AMA)	Mitosis	Intracellular Ca increase?	8
Rat superior cervical ganglion cells	Activation	20-Hz electrical stimulation, dimethylphenylpipezazinium, veratridine	30
Rat pheochromocytoma cells PC-12	Activation	Depolarizing K^+	31
Bovine adrenal chromaffin cells	Activation	Acetylcholine, nicotine, veratridine, depolarizing K^+, ionophore A23187, forscolin	32
Bovine aortic endothelial cells	Activation	ATP, bradykinin, ionophore A23187	33
Human umblical vein endothelial cells (HUVEC)	Activation	Thrombin, histamine, ionomycin	34
Rat pituitary cells GH_3	Activation	Thyrotropin-releasing hormone, depolarizing K^+ scorpion venom toxin Ca^{2+}, ionophore A23187, Ba^{2+}, 4-aminopyridine, ethylammonium ion, valinomycin, oligomycin plus antimycin	35[a]

[a]According to its properties, the 97-kDa polypeptide described in ref. 35 seems to be identical to eEF-2.

transient increase in cytoplasmic Ca^{2+} concentration is usually observed upon treatment of quiescent cells with various mitogens.[36] Since eEF-2 kinase is present in all cultivated mammalian cells studied to date, one can assume that transient phosphorylation of eEF-2 is a common response of quiescent cells to mitogen action. Hence, mitogenic stimulation may be accompanied by transient inhibition of protein synthesis, because, as we discussed before, eEF-2 becomes inactive after phosphorylation. This seems paradoxical because it is well known that mitogenic stimulation of quiescent cells results in the activation of protein synthesis.[37] The explanation is simple: the stimulation of protein synthesis is usually observed several hours after mitogen treatment. To our knowledge there are no reports in which the rate of protein synthesis is measured during the first minutes after mitogen stimulation.

At the same time there is indirect evidence suggesting that the initial effect of growth factors results in transient inhibition of protein synthesis. The evidence came from studies in which the inhibition of protein synthesis in quiescent cells with cycloheximide, anisomycin, or puromycin mimics some of the effects of growth factors. One of the earliest observations was that treatment of serum-deprived 3T3 cells with cycloheximide stimulates uridine incorporation in a manner similar to that in the case of serum treatment.[38] Subsequently, several reports demonstrated that inhibitors of protein synthesis elicit in quiescent cells many of the responses characteristic of initiation of proliferation (such as increase in the rate of putrescine transport into the cells[39] or stimulation of ribosomal protein S_6 phosphorylation[40–42]). The most interesting observation is that inhibitors of protein synthesis superinduce and can often induce the expression of a number of genes called immediate-early genes or primary response genes[43–52] (see ref. 53 for review). These genes include the protooncogenes *c-fos*, c-*jun*, and c-*myc* and many other genes whose expression seems to be necessary for the induction of the proliferative state.

Certainly, the transient inhibition of protein synthesis is not the only prerequisite for the induction of cell proliferation. In many types of cells activation of protein kinase C or alteration in cAMP concentration is necessary for the initiation of proliferation.[36] There are also examples, however, where a transient inhibition of protein synthesis is sufficient to induce proliferation in the absence of any other stimuli. It was shown in quiescent Swiss 3T3 cells[54] and NIH 3T3 cells[55] that incubation with cycloheximide or puromycin could fully substitute for platelet-derived growth factor in stimulation of cells to enter the S phase of the cell cycle. While in the case of Swiss 3T3 cells other growth factors were necessary to induce DNA synthesis,[54] in the case of NIH 3T3 cells incubation with cycloheximide or puromycin for 45 min was enough to make cells enter the S phase[55] (Fig. 3).

All these results strongly suggest that cells are maintained in a quiescent state due to some process which is dependent on continuous protein synthesis; temporary inhibition of translation results in disruption of this process and transition of cells into the proliferative state.

In light of the above observations, the role of transient eEF-2 phosphorylation and resultant transient inhibition of protein synthesis as an immediate result of mitogenic stimulation becomes more understandable. This could be a mechanism for disrupting a process which maintains cells in the quiescent state.

It is unclear, however, why inhibition of protein synthesis results in the stimulation of cell proliferation. Several hypotheses can be proposed. The most straightforward idea is that the quiescent state of the cell is maintained by the continuous synthesis of short-lived proteins which prevent proliferative events. For example, transcription of immediate-early genes in quiescent cells may be blocked by a short-lived repressor. If this is the case, the protein synthesis inhibition will lead to the disappearance of such a repressor and consequently to

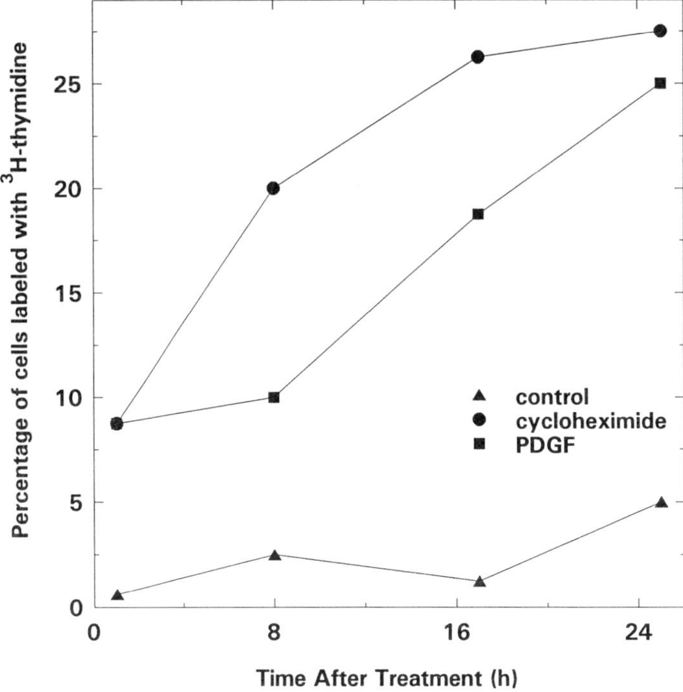

Figure 3. Stimulation of DNA synthesis (which means entrance into the S phase) in quiescent NIH 3T3 cells by incubation with cycloheximide (for 45 min) in comparison with platelet-derived growth factor (PDGF). Experimental details can be found in ref. 55. Data for this figure were kindly provided by Dr. Igor Rosenwald, MGH Cancer Center, Charlestown, Massachusetts.

the activation of transcription of these genes. Similarly, short-lived proteins may exist[56] which make certain "proliferative" mRNAs unstable in quiescent cells; transient inhibition of protein synthesis will result in stabilization and increased concentration of these mRNAs.

The idea of regulating gene expression through short-lived proteins is very attractive, especially because it is now known that extremely short-lived proteins with half-lives of several minutes do exist in the cell.[57] As an example, α 2-repressor in yeast has a half-life of about 5 min.[58] Short-lived proteins responsible for maintaining cells in the quiescent state, however, have not yet been identified.

Transient inhibition of protein synthesis can also activate translation of some mRNAs. It has been reported that in some cases the incubation of cells with cycloheximide can induce the translation of several mRNAs.[59–62] This paradoxical effect of protein synthesis inhibition can be due either to the disappearance of short-lived repressors of translation or to the low affinity of some mRNAs for initiation factors. According to the model of Thach and colleagues,[63–66] translation of such "weak" mRNAs can be stimulated by overall inhibition of the elongation rate because this condition will result in mobilization of the limiting initiation factors from "strong" mRNAs having high affinity for the factors.

Also, it cannot be excluded that inhibition of protein synthesis may result in the generation of a small molecule (like ppGpp, generated in response to amino acid starvation in bacteria) which can serve as a direct effector of transcription or translation.

Then general hypothetical mechanism by which mitogenic stimulation could lead to the induction of the proliferative state through eEF-2 phosphorylation is depicted in Fig. 4. Though the exact mechanism is speculative, the notion that the transient inhibition of elongation may result in the induction of the proliferative state seems attractive.

11. PHOSPHORYLATION OF eEF-2 DURING MITOSIS

When the degree of eEF-2 phosphorylation was measured at different phases of the cell cycle in transformed human amnion cells, it was found that eEF-2

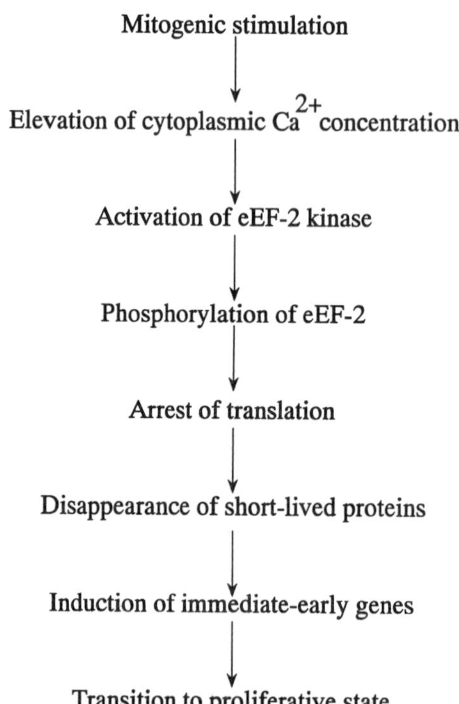

Figure 4. A scheme explaining how mitogens can induce the transition of the quiescent cell into the proliferative state through eEF-2 phosphorylation.

becomes phosphorylated only during mitosis.[8] Since it is well known that protein synthesis in mitosis is inhibited by 50–80%,[8,67–77] and also that a transient increase in intracellular Ca^{2+} concentration during mitosis takes place,[78–82] it was suggested that Ca^{2+}/calmodulin-dependent phosphorylation of eEF-2 is the mechanism of protein synthesis inhibition in this case.[8] This suggestion is in line with data on synchronous culture of HeLa cells,[74] where an accumulation of large polyribosomes was observed during mitosis.

There are other reports,[70–72,75–77] however, that protein synthesis during mitosis is inhibited at the stage of initiation, but not elongation. The reason for the discrepancy seems to be that the mechanism of protein synthesis inhibition at different subphases of mitosis is also different. Transient Ca^{2+} increase during mitosis is usually observed at the metaphase–anaphase transition[78–82] and consequently eEF-2 phosphorylation and inhibition of protein synthesis at the stage of elongation should occur at anaphase but not at metaphase. Indeed, in all reports where inhibition of protein synthesis during mitosis was found to occur at the stage of initiation, measurements were made in cells at metaphase.[70–72,75–77] When protein synthesis rates were measured by pulse-labeling with radioactive amino acids with subsequent counting of radioactivity incorporated into proteins by autoradiography of fixed cells in different subphases of mitosis, it was found that protein synthesis was inhibited in both metaphase and anaphase, and a stronger inhibition was observed in anaphase.[67,68] Thus protein synthesis is inhibited at metaphase probably due to the inhibition of initiation by unknown mechanisms and more strongly at anaphase due to inhibition of elongation through phosphorylation of eEF-2.

12. WHAT IS THE ROLE OF eEF-2 PHOSPHORYLATION AND PROTEIN SYNTHESIS INHIBITION DURING MITOSIS?

Unfertilized eggs provide an excellent model for the study of the metaphase–anaphase transition. Eggs of many animals are blocked in metaphase before fertilization. Eggs of vertebrates are blocked in metaphase II of meiosis before fertilization, while eggs of many invertebrates are blocked in metaphase I. Fertilization of the eggs results in the transition from metaphase into anaphase, completion of meiosis, and subsequent cleavage divisions.

Unfortunately, no one, to our knowledge, has measured the rate of protein synthesis within minutes after fertilization, i.e., at the beginning of anaphase. Nonetheless, there are reasons to believe that fertilization is accompanied by phosphorylation of eEF-2 and transient inhibition of protein synthesis. First of all, eEF-2 kinase is present in sufficiently high concentration in metaphase II-arrested eggs of *Xenopus*[83] and in metaphase I-arrested eggs of different invertebrates (H. Abdelmajid, P. Guerrier and A. G. Ryazanov, unpublished observations). Second, fertilization is accompanied by a drastic transient increase in Ca^{2+} concentration.[84–86] Third, artificial increase in cytoplasmic Ca^{2+} concentration caused by ionophores or by direct injection of Ca^{2+} (refs. 87 and 88) results in release from metaphase block. Fourth, and most important, metaphase block can be released by inhibition of protein synthesis.[89–97] It is tempting to suggest that parthenogenetic activation of metaphase-arrested eggs by protein synthesis inhibitors mimics the phosphorylation of eEF-2 and the transient inhibition of protein synthesis that occurs during natural fertilization.

Why does transient inhibition of protein synthesis activate eggs? It is known that the metaphase-arrested state is maintained because of the high activity of MPF (maturation-promoting factor), which is a complex of two proteins, $p34^{cdc2}$ protein kinase and cyclins (see ref. 98 for review). Fertilization results in the degradation of cyclins, which inactivates MPF and consequently leads to the transition into anaphase and subsequent development.[98] Inhibition of protein synthesis may result in the release of metaphase block because it leads to the degradation of cyclins. This can be due to the fact that either cyclins are relatively short-lived proteins or, more probably, some short-lived proteins in metaphase-arrested eggs negatively regulate cyclin degradation. In any case, Ca^{2+}/calmodulin-dependent phosphorylation of eEF-2 during fertilization can result in transient inhibition of protein synthesis, which would facilitate cyclin degradation through depletion of a putative cyclin protease inhibitor.

Since the general mechanism of cell cycle control is the same in eggs and somatic cells,[98] it is reasonable to believe that inhibition of protein synthesis during anaphase in somatic mammalian cells plays the same role as during egg fertilization. In other words, inhibition of protein synthesis in anaphase may result in the disappearance of short-lived proteins (which negatively regulate cyclin degradation), leading to the induction of the G_1 state.

13. eEF-2 PHOSPHORYLATION AND THE REGULATION OF ENTRANCE INTO AND EXIT FROM THE CELL CYCLE

As we discussed above, there are two principal cases when phosphorylation of eEF-2 takes place *in vivo*: during the transition from the quiescent into the proliferative state, i.e., G_0–G_1 transition, and during the transition from metaphase to anaphase, which can be considered as the M–G_1 transition. Since in both cases the cell goes into the G_1 phase of the cell cycle, it can be suggested that in both cases phosphorylation of eEF-2 and transient inhibition of the protein synthesis switch are similar mechanisms.

Thus, we believe that eEF-2 phosphorylation is obligatory for exit from mitosis into the G_1 phase. Then, it phosphorylation of eEF-2 does not occur during mitosis, the cell will go into the G_0 phase, or the quiescent state. In other words, we suggest that during the metaphase–anaphase transition the cell makes the decision of whether to continue proliferating or to enter a quiescent state. If during the metaphase–anaphase transition the Ca^{2+} concentration increases, phosphorylation of eEF-2 is induced and inhibition of translation takes place; the cell will go into the G_1 phase. If these events do not occur, the cell will go into the G_0 phase.

This idea is supported by recent experiments in which Ca^{2+} transients during mitosis were studied under different cultivation conditions of cells. It was observed that Ca^{2+} transients during mitosis occur only when both serum and Ca^{2+} are present in the medium.[80,82] When either serum or Ca^{2+} concentration was reduced in the medium, no Ca^{2+} transients during mitosis were found, even though mitosis proceeded normally. It is clear from these experiments that Ca^{2+} transients are not necessary for mitosis itself, but they also indicate that Ca^{2+} transients are important for continuation of proliferation, because removal of serum drives cells into the quiescent state (G_0 phase). In the absence of Ca^{2+} transients, eEF-2 is not phosphorylated. Thus, one can expect that upon removal of serum there will be no transient inhibition of elongation during anaphase.

Thus phosphorylation of eEF-2 seems to be the mechanism regulating exit from and entrance into the cell cycle rather than the cycle itself (see Fig. 5). Our hypothesis is very much in line with, and provides a molecular mechanism, for the theory of quiescence–proliferation regulation developed by Epifanova and Polunovsky.[99] Their theory is based on cell fusion experiments, which demonstrated that upon fusion of a quiescent cell with a proliferating cell, the cytoplasm from the quiescent cell inhibited DNA synthesis in the proliferating cell.[100] Moreover, they found that this inhibitory effect could be abolished by incubating the quiescent cell with cycloheximide prior to fusion,[100] suggesting that an unstable protein is present in the cytoplasm of quiescent cells which inhibits cell proliferation. On the basis of these and other experiments they proposed that the transition between quiescent and proliferative states and vice versa is controlled by the interplay of endogenous inhibitors and exogenous stimulators of proliferation.[99] According to their theory, during the cell cycle, the cell accumulates endogenous inhibitor of

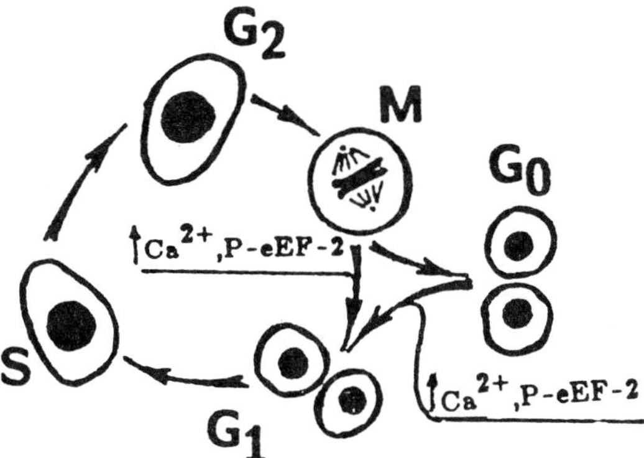

Figure 5. A scheme explaining how eEF-2 phosphorylation can regulate exit from and entrance into the cell cycle.

proliferation, which is most probably an unstable protein. This protein reaches a certain threshold level at the end of the cell cycle and, if it is not destroyed, the cell will go into the quiescent state. On the other hand, if it is destroyed by some mechanism, which is dependent on exogenous stimulators (growth factors), the cell will go into the G_1 phase and continue to proliferate.[99]

Clearly, our idea concerning the role of eEF-2 phosphorylation during mitosis fits very well with this theory, because it explains how growth factors induce the destruction of the endogenous inhibitor of proliferation at the metaphase–anaphase transition.

While many growth factors are well characterized, almost nothing is known about endogenous inhibitors of proliferation. Candidates for the role of such inhibitors are mitotic cyclins because they accumulate during the cell cycle and are destroyed at the end of mitosis. Unfortunately, no information about the level of mitotic cyclins in quiescent cells is available in the literature. If mitotic cyclins are identical to inhibitors of proliferation, one can expect that they should be present in a relatively high concentration in quiescent cells.

14. POSSIBLE ROLE OF eEF-2 PHOSPHORYLATION IN REGULATION OF CELL DIFFERENTIATION

If the absence of eEF-2 phosphorylation commits a cell to exit from the cell cycle and enter the quiescent state, this mechanism may also be involved in the regulation of differentiation. It is well known that during terminal differentiation the cells leave the cell cycle usually through a G_0-like state. For example, when nerve growth factor (NGF) induces differentiation of PC-12 cells into neuronlike cells, it arrests the cells in a G_0-like state, called the G_D state.[101] According to our idea, in order to go into the quiescent state there should be no significant inhibition of protein synthesis during anaphase. If this is correct, we can expect that NGF will prevent eEF-2 phosphorylation during mitosis. This appears to be the case. One of the earliest effects of NGF on PC-12 cells is the downregulation of the eEF-2 kinase.[102–107] Recently we found that this downregulation of eEF-2 kinase results in a decrease of eEF-2 phosphorylation during mitosis.[108] In addition, the protein synthesis rate during mitosis was significantly higher in NGF-treated cells than in control cells.[108] These results suggest that NGF arrests PC-12 cells in the G_D state because it prevents the phosphorylation of eEF-2 and the inhibition of protein synthesis that occurs normally during mitosis.

15. eEF-2 PHOSPHORYLATION IN NEURONS AND MECHANISMS OF MEMORY

As can be seen in Table I, eEF-2 phosphorylation is observed not only during mitosis and the G_0–G_1 transition, but also during the activation of various types of cells. The role of eEF-2 phosphorylation in these cases may be similar to the role of eEF-2 phosphorylation during the G_0–G_1 transition: it leads to transient inhibition of protein synthesis, which is proposed to result in the disappearance of short-lived proteins or the destruction of some other processes dependent on continuous protein synthesis. Thus, it eventually leads to the induction of a new program of gene expression.

One of the most intriguing possible roles of eEF-2 phosphorylation is its involvement in the activation of neurons. It has been shown that the activation of neurons by membrane depolarization or neurotransmitters is accompanied by the induction of transcription of the immediate-early genes—the same genes that are induced during the G_0–G_1 transition.[109–114] It is believed that these changes in gene expression in response to trans-synaptic activation mediate the formation of long-term memory.[112–116] Ca^{2+} and calmodulin play the central role in linking trans-synaptic signals (millisecond range) with the induction of a new program of gene expression (which can last for hours, days, or years).[109,114,117,119]

The most intensively studied example of gene induction in the nervous system is the stimulation of transcription of the c-*fos* protooncogene in undifferentiated or differentiated PC-12 cells.[109–114,117–120] According to Sheng *et al.*,[118,119] Ca^{2+} induces the transcription of c-*fos* by activating Ca^{2+}/calmodulin-dependent protein kinases I and II, which phosphorylate the transcription factor CREB. Phosphorylation of CREB stimulates it and results in the activation of c-*fos* transcription.[118,119] Mutation in CREB which makes it unable to undergo phosphorylation strongly reduces the response of c-*fos* to the increase of Ca^{2+} concentration. Nevertheless, the response is not abolished completely.[119] This suggests that there are other mechanisms by which Ca^{2+} induces transcription of c-*fos*. Furthermore, it has been found recently[120] that in a mutant of PC-12 cells which has no active cAMP-dependent protein kinase, but has normal amounts of Ca^{2+}/calmodulin-dependent protein kinases I and II, an increase of Ca^{2+} concentration in the cytoplasm does not induce c-*fos* transcription.

eEF-2 phosphorylation is a good candidate for the mechanism of induction of c-*fos* and other immediate-early genes in neurons. Transient phosphorylation of eEF-2 in response to stimulation in PC-12 cells[31] and in rat superior cervical ganglion cells[30] has been reported and stimulation of immediate-early gene transcription by protein synthesis inhibition is well documented.[43–52]

16. eEF-2 PHOSPHORYLATION AND PROGRAMMED CELL DEATH

One of the striking features of eEF-2 phosphorylation is that eEF-2 represents the most prominent phosphorylatable protein in extracts from animal cells (see Fig. 1). At the same time in intact cells labeled with ^{32}P, phosphorylation of eEF-2 is not most prominent even under conditions where Ca^{2+} concentration in the cytoplasm is high. The difference between lysed and intact cells suggests that physical destruction of the cell induces eEF-2 phosphorylation. It is unclear, however, at which stage of the cell lysis the phosphorylation of eEF-2 is activated. From the data discussed below it is tempting to suggest that this phosphorylation is activated very early during the process of cell lysis and, moreover, it can be a part of the mechanism of programmed cell death.

Programmed cell death, or apoptosis, is the process involved in the natural turnover of cells as well as in the elimination of unneeded cells in the organism (see refs. 121 and 122 for review). A characteristic feature of apoptosis is the fragmentation of internucleosomal DNA into units of about 200 base pairs. In all cases studied, a sustained increase in the cytoplasmic Ca^{2+} concentration appears to serve as a common signal for the initiation of apoptosis.[123,124] In addition, inhibitors of calmodulin have been shown to prevent apoptosis.[124] These results permit one to suggest that a sustained phosphorylation of eEF-2 takes place in the early phase of programmed cell death and can be a part of a mechanism which eventually leads to DNA fragmentation. This model is supported by recent studies which have demonstrated that the diphtheria toxin, which kills cells by ADP-ribosylation of eEF-2, induces typical apoptosis in cells.[125,126] Initially this effect was attributed to the nuclease activity of diphtheria toxin.[125,126] More recently, the existence of such nuclease was questioned,[127] suggesting that DNA fragmentation in diphtheria toxin-treated cells is somehow related to inactivation of eEF-2.

17. CONCLUSION

Initially, when the phosphorylation of eEF-2 was discovered, we thought that this mechanism would provide an explanation for the different cases of translational control at the level of polypeptide chain elongation. Curiously, in the vast majority of cases where regulation of elongation rate is well documented (reviewed in refs. 128 and 129), phosphorylation of eEF-2 is definitely not implicated, because this phosphorylation is very transient and usually does not last for more than a few minutes. This fact, nevertheless, does not mean that phosphorylation of eEF-2 is not important.

Studies of eEF-2 phosphorylation lead us to the formulation of a new concept, which postulates the existence of stable physiological states of the cell, determined by a specific pattern of synthesized proteins (expressed genes). The main idea is that in any particular physiological state of the cell, the pattern of synthesized proteins is *self-maintained*, which ensures the stability of the physiological state. Switching into a new program (new pattern of protein synthesis) needs a transient shutoff of the protein synthesis. Special physiological signaling mechanisms exist for this purpose. Different hormones and growth factors induce a transient increase in the cytoplasmic Ca^{2+} concentration which eventually results in a transient arrest of translation due to Ca^{2+}-dependent phosphorylation of eEF-2. Arrest of translation disrupts a self-maintained pattern of synthesized proteins and thus allows the expression of new genes and the establishment of a new program of gene expression. We propose that reprogramming through transient translational arrest is a common mechanism which is involved when the cell makes a transition from one physiological state into another, like the transition of the cell from the quiescent into the proliferative state, or the transition of an unfertilized egg into a fertilized one.

One can ask why this arrest takes place at the level of translational elongation, but not at the level of transcription or translational initiation. We think that there is a strong reason for this, because only at the level of translational elongation is it possible to arrest synthesis *instantly*.

Transient inhibition of protein synthesis is involved in the regulation of so many different processes that it probably can be considered as a separate signal transduction pathway analogous to protein kinase C- or cAMP-dependent pathways. We suggest calling it the "translational arrest-dependent pathway."

REFERENCES

1. Ryazanov, A. G., 1987, Ca^{2+}/calmodulin-dependent phosphorylation of elongation factor 2, *FEBS Lett.* **214:**331–334.
2. Palfrey, H. C., 1983, Presence in many mammalian tissues of an identical major cytosolic substrate (Mr 100 000) for calmodulin-dependent protein kinase, *FEBS Lett.* **157:**183–190.
3. Nairn, A. C., Bhagat, B., and Palfrey, H. C., 1985, Identification of calmodulin-dependent protein kinase III and its major Mr 100,000 substrate in mammalian tissues, *Proc. Natl. Acad. Sci. USA* **82:**7939–7943.
4. Ryazanov, A. G., Natapov, P. G., Shestakova, E. A., Severin, F. F., and Spirin, A. S., 1988, Phosphorylation of the elongation factor 2: The fifth Ca^{2+}/calmodulin-dependent system of protein phosphorylation, *Biochimie* **70:**619–626.
5. Ovchinnikov, L. P., Motuz, L. P., Natapov, P. G., Averbuch, L. J., Wettenhall, R. E. H., Szyszka, R., Kramer, G., and Hardesty, B., 1990, Three phosphorylation sites in elongation factor 2, *FEBS Lett.* **275:**209–212.
6. Price, N. T., Redpath, N. T., Severinov, K. V., Campbell, D. G., Russell, J. M., and Proud, C. G., 1991, Identification of the phosphorylation sites in elongation factor-2 from rabbit reticulocytes, *FEBS Lett.* **282:**253–258.
7. Palfrey, H. C., Nairn, A. C., Muldoon, L. L., and Villereal, M. L., 1987, Rapid activation of calmodulin-dependent protein kinase III in mitogen-stimulated human fibroblasts. Correlation with intracellular Ca^{2+} transients, *J. Biol. Chem.* **262:**9785–9792.
8. Celis, J. E., Madsen, P., and Ryazanov, A. G., 1990, Increased phosphorylation of elongation factor 2 during mitosis in transformed human amnion cells correlates with a decreased rate of protein synthesis, *Proc. Natl. Acad. Sci. USA* **87:**4231–4235.
9. Donovan, M. G., and Bodley, J. W., 1991, *Saccharomyces cerevisiae* elongation factor 2 is phosphorylated by an endogenous kinase, *FEBS Lett.* **291:**303–306.
10. Tuazon, P. T., Merrick, W. C., and Traugh, J. A., 1989, Comparative analysis of phosphorylation of translational initiation and elongation factors by seven protein kinases, *J. Biol. Chem.* **264:** 2773–2777.
11. Kim, Y. W., Kim, C. W., Kang, K. R., Byun, S. M., and Kang, Y.-S., 1991, Elongation factor 2 in chick embryo is phosphorylated on tyrosine as well as serine and threonine, *Biochem. Biophys. Res. Commun.* **175:**400–406.
12. Redpath, N. T., and Proud, C. G., 1989, The tumour promoter okadaic acid inhibits reticulocyte-

lysate protein synthesis by increasing the net phosphorylation of elongation factor 2, *Biochem. J.* **262:**69–75.

13. Gschwendt, M., Kittstein, W., Mieskes, G., and Marks, F., 1989, A type 2A protein phosphatase dephosphorylates the elongation factor 2 and is stimulated by the phorbol ester TPA in mouse epidermis *in vivo*, *FEBS Lett.* **257:**357–360.
14. Redpath, N. T., and Proud, C. G., 1990, Activity of protein phosphatases against initiation factor-2 and elongation factor-2, *Biochem. J.* **272:**175–180.
15. Nairn, A. C., and Palfrey, H. C., 1987, Identification of the major Mr 100,000 substrate for calmodulin-dependent protein kinase III in mammalian cells as elongation factor-2, *J. Biol. Chem.* **262:**17299–17303.
16. Gillette, R., Gillette, M., Lipeski, L., and Connor, J., 1990, pH-sensitive, Ca^{2+}/calmodulin-dependent phosphorylation of unique protein in molluscan nervous system, *Biochim. Biophys. Acta* **1036:**207–212.
17. Ernst, V., Levin, D. H., and London, I. M., 1979, *In situ* phosphorylation of the α subunit of eukaryotic initiation factor 2 in reticulocyte lysates inhibited by heme deficiency, double-stranded RNA, oxidized glutathione, or the heme-regulated protein kinase, *Proc. Natl. Acad. Sci. USA* **76:**2118–2122.
18. Sitikov, A. S., Simonenko, P. N., Shestakova, E. A., Ryazanov, A. G., and Ovchinnikov, L. P., 1988, cAMP-dependent activation of protein synthesis correlates with dephosphorylation of elongation factor 2, *FEBS Lett.* **228:**327–331.
19. Nilsson, A., Carlberg, U., and Nygard, O., 1991, Kinetic characterization of the enzymatic activity of the eEF-2-specific Ca^{2+}- and calmodulin-dependent protein kinase III purified from rabbit reticulocytes, *Eur. J. Biochem.* **195:**377–383.
20. Knight, S. A. B., Kohr, W., and Korc, M., 1991, Manganese-stimulated phosphorylation of a rat pancreatic protein: Identity with elongation factor 2, *Biochim. Biophys. Acta* **1092:**196–204.
21. Nygard, O., Nilsson, A., Carlberg, U., Nilsson, L., and Amons, R., 1991, Phosphorylation regulates the activity of the eEF-2-specific Ca^{2+}- and calmodulin-dependent protein kinase III, *J. Biol. Chem.* **266:**16425–16430.
22. Shestakova, E. A., and Ryazanov, A. G., 1987, Influence of elongation factor 2 phosphorylation on its activity in the cell-free translation system, *Dokl. Akad. Nauk USSR* **297:**1495–1498 [in Russian].
23. Ryazanov, A. G., Shestakova, E. A., and Natapov, P. G., 1988, Phosphorylation of elongation factor 2 by EF-2 kinase affects rate of translation, *Nature* **334:**170–173.
24. Kohno, K., Uchida, T., Ohkubo, H., Nakanishi, S., Nakanishi, T., Fukui, T., Ohtsuka, E., Ikehara, M., and Okada, Y., 1986, Amino acid sequence of mammalian elongation factor 2 deduced from the cDNA sequence: Homology with GTP-binding proteins, *Proc. Natl. Acad. Sci. USA* **83:**4978–4982.
25. Ryazanov, A. G., and Davydova, E. K., 1989, Mechanism of elongation factor 2 (EF-2) inactivation upon phosphorylation. Phosphorylated EF-2 is unable to catalyze translocation, *FEBS Lett.* **251:**187–190.
26. Carlberg, U., Nilsson, A., and Nygard, O., 1990, Functional properties of phosphorylated elongation factor 2, *Eur. J. Biochem.* **191:**639–645.
27. Wong, W. L., Brostrom, M. A., and Brostrom, C. O., 1991, Effects of Ca^{2+} and ionophore A23187 on protein synthesis in intact rabbit reticulocytes, *Int. J. Biochem.* **23:**605–608.
28. Brostrom, C. O., and Brostrom, M. A., 1990, Calcium-dependent regulation of protein synthesis in intact mammalian cells, *Annu. Rev. Physiol.* **52:**577–590.
29. Levenson, R. M., and Blackshear, P. J., 1989, Insulin-stimulated protein tyrosine phosphorylation in intact cells evaluated by giant two-dimensional gel electrophoresis, *J. Biol. Chem.* **264:** 19984–19993.
30. Cahill, A. L., Applebaum, R., and Perlman, R. L., 1988, Phosphorylation of elongation factor 2 in the rat superior cervical ganglion, *Neurosci. Lett.* **84:**345–350.

31. Nairn, A. C., Nichols, R. A., Brady, M. J., and Palfrey, H. C., 1987, Nerve growth factor treatment or cAMP elevation reduces Ca^{2+}/calmodulin-dependent protein kinase III activity in PC12 cells, *J. Biol. Chem.* **262:**14265–14272.
32. Haycock, J. W., Browning, M. D., and Greengard, P., 1988, Cholinergic regulation of protein phosphorylation in bovine adrenal chromaffin cells, *Proc. Natl. Acad. Sci. USA* **85:**1677–1681.
33. Demolle, D., Lecomte, M., and Boeynaems, J.-M., 1988, Pattern of protein phosphorylation in aortic endothelial cells. Modulation by adenine nucleotides and bradykinin, *J. Biol. Chem.* **263:** 18459–18465.
34. Mackie, K. P., Nairn, A. C., Hampel, G., Lam, G., and Jaffe, E. A., 1989, Thrombin and histamine stimulate the phosphorylation of elongation factor 2 in human umbilical vein endothelial cells, *J. Biol. Chem.* **264:**1748–1753.
35. Drust, D. S., and Martin, T. F. J., 1982, Thyrotropin-releasing hormone rapidly and transiently stimulates cytosolic calcium-dependent protein phosphorylation in GH_3 pituitary cells, *J. Biol. Chem.* **257:**7566–7573.
36. Rozengurt, E., 1986, Early signals in the mitogenic response, *Science* **234:**161–166.
37. Rudland, P. S., and Jimenez De Asua, L., 1979, Action of growth factors in the cell cycle, *Biochim. Biophys. Acta* **560:**91–133.
38. Hershko, A., Mamont, P., Shields, R., and Tomkins, G. M., 1971, "Pleiotropic response," *Nature New Biol.* **232:**206–211.
39. Pohjanpelto, P., 1976, Cycloheximide elicits in human fibroblasts a response characteristic for initiation of cell proliferation, *Exp. Cell Res.* **102:**138–142.
40. Gressner, A. M., and Wool, I. G., 1974, The stimulation of the phosphorylation of ribosomal protein S6 by cycloheximide and puromycin, *Biochem. Biophys. Res. Commun.* **60:**1482–1490.
41. Krieg, T., Hofsteenge, J., and Thomas, G., 1988, Identification of the 40S ribosomal protein S6 phosphorylation sites induced by cycloheximide, *J. Biol. Chem.* **263:**11473–11477.
42. Price, D. J., Nemenoff, R. A., and Avruch, J., 1989, Purification of a hepatic S6 kinase from cycloheximide-treated rats, *J. Biol. Chem.* **264:**13825–13833.
43. Kelly, K., Cochran, B. H., Stiles, C. C., and Leder, P., 1983, Cell-specific regulation of the *c-myc* gene by lymphocyte mitogens and platelets derived growth factor, *Cell* **35:**603–610.
44. Makino, R., Hayashi, K., and Sugimura, T., 1984, *c-myc* transcript is induced in rat liver at a very early stage of regeneration or by cycloheximide treatment, *Nature* **310:**697–698.
45. Elder, P. K., Schmidt, L. J., Ono, T., and Getz, M. J., 1984, Specific stimulation of actin gene transcription by epidermal growth factor and cycloheximide, *Proc. Natl. Acad. Sci. USA* **81:** 7476–7480.
46. Rittling, S. R., Gibson, C. W., Ferrari, S., and Baserga, R., 1985, The effect of cycloheximide on the expression of cell cycle dependent genes, *Biochem. Biophys. Res. Commun.* **132:**327–335.
47. Lau, L. F., and Nathans, D., 1985, Identification of a set of genes expressed during the G0/G1 transition of cultured mouse cells, *EMBO J.* **4:**3145–3151.
48. Greenberg, M. E., Hermanowski, A. L., and Ziff, E. B., 1986, Effect of protein synthesis inhibitors on growth factor activation of *c-fos*, *c-myc*, and actin gene transcription, *Mol. Cell. Biol.* **6:**1050–1057.
49. Sobczak, J., Mechti, N., Tournier, M.-F., Blanchard, J.-M., and Duguet, M., 1989, *c-myc* and *c-fos* gene regulation during mouse liver regeneration, *Oncogene* **4:**1503–1508.
50. Mahadevan, L. C., and Edwards, D. R., 1991, Signalling and superinduction, *Nature* **349:**747–748.
51. Wisdom, R., and Lee, W., 1991, The protein-coding region of *c-myc* mRNA contains a sequence that specifies rapid mRNA turnover and induction by protein synthesis inhibitors, *Genes Dev.* **5:** 232–243.
52. Messina, J. L., 1990, Insulin's regulation of *c-fos* gene transcription in hepatoma cells, *J. Biol. Chem.* **265:**11700–11705.
53. Herschman, H. R., 1991, Primary response genes induced by growth factors and tumor promoters, *Annu. Rev. Biochem.* **60:**281–319.

54. Kaczmarek, L., Surmacz, E., and Baserga, R., 1986, Cycloheximide or puromycin can substitute for PDGF in inducing cellular DNA synthesis in quiescent 3T3 cells, *Cell Biol. Int. Rep.* **10:**455–463.
55. Epifanova, O. I., Rosenwald, I. B., and Makarova, G. F., 1990, Growth factors and endogenous control of cell proliferation, *Acta Histochem. Suppl.* **39:**211–214.
56. Koeller, D. M., Horowitz, J. A., Casey, J. L., Klausner, R. D., and Harford, J. B., 1991, Translation an the stability of mRNAs encoding the transferrin receptor and *c-fos*, *Proc. Natl. Acad. Sci. USA* **88:**7778–7782.
57. Bachmair, A., Finley, D., and Varshavsky, A, 1986, *In vivo* half-life of a protein is a function of its amino-terminal residue, *Science* **234:**179–186.
58. Hochstrasser, M., and Varshavsky, A., 1990, *In vivo* degradation of a transcriptional regulator: The yeast α 2 repressor, *Cell* **61:**697–708.
59. Lee, G. T.-Y., and Engelhardt, D. L., 1979, Peptide coding capacity of polysomal and non-polysomal messenger RNA during growth of animal cells, *J. Mol. Biol.* **129:**221–233.
60. Bergmann, I. E., Cereghini, S., Georghegan, T., and Brawerman, G., 1982, Functional characteristics of untranslated messenger ribonucleoprotein particles from mouse sarcoma ascites cell. Possible relation to the control of messenger RNA utilization, *J. Mol. Biol.* **156:**567–582.
61. Sorrentino, V., Battistini, A, Curatola, A. M., Di Francesco, P., and Rossi, G. B., 1985, Induction and/or selective retention of proteins in mammalian cells exposed to cycloheximide, *J. Cell. Physiol.* **125:**313–318.
62. Walden, W. E., and Thach, R. E., 1986, Translational control of gene expression in a normal fibroblast. Characterization of a subclass of mRNAs with unusual kinetic properties, *Biochemistry* **25:**2033–2041.
63. Walden, W. E., Godefroy-Colburn, T., and Thach, R. E., 1981, The role of mRNA competition in regulating translation. I. Demonstration of competition *in vivo*, *J. Biol. Chem.* **256:**11739–11746.
64. Brendler, T., Godefroy-Colburn, T., Carlill, R. D., and Thach, R. E., 1981, The role of mRNA competition in regulating translation. II. Development of a quantitative *in vitro* assay, *J. Biol. Chem.* **256:**11747–11754.
65. Brendler, T., Godefroy-Colburn, T., Yu, S., and Thach, R. E., 1981, The role of mRNA competition in regulating translation. III. Comparison of *in vitro* and *in vivo* results, *J. Biol. Chem.* **256:**11755–11761.
66. Godefroy-Colburn, T., and Thach, R. E., 1981, The role of mRNA competition in regulating translation. IV. Kinetic model, *J. Biol. Chem.* **256:**11762–11773.
67. Prescott, D. M., and Bender, M. A., 1962, Synthesis of RNA and protein during mitosis in mammalian tissue culture cells, *Exp. Cell Res.* **26:**260–268.
68. Konrad, C. G, 1963, Protein synthesis and RNA synthesis during mitosis in animal cells, *J. Cell Biol.* **19:**267–277.
69. Johnson, T. C., and Holland, J. J., 1965, Ribonucleic acid and protein synthesis in mitotic HeLa cells, *J. Cell Biol.* **27:**565–574.
70. Scharff, M. D., and Robbins, E., 1966, Polyribosome disaggregation during metaphase, *Science* **151:**992–995.
71. Steward, D. L., Shaeffer, J. R., and Humphrey, R. M., 1968, Breakdown and assembly of polyribosomes in synchronized Chinese hamster cells, *Science* **161:**791–793.
72. Fran, H., and Penman, S., 1970, Regulation of protein synthesis in mammalian cells. II. Inhibition of protein synthesis at the level of initiation during mitosis, *J. Mol. Biol.* **50:**655–670.
73. Mano, Y., 1970, Cytoplasmic regulation and cyclic variation in protein synthesis in the early cleavage stage of the sea urchin embryo, *Dev. Biol.* **22:**433–460.
74. Eremenko, T., and Volpe, P., 1975, Polysome translational state during the cell cycle, *Eur. J. Biochem.* **52:**203–210.
75. Tarnowka, M. A., and Baglioni, C., 1979, Regulation of protein synthesis in mitotic HeLa cells, *J. Cell. Physiol.* **99:**359–368.

76. Bonneau, A.-M., and Sonenberg, N., 1987, Involvement of the 24-kDa cap-binding protein in regulation of protein synthesis in mitosis, *J. Biol. Chem.* **262:**11134–11139.
77. Kanki, J. P., and Newport, J. W., 1991, The cell cycle dependence of protein synthesis during *Xenopus laevis* development, *Dev. Biol.* **146:**198–213.
78. Poenie, M., Alderton, J., Steinhardt, R., and Tsien, R., 1986, Calcium rises abruptly and briefly throughout the cell at the onset of anaphase, *Science* **223:**886–889.
79. Ratan, R. R., Maxfield, F. R., and Shelanski, M. L., 1988, Long-lasting and rapid calcium changes during mitosis, *J. Cell Biol.* **107:**993–999.
80. Tombes, R. M., and Borisy, G. G., 1989, Intracellular free calcium and mitosis in mammalian cells: Anaphase onset is calcium modulated, but is not triggered by a brief transient, *J. Cell Biol.* **109:**627–636.
81. Hepler, P. K., 1989, Calcium transients during mitosis: Observations in flux, *J. Cell Biol.* **109:** 2567–2573.
82. Kao, J. P. Y., Alderton, J. M., Tsien, R. Y., and Steinhardt, R.A., 1990, Active involvement of Ca^{2+} in mitotic progression of Swiss 3T3 fibroblasts, *J. Cell Biol.* **111:**183–196.
83. Severinov, K. V., Melnikova, E. G., and Ryazanov, A. G., 1990, Down-regulation of the translation elongation factor 2 kinase in *Xenopus laevis* oocytes at the final stages of oogenesis, *New Biol.* **2:**887–893.
84. Cuthbertson, K. S. R., Whittingham, D. G., and Cobbold, P. H., 1981, Free Ca^{2+} increases in exponential phases during mouse oocyte activation, *Nature* **294:**754–757.
85. Busa, W. B., and Nuccitelli, R, 1985, An elevated free cytosolic Ca^{2+} wave follows fertilization in eggs of the frog, *Xenopus laevis*, *J. Cell Biol.* **100:**1325–1329.
86. Kubota, H. Y., Yoshimoto, Y., Yoneda, M., and Hiramoto, Y., 1987, Free calcium wave upon activation in *Xenopus* eggs, *Dev. Biol.* **119:**129–136.
87. Steinhardt, R. A., Epel, D., Carroll, E. J., and Yanagimachi, R., 1974, Is calcium ionophore a universal activator for unfertilized eggs? *Nature* **252:**41–43.
88. Fulton, B. P., and Whittingham, D. G., 1978, Activation of mammalian oocytes by intracellular injection of calcium, *Nature* **273:**149–151.
89. Siracusa, G., Whittingham, D. G., Molinaro, M., and Vivarelli, E., 1978, Parthenogenetic activation of mouse oocytes induced by inhibitors of protein synthesis, *J. Embryol. Exp. Morphol.* **43:**157–166.
90. Clarke, H. J., and Masui, Y., 1983, The induction of reversible and irreversible chromosome decondensation by protein synthesis inhibition during meiotic maturation of mouse oocytes, *Dev. Biol.* **97:**291–301.
91. Clarke, H. J., Rossant, J., and Masui, Y., 1988, Suppression of chromosome condensation during meiotic maturation induces parthenogenetic development of mouse oocytes, *Development* **104:** 97–103.
92. Kim, H., and Schuetz, A. W., 1991, Regulation of parthenogenetic activation of metaphase II mouse oocytes by pyruvate, *J. Exp. Zool.* **257:**375–385.
93. Zampetti-Bosseler, F., Huez, G., and Brachet, J., 1973, Effects of several inhibitors of macromolecule synthesis upon maturation of marine invertebrate oocytes, *Exp. Cell Res.* **78:**383–393.
94. Neant, I., and Guerrier, P., 1988, Meiosis reinitiation in the mollusc *Patella vulgata*. Regulation of MPF, CSF and chromosome condensation activity by intracellular pH, protein synthesis and phosphorylation, *Development* **102:**505–516.
95. Dube, F., and Dufresne, L., 1990, Release of metaphase arrest by partial inhibition of protein synthesis in blue mussel oocytes, *J. Exp. Zool.* **256:**323–332.
96. Guerrier, P., Neant, I., Colas, P., Dufresne, L., Saint-Pierre, J., and Dube, F., 1990, Protein synthesis and protein phosphorylation as regulators of MPF activity, in: *Mechanism of Fertilization, Plants to Humans* (B. Dale, ed.), pp. 79–100, Springer-Verlag, Berlin.
97. Van Loon, A. E., Colas, P., Goedemans, H. J., Neant, I., Dalbon, P., and Guerrier, P., 1991, The role of cyclins in the maturation of *Patella vulgata* oocytes, *EMBO J.* **10:**3343–3349.

98. Hunt, T., 1989, Maturation promoting factor, cyclin and the control of M-phase, *Cur. Opin. Cell Biol.* **1:**268–274.
99. Epifanova, O. I., and Polunovsky, V. A., 1986, Cell cycle controls in higher eukaryotic cells: Resting state or a prolonged G_1 period? *J. Theor. Biol.* **120:**467–477.
100. Polunovsky, V. A., Setkov, N. A., and Epifanova, O. I., 1983, Onset of DNA replication in nuclei of proliferating and resting NIH 3T3 fibroblasts following fusion, *Exp. Cell Res.* **146:**377–383.
101. Rudkin, B. B., Lazarovici, P., Levi, B.-Z., Abe, Y., Fujita, K., and Guroff, G., 1989, Cell cycle-specific action of nerve growth factor in PC12 cells: Differentiation without proliferation, *EMBO J.* **8:**3319–3325.
102. End, D., Hanson, M., Hashimoto, S., and Guroff, G., 1982, Inhibition of the phosphorylation of a 100,000-dalton soluble protein in whole cells and cell-free extracts of PC12 pheochromocytoma cells following treatment with nerve growth factor, *J. Biol. Chem.* **257:**9223–9225.
103. End, D., Tolson, N., Hashimoto, S., and Guroff, G., 1983, Nerve growth factor-induced decrease in the cell-free phosphorylation of a soluble protein in PC12 cells, *J. Biol. Chem.* **258:**6549–6555.
104. Togari, A., and Guroff, G., 1985, Partial purification and characterization of a nerve growth factor-sensitive kinase and its substrate from PC12 cells, *J. Biol. Chem.* **260:**3804–3811.
105. Hama, T, Huang, K.-P., and Guroff, G. 1986, Protein kinase C as a component of a nerve growth factor-sensitive phosphorylation system in PC12 cells, *Proc. Natl. Acad. Sci. USA* **83:**2353–2357.
106. Koizumi, S., Ryazanov, A. G., Hama, T., Chen, H.-C., and Guroff, G., 1989, Identification of Nsp100 as elongation factor 2 (EF-2), *FEBS Lett.* **253:**55–58.
107. Brady, M. J., Nairn, A. C., Wagner, J. A. and Palfrey, H. C., 1990, Nerve growth factor-induced down-regulation of calmodulin-dependent protein kinase III in PC 12 cells involves cyclic AMP-dependent protein kinase, *J. Neurochem.* **54:**1034–1039.
108. Ryazanov, A. G., Prisyazhnoy, V. S., Kindbeiter, K., Diaz, J.-J., Madjar, J.-J., Abdelmajid, H., and Rudkin, B. B., 1992, Does down-regulation of elongation factor 2 kinase mediate the automitogenic effect of nerve growth factor? *J. Cell. Biochem. Suppl.* **16B:**156.
109. Morgan, J. I., and Curran, T., 1986, Role of ion flux in the control of *c-fos* expression, *Nature* **322:**552–555.
110. Greenberg, M. E., Ziff, E. B., and Green, L. A., 1986, Stimulation of neuronal acetylcholine receptors induces rapid gene transcription, *Science* **234:**80–83.
111. Morgan, J. I, Cohen, D. R., Hempstead, J. L., and Curran, T., 1987, Mapping patterns of *c-fos* expression in the central nervous system after seizure, *Science* **237:**192–197.
112. Curran, T., and Morgan, J. I., 1987, Memories of fos, *Bioessays* **7:**255–258.
113. Sheng, M., and Greenberg, M. E., 1990, The regulation and function of *c-fos* and other immediate early genes in the nervous system, *Neuron* **4:**477–485.
114. Morgan, J. I., and Curran, T., 1991, Proto-oncogene transcription factors and epilepsy, *Trends Pharmacol.* **12:**343–349.
115. Goelet, P., Castellucci, V. F., Schacher, S., and Kandel, E. R., 1986, The long and the short of long-term memory—A molecular framework, *Nature* **322:**419–422.
116. Black, I. B., Adler, J. E., Dreyfus, C. F., Friedman, W. F., LaGamma, E. F., and Roach, A. H., 1987, Biochemistry of information storage in the nervous system, *Science* **236:**1263–1268.
117. Morgan, J. I., and Curran, T., 1988, Calcium as a modulator of the immediate-early gene cascade in neurons, *Cell Calcium* **9:**303–311.
118. Sheng, M., McFadden, G., and Greenberg, M. E., 1990, Membrane depolarization and calcium induce *c-fos* transcription via phosphorylation of transcription factor CREB, *Neuron* **4:**571–582.
119. Sheng, M., Thompson, M. A., and Greenberg, M. E., 1991, CREB: A Ca^{2+}-regulated transcription factor phosphorylated by calmodulin-dependent kinases, *Science* **252:**1427–1430.
120. Ginty, D. D., Glowacka, D., Bader, D. S., Hidaka, H., and Wagner, J. A., 1991, Induction of immediate early genes by Ca^{2+} influx requires cAMP-dependent protein kinase in PC-12 cells, *J. Biol. Chem.* **266:**17454–17458.

121. Duvall, E., and Wyllie, A. H., 1986, Death and the cell, *Immunol. Today* **7:**115–119.
122. Boobis, A. R., Fawthrop, D. J., and Davies, D. S., 1989, Mechanisms of cell death, *Trends Pharmacol. Sci.* **10:**275–280.
123. Orrenius, S., McConkey, D. J., Bellomo, G., and Nicotera, P., 1989, Role of Ca^{2+} in toxic cell killing, *Trends Pharmacol. Sci.* **10:**281–285.
124. McConkey, D. J., Nicotera, P., Hartzell, P., Bellomo, G., Wyllie, A. H., and Orrenius, S., 1989, Glucocorticoids activate a suicide process in thymocytes through an elevation of cytosolic Ca^{2+} concentration, *Arch. Biochem. Biophys.* **269:**365–370.
125. Chang, M. P., Bramhall, J., Graves, S., Bonavida, B., and Wisnieski, B. J., 1989, Internucleosomal DNA cleavage precedes diphtheria toxin-induced cytolysis. Evidence that cell lysis is not a simple consequence of translation inhibition. *J. Biol. Chem.* **264:**15261–15267.
126. Chang, M. P., Baldwin, R. L., Bruce, C., and Wisnieski, B. J., 1989, Second cytotoxic pathway of diphteria toxin suggested by nuclease activity, *Science* **246:**1165–1168.
127. Bodley, J. W., Johnson, V. G., Wilson, B. A., Blanke, S. R. Murphy, J. R., Pappenheimer, A. M., Jr., Collier, R. J., Lessnick, S. L., Bruce, C., Baldwin, R. L., Chang, M. P., Nakmura, L. T., and Wisnieski, B. J., 1990, Does diphtheria toxin have nuclease activity? *Science* **250:**832–838.
128. Ryazanov, A. G., Rudkin, B. B., and Spirin, A. S., 1991, Regulation of protein synthesis at the elongation stage. New insights into the control of gene expression in eukaryotes, *FEBS Lett.* **285:** 170–175.
129. Spirin, A. S, and Ryazanov, A. G., 1991, Regulation of elongation rate, in: *Translation in Eukaryotes* (H. Trachsel, ed.), pp. 321–346, CRC Press, Boca Raton, Florida.

Chapter 22

A Coupled Translation–Transcription Cell-Free System

Katherine T. Schmeidler-Sapiro and Joseph Ilan

1. INTRODUCTION

In many systems a correlation can be drawn between an increase in the translational capacity of cells and their subsequent embryological development, growth, or response to hormone or other stimulus. In order to distinguish a causal relationship from a coincidental one between the induction of specific protein products and the preceding ribosome accumulation, one would like to be able to separate the events. Such intervention is not feasible *in vivo*. Therefore, a cell-free system was developed in which isolated intact cockerel liver nuclei were transcriptionally active, in the presence or absence of fractionated or unfractionated active translational machinery isolated from rabbit reticulocytes.

A common feature of most biological systems that can be induced to produce large amounts of newly synthesized protein is an increase in cellular ribosomal content prior to accumulation of the specific messenger RNA (mRNA). For instance, as reviewed by Liao,[1] in chickens the increased protein synthetic rate observed after *in vivo* estradiol treatment is preceded by the appearance of

KATHERINE T. SCHMEIDLER-SAPIRO • Department of Biological Sciences, California State University—Long Beach, Long Beach, California 90840. JOSEPH ILAN • Institute of Pathology, Case Western Reserve University School of Medicine, Cleveland, Ohio 44106.

Translational Regulation of Gene Expression 2, edited by Joseph Ilan. Plenum Press, New York, 1993.

hormone-induced mRNA, which, in turn, is preceded by a surge of rRNA synthesis. We have shown that estrogen injection into cockerels or immature female chickens results in a large increase in liver ribosomal content. The specifically induced vitellogenin mRNA starts to appear in polysomes 1 day following estradiol administration—after a substantial amount of newly synthesized ribosomes have accumulated.[2] Increased rates of rRNA synthesis that precede accumulation of specific protein are characteristic of many systems, such as prostate response to *in vivo* testosterone administration,[1] ovalbumin synthesis by oviduct in response to estradiol,[2] lens regeneration in which there is extensive ribosome accumulation preceding crystallin mRNA synthesis,[3] and during periods of rapid growth in response to nutritional shifts, return to permissive temperature, or growth factor activity.[4–10] The converse is also observed: prostatic rRNA synthesis and translational capacity decrease rapidly after castration;[1] one of the earliest functions lost during programmed cell death is ribosome synthesis and accumulation.[11,12]

In early embryogenesis, later differentiation, and up- and down-regulatory shifts in response to nutritional or hormonal changes, a temporal correlation between increased translational capacity and subsequent gene induction has been well documented in many diverse systems. A direct causal relationship, however, has not been proved. A direct effect of the translational machinery on the transcription of specific genes is clearly established in prokaryotes; attenuation of the *trp* operon is one example.[13] Attenuation, however, depends on direct linkage of translation and transcription. Such a model seems less feasible for eukaryotes, where transcription occurs within the nuclear envelope and translation in the cytosolic compartment. Furthermore, the initial eukaryotic transcript typically undergoes extensive posttranscriptional modification and association with ribonucleoproteins (RNPs) before and during transport to the cytoplasm for translation. Nonetheless, eukaryotes may have evolved some mechanism to preserve transcriptional sensitivity to the state of translational activity. While transcription is clearly dependent on the prior translation of the requisite enzymes, these tend to be relatively long-lived. Regulation mediated by controlling the availability of these components would therefore not be very finely tuned to changes in a cell's requirements.

In many systems an increase in cellular translational capacity precedes or is concurrent with a qualitative change in gene expression. While such temporal correlation between increased translational capacity and subsequent "gene induction" has been well documented, no direct causal relationship has been proved. Cytoplasmic factors can be instrumental in causing *de novo* synthesis of gene products or initiation of nuclear programs. The question posed here is whether the translational capacity of the environment in which a nucleus finds itself affects its transcription either qualitatively or quantitatively.

Therefore, we developed a joint cell-free system with which the effect of active translational machinery and its components on transcription could be investigated. The requirements for such a system are that a medium must be found which allows activity of these two systems so that they may be incubated together and allowed to interact. This is not trivial, since the conditions usually employed for *in vitro* translation differ from those of transcription, and some requirements might be mutually exclusive. Activity must persist for sufficient time to permit responses to any interaction which might occur. The sources of the two systems should be sufficiently disparate to avoid inclusion of other cytoplasmic factors which might directly affect gene expression. Products must be measurable. Finally, the isolated transcriptional and translational systems should function similarly *in vivo*.

2. A FRACTIONATED TRANSLATIONAL CELL-FREE SYSTEM DERIVED FROM RABBIT RETICULOCYTES

2.1. Whole-Lysate Preparation and Characterization

Rabbit reticulocyte lysate was chosen as a convenient source of translational machinery because it satisfied several criteria. Reticulocytes are immature red blood cells. In mammals, those isolated from the peripheral circulation have lost their nuclei but are still actively engaged in translation using residual, long-lived mRNA. Reticulocyte lysate has little or no endogenous nuclease activity under normal conditions. This is vital in the development of a coupled transcription–translation system since only minute quantities of nascent or *de novo* initiated mRNA are expected to be provided to the lysate by the isolated nuclei *in vitro* during the short incubation period. The reticulocyte cell-free system is capable of initiating and terminating translation of exogenously supplied mRNA[14,15] and even the translation of long mRNAs such as myosin,[16] vitellogenin,[17] and silk fibroin.[18] Moreover, reticulocyte lysate can be fractionated and manipulated without substantial loss of activity, and has been extensively investigated by others.[19–21]

2.1.1. Isolation of Reticulocytes

Reticulocytosis was achieved in rabbits by daily bleeding of 35–60 ml/day. On the fourth or fifth day, when the packed cell volume (PCV) of the blood was 20–25%, 0.5 ml of 100 mg/ml iron dextran (NONEMIC) was administered intramuscularly, and the animals were rested 2 days. Thereafter, 40–50 ml of blood was collected daily into 0.3 ml of sodium heparin (1000 units/ml in normal saline), on ice. PCV was monitored daily. When PCV dropped below 25%, iron dextran was administered as above and the animals were rested for 2–3 days before bleeding was resumed.

2.1.2. Preparation of the Reticulocyte Lysate

Lysate was prepared essentially as described by Hunt *et al.*[20] except that cells were lysed into 10 mM DTT. Briefly, cells collected after gentle centrifugation and removal of serum and buffy coat layers were washed five times in reticulocyte wash buffer (0.13 M NaCl, 5 mM KCl, 7.5 mM $MgCl_2$). For cell lysis, the packed cells were resuspended in one volume of 10 mM DTT and mixed vigorously, and the supernatant was collected after centrifugation for 15 min at 1600 × *g*. Aliquots of this lysate supernatant were frozen under nitrogen and stored at −70°C. All steps were carried out at 0–4°C. Incubation of the lysate and analysis of products are described below.

When reticulocytosis was achieved as described above, the first 3 days of the second week were found to yield the most active lysates (Fig. 1). Reticulocytosis during this period was 45–65%, with PCV of 22–40%. Failure to administer iron dextran resulted in the death of the animals when anemia was induced to this extent. Although a higher percent of reticulocytes could be achieved by longer maintenance of anemia and additional iron dextran supplements (essentially

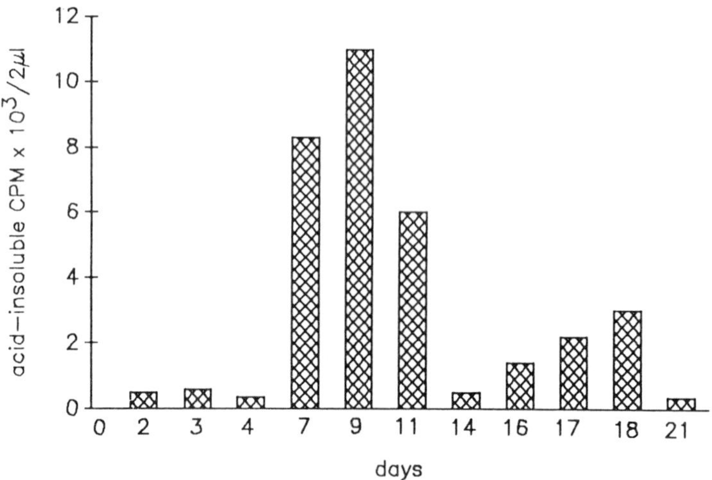

Figure 1. Rates of protein synthesis in lysates prepared on different days of bleeding regimen. Bleeding regimen and lysate incubations as described in the text. Iron dextran was administered after bleeding on days 4, 11, and 18. Lysate activity was assayed in 20-μl samples containing 0.4 volume lysate prepared, as described, immediately after blood was collected on each day. Samples were incubated at 28°C as described in the text, with 100 μCi ^{3}H-leucine, final concentration 10μM; aliquots were taken at timed intervals for scintillation counting. The 15-min values shown here were within the linear period of incorporation for all samples. Otherwise identical samples were incubated with hemin. Replicates +/− hemin fell within the same range; since hemin had no effect, these data are not presented here.

repeating the procedure for a second week), lysates prepared from these collections were not as active. We did not observe the precipitous loss of activity after 5–10 min which has been reported by others,[19–21] nor were lysates prepared in this manner hemin-dependent for their long-term activity. These observations held consistently for every lysate preparation, regardless of when during the bleeding regimen the cells were harvested. It may be that phenylhydrazine treatment to induce reticulocytosis reduces the available iron and/or hemin, so that cells prepared from phenylhydrazine-treated animals are iron-depleted. Since rabbits used in these studies were supplemented with iron dextran before reticulocytes were collected, endogenous hemin levels may have been sufficient to support linear rates of incorporation over longer periods without activating the heme-dependent inhibitor (HDI) and thus the requirement for added hemin. This has not been tested directly.

2.2. Preparation and Characterization of a Fractionated Reticulocyte Lysate

Since the reticulocyte lysate has such high endogenous mRNA activity, small amounts of added mRNA are at a competitive disadvantage and their translation is masked. Therefore it is necessary to inactivate or remove the endogenous mRNA. Typically, endogenous message is digested by calcium-dependent micrococcal endonuclease.[19] Excess ethylene-glycol-bis(2-aminoethylether)-N,N′-tetraacetic acid (EGTA) is then introduced to chelate the Ca^{2+} and thus inactivate the nuclease. This approach is not appropriate for a translation–transcription coupled cell-free system for several reasons. First of all, residual levels of endonuclease activity, while insignificant when large quantities of exogenous mRNA are added in typical experimental designs, may destroy the minute quantities of *de novo* synthesized mRNAs produced by nuclei in a coupled system. Second, normal nuclear transcription is Ca^{2+}-dependent. Thus, excess EGTA might interfere with appropriate nuclear transcriptional activity. Third, calcium has been shown to be important in many cytoskeletal and membrane interactive processes. Thus, one would expect the Ca^{2+} might be important in a physiological interplay between cytoplasmic components and transcriptional regulation within intact nuclei. Care was taken to prepare nuclei gently, so that envelope structures which might be involved in cytoplasmic interaction might remain intact and functional. These putative interactions with translational apparatus components may be via ribosomes attached to the outer portion of the nuclear envelope, or other attached cytoskeletal elements. Calcium chelation might well disrupt these structures or their ability to interact. Finally, introduction of intact nuclei into a reticulocyte lysate compartmentalizes what had been a soluble system. It is not immediately apparent that these membranous compartments, or the large additional quantities of protein, DNA, and other macromolecules, might not effectively separate calcium from EGTA, thus conceivably reactivating the micrococcal endonuclease.

Clearly this possible eventuality must be avoided. Therefore a message-depleted system was prepared by centrifugation, and without the use of nuclease.

Briefly, the lysate was separated into three major translational fractions: ribosomes, mRNP, and the soluble fraction, which presumably included tRNAs, amino acids, cofactors such as ATP and GTP, soluble enzymes, etc. (Fig. 2). Differential centrifugation sedimented ribosomes and polysomes, thus segregating the soluble factors and transfer RNA (tRNA). Messenger RNPs were dissociated from the ribosomes and polysomes during incubation of the complete system for protein synthesis in the cold (cold runoff), followed by differential centrifugation to sediment the ribosomal fraction. This segregates the mRNP fraction to the supernatant. This procedure results in a reticulocyte cell-free system, crudely fractionated into its major translational components, which lends itself to the

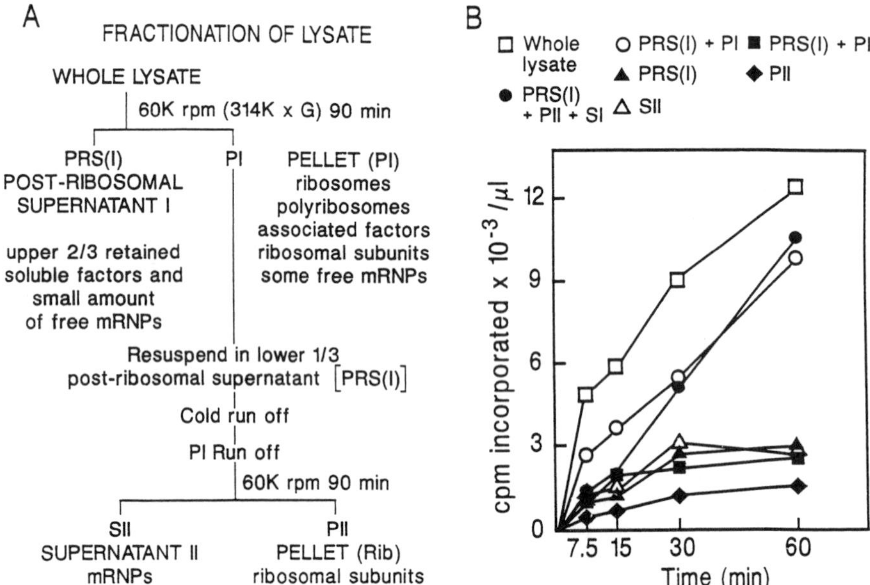

Figure 2. Fractionation of the reticulocyte lysate. **(A)** Flow diagram of the fractionation protocol. **(B)** Protein synthetic activities of the lysate fractions. The methods for preparation of reticulocytes, lysate, and lysate fractionation are described in the text. Incubations were carried out in 20 µl at 28°C. The amount of each fraction added was equivalent to its part of the whole unfractionated lysate which would be added (0.4 total sample volume). The final concentration of ^{35}S-methionine was 12 µM (1.4 mCi/ml). The incubation mixture also included 4 mM DTT, 25 mM creatine phosphate, 50 µg/ml creatine phosphokinase, 2 mM $MgCl_2$, 80 mM KCl, 1.6 mM ATP, 0.2 mM GTP, and 50 µM each of the other 19 amino acids. Timed aliquots were withdrawn, spotted onto Whatman GF/A filters, and placed into ice-cold 10% trichloroacetic acid (TCA) in acetone. The filters were rinsed thoroughly in cold 5% TCA in 0.1 M methionine (5% TCA-met), heated at 90°C for 15 min, then rinsed again in cold 5% TCA-met. Filters were then rinsed twice in acetone and air-dried, and precipitated radioactivity was measured by liquid scintillation counting. Duplicate aliquots were taken at each time point.

analysis of a coupled transcription–translation system. It has not been treated with nuclease or any inhibitors. Moreover, the reconstituted system retains 60% of the original crude lysate protein synthetic activity.

High-speed centrifugation of reticulocyte lysates has been reported to activate HDI in reticulocytes harvested from rabbits made reticulocytotic with phenylhydrazine.[20,21] Most reticulocyte lysates must be supplemented with hemin at some time during their preparation, or immediately before use, to prevent activation of HDI and permit linear translation rates for longer than about 5 min. In our system, however, HDI was not active before or after centrifugation, and our lysate activity was not dependent on added heme.

2.2.1. Fractionation of the Reticulocyte Lysate

Thawed aliquots of lysate were centrifuged for 100 min at 314,000 $\times$ $g_{(max)}$ at 4°C to pellet ribosomes and polysomes. The upper two-thirds (approximately) of the postribosomal supernatant (PRS-I) was removed by pipette and refrozen under nitrogen. The ribosomal pellet (P-I) was resuspended in the remaining one-third of the PRS-I and brought to one-half of the original volume using concentrated lysate incubation medium, so that the final concentrations of the medium components were as described below for translation alone. Included in this were creatine phosphate (CP), creatine phosphokinase (CPK), ATP, GTP, $MgCl_2$, KCl, and amino acids other than those that might be used as label: leucine and methionine.

This suspension was incubated for 1 hr at 10°C to allow ribosomes to run off the message but not to reinitiate because of the low temperature. After the incubation the mixture was centrifuged as above. The supernatant (PRS-II) was drawn off quantitatively, frozen, and stored for subsequent reconstitution of the system. The pelleted runoff ribosomes (P-II) were rinsed gently several times with cold sterile 1 mM $MgCl_2$ and resuspended to one-fourth the original lysate volume in 1 mM $MgCl_2$ by gentle vortexing. During each preparation small portions of each fraction were taken before and after each step and were used in the cell-free translation system to monitor loss of activity due to each manipulation.

2.2.2. Incubation of Reticulocyte Lysate

Incubations were carried out in 20 μl at 28°C in incubation medium as shown in Table I. Each component of the medium was titrated to achieve maximal incorporation rates for extended periods of time. When fractionated lysate components were used, the amount of each fraction was equivalent to its part in the whole, unfractionated lysate. For liquid scintillation counting of incorporated amino acids, timed aliquots were withdrawn, spotted onto Whatman GF/A filters, and placed into cold 10% trichloroacetic acid (TCA) in acetone to bleach the hemoglobin. Filters were washed in 5% TCA in 0.1 M methionine or leucine,

Table I. Conditions for Optimized Amino Acid and/or Nucleotide Incorporation in Reticulocyte Lysate, Isolated Nuclei, and Combined System

Component	Lysate	Nuclei	Combined system
Dithiothreitol (mM)	4	3.5	3.5
Creatine phosphate (mM)	25	25	25
Creatine phosphokinase (μg/ml)	50	50	50
$MgCl_2$ (mM)	2.0	2.0	2.0
KCl (mM)	80	80	80
$(NH_4)_2SO_4$ (mM)	—	300	300
ATP (mM)	1.6	1.65	1.65
GTP (mM)	0.2	0.125	0.125
UTP, CTP (mM)	—	0.125	0.125
$MnCl_2$ (mM)	—	0.5	0.5
NaF (mM)	—	3.0	3.0
Glycerol (%)	—	15–20	15–20
HEPES[a], pH 7.8 (mM)	—	50	50
Tris (mM)	—	4.0	3–4
Buffer pH	None	7.8	7.8
19 amino acid (μm)	50	—	50
Labeled nucleotide triphosphates (mM)	—	0.125	0.125
Labeled amino acids (μm)	20	—	20

[a]HEPES: N-2-hydroxyethylpiperazine-N′-2-ethane sulfonic acid

depending on which amino acid was radioactively labeled. This effectively reduced background binding levels. Samples were heated to 90°C in 5% TCA for 15 min to hydrolyse aminoacyl-tRNA. Filters were then rinsed once again in 5% TCA, then twice in acetone, air-dried, and counted. Duplicate aliquots were taken at each time point. Standard conditions were used for sodium dodecyl sulfate–polyacrylamide gel electrophoresis (SDS–PAGE).

2.2.3. Characterization of the Fractionated Lysate

Most fractionated cell-free systems have 1–10% or less of the unfractionated lysate activity. The activity of our fractionated system, however, when reconstituted, approached the activity of the original cell lysate. Lysates were fractionated by centrifugation and the ribosomes were run off in the cold (Fig. 2). Individual fractions had little activity, while activity of the reconstituted system approached that of the unfractionated lysate. The fractionated system, as well as the original complete lysate, are capable of linear incorporation for at least 60 min. In other experiments, incorporation of labeled amino acids into protein continued linearly for 3–6 hr. About 40–50% of the lysate activity is lost as a result of the first centrifugation step. This can be seen by comparing the protein synthetic activities of the whole lysate and the first high-speed postribosomal supernatant (PRS-I) combined with the first high-speed postribosomal pellet (P-I) (Fig. 2B). This combination reconstitutes all of the components of the complete system, and demonstrates the effect of the centrifugation itself on activity. No further loss of activity was apparent after subsequent cold runoff and the second centrifugation. The strategy employed to fractionate the reticulocyte lysate and to isolate the various functional components is depicted in Fig. 2A. Residual activity in PRS-I and the second high-speed postribosomal supernatant (PRS-II), which should contain mRNP, tRNA, and soluble factors, is presumably due to residual ribosomal subunits not pelleted by the centrifugation. The activity of the sedimented ribosomes obtained after cold runoff (P-II) may be due to mRNPs which were not completely dissociated from the ribosomes by the runoff procedure, from large mRNPs pelleted with the ribosomal subunits, or from mRNPs which might be nonspecifically associated with the ribosomes and thus co-pelleted. Simple summation of the activities of the individual constituent fractions cannot account for the final activity of the reconstituted system. Thus, without the use of nuclease, we have achieved, by combining PRS-I and P-II, an mRNP-depleted translational system dependent upon addition of exogenously supplied mRNP (or PRS-II).

3. A TRANSCRIPTIONAL CALL-FREE SYSTEM OF INTACT COCKEREL LIVER NUCLEI

In order to explore a possible role of the translational machinery or its individual components in regulation-specific transcriptional activities, chicken liver nuclei were used. Nuclei from avian liver would seem to lend themselves to this kind of analysis, since a single *in vivo* treatment of young cockerels with estradiol brings about vast induction of vitellogenin and very low-density lipoprotein (VLDL) following accumulation of ribosomes, and for this reason we have studied many aspects of this system.[2,22–27] These livers commit 10% of their protein synthetic capacity to produce albumin constitutively, regardless of the hormonal status of the animal. Thus, in avian liver we have a tissue capable of producing induced product in large quantity while maintaining normal "housekeeping" functions. Before hormone treatment, vitellogenin and VLDL are not detectable. Among the constitutive and induced gene products are a few readily identifiable polypeptides synthesized in great abundance. Since the putative interactions of cytoplasmic components with nuclei might be mediated by any component of the nuclear matrix or envelope, a crude nuclear preparation was used for these studies. No detergent was used at any time, to ensure that the entire nuclear envelope would be preserved.

3.1. Preparation of the Cell-Free Transcription System

3.1.1. Isolation of Cockerel Liver Nuclei

Liver nuclei were isolated from young cockerels after the method of Ilan and Taubert.[28,29] Briefly, after removal of the gall bladder from the exsanguinated cockerel, the liver was rinsed with ice cold nuclei homogenization buffer (NHB) containing 10 mM Tris, 3 mM $MgCl_2$, 0.32 M sucrose, and 5 mM dithiothreitol (DTT), pH 7.8. It was quickly diced and homogenized in NHB with a Dounce homogenizer using a loose pestle. A crude nuclear pellet was obtained by centrifugation for 10 min at 800 $\times$ g, and was resuspended in nuclei resuspension buffer (NRB)-I (10 mM Tris, 1 mM $MgCl_2$, 2.4 M sucrose, and 5 mM DTT, pH 7.8). This suspension was centrifuged at 23,000 $\times$ g for 120 min to pellet the nuclei. The buffer was decanted and the nuclear pellet resuspended in NRB-II (10 mM Tris, 4 mM $MgCl_2$, 0.25 M sucrose, and 5 mM DTT, pH 7.8) by gentle vortexing. The nuclei were rinsed 2–3 times with NRB-II by resuspension and centrifugation at 1000 $\times$ g for 10 min. The final suspension was in NRB-II containing 50% glycerol. Aliquots were frozen under nitrogen and stored at −70°C until used. All procedures were carried out at 4°C. Nuclei were counted using a hemocytometer.

3.1.2. Incubation of Nuclei

Incubations were carried out at 28°C in incubation medium as shown in Table I. Each component of the medium was titrated to achieve maximal incorporation rates for extended periods of time. Liquid scintillation counting of the timed aliquots was performed essentially as described for the reticulocyte lysate except that no TCA–acetone rinse was used, and filters were always kept in ice-cold TCA.

3.2. Transcriptional Activity of the Cell-Free System

Nuclei continue to incorporate labeled precursor into RNA for at least 3 hr. Figure 3 shows a typical time course for nuclei incubated under the standard conditions (Table I). Incubations for longer times resulted in incorporation continuing at similar rates. Throughout the course of the incubation, approximately 90% of the labeled RNA was released into the supernatant (Fig. 3). This was demonstrated by pelleting nuclei by centrifugation after the incubation, and assaying the pellet and the supernatant for radioactivity. If the nuclei did lyse during an incubation, the suspension gelled rapidly. Since nuclear counts remained unchanged over the course of even long incubations, lysis of a relatively small proportion of nuclei was apparently able to gel the solution, and serves as an extremely sensitive gauge for nuclear breakage. Thus, release of radioactive macromolecules represents transport rather than decay of nuclear integrity. Incorporation of nucleotides into RNA by isolated nuclei is dependent on addition of all

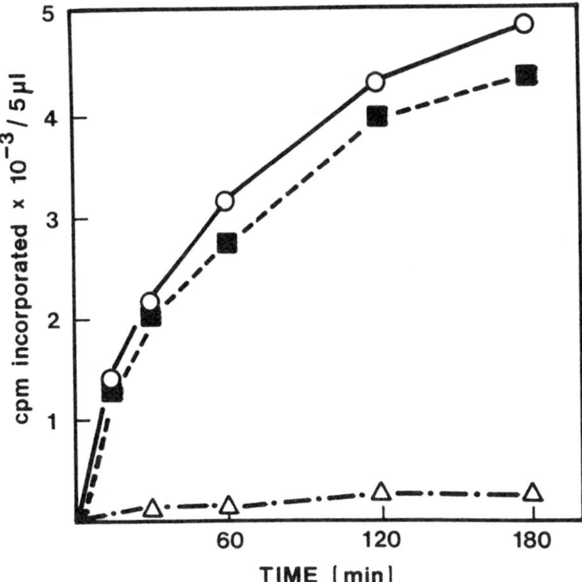

Figure 3. Transcriptional activity of isolated cockerel liver nuclei. Incorporation of ^{3}H-GTP by isolated nuclei, as measured by cold TCA precipitation, was carried out for 180 min in 200 µl at 28°C. Each sample contained 5.4×10^4 nuclei. The mixture also contained 80 µCi ^{3}H-GTP, 3.5 mM DTT, 25 mM creatine phosphate, 50 mg/ml creatine phosphokinase, 2 mM $MgCl_2$, 80 mM KCl, 300 mM $(NH_4)_2SO_4$, 0.5 mM $MnCl_2$, 3.0 mM NaF, 20% (v/v) glycerol, 4 mM Tris-HCl (pH 7.4), and 50 mM HEPES (pH 7.4). Duplicate 5µl aliquots were removed at the times indicated and spotted onto paper filters. These were placed into ice-cold 10% TCA in 0.1 M sodium phosphate (TCA-PO_4), rinsed thoroughly in cold 5% TCA-PO_4, rinsed twice in acetone, and air-dried, and precipitated radioactivity was measured by liquid scintillation counting. Duplicate aliquots were taken at each time point. Open triangles, no added ATP, CTP, or UTP; open circles 0.125 mM NTPs; closed squares, 1.25 mM NTPs.

four nucleotide triphosphates (NTPs). Regardless of which NTP is labeled, a very high-specific-activity labeled precursor could be rate limiting. Therefore, unlabeled NTP of the labeled species was also added to the incubation medium (Table I).

4. A COUPLED TRANSCRIPTION–TRANSLATION CELL-FREE SYSTEM

4.1. Incubation Medium

To achieve a coupled incubation system for nuclei and lysate, the conditions for maximal incorporation by the two individual systems were compared. Concentration curves for each component were generated for each system separately and maxima determined. In some cases, a fairly wide range of concentrations yielded equivalent incorporation time courses and maxima; in other cases the concentration maxima were fairly sharp. Where conditions differed, attempts were made to reconcile the two media. Differences between the two systems may be divided into three categories: (1) those where the difference in concentration of a component was small, or the concentration maximum in one system was within the acceptable range for the other medium; (2) those where, in standard procedures, one but not the other medium contained the specific component; and (3) those where the two systems had very different requirements for the component. Concentrations of several components did not have distinct maxima. Therefore, while a particular concentration had been chosen as "optimal," there might be a considerable range through which differences in activity were minimal. A compromise concentration was chosen so that it would allow as close as possible to maximal incorporation for both systems. Optimal concentrations for each system separately and for the combined system are shown in Table I.

Where concentration maxima for the two systems coincided or overlapped, a compromise concentration was chosen. Those media components ordinarily found in one but not the other system were tested for compatibility. In most cases a compromise condition was readily found. There were a few components that posed more difficulty, however. For example, addition of HEPES or Tris to the lysate proved very inhibitory; activity dropped precipitously in even 4 mM Tris, and with greater than 25 mM HEPES. In contrast, added buffer appeared to stabilize nuclear incorporation in long incubations, and slightly increased the rate of incorporation as well. The most difficult paradox was monovalent cation addition. KCl or NH_4OAc, 100 mM or greater, inhibited nuclear transcription, while concentrations of $(NH_4)_2SO_4$ lower than 150 mM were also inhibitory; 300 mM $(NH_4)_2SO_4$ was quite effective. However, KCl was well tolerated, even required by the reticulocyte lysate, while ammonium salts were severely inhibitory. Lysate activity in 100 mM $(NH_4)_2SO_4$ was one-third that in 50 mM, and even 50 mM NH_4OAc inhibited translation effectively. To our surprise, these apparently irreconcilable differences in salt and buffer requirements and tolerances were ameliorated to a large extent when the two systems were combined. In what would otherwise have been an inhibitory concentration of 300 mM $(NH_4)_2SO_4$, proteins were labeled *in vitro* at easily measurable rates, albeit at lower rates than in comparable incubations in low salt. These results suggest that compartmentalization is occurring *in vitro*, so that the average concentration of a given medium component may not be its effective concentration at a given locale. Precursors are used preferentially from different cellular and extracellular pools *in vivo*;[25] it is possible that similar mechanisms of compartmentalization can be established *in vivo* when intact organellar systems are employed.

4.2. Protein Synthesis in the Coupled System

Figure 4 depicts a fluorograph of an SDS–polyacrylamide gel showing ^{35}S-labeled protein products of the reticulocyte lysate incubated in the presence of the

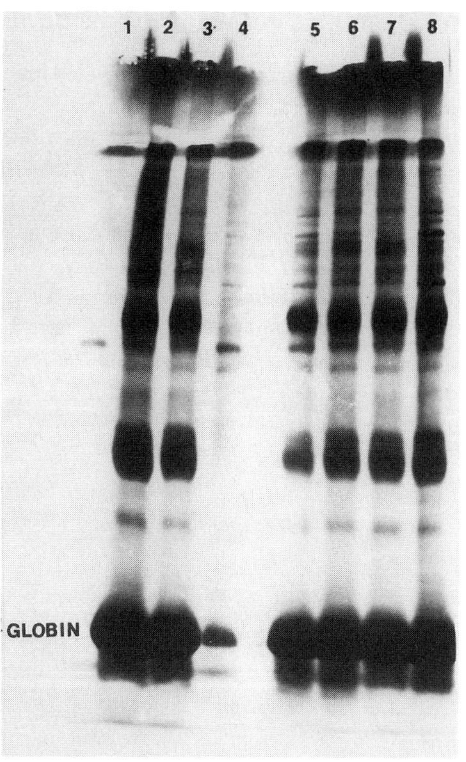

Figure 4. Protein synthesized in the coupled transcription–translation cell-free system. Incubation conditions were as described for Fig. 3, except that 1.1 mCi/ml ^{35}S-methionine (20μM) and 50 μM of the other 19 amino acids were also included. Incubations were carried out in a total volume of 150 μl, with 7.8×10^5 nuclei/tube. Duplicate 1-μl aliquots were removed at timed intervals to monitor incorporation of ^{35}S-methionine into TCA-precipitable protein, as described for Fig. 2; these showed essentially linear incorporation for 3 hr. At the end of the 3 hr of incubation, 7 μl was removed from each sample and processed for SDS–PAGE and fluorography.[19] The amount of each fraction loaded onto the gel was equivalent to its part in the whole lysate. Stained molecular weight markers were incubated in these gels for estimation of molecular weights; these are not shown in the fluorograph. Lane 1: Nuclei only (no added lysate components). Lane 2: Nuclei + whole reticulocyte lysate. Lane 3: Nuclei + PRS-I. Lane 4: Nuclei + P-II. Lane 5: Nuclei + PRS-II. Lane 6: Nuclei + PRS-I + PRS-II. Lane 7: Nuclei + PRS-I + P-II + PRS-II. Lane 8: Whole reticulocyte lysate (no nuclei included).

nuclei. Equal amounts of sample, containing different amounts of label, were run in each lane. The fluorograph was overexposed to visualize bands in lanes 1 and 4; therefore intensity does not accurately reflect quantity in the darker bands. The reticulocyte lysate prepared by bleeding, rather than by phenylhydrazine treatment, synthesizes a large spectrum of proteins in addition to globin. Shorter exposure times of the same gel show clear bands which represent only the predominant labeled species: mostly globin and some other reticulocyte polypeptides. Even under these conditions, a much more diverse population of polypeptides was labeled than is typical of commercial lysates. In the context of these experiments, however, the primary goal was to visualize bands too light to be seen without overexposure. The synthetic activity of lysate alone is shown in lane 8. Addition of nuclei to the complete lysate (lane 2) does not measurably alter protein synthesis by the lysate. When the ribosomal subunits, fraction P-II, are incubated with nuclei (lane 4) a few bands are apparent. They clearly represent globin and other lysate products, most probably resulting from residual lysate mRNP in the P-II fraction. Other bands in lane 4 do not seem to be predominant among the lysate products, and may, as is shown below for specific polypeptides, represent translational products directed by message contributed by the isolated nuclei.

4.3. Transcriptional Activity in the Coupled System

Isolated nuclei incorporate labeled precursor into RNA proportionately for at least 3 hr. The rate of RNA synthesis is dependent on temperature and the addition of all four NTPs. Nuclear transcriptional rates in the coupled system are identical to those shown in Fig. 3 and the curves can be superimposed. Incorporation is inhibited by actinomycin D and α-amanitin (Fig. 5A). RNA labeled and released by isolated nuclei resembles the transcriptional products of liver nuclei *in vivo* (Fig. 5B). The pattern of RNA labeled *in vitro* by nuclei alone (Fig. 5B, lane 1) is very similar to that of liver polysomal RNA labeled *in vivo* (lane 2) and to mass liver polysomal RNA (stained, lane 3). Major *in vitro* products include 18S and 28S RNAs, comigrating with mature, processed rRNAs, as well as diverse

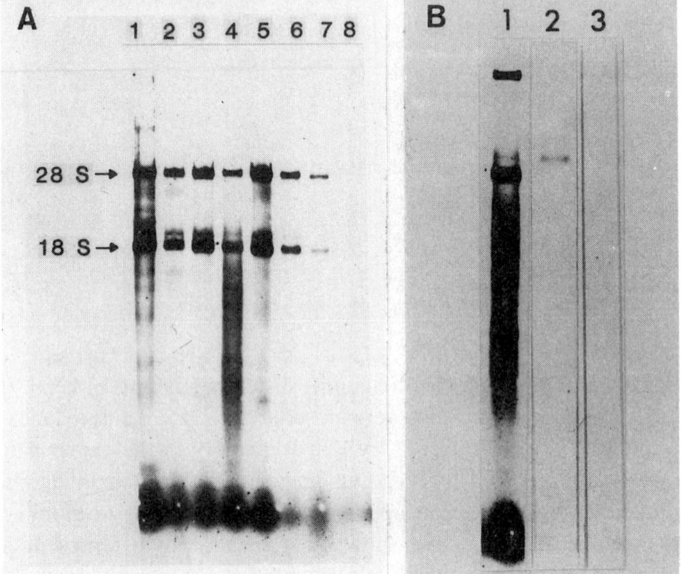

Figure 5. (**A**) Effects of actinomycin D and α-amanitin on transcription by nuclei in the coupled transcription–translation cell-free system. Incubation conditions were as described for Fig. 4, except that whole lysate was used for the coupled system in each sample. RNA synthesis inhibitors were added as indicated. Nuclei were incubated for 3 hr as described in Fig. 3. (**B**)Comparison of labeled RNA synthesized *in vitro* by isolated nuclei and liver polysomal RNA labeled *in vivo*. Nuclei were incubated for 3 hr as described for Fig. 3, except that ^{3}H-UTP replaced ^{3}H-GTP. At the end of the incubation, nuclei were pelleted by brief centrifugation and RNA was purified from the postnuclear supernatant by phenol extraction. The RNA was electrophoresed on a formaldehyde–agarose gel and subjected to fluorography. Liver polysomal RNA was labeled *in vivo* by injection 1 mCi of ^{3}H-uridine in 1 ml sterile water intraperitoneally before isolating polysomes. Liver polysomes and polysomal RNA were isolated as previously described[2] and the isolated RNA treated as above. Lane 1: RNA from isolated nuclei, labeled *in vitro*; lane 2: polysomal RNA labeled *in vivo*; lane 3: ethidium bromide stain pattern of lane 2 RNA.

intermediate-sized RNA species which comigrate with native nonribosomal polysomal RNAs. Transfer RNA (4S) as well as 5S ribosomal RNA are also among the transcriptional products. These gel patterns are unlikely to be due to RNase activity, since, when prelabeled RNA was incubated with nuclei and ribosomes under otherwise similar conditions, there was no apparent degradation.

When nuclei were incubated with lysate fractions, the total amount of radioactivity incorporated did not change. The pattern of labeled RNA species, however, did change (Fig. 6A). Nuclei incubated in buffer alone incorporated label which migrated as 18S and 28S rRNAs, diverse RNA species mostly between approximately 7S and 18S in size, and 5S and 4S species (Fig. 6A, lane 1). The size range of the diverse 7–18S RNA species is consistent with that of mRNA. Nuclei incubated with P-II, the runoff ribosome fraction, which consists primarily of ribosomal subunits and is to a large extent devoid of mRNA, show a high rate of incorporation into these diverse RNA species and relatively less into rRNAs (Fig. 6A, lane 4). In contrast, nuclei incubated with PRS-II, the mRNP lysate fraction, show little if any label in the putative mRNA and an increased relative rate of labeling in rRNA (Fig. 6A, lane 5). Nuclei incubated in the presence of any lysate fraction or mixture of fractions which included supernatant-containing mRNP show this phenomenon (Fig. 6A, lanes 2, 3, and 5–7).

4.4. Characterization of RNAs Labeled *in Vitro* in the Coupled System

The nature of these labeled diverse RNA species was analyzed further. Nuclei were incubated with ^{3}H-NTP, without translational system components (Fig. 6B, lane A), or in the presence of 0.1 μg/ml α-amanitin (Fig. 6B, lane C). At this concentration, α-amanitin inhibits RNA poly-II activity. Alternatively, nuclear RNA was labeled in incubation medium containing runoff ribosomal subunits (P-II) in the presence or absence of α-amanitin (Fig. 6B, lanes B and D, respectively). The nuclei were pelleted by brief centrifugation, so that only the RNA released from the nuclei was analyzed. Previous experiments indicated that at least 90% of labeled products were released from the nuclei under these conditions. This RNA was phenol-extracted, then bound to oligo(dT) cellulose in 0.5 M NaCl. Poly(A)$^{+}$ RNA eluted with distilled water was separated on a standard formaldehyde agarose gel and submitted to fluorography (Fig. 6B). Lanes A and B contain poly(A)$^{+}$ RNA labeled in the absence of α-amanitin. Significantly, more poly(A)$^{+}$ RNA is labeled by nuclei incubated with ribosomes (Fig. 6B, lane B) than in their absence (lane A). Addition of α-amanitin greatly reduced the labeling of poly(A)$^{+}$ RNA, and completely abolishes the effect of ribosomal addition to the nuclei (lanes C and D).

Thus, three kinds of labeled RNA were identified after incubation of isolated nuclei in this transcription–translation coupled cell-free system. One comigrates with mature cytoplasmic 18S and 28S rRNAs, indicating correct processing of rRNA. The second migrates with 5S and 4S RNA species. The third in polyadenyl-

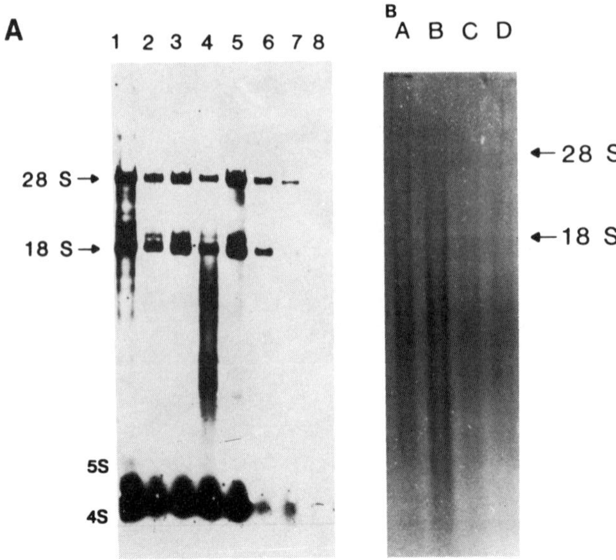

Figure 6. Effects of individual components of the fractionated translation system on specific transcriptional activity in the coupled system. **(A)** Transcriptional specificity is dependent on the presence of individual translational components. Incubation conditions and the addition of ^{3}H-GTP were as described in Fig. 3. For each experiment 7.8×10^5 nuclei were incubated for 3 hr at 28°C in a final volume of 150 μl. Lysate fractions were added as indicated below. The amount of each lysate fraction was equivalent to its relative proportion in the volume of whole lysate which would have been used (0.4×150 μl). After 3 hr of incubation, the nuclei were pelleted by brief centrifugation (5 min, 12,000 $\times$ *g*). RNA was phenol-extracted from 40 μl of the supernatant and subjected to electrophoresis on a standard formaldehyde–agarose gel, and the labeled RNA was visualized by fluorography. Lane 1: Nuclei only (no added lysate components). Lane 2: Nuclei + whole reticulocyte lysate. Lane 3: Nuclei + PRS-I. Lane 4: Nuclei + P-II. Lane 5: Nuclei + PRS-II. Lane 6: Nuclei + PRS-I + P-II. Lane 7: Nuclei + PRS-I + P-II + PRS-II. Lane 8: Whole reticulocyte lysate (no nuclei included). **(B)** Nature of RNA species transcribed and transported by isolated nuclei. Nuclei were incubated with ^{3}H-NTPs for 3 hr in the presence or absence of 0.1 μg/ml α-amanitin and/or runoff ribosomes (P-II fraction). Following the incubation, nuclei were pelleted by brief centrifugation, and the supernatant was phenol-extracted. Extracted RNA was bound to oligo(dT) cellulose in the presence of 0.5 M NaCl, eluted in distilled water, and electrophoresed on a formaldehyde–agarose gel as in **A**. Lane A: −α-amanitin; −P-II. Lane B: −α-amanitin; +P-II. Lane C: +α-amanitin; −P-II. Lane D: +α-amanitin; +P-II. **(C)** Identification of labeled preproalbumin transcripts transported from the nuclei to the incubation medium. Incubation conditions were as described in panel 2, lane 4, in the presence of ribosomes to maximize transcription of the 7–18S RNA species. Ten μCi of [α-^{32}P]GTP was added to each sample. The postnuclear RNA was phenol-extracted and electrophoresed as described in panel **A**. Lane 1: Autoradiograph of the gel. Lane 2: Autoradiograph of the material in lane 1 after transfer blotting onto nitrocellulose strips impregnated with plasmid containing cloned chicken serum albumin cDNA sequences. The transfer blotting was carried out under stringent conditions as described.[26] Lane 3: Hybrid selection as in lane 2, except that the nitrocellulose strips were impregnated with plasmid containing cloned chicken apoVLDL-II cDNA sequences. **(D)** Translational products of hybrid-selected RNA. The hybrid selection was carried out as described[26] and in panel B, except that no label was used. Estradiol was administered to cockerels 1 day before sacrifice in order to induce apoVLDL-II mRNA expression. After hybrid selection the RNA was eluted and used as a template for protein synthesis in a commercially obtained, nuclease-treated, reticulocyte lysate, using ^{35}S-methionine. The translational products were subjected to SDS–polyacrylamide gel electrophoresis, followed by fluo-

ated, suggesting transcription, processing, and transport of mRNA. The third category of RNA fulfills the definition of mRNA by two criteria. First, they are cytoplasmic polyadenylated RNA species. Second, their synthesis is inhibited by α-amanitin in concentrations which have little inhibitory effect on the synthesis and accumulation of 18S and 28S rRNAs or 5S rRNAs or 4S tRNAs.

4.5. Identification of a Specific mRNA Species among the *in Vitro* Synthesized Products of the Coupled System

Since albumin mRNA comprises about 10% of total hepatocyte message, it can be used to follow the synthesis, processing, and transport of mRNA by hepatic nuclei. Nuclei were incubated in the presence of [α-^{32}P]GTP and ribosomes. These conditions result in large increases in the synthesis rates of 7–18S polyadenylated RNA species. The RNA that was transported into the postnuclear supernatant was isolated and electrophoresed on an agarose gel as shown in Fig. 6A, lane 4. It was blotted onto nitrocellulose strips impregnated with plasmid containing albumin cDNA, under conditions for high-stringency hybridization selection as previously reported.[26] The strips were washed with hybridization buffer and autoradiographed (Fig. 6C, lane 2). There is only one apparent radio-

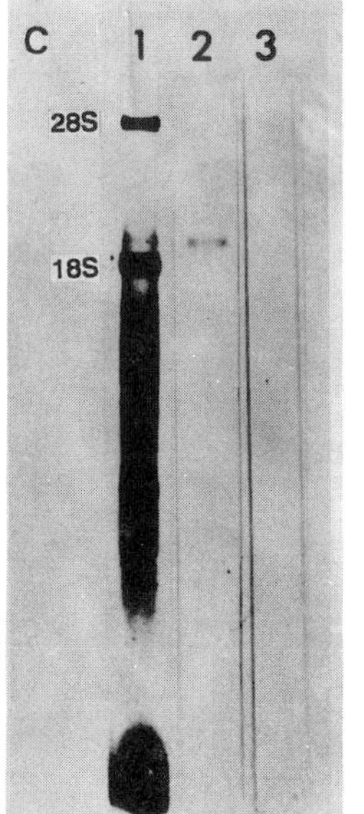

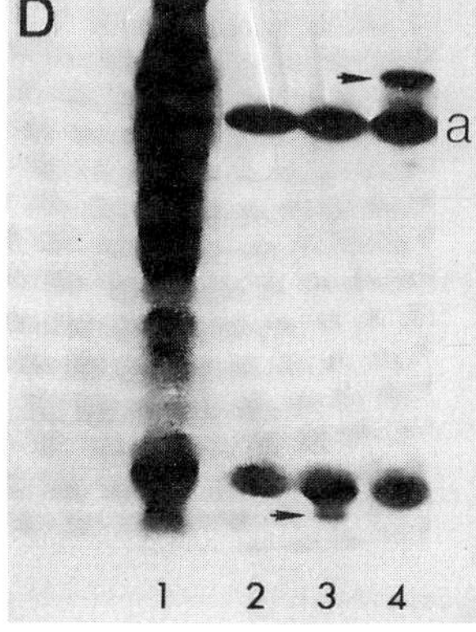

rography.[26] The prominent band below preproalbumin (a) is an artificial band generated by some lots of ^{35}S-methionine. Lane 1: Translational products of total poly(A)$^+$ RNA before hybrid selection. Lane 2: Carrier tRNA only. Lane 3: Hybrid-selected apoVLDL-II RNA. Lane 4: Hybrid-selected preproalbumin mRNA.

active band, slightly above the 18S rRNA marker. It has a molecular mass of 850 kDa. This molecular mass corresponds to approximately 2600 nucleotides, which is in agreement with the reported blots of chicken liver poly(A)$^+$ RNA probed with chicken albumin cDNA to detect preproalbumin transcripts.[30] As a control, strips impregnated with plasmid DNA harboring apoVLDL II cDNA were used (Fig. 6C, lane 3). The strips are highly selective in hybridization of apoVLDL II mRNA, as we have previously shown.[26] ApoVLDL mRNA is expressed *in vivo*, however, only in estradiol-treated animals. Figure 6C, lane 3, shows RNA produced by nuclei taken from cockerels not treated with estradiol. Therefore, apoVLDL II mRNA is not expected to be expressed by these nuclei. Indeed, no apparent transfer could be detected.

The same was not observed using hybridization selection of poly(A)$^+$ RNA obtained from cockerels 3 days after estradiol administration. The blood of such cockerels shows high levels of vitellogenin, indicating apoVLDL II induction.[2] Figure 6D shows the specificity of the cDNA impregnated in the nitrocellulose prior to the blot transfer. Poly(A)$^+$ RNA was prepared and hybrid-selected with nitrocellulose strips impregnated with albumin cDNA or apoVLDL II cDNA. The hybridized RNA was eluted and translated in a commercially-obtained, nuclease-treated reticulocyte lysate in the presence of yeast carrier RNA (200 μl/ml) and ^{35}S-methionine. Aliquots were electrophoresed on a 20% SDS–polyacrylamide gel.[26] Figure 6D, lane 4 depicts the translational products directed by RNA eluted from the nitrocellulose strips impregnated with albumin cDNA. In addition to globin, albumin is the only apparent translation product. When apoVLDL II cDNA was impregnated into the nitrocellulose strips and used for hybrid selection, only the apoVLDL II translation product is apparent in addition to globin (Fig. 6D, lane 3 arrowhead). The strong autoradiogram signal below the albumin (arrow) is the nonspecific ^{35}S-methionine signal which always appears on the gels even after incubation in the absence of lysate in the medium. Thus, impregnation of nitrocellulose with either cDNA is a highly specific method to select specific complementary sequences.

4.6. RNA May Be Synthesized, Processed, Transported, and Translated in the Coupled Cell-Free System

From these experiments it appears that the presence of rabbit reticulocyte runoff ribosomes in the cockerel liver nuclei incubation medium increases the *in vitro* synthesis of putative mRNAs. Among these, albumin could be specifically identified. Moreover, the message appears to be intact, with the correct cytoplasmic molecular mass. These nuclei are apparently able to transcribe and correctly process, as well as transport, albumin mRNA to the medium. It is not clear whether these nuclei can initiate RNA synthesis. Since the incorporation of radioisotope precursors into RNA by isolated nuclei increases proportionately for many hours, however, we assume that reinitiation of RNA transcription is most likely taking place. Furthermore, the ability of these isolated nuclei to transcribe different products under different conditions is consistent with the notion that transcription is initiating *in vitro*.

There is supportive evidence for the notion that the putative mRNA is not only transcribed, processed, and transported, but is also translated *in vitro*. This translation of the newly transcribed message is evident from the following experiment. The coupled transcription–translation cell-free system was incubated in the presence of α-labeled ^{32}P-NTPs as shown in Fig. 6A, lane 4. These incubation conditions included ribosomes and favor incorporation of label into putative mRNA. After incubation for 1 hr, the nuclei were separated from the incubation medium by brief centrifugation at 900 × *g* at 4°C. The presence of polysomes in the incubation medium was examined by sucrose gradient centrifugation analysis as we have previously described.[31] An aliquot of the same medium was treated with puromycin in order to disrupt polysomes, and analyzed in the same way.[31] In both cases unlabeled cockerel or rat liver polysomes were added as carrier as well as visual polysomal markers.[32] It is clear that the radioactivity is associated with the polysomal fraction (Fig. 7A) and that it is completely moved into postribosomal portion by puromycin treatment prior to fractionation on the sucrose gradient (Fig. 7B).

It is worth noting that the puromycin does not shift the radiolabeled RNA

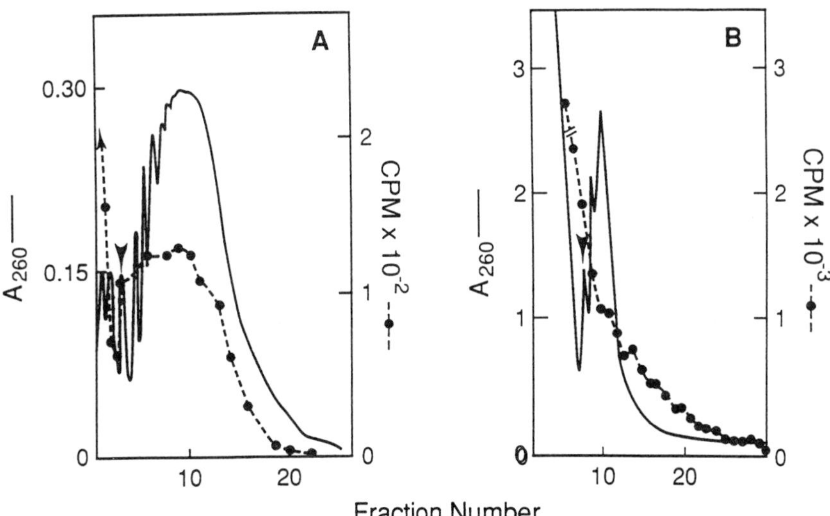

Figure 7. Polysome formation in the coupled system. Incubations were carried out as described in Fig. 5A. After 3 hr the nuclear incubation medium was separated from the nuclei by brief centrifugation. The solid line of absorbance was generated from rat liver polysomes which served as a carrier and marker.[32] **(A)** The incubation medium was fractionated on a 20–50% sucrose gradient as we described.[2] **(B)** The incubation medium was treated with 50 μg/ml puromycin prior to fractionation.[31]

toward the top of the gradient as much as absorbance appears to shift. Puromycin treatment disrupts polysomes, dissociating the ribosomal subunits and releasing veritable polysomal mRNP. Liver polysomes are comprised of ribosomes and mRNP. Unlike rabbit reticulocytes prepared by phenylhydrazine-induced anemia, in which globin mRNPs overwhelmingly predominate in the mRNP population, liver mRNPs are diverse and many are quite large. Therefore, instead of floating high in the gradient as one would expect for the relatively small globin mRNP, liver mRNPs are found well into a sucrose gradient, often having migrated further than ribosomal subunits. Since the label is in the mRNP, it is not unexpected to see its distribution as depicted in Fig. 7. Absorbance represents almost entirely ribosomal particles, as they are the predominant form by mass, and clearly this pattern shows a classical response to puromycin. Thus, these data are consistent with our suggestion that the labeled RNA in these RNPs represents cockerel liver mRNA. Furthermore, it serves as an internal control suggesting that there is not appreciable RNase activity in these preparations, since large mRNAs would be expected to be acutely sensitive to RNase.

These results may be taken as supportive evidence that under the above conditions the nuclei not only transcribed, processed, and transported mRNA into the medium, but that these *in vitro* transcribed messages are actively participating in translation in the coupled system. This interpretation is strengthened by the observation that when the transcription–translation coupled system is incubated in the presence of ^{35}S-methionine and the labeled proteins are analyzed by SDS–PAGE, a prominent band that comigrates with preproalbumin is apparent (Fig. 4, arrow). The appearance of labeled preproalbumin is probably due to two sources of mRNA. First, preexisting mRNA, ribosomes, and other components of the translational apparatus together with some rough endoplasmic reticulum (RER) are presumably still attached to the nuclear envelope, since no detergent was used during preparation. Detergent treatment was avoided because after such treatment nuclei lose the transcriptional properties described above and instead transcribe unidentified RNA products which do not resemble the *in vivo* pattern. Although the nuclear isolation protocol was developed to minimize damage to the outer portion of the nuclear envelope, there did not appear to be extensive carryover of RER membranes associated in the nuclear preparation, since *in vitro*-labeled albumin migrates as preproalbumin, with its leader portion intact. In the presence of functional RER the amino-terminal leader would be cleaved, leaving only proalbumin.

The existence of residual albumin mRNA on the isolated nuclei is apparent from Fig. 4, lane 1, in which nuclei were incubated without translational components, but rather with buffer alone. These translational products also provide an estimate of the background activity of preproalbumin synthesis directed by preexisting mRNA and the activity of ribosomes carried over from the nuclear preparation. When liver nuclei and rabbit reticulocyte ribosomes are included in the incubation medium together (Fig. 4, lane 4), however, preproalbumin synthesis appears to be severalfold higher than this background. This difference may represent *in vitro* expressed preproalbumin message. Taken together, these results indicate that the coupled transcription–translation cell-free system is functional in that it is able to transcribe, process, transport, and translate at least one defined protein, namely preproalbumin.

5. CONCLUSIONS

Our experiments strongly suggest that there is a continuous communication between the protein synthetic apparatus and nuclei. The latter can respond to the cytoplasmic environment and the nature of the translational milieu, and regulate transcription by switching activity from one RNA polymerase to another. This is suggested by the observation that while the rate of incorporation into total RNA did not vary significantly when nuclei were incubated with different components of the translational apparatus, the relative proportions of labeled RNA species did vary with the conditions. Interactions between translational components and transcription factors may be essential for the regulation of nuclear-specific transcription. Such interaction may influence *trans*-acting element(s) affecting RNA polymerase I, II, and III (pol I, II, and III).

The nuclei in this study were isolated from avian liver and therefore are expected to be committed to transcribe the repertoire of avian hepatocytes. Hence, if the cell-free system employs the right conditions, one would expect the system to mimic the behavior of avian hepatocyte transcriptional activity *in vivo*. This prediction is supported by the results of our experiments.

When ribosomes or mRNP, labeled either in their RNA component or their protein fraction, are incubated with nuclei, the label is immediately found associated with the nuclei (results not shown). That similar results are seen whether the label is in RNA or protein suggests (but does not prove) that the RNP particles remain essentially intact. This association was measured by separating the nuclei from the incubation medium by brief centrifugation through a 1 M sucrose cushion. These observations, however, do not distinguish whether the label is attached outside to the nuclear envelope or has entered the nuclei and has become attached to chromatin.

Spirin and co-workers[33] reported cell-free translation systems from *Escherichia coli* or wheat germ capable of producing polypeptides in high yield and active for 10–20 hr. Their system differs from ours in that it is based on continuous flow of the feeding buffer, which contains amino acids, ATP, and GTP, as well as a continuous removal of polypeptide products. A similar continuous-flow cell-free system capable of producing polypeptides in high yield was reported recently for reticulocyte lysate.[34] An elegant transcription–translation coupled system using a DNA-free *E. coli* extract and added plasmid DNA, employing the continuous-flow approach, was recently described.[35] Previous attempts to couple transcription and translation in a cell-free system using *E. coli*, rabbit reticulocytes, or wheat germ[36,37] were hindered by the cessation of translation after 30–60 min of incubation. Moreover, these systems used purified DNA rather than nuclei, thereby precluding the possibility of exploring the interrelationship between the protein-synthesizing machinery apparatus and nuclear activity, or of studying the mechanisms underlying cytoplasmic factors involved in the regulation of nuclear activity.

Even though our coupled system is not based on continuous flow, it is quite efficient in that it is able to synthesize RNA and protein in high yields for hours. Incorporation of amino acids into hot TCA precipitate or nucleotides into RNA was continuous and proportional for 6–12 hr. Since the rabbit reticulocyte lysate had not been dialyzed, it contained relatively high concentrations of amino acids and therefore was able to carry out a significant amount of protein synthesis over a prolonged period of time. To ensure this, and that the amino acid balance did not overwhelmingly favor globin synthesis,[38] the medium was also supplemented with amino acids. In translation experiments the specific activity of the labeled amino acid was low so that its concentration would not be rate limiting. The concentrations of NTP in the incubation medium were also high. Since the specific activity of the added labeled NTP is known (and assuming low amounts carried over in the nuclei), we can calculate an estimate of total RNA produced. This amounts to 5 μg RNA per 10^6 nuclei per 3 hr incubation. The reaction mixture typically contained about 10^5 nuclei, so the approximate yield per reaction tube was 0.5 μg RNA. It should be noted that if significant amounts of NTP were in the isolated nuclei, this would be an underestimate.

More than 90% of the transcriptional product is processed and transported to the incubation medium. If the incubation medium is enriched with mRNP, the main transcriptional products are processed to 18S and 28S rRNAs. In such experiments, if the postnuclear supernatant is analyzed by sucrose gradient centrifugation, the label comigrates with 80S ribosomal markers. This implies that intact ribosomes have been formed, including the *in vitro* synthesized rRNAs. This is not surprising, as liver nuclei are known to contain large reservoirs of ribosomal proteins.[39] Ribonucleoproteins have been shown to become associated with nascent RNA cotranscriptionally.[40] The formation of intact ribosomal subunits and ribosomes, however, indicates that 5S rRNA has also been synthesized and processed. Therefore, in the presence of mRNP, the major transcriptional products are those of pol I and pol III. Conversely, when excess ribosomes, depleted of mRNP, are included in the coupled transcription–translation system, the major transcriptional products are those of pol II, namely mRNAs.

These diverse RNA products satisfy three criteria to be identified as mRNAs. First, much of this material is polyadenylated. Second, it shows characteristic α-amanitin sensitivity. Third, and most important, specific *in vitro*-labeled mRNA can be detected by hybrid selection as a transcriptional product of cockerel nuclei. This hybrid-selected mRNA migrates to a spot which coincides with the known transcriptional product marker of cloned chicken preproalbumin mRNA.[30] Moreover, upon translation of the hybrid-selected albumin mRNA in a reticulocyte cell-free system, the only translational product (besides globin) is preproalbumin. When poly A^+ RNA were prepared from livers obtained from cockerels treated *in vivo* with estradiol and the induced mRNA for apoVLDL was hybrid-selected, apoVLDL polypeptide was translated. These results indicate that in the coupled system, the presence of ribosomes increases the relative activity of RNA pol II. Furthermore, when the coupled system is incubated in the presence of radioactive amino acids, labeled albumin can be identified among the translational products. The controls in these experiments indicate that the synthesis of labeled albumin cannot be attributed entirely to residual albumin mRNA that was carried along with the isolated nuclei, but resulted from the *in vitro* transcription, processing, and transport of preproalbumin transcripts. Whether these transcripts are initiated *in vitro* is not known, but its seems likely that at least some initiation is taking place during such a long period of nearly linear incorporation.

Thus, the coupled system closely resembles *in vivo* activities. Since it is a cell-free preparation, it may not faithfully represent *in vivo* behavior. Nonetheless, our results suggest that the cell-free system reflects *in vivo* activities and interactions. For example, it has been observed that when rats are starved, the ribosomal content of their livers is rapidly depleted. Albumin mRNP is no longer found in polysomes. Following starvation, the residual polysomes that survived are engaged in the translation of mRNA needed for housekeeping enzymes and survival of the hepatocytes. The albumin mRNP is not degraded, but rather is maintained in the postribosomal cytoplasmic fraction. When such fasted rats are refed, new mRNA is not synthesized to any appreciable extent. Instead, extensive ribosome synthesis takes place and the preexisting albumin mRNPs are reutilized to form new polysomes.[41] These *in vivo* observations are consistent with the behavior of our cell-free transcription–translation coupled system.

Our experiments clearly indicate that individual components of the translational apparatus are involved in the regulation of the different RNA polymerase activities. When mRNPs are included in the coupled transcription–translation cell-free system, the main transcriptional products are 28S and 18S ribosomal rRNAs—products of pol I. In contrast, when intact mRNP-depleted ribosomal subunits are included in the coupled system, the main transcriptional products are RNA species whose characteristics are consistent with mRNA—products of pol II. This interplay between the translational and transcriptional systems implies the existence of *trans*-acting factor(s) in common, and presumably found associated with ribosomes and/or mRNPs. Such a putative factor must be present in restricted amounts and be a rate-limiting component for either pol I or pol II activity. It has recently been shown that RNA polymerase I and II promoters share interchangeable enhancers.[42] This common enhancer is a binding site for the transcriptional activators general regulatory factor 2 and the autonomously replicating sequence-binding factor I, a thymidine-rich element. When a particular form of pol I-enhancer was placed in front of a pol II promoter, transcription from that promoter was increased 43-fold comparable to the effect of a powerful pol II activator such as Gal4. Conversely, when two copies of the thymidine-rich element from a pol II enhancer were placed upstream to a pol I promoter, transcription was stimulated 38-fold. This functional reciprocity of pol I and II enhancers indicates similarities in the mechanisms of transcriptional activation.[42]

Several models could be proposed. Drawing an analogy to the *Xenopus* oocyte 5S RNA binding factor TF III,[43] free ribosomes might compete with the pol I promoter for a *trans*-acting factor. Thus, the presence of excess free ribosomes could reduce the activity of pol I. If the same or a different *trans*-acting factor also could stimulate pol II promoters, and these had a higher binding affinity, added ribosomes would introduce more of this factor and thus stimulate pol II activity. Conversely, mRNP might bind either of these putative factors, changing the competitive milieu, and shifting the transcription balance toward pol I activity.

Another possible model would invoke the requirement for a *trans*-acting protein duplex: one member pol I- or pol II-specific, and the other held in common. The "default" situation where nuclei alone, with no added ribosomes or mRNP, transcribe predominately rRNA occurs because in the pol I and pol II competition for the common *trans*-acting factor, pol I out competes pol II. Ribosomes compete with pol I promoters for the pol I-specific factor, however; thus, addition of ribosomes enhances the relative activity of pol II. In the same manner, additional mRNP competes for the pol II-specific factor, allowing pol I access to virtually all of the common *trans*-acting factor. The existence of the common, rate-limiting *trans*-acting factor is inferred from the observation that total RNA synthesis remains constant in all cases, only the relative activities of pol I and II are altered. Were there only promoter-specific factors, one would expect addition of ribosomes to reduce pol I activity without affecting pol II, and vice versa. Moreover, with a single *trans*-acting factor held in common, it would be difficult to explain the differential effect.

The results obtained with the transcription–translation coupled cell-free system indicate formation of intact functional ribosomes *in vitro*, suggesting coordinated regulation of RNA polymerases I and III, since 5S rRNA is a pol III product. This is known to obtain *in vivo*. Moreover, this coordination involves two different nuclear loci, since pol I is localized to the nucleolus and pol III is nucleoplasmic. RNA pol III differs from pol I in that the control region of the 5S RNA gene is within the coding region.[44] In this classic example, the sequences essential for transcription are closer to the 3′ end than the 5′ end of the gene. Other sequence elements required and sufficient for transcription of the adenovirus-encoded VA1 gene and several tRNA genes, all products of pol III, have also been shown to reside within the transcribed region.[45] Crucial elements of pol III promoters, however can also reside external to the transcribed region.[46]

Many diverse pol III-transcribed genes exhibit conserved sequence elements, including a TATA-like box.[47] Such an element is generally considered typical of promoters for pol II transcription. Point mutations within the TATA-like sequence of the 7SK gene dramatically reduce its ability to act as a pol III template,[47] indicating that this sequence is indeed a pol III promoter element. These results imply that a TATA-like binding factor(s) that function in pol II transcription is also involved in pol III initiation. This is consistent with our observations, and reinforces the concept that different eukaryotic polymerases may use common transcription factors. The notion of common transcription factors is not difficult to imagine, since the three polymerase classes also share common enzyme subunits.[48] Moreover, enhancers, once thought to be restricted to pol II transcription systems, have now been shown to augment transcription by all three classes of RNA polymerase. It was shown that the promoters of the U6 and 7SK genes, which are transcribed by poly II, contain sequence elements characteristic of pol III promoters. These genes can also be transcribed by pol III. U6 and 7SK are typical of pol III genes. They exhibit a characteristic α-amanitin sensitivity and the presence of a thymidine (T)-rich termination signal. Furthermore, these genes are transcribed in extracts lacking pol II activity, and their transcription is inhibited by 5S or VA genes, which presumably compete for pol III.[47,49–52] While it is generally accepted that various modes of pol III transcription involve completely unique accessory proteins, serious consideration must also be given to the possibility that the same core proteins are shared by all three polymerases.

Our results strongly suggest competition among the three classes of eukaryotic RNA polymerases for one or more *trans*-acting factors. The coupled transcription–translation cell-free experimental system can be used to investigate many questions under conditions which more closely resemble the *in vivo* state. All three polymerases appear to be active, and their relative activities are interdependent. Their relative activities respond to alterations in the extranuclear environment—specifically, the availability of components of the protein synthetic apparatus and the products of the three polymerases. Thus, this system provides experimental opportunities to explore the regulatory functions of the various components of the transcriptional–translational apparatus: their interdependence and what factors come into play. This experimental system allows us to ask questions which are obscured in other systems which may be either too purified to demonstrate these complex interactions or too complex to manipulate with any clarity. It has provided us with a tantalizing taste and perhaps the first glimpse and direction for exploring a common denominator in the regulation of the transcriptional action of the three RNA polymerases in a readily manipulated cell-free system which closely resembles the *in vivo* situation.

Using cell-free preparations to investigate cellular processes raises the specter of artifacts. While this cannot be avoided completely, several precautions have been taken. Nuclei were prepared gently, to preserve as much as possible the integrity of the nuclear envelope. Reticulocytosis was induced by bleeding rather than phenylhydrazine treatment, and the derived lysates were not nuclease-treated. The patterns of nuclear transcripts and lysate polypeptide products both closely resemble the *in vivo* products of their respective sources, suggesting that the *in vitro* systems closely resembled their *in vivo* counterparts. Furthermore, the observations of apparent interaction between these disparate components in the coupled system lends further credibility to this interpretation. The nuclei clearly compartmentalized the milieu, and their presence altered the conditions experienced by the soluble translational system. This is especially evident for those medium supplements, such as ammonium sulfate, which were required by nuclei and were toxic for the lysate when incubated alone, but were almost completely innocuous for translation in the coupled system. The interactions described here are complex, in the sense that many components, prepared separately and combined *in vitro*, must have reconstituted an integrated and structurally interrelated system. On the other hand, the hypothesis supported by these data, that the available protein synthetic machinery modulates nuclear transcription, is a simple and unifying concept, and might satisfy William of Ockham.

ACKNOWLEDGMENTS. This work has been supported by a grant from the NIH to Joseph Ilan, and NIH and American Diabetes Association grants to Judith Ilan.

REFERENCES

1. Liao, S., 1975, Cellular receptors and mechanisms of action of steroid hormones, *Int. Rev. Cytol* **41:**93–172.
2. Bast, R. E., Garfield, S. A., Gehrke, L., and Ilan, J, 1977, Coordination of ribosome content and polysome formation during estradiol stimulation of vitellogenin synthesis in immature male chick livers, *Proc. Natl. Acad. Sci. USA* **74:**3133–3137.
3. Brachet, J., 1975, Nucleocytoplasmic interactions in cell differentiation, in: *Molecular Biology of Nucelocytoplasmic Relationships*, Volume 1, (S. Puiseux-Dao, ed.), pp. 187–201, Elsevier, Amsterdam.
4. Champney, W. S., 1977, Kinetics of ribosome synthesis during a nutritional shift-up in *Escherichia coli* K-12, *Mol. Gen. Genet.* **152:**259–266.
5. Wolf, S. F., and Schlesinger, D., 1977, Nuclear metabolism of ribosomal RNA in growing, methionine-limited, and ethionine-treated HeLa cells, *Biochemistry* **16:**2783–2791.
6. DePhilip, R. M., Chadwick, D. E., Ignotz, R. A., Lynch, W. E., and Lieberman, I., 1979, Rapid stimulation by insulin of ribosome synthesis in cultured chick embryo fibroblasts, *Biochemistry* **18:**4812–4817.
7. Benoff, S., and Nadel-Ginard, B., 1979, Cell-free translation of mammalian myosin heavy-chain messenger ribonucleic acid from growing and fused L6E9 myoblasts, *Biochemistry* **18:**494–500.
8. Sturani, E., Constantini, M. G., Zippel, R., and Alberghina, F. A. M., 1976, Regulation of RNA synthesis in *Neurospora crassa*. An analysis of a shift-up, *Exp. Cell. Res.* **99:**245–252.
9. Hallberg, R. L., and Bruns, P. J., 1976, Ribosome biosynthesis in *Tetrahymena pyriformis*: Regulation in response to nutritional changes, *J. Cell Biol.* **71:**383–394.
10. Bailey, R. P., Vrooman, M. J., Sawai, Y., Tsukada, K., Short, J., and Leiberman, I., 1976, Amino acids and control of nucleolar size, the activity of RNA polymerase I, and DNA synthesis in liver, *Proc. Natl. Acad. Sci. USA* **73:**3201–3205.
11. Protzel, A., Sridhara,S., and Levenbook, L., 1076, Ribosomal replacement and degradation during metamorphosis of the blowfly, *Calliphora vicina*, *Insect Biochem.* **6:**571–578.

12. Mishima, Y., Matsui, T., and Muramatsu, M., 1979, The mechanism of decrease in nucleolar RNA synthesis by protein synthesis inhibition, *J. Biochem.* **85:**807–818.
13. Hatfield, G. W., and Sharp, J. A., 1987, Translational control of transcription termination in prokaryotes, in: *Translational Regulation of Gene Expression* (J. Ilan, ed.), pp. 447–471, Plenum Press, New York.
14. Ilan, J., and Ilan, J., 1976, Requirement for homologous rabbit reticulocyte initiation factor 3 for initiation of α- and β-globin mRNA translation in a crude protozoal cell-free system, *J. Biol. Chem.* **251:**5718–5725.
15. Ilan, J., and Ilan, J., 1978, Translation of maternal messenger ribonucleoprotein particles from sea urchin in a cell free system from unfertilized eggs and product analysis, *Dev. Biol.* **66:**375–385.
16. DePhilip, R. M., Rudert, W. A., and Lieberman, I., 1980, Preferential stimulation of ribosomal protein synthesis by insulin and in the absence of ribosomal and messenger ribonucleic acid formation, *Biochemistry* **19:**1662–1669.
17. Shapiro, D. J., and Baker, H., 1977, Purification and characterization of *Xenopus laevis* vitellogenin messenger RNA, *J. Biol. Chem.* **252:**5244–5250.
18. Lizardi, P. M., Mahdavi, V., Schields, D., and Candelas, G., 1979, Discontinuous translation of silk fibroin in a reticulocyte cell-free system and in intact silk gland cells, *Proc. Natl. Acad. Sci. USA* **76:**6211–6215.
19. Pelham, H. R. B., and Jackson, R. J., 1976, An efficient mRNA-dependent translation system from reticulocyte lysate, *Eur. J. Biochem.* **67:**249–256.
20. Hunt, T., Vanderhoff, G., and London, I. M., 1972, Control of globin synthesis: The role of heme, *J. Mol. Biol.* **66:**471–481.
21. Hunt, T., 1980, The initiation of protein synthesis, *TIBS* **5:**178–181.
22. Johnson, T. R., and Ilan, J., 1982, Proteins associated with poly(A)$^+$ RNA of cockerel liver: Effects of estradiol stimulation, *Proc. Natl. Acad. Sci. USA* **79:**4088–4092.
23. Gehrke, L., Bast, R. E., and Ilan, J., 1981, An analysis of rates of polypeptide elongation in avian liver explants following *in vivo* estrogen treatment. I. Determination of average rates of polypeptide chain elongation, *J. Biol. Chem.* **256:**2514–2521.
24. Gehrke, L., Bast, R. E., and Ilan, J., 1981, An analysis of rates of polypeptide elongation in avian liver explants following *in vivo* estrogen treatment. II. Determination of the specific rates of elongation of serum albumin and vitellogenin nascent chains, *J. Biol. Chem.* **256:**2522–2530.
25. Gehrke, L., and Ilan, J., 1983, Preferential utilization of exogenously supplied leucine for protein synthesis in estradiol-induced and uninduced cockerel liver explants, *Proc. Natl. Acad. Sci. USA* **80:**3274–3278.
26. Johnson, T. R., and Ilan, J., 1985, Hybrid selection of messenger ribonucleoprotein for serum albumin: Analysis of specific message-bound proteins, *Proc. Natl. Acad. Sci. USA* **82:**7327–7329.
27. Boehm, K. D., Hood, R. L., and Ilan, J., 1988, Induction of vitellogenin in primary monolayer cultures of cockerel hepatocytes, *Proc. Natl. Acad. Sci. USA* **85:**3450–3454.
28. Ilan, J., and Taubert, H.-D., 1968, Ribonucleic acid polymerase activity in isolated uterine nuclei from cow and rat, *Gynaecologia* **165:**45–52.
29. Taubert, H.-D., and Ilan, J., 1968, The effect of ammonium ion upon ribonucleic acid polymerase and guanidine triphosphatase activity in isolated bovine endometrial nuclei, *Experientia* **23:** 706–710.
30. Gordon, J. I., Burns, A. T., Christmann, J. L., and Deeley, R. G., 1978, Cloning of a double-stranded cDNA that codes for a portion of chicken preproalbumin. A general method for isolating a specific DNA sequence from partially purified mRNA, *J. Biol. Chem.* **253:**8629–8639.
31. Ilan, J., and Ilan, J., 1981, Preferential channeling of exogenously supplied methionine into protein by sea urchin embryos, *J. Biol. Chem.* **256:**2830–2834.
32. Bast, R. E., Singer, M., and Ilan, J., 1979, Nerve-dependent changes in content of ribosomes, polysomes, and nascent peptides in newt limb regenerates, *Dev. Biol.* **70:**13–26.

33. Spirin, A. S., Baranov, V. I., Ryabova, L. A., Ovodov, S. Y., and Alakhov, Y. B., 1988, A continuous cell-free translation system capable of producing polypeptides in high yield, *Science* **242:**1162–1164.
34. Ryabova, L. A., Ortlepp, S. A., and Baranov, V. I., 1989, Preparative synthesis of globin in a continuous cell-free translation system from rabbit, *Nucleic Acids Res.* **17:**4412.
35. Baranov, V. I., Morozov, I. Yu., Ortlepp, S. A., and Spirin, A. S., 1989, Gene expression in a cell-free system on the preparative scale, *Gene* **84:**463–466.
36. Zubay, G., 1973, *In vitro* synthesis of protein in microbial systems, *Annu. Rev. Genet.* **7:**267–287.
37. Roberts, B. E., and Paterson, B. M., 1973, Efficient translation of tobacco mosaic virus RNA and rabbit globin 9S mRNA in a cell-free system from commercial wheat germ, *Proc. Natl. Acad. Sci. USA* **70:**2330–2334.
38. Smith, D. W. E., 1975, Reticulocyte transfer RNA and hemoglobin synthesis, *Science* **190:** 529–535.
39. Wu, R. S., Kumar, A., and Warer, J. R., 1971, Ribosome formation is blocked by camptothecin, a reversible inhibitor of RNA synthesis, *Proc. Natl. Acad. Sci. USA* **68:**3009–3014.
40. Beyer, A. L., Miller, O. L., Jr., and McKnight, S. L., 1980, Ribonucleoprotein structure on nascent hnRNAs is non-random and sequence-dependent, *Cell* **20:**75–84.
41. Yap, S. H., Strair, R. K., and Shafritz, D. A., 1978, Effect of a short term fast on the distribution of cytoplasmic albumin messenger ribonucleic acid in rat liver. Evidence for formation of free albumin messenger ribonucleoprotein particles, *J. Biol. Chem.* **253:**4944–4950.
42. Lorch, Y., Lue, N. F., and Kornberg, R. D., 1990, Interchangeable RNA polymerase I and II enhancers, *Proc. Natl. Acad. Sci. USA* **87:**8202–8206.
43. Pelham, H. R. B., and Brown, D. D., 1980, A specific transcription factor that can bind the 5S RNA gene or 5S RNA, *Proc. Natl. Acad. Sci. USA* **77:**4170–4174.
44. Sakonju, S., Bogenhage, D. F., and Brown, D. D., 1980, A control region in the center of the 5S RNA gene directs specific initiation of transcription: I. The 5′ border of the region, *Cell* **19:**13–25.
45. Ciliberto, G., Castagnoli, L., and Cortese, R., 1983, Transcription by RNA polymerase III, *Curr. Top. Dev. Biol.* **18:**59–88.
46. Soliner-Webb, B., 1988, Surprises in polymerase III transcription, *Cell* **52:**153–154.
47. Murphy, S., Tripodi, M., and Melli, M., 1986, A sequence upstream from the coding region is required for the transcription of the 7SK RNA genes, *Nucleic Acids Res.* **14:**9243–9260.
48. Sentenac, A., 1985, Eukaryotic RNA polymerases, *CRC Crit. Rev. Biochem.* **18:**31–90.
49. Kunkel, G. R., Maser, R. L., Calvet, J. P., and Pederson, T., 1986, U6 small nuclear RNA is transcribed by RNA polymerase III, *Proc. Natl. Acad. Sci. USA* **83:**8575–8579.
50. Reddy, R., Henning, D., Das, G., Harless, M., and Wright, D., 1987, The capped U6 small nuclear RNA is transcribed by RNA polymerase III, *J. Biol. Chem.* **262:**75–81.
51. Carbon, P., Murgo, S., Ebel, J. P., Krol, A., Tebb, G., and Mattaj, L. W., 1987, A common octamer motif binding protein is involved in the transcription of U6 snRNA by RNA polymerase III and U2 snRNA by RNA polymerase II, *Cell* **51:**71–79.
52. Margottin, F., Dujardin, G., Gerard, M., Egly, J.-M., Huet, J., and Sentena, A., 1990, Participation of the TATA factor in transcription of the yeast U6 gene by RNA polymerase C, *Science* **251:** 424–426.

Index